This series reports new developments in the fields of control and information sciences—quickly, informally and at a high level. The type of material considered for publication includes:

1. Preliminary drafts of monographs and advanced textbooks
2. Lectures on a new field, or presenting a new angle on a classical field
3. Research reports
4. Reports of meetings, provided they are

   (a) of exceptional interest and
   (b) devoted to a specific topic. The timeliness of subject material is very important.

Indexed by EI-Compendex, SCOPUS, Ulrich's, MathSciNet, Current Index to Statistics, Current Mathematical Publications, Mathematical Reviews, IngentaConnect, MetaPress and Springerlink.

Lecture Notes in Control and Information Sciences

Volume 493

Romain Postoyan · Paolo Frasca · Elena Panteley ·
Luca Zaccarian
Editors

Hybrid and Networked Dynamical Systems

Modeling, Analysis and Control

Springer

Editors
Romain Postoyan
Department Contrôle Identification Diagnostic
CNRS, Université de Lorraine, CRAN
Vandoeuvre-lès-Nancy, France

Elena Panteley
MODESTY
L2S CNRS, CentraleSupélec
Gif-sur-Yvette, France

Paolo Frasca
DANCE INRIA
Gipsa-lab, CNRS
Saint Martin d'Hères, Isère, France

Luca Zaccarian
MAC
LAAS-CNRS
Toulouse, France

Dipartimento di Ingegneria Industriale
University of Trento
Trento, Italy

ISSN 0170-8643 ISSN 1610-7411 (electronic)
Lecture Notes in Control and Information Sciences
ISBN 978-3-031-49554-0 ISBN 978-3-031-49555-7 (eBook)
https://doi.org/10.1007/978-3-031-49555-7

Mathematics Subject Classification: 93

This Springer imprint is published by the registered company Springer Nature Switzerland AG
The registered company address is: Gewerbestrasse 11, 6330 Cham, Switzerland

Preface

The chapters of this book reflect the presentations given during the workshop on "Hybrid and Networked Dynamical Systems", which was held on June 20–22, 2022, at the ENSEEIHT (Toulouse, France) within the framework of the HANDY project funded by the French National Research Agency (ANR). The workshop was an opportunity to present recent advances in the fields of networked systems, hybrid dynamical systems and beyond, by renown experts of the field. In view of the quality of talks and the overall success of the workshop, it was a natural next step to make the material available to a broader audience, which led to the present book. We wish to thank all the speakers of the workshop and the contributors to this volume for their constant support and encouragement during both the organization of the workshop and the editorial work for the preparation of this volume.

Nancy, France
Grenoble, France
Gif-sur-Yvette, France
Toulouse, France/Trento, Italy
October 2023

Romain Postoyan
Paolo Frasca
Elena Panteley
Luca Zaccarian

Acknowledgements

The editors would like to thank the French National Research Agency (ANR) for their support via the grant "Hybrid And Networked Dynamical sYstems" (HANDY), number ANR-18-CE40-0010. This grant was a unique opportunity to consolidate existing collaborations and to create new synergies among the partners involved, which has led to numerous contributions in control theory, and more specifically in the development of novel hybrid techniques for representing and controlling collective behavior in multi-agent systems, some of which are presented in this book. This project also offered the members the possibility to increase their collaboration potential, whose outcome is already reflected by new collaborations and follow-up funding proposals.

The editors would also like to thank the ENSEEIHT (Toulouse, France) for hosting the workshop on "Hybrid and Networked Dynamical Systems" in June 2022, as well as Pauline Kergus and Simone Mariano for their priceless help with the workshop organization.

Contents

Part II Hybrid Techniques

Contributors

Georges Aazan Université Paris-Saclay, CNRS, ENS Paris-Saclay, Laboratoire Méthodes Formelles, Gif-sur-Yvette, France

Vincent Andrieu CNRS and LAGEPP UMR CNRS, Université Claude Bernard Lyon, Université de Lyon, Villeurbanne, France

Daniele Astolfi CNRS and LAGEPP UMR CNRS, Université Claude Bernard Lyon, Université de Lyon, Villeurbanne, France

Pauline Bernard Centre Automatique et Systèmes, Mines Paris, Université PSL, Paris, France

M. Kanat Camlibel Johann Bernoulli Institute for Mathematics and Computer Science, and Artificial Intelligence, University of Groningen, Groningen, The Netherlands

Andrea Cristofaro Dipartimento di Ingegneria Informatica, Automatica e Gestionale A. Ruberti, Sapienza University of Rome, Rome, Italy

Jamal Daafouz Université de Lorraine, CNRS, CRAN, Nancy, France

Claudio De Persis Engineering and Technology Institute Groningen, University of Groningen, Groningen, The Netherlands

Carla De Souza Leyfa Measurement une filiale SNCF, Aucamville, France

Francesco D'Orazio Dipartimento di Ingegneria Informatica, Automatica e Gestionale A. Ruberti, Sapienza University of Rome, Rome, Italy

Arthur Scolari Fagundes Universidade Federal do Rio Grande do Sul (UFRGS), Porto Alegre, RS, Brazil

Paolo Frasca CNRS, Inria, Grenoble INP, GIPSA-lab, University Grenoble Alpes, Grenoble, France

Germain Garcia LAAS-CNRS, University of Toulouse, INSA, Toulouse, France

Antoine Girard Université Paris-Saclay, CNRS, CentraleSupélec Laboratoire des Signaux et Systèmes, Gif-sur-Yvette, France

João Manoel Gomes da Silva Jr. Universidade Federal do Rio Grande do Sul (UFRGS), Porto Alegre, RS, Brazil

Lorenzo Govoni Dipartimento di Ingegneria Informatica, Automatica e Gestionale A. Ruberti, Sapienza University of Rome, Rome, Italy

Luca Greco Université Paris-Saclay, CNRS, CentraleSupélec, Laboratoire des signaux et systèmes, Gif-sur-Yvette, France

Meichen Guo Delft Center for Systems and Control, Delft University of Technology, Delft, The Netherlands

Julien M. Hendrickx ICTEAM Institute, Louvain-la-Neuve, Belgium

Mónika Józsa Department of Engineering, Control, University of Cambridge, Cambridge, UK

Marc Jungers Université de Lorraine, CNRS, CRAN, Nancy, France

Antonio Loría Laboratoire des signaux et systèmes, CNRS, Gif-sur-Yvette, France

Lorenzo Marconi Department of Electrical, Electronic, and Information Engineering “Guglielmo Marconi”, University of Bologna, Bologna, Italy

Paolo Mason Université Paris-Saclay, CNRS, CentraleSupélec, Laboratoire des signaux et systèmes, Gif-sur-Yvette, France

Mattia Mattioni Dipartimento di Ingegneria Informatica, Automatica e Gestionale A. Ruberti, Sapienza University of Rome, Rome, Italy

Charles Monnoyer de Galland ICTEAM Institute, Louvain-la-Neuve, Belgium

Emmanuel Nuño Department of Computer Science at the University of Guadalajara, Guadalajara, Mexico

Elena Panteley Université Paris-Saclay, CNRS, CentraleSupélec, Laboratoire des signaux et systèmes, Gif-sur-Yvette, France

Mihály Petreczky University of Lille, CNRS, Centrale Lille, UMR CRIStAL, Lille, France

Romain Postoyan Université de Lorraine, CNRS, CRAN, Nancy, France

Christophe Prieur University Grenoble Alpes, CNRS, Grenoble-INP, Grenoble, France

Isabelle Queinnec LAAS-CNRS, University of Toulouse, CNRS, Toulouse, France

Esteban Restrepo CNRS, University of Rennes, Inria, IRISA, Rennes, France

Pierre Riedinger Université de Lorraine, CNRS, CRAN, Nancy, France

Sophie Tarbouriech LAAS-CNRS, University of Toulouse, CNRS, Toulouse, France

Pietro Tesi Department of Information Engineering, University of Florence, Florence, Italy

Gia Quoc Bao Tran Centre Automatique et Systèmes, Mines Paris, Université PSL, Paris, France

Flora Vernerey Université de Lorraine, CNRS, CRAN, Nancy, France

Renato Vizuete ICTEAM Institute, Louvain-la-Neuve, Belgium

Luca Zaccarian MAC, LAAS-CNRS, CNRS, University of Toulouse, Toulouse, France; Dipartimento di Ingegneria Industriale, University of Trento, Trento, Italy

Chapter 1
Introduction

Romain Postoyan, Paolo Frasca, Elena Panteley, and Luca Zaccarian

Abstract This chapter serves as introduction to the book and to the contributions therein. We provide some relevant background on hybrid and networked dynamical systems and summarize the contribution of each chapter, while illustrating their synergies and connecting themes. The broad spectrum of the contributed chapters is organized into three main thematic areas: Networked Systems; Hybrid Techniques; and Emerging Trends and Approaches for Analysis and Design.

Networked dynamical systems are ubiquitous in our everyday life. From energy grids and fleets of robots or vehicles to social networks, the same scenario arises in each case: dynamical units interact locally to achieve a global task. When considering a networked system as a whole, very often continuous-time dynamics are affected by instantaneous changes, called jumps, leading to the so-called hybrid dynamical systems. These jumps may come from (i) the intrinsic dynamics of the nodes, like in multimedia delivery with fixed rate encoding, e.g., [8, 9], (ii) the intrinsic degrees of freedom of the control actions, like in power converters within energy grids, e.g., [2, 3, 10, 29] or valve-based controllers [24] that switch between several modes of operation, (iii) the creation/loss of links or the addition/removal of nodes, like

R. Postoyan (✉)
Université de Lorraine, CNRS, CRAN, 54000 Nancy, France
e-mail: romain.postoyan@univ-lorraine.fr

P. Frasca
CNRS, Inria, Grenoble INP, GIPSA-lab, University Grenoble Alpes, 38000 Grenoble, France
e-mail: paolo.frasca@gipsa-lab.fr

E. Panteley
Laboratoire des signaux et systèmes, CNRS, Gif-sur-Yvette, France
e-mail: elena.panteley@cnrs.fr

L. Zaccarian
MAC, LAAS-CNRS, CNRS, University of Toulouse, Toulouse, France
e-mail: zaccarian@laas.fr

Dipartimento di Ingegneria Industriale, University of Trento, Trento, Italy

R. Postoyan et al. (eds.), *Hybrid and Networked Dynamical Systems*,
Lecture Notes in Control and Information Sciences 493,
https://doi.org/10.1007/978-3-031-49555-7_1

in social networks where opinion dynamics take place, e.g., [7, 12, 18–20]. These control applications call upon the development of powerful analysis and design tools capable of handling the multi-agent interaction patterns combined with the above-mentioned hybrid nature of the dynamics.

Both hybrid systems and networked systems have been extensively studied during the last two decades. On the one hand, methods to deal with hybrid dynamics have witnessed a fast growth due to the emergence of powerful modeling and analytical tools, see, e.g., [13, 16, 17]. However, when dealing with such hybrid dynamics, the existing results typically require that a single global model be specified, which may be an obstacle when dealing with networked systems. More work needs to be done to develop tailored tools when investigating networked systems exhibiting hybrid dynamics. On the other hand, research in networked systems has a long history in many fields of science and engineering: in the years 2000, the interest of control researchers in networked systems, also called multi-agents systems, exploded thanks to the pioneering works of [11, 14, 22, 30], which unraveled the fundamental links between algebraic graph theory and stability in this context. Existing works mostly concentrate on simple node dynamics, often single or double integrators, with the exception of the passivity-based approach put forward in [1] or the recent works in, e.g., [15, 23] for instance, which allow dealing with more general nonlinear heterogeneous models. Despite these two bodies of work about hybrid systems and networked systems, works addressing hybrid networked dynamical systems are scarcer: we could cite among others [27] for sampled-data systems, [4, 6, 21, 25, 28] for networked control systems, [5, 26] for networks with hybrid nodes. Yet, hybrid phenomena in multi-agent systems remain largely unexplored despite their fundamental importance.

In this context, the aim of this book is to bring together recent, exciting contributions in the field of hybrid, networked dynamical systems with an opening to related emergent topics. Although each chapter is self-contained, the book has been organized in such a way that theme-related chapters are grouped together. The book is thus organized into three parts, which all address key aspects of the topic.

The chapters in Part I deal with networked systems and focus on two main challenges of the field: distributed control and estimation. Chapters 2 and 3 present distributed control design tools to enforce synchronization or consensus properties within a network of nonlinear dynamical agents using distributed, local rules. Chapter 4 in Part I is devoted to recovering the communication graph given only output information from the nodes.

Part II presents a range of novel hybrid techniques for analysis, control, and estimation purposes. Chapter 5 presents recent advances in the study of multi-consensus for classes of hybrid multi-agent systems. Chapter 6 surveys recent results on the observer design for hybrid dynamical systems with known jump times. Chapter 7 introduces tools to analyze stability properties of discrete-time switched linear systems driven by switching signals generated by an important class of automata, namely, deterministic Büchi automata, where the alphabet coincides with the modes of the switched system. Chapters 8 and 9 of Part II present hybrid control design techniques dedicated to uncertain systems and systems subject to input saturation.

Finally, Part III is devoted to emerging tools for analysis and design. In particular, Chapt. 10 provides a survey of recent trends in the analysis of open multi-agent systems, which are characterized by their time-varying number of agents, and discusses three applications of major importance: consensus, optimization and epidemics. Chapter 11 provides insights on how event-triggering control techniques can be used to reduce the computational effort required by neural network controllers, which are currently attracting a lot of attention within the control community. Chapter 12 explains how to (locally) stabilize nonlinear dynamical systems by only relying on available input-state data, and how to estimate the corresponding region of attraction. Finally, Chapt. 13 surveys recent results on harmonic control, which shed new light on the rigorous, frequency control of general dynamical systems.

We believe that the variety of topics covered in this book and the overall tutorial writing style that many of the authors have used will render this book a pleasant reading both for experts in the field and for young researchers who seek for an intuitive understanding of the main recent highlights in these research areas.

Acknowledgements This work was supported by the "Agence Nationale de la Recherche" (ANR) under Grant HANDY ANR-18-CE40-0010.

References

1. Bai, H., Arcak, M., Wen, J.: Cooperative Control Design: A Systematic, Passivity-Based Approach. Springer Science & Business Media (2011)
2. Baldi, S., Papachristodoulou, A., Kosmatopoulos, E.B.: Adaptive pulse width modulation design for power converters based on affine switched systems. Nonlinear Anal. Hybrid Syst. **30**, 306–322 (2018)
3. Beneux, G., Riedinger, P., Daafouz, J., Grimaud, L.: Adaptive stabilization of switched affine systems with unknown equilibrium points: application to power converters. Automatica **99**, 82–91 (2019)
4. Berneburg, J., Nowzari, C.: Robust dynamic event-triggered coordination with a designable minimum inter-event time. IEEE Trans. Autom. Control **66**(8), 3417–3428 (2020)
5. Casadei, G., Isidori, A., Marconi, L.: About disconnected topologies and synchronization of homogeneous nonlinear agents over switching networks. Int. J. Robust Nonlinear Control **28**(3), 901–917 (2018)
6. Ceragioli, F., De Persis, C., Frasca, P.: Discontinuities and hysteresis in quantized average consensus. Automatica **47**(9), 1916–1928 (2011)
7. Ceragioli, F., Frasca, P., Piccoli, B., Rossi, F.: Generalized solutions to opinion dynamics models with discontinuities. In: Crowd Dynamics, Modeling and Social Applications in the Time of COVID-19, vol. 3, pp. 11–47. Springer, Berlin (2021)
8. Cofano, G., De Cicco, L., Mascolo, S.: Modeling and design of adaptive video streaming control systems. IEEE Trans. Control Netw. Syst. **5**(1), 548–559 (2016)
9. Dal Col, L., Tarbouriech, S., Zaccarian, L., Kieffer, M.: A consensus approach to PI gains tuning for quality-fair video delivery. Int. J. Robust Nonlinear Control **27**(9), 1547–1565 (2017)

10. Deaecto, G.S., Geromel, J.C., Garcia, F.S., Pomilio, J.A.: Switched affine systems control design with application to DC-DC converters. IET Control Theory Appl. **4**(7), 1201–1210 (2010)
11. Fax, J.A., Murray, R.M.: Information flow and cooperative control of vehicle formations. IEEE Trans. Autom. Control **49**(9), 1465–1476 (2004)
12. Frasca, P., Tarbouriech, S., Zaccarian, L.: Hybrid models of opinion dynamics with opinion-dependent connectivity. Automatica **100**, 153–161 (2019)
13. Goebel, R., Sanfelice, R.G., Teel, A.R.: Hybrid Dynamical Systems. Princeton University Press (2012)
14. Jadbabaie, A., Lin, J., Morse, A.S.: Coordination of groups of mobile autonomous agents using nearest neighbor rules. IEEE Trans. Autom. Control **48**(6), 988–1001 (2003)
15. Lee, J.G., Trenn, S., Shim, H.: Synchronization with prescribed transient behavior: heterogeneous multi-agent systems under funnel coupling. Automatica **141**, 110276 (2022)
16. Liberzon, D.: Switching in Systems and Control. Springer, Berlin (2003)
17. Lygeros, J., Tomlin, C., Sastry, S.: Hybrid systems: modeling, analysis and control. In: Electronic Research Laboratory, University of California, Berkeley, CA, Technical Report UCB/ERL M, vol. 99, p. 6 (2008)
18. Mariano, S., Morărescu, I.-C., Postoyan, R., Zaccarian, L.: A hybrid model of opinion dynamics with memory-based connectivity. IEEE Control Syst. Lett. **4**(3), 644–649 (2020)
19. Morărescu, I.-C., Girard, A.: Opinion dynamics with decaying confidence: application to community detection in graphs. IEEE Trans. Autom. Control **56**(8), 1862–1873 (2011)
20. Morărescu, I.-C., Martin, S., Girard, A., Muller-Gueudin, A.: Coordination in networks of linear impulsive agents. IEEE Trans. Autom. Control **61**(9), 2402–2415 (2016)
21. Nowzari, C., Garcia, E., Cortés, J.: Event-triggered communication and control of networked systems for multi-agent consensus. Automatica **105**, 1–27 (2019)
22. Olfati-Saber, R., Murray, R.M.: Consensus problems in networks of agents with switching topology and time-delays. IEEE Trans. Autom. Control **49**(9), 1520–1533 (2004)
23. Panteley, E., Loría, A.: Synchronization and dynamic consensus of heterogeneous networked systems. IEEE Trans. Autom. Control **62**(8), 3758–3773 (2017)
24. Panzani, G., Colombo, T., Savaresi, S.M., Zaccarian, L.: Hybrid control of a hydro-pneumatic tractor suspension. In: IEEE Conference on Decision and Control, pp. 250–255. Melbourne, Australia (2017)
25. De Persis, C., Postoyan, R.: A Lyapunov redesign of coordination algorithms for cyber-physical systems. IEEE Trans. Autom. Control **62**(2), 808–823 (2017)
26. Phillips, S., Sanfelice, R.G.: On asymptotic synchronization of interconnected hybrid systems with applications. In: American Control Conference, Seattle, U.S.A., pp. 2291–2296 (2017)
27. Poveda, J.I., Teel, A.R.: Hybrid mechanisms for robust synchronization and coordination of multi-agent networked sampled-data systems. Automatica **99**, 41–53 (2019)
28. Scheres, K.J.A., Dolk, V.S., Chong, M.S., Postoyan, R., Heemels, W.P.M.H.: Distributed periodic event-triggered control of nonlinear multi-agent systems. IFAC-PapersOnLine **55**(13), 168–173 (2022)
29. Torquati, L., Sanfelice, R., Zaccarian, L.: A hybrid predictive control algorithm for tracking in a single-phase DC/AC inverter. In: IEEE Conference on Control Technology and Applications, pp. 904–909. Kohala Coast (HI), USA (2017)
30. Tsitsiklis, J., Bertsekas, D., Athans, M.: Distributed asynchronous deterministic and stochastic gradient optimization algorithms. IEEE Trans. Autom. Control **31**(9), 803–812 (1986)

Part I
Networked Systems: Control and Estimation

Chapter 2
Contracting Infinite-Gain Margin Feedback and Synchronization of Nonlinear Systems

Daniele Astolfi and Vincent Andrieu

Abstract In this chapter, we study the problem of synchronization of homogeneous multi-agent systems in which the network is described by a connected graph. The dynamics of each single agent is described by an input-affine nonlinear system. We propose a feedback controller based on contraction analysis and Riemannian metrics. The starting point is the existence of a solution to a specific partial differential inequality which generalizes, in the nonlinear context, the well-known algebraic Riccati inequality. Further, we provide new results that allow to relax the Killing vector property so that to obtain a less stringent solution to be approximated with numerical methods. The proposed approach allows to design an infinite-gain margin feedback which is the fundamental ingredient to solve the problem of synchronization. We show that the synchronization problem is solved between two agents and we conjecture that the result holds for any connected graph. Simulations for a connected directed graph of Duffing oscillators corroborate the conjecture.

2.1 Introduction

Control of multi-agent systems [1] is a topic that attracted a lot of attention in control community due to its numerous applications in many fields, such as in power electronics, e.g., [2, 3]. One fundamental class of problems to be solved is the synchronization problem, that is, the problem of designing a distributed feedback law (i.e., that uses only information of agents that communicate) so that all the agents converge to a common trajectory. Such a problem has been completely solved for linear systems, see, e.g., [4–6] for homogeneous continuous-time dynamics. The key ingredient is to design a so-called diffusive coupling which weights the

D. Astolfi (✉) · V. Andrieu
CNRS and LAGEPP UMR CNRS 5007, Université Claude Bernard Lyon 1, Université de Lyon, F-69100 Villeurbanne, France
e-mail: daniele.astolfi@univ-lyon1.fr

V. Andrieu
e-mail: vincent.andrieu@univ-lyon1.fr

R. Postoyan et al. (eds.), *Hybrid and Networked Dynamical Systems*,
Lecture Notes in Control and Information Sciences 493,
https://doi.org/10.1007/978-3-031-49555-7_2

distance between any two communicating trajectories. Furthermore, the feedback gain is designed via the use of algebraic Riccati equations that allow for an infinite-gain margin feedback that allow for an infinite-gain margin feedback [7–9], that is, a feedback that can be multiplied by any positive number and is still a "good feedback". In turn, such an infinite-gain margin law is a robust control allowing to cope with the fact that the Laplacian matrix has a potential different impact on the diffusive coupling terms acting on each agent. For nonlinear dynamics, however, the synchronization problem is still open, due to the intrinsic difficulties arising from nonlinearities. Remarkable results have been obtained when the dynamics are passive [10], in Lur'e form [11], or controllable via high-gain feedback [12]. Contraction tools have been also employed in [13–15], with possibly non-smooth dynamics [16] or integral diffusive coupling [17]. Simple diffusive couplings containing integral actions have been proposed in [18] and some researchers also started studying the problem of synchronization of heterogeneous dynamics [19]. Yet, for nonlinear input-affine systems, the theory is still non-exhaustive.

The objective of this work is to propose further and new results in the theory of synchronization of multi-agent systems described by nonlinear dynamics building up on the set of recent results [20–22]. To this end, we rely on contraction theory and Riemannian metric analysis [23–26]. Similar to the Control Contraction Metric conditions in [27], we propose a new set of sufficient conditions providing the existence of a contracting infinite-gain margin law [8]. This set of conditions, already proposed in [8, 20], is based on the existence of a solution to a partial differential equation generalizing the algebraic Riccati equation for stabilization. Furthermore, a Killing vector condition must be satisfied. As a consequence, in this chapter, we relax such a Killing vector condition, allowing for more treatable conditions when one is interested in solving such a partial differential equation with numerical tools, such as neural networks, see, e.g., [21]. Based on the aforementioned new conditions, we show the existence of a new nonlinear infinite-gain margin law for input-affine nonlinear systems. Then, we show that the problem of synchronization between two agents can be solved by a diffusive coupling directly obtained from the previous conditions. We also conjecture that the same type of control law can solve a synchronization problem for any communication graph. Finally, we illustrate the new conditions on the synchronization problem in which each agent is described by a Duffing oscillators and the graph is direct and connected.

2.2 Preliminaries

2.2.1 Notation

Given a vector field $f : \mathbb{R}^n \to \mathbb{R}^n$ and a 2-tensor $P : \mathbb{R}^n \to \mathbb{R}^{n\times n}$ both C^1, we indicate with $L_f P(x)$ the Lie derivative of the tensor P along f defined as

$$L_f P(x) := \mathfrak{d}_f P(x) + P(x)\tfrac{\partial f}{\partial x}(x) + \tfrac{\partial f}{\partial x}^\top(x) P(x)\,,$$
$$\mathfrak{d}_f P(x) := \lim_{h\to 0} \frac{P(X(x,t+h)) - P(x)}{h}\,,$$

where $X(x,t)$ is the solution of the initial value problem

$$\tfrac{\partial}{\partial t} X(x,t) = f(X(x,t)), \quad X(x,0) = x,$$

for all $t \geq 0$. Note that $L_f P(x)$ can be equivalently expressed as

$$L_f P(x) = \lim_{h\to 0} \frac{(I + h\frac{\partial f}{\partial x}(x))^\top P(x + hf(x))(I + h\frac{\partial f}{\partial x}(x)) - P(x)}{h}\,,$$

with coordinates

$$(L_f P(x))_{i,j} = \sum_k \left[2P_{ik}\frac{\partial f_k}{\partial x_j}(x) + \frac{\partial P_{ij}}{\partial x_k}(x) f_k(x) \right].$$

Given the tensor P any two elements $x_1, x_2 \in \mathbb{R}^n$, let $\gamma : [0,1] \mapsto \mathbb{R}^n$ as any C^1 path such that $\gamma(0) = x_1$ and $\gamma(1) = x_2$. We define the length of the curve γ as

$$\ell_P(\gamma) := \int_0^1 \sqrt{\frac{d\gamma}{ds}(s)^\top P(\gamma(s)) \frac{d\gamma}{ds}(s)}\, ds\,. \tag{2.1}$$

The Riemannian distance between x_1 and x_2 is then defined as the infimum of the length among all the possible piecewise C^1 paths γ, namely

$$\mathrm{d}_P(x_1, x_2) := \inf_\gamma \{\ell_P(\gamma)\}\,.$$

For more details on Riemannian analysis, we refer to [28] and references therein or to [29].

2.2.2 *Highlights on Graph Theory*

In a general framework, a communication graph is described by a triplet $\mathcal{G} = \{\mathcal{V}, \mathcal{G}, \mathcal{A}\}$ in which $\mathcal{V} = \{v_1, v_2, \ldots, v_N\}$ is a set of $N \subset \mathbb{N}$ *vertexes* (or nodes), $\mathcal{G} \subset \mathcal{V} \times \mathcal{V}$ is the set of *edges* e_{jk} that models the interconnection between the vertexes with the flow of information from vertex j to vertex k *weighted* by the (k, j)-th entry $a_{kj} \geq 0$ of the *adjacency matrix* $\mathcal{A} \in \mathbb{R}^{N\times N}$. We denote by $\mathcal{L} \in \mathbb{R}^{N\times N}$ the *Laplacian matrix* of the graph, defined as

$$\ell_{kj} = -a_{kj} \text{ for } k \neq j, \qquad \ell_{kj} = \sum_{i=1}^{N} a_{ki} \text{ for } k = j,$$

where $\ell_{j,k}$ is the (j,k)-th entry of $\mathcal{L}$. We denote with $\mathcal{N}_i$ the set of in-neighbors of node i, i.e., the set $\mathcal{N}_i := \{j \in \{1, \ldots, N\} \mid e_{ji} \in \mathcal{G}\}$. A time-invariant graph is said to be *connected* if and only if $\mathcal{L}$ has only one trivial eigenvalue $\lambda_1(\mathcal{L}) = 0$ and all other eigenvalues $\lambda_2(\mathcal{L}), \ldots, \lambda_N(\mathcal{L})$ have strictly positive real parts (see [30]).

2.2.3 Problem Statement

We consider a network of N agents in which each agent dynamics is described by an input-affine nonlinear ODE of the form

$$\dot{x}_i = f(x_i) + g(x_i)u_i, \qquad i = 1, \ldots, N, \tag{2.2}$$

where $x_i \in \mathbb{R}^n$ is the local state of node i and $u_i \in \mathbb{R}^m$ is its control input. We denote the state of the entire network as

$$\mathbf{x} := \text{col}\{x_1^\top, \ldots, x_N^\top\}^\top \in \mathbb{R}^{Nn}. \tag{2.3}$$

Furthermore, we denote with $X_i(x_i^\circ, t)$ the trajectory of agent i starting from the initial condition $x_i^\circ(0)$ evaluated at time $t \geq 0$, and with $\mathbf{X}(\mathbf{x}^\circ, t)$ the trajectory of the entire network (2.3) evaluated at initial condition $\mathbf{x}^\circ \in \mathbb{R}^{Nn}$ at time $t \geq 0$. Our synchronization objective is to design a nonlinear diffusive coupling, namely a distributed feedback control law of the form

$$u_i = \sum_{j \in \mathcal{N}_i} a_{ij} \Big[\varphi(x_j) - \varphi(x_i)\Big] = -\sum_{j=1}^{N} \ell_{ij}\varphi(x_j) \tag{2.4}$$

for all $i = 1, \ldots, N$, for some C^1 function $\varphi : \mathbb{R}^n \to \mathbb{R}^m$, that stabilizes the dynamics (2.2) on the so-called *synchronization manifold* $\mathcal{D}$ defined as

$$\mathcal{D} := \{\mathbf{x} \in \mathbb{R}^{Nn} \mid x_i = x_j, \quad \text{for all } i, j \in \{1, \ldots, N\}\}, \tag{2.5}$$

where the states of all the agents of the network agree with each other. By construction, the i-th agent uses only the information x_j of its neighborhoods $j \in \mathcal{N}_i$ and its own local information x_i. Furthermore, the control action u_i is equal to zero on the synchronization manifold. In other words, when consensus is achieved, no correction term is needed for each individual agent. As a consequence, stabilizing all the agents on a desired equilibrium point is generally not a valid solution in such a framework. We formalize our synchronization problem as follows.

Problem 2.2.1 (Network synchronization) *Let the function φ be such that the manifold $\mathcal{D}$ defined in* (2.5) *is globally uniformly exponentially stable for the closed-loop system*

$$\dot{x}_i = f(x_i) - \sum_{j \in \mathcal{N}_i} \ell_{ij} \varphi(x_j), \qquad i = 1, \ldots, N,$$

namely, there exist positive constants k and $\lambda > 0$ such that, for all $\mathbf{x}^\circ$ in $\mathbb{R}^{Nn}$ and for all $t \geq 0$ in the time domain of existence of solutions $\mathcal{T} \subseteq \mathbb{R}$, we have

$$|\mathbf{X}(\mathbf{x}^\circ, t)|_{\mathcal{D}} \leq k \exp(-\lambda t)\, |\mathbf{x}^\circ|_{\mathcal{D}}. \tag{2.6}$$

Then, we say that the distributed feedback control law (2.4) *solves the* global exponential synchronization problem *for the network* (2.3).

2.2.4 From Linear to Nonlinear Synchronization

To better contextualize the results of our paper, it is useful to recall some important aspects on synchronization of homogeneous network of linear systems, see, e.g., [6, Chap. 5]. Consider a network where each agent is described by

$$\dot{x}_i = Ax_i + Bu_i, \qquad i = 1, \ldots N, \tag{2.7}$$

where $x \in \mathbb{R}^n$ is the state, $u \in \mathbb{R}^m$ is the control action, and A, B are matrices of appropriate dimension. The following result holds, see [6, Proposition 5.2].

Proposition 2.2.2 (Synchronization of linear systems) *Consider a connected network $\mathcal{G} = \{\mathcal{V}, \mathcal{E}, \mathcal{A}\}$ where each agent is described by* (2.7). *Assume there exists a matrix K such that the matrix $A - \lambda_{\mathcal{L}} BK$ is Hurwitz for all $\lambda_{\mathcal{L}} \in \operatorname{spec}\{\mathcal{L}\} \setminus \{0\}$. Then the network in closed loop with the distributed control law*

$$u_i = K \sum_{j \in \mathcal{N}_i} a_{ij}(x_j - x_i) = -K \sum_{j=1}^{n} \ell_{ij} x_j \tag{2.8}$$

achieves synchronization, i.e., $\lim_{t \mapsto +\infty} \left| x_i(t) - x_j(t) \right| = 0$ *for all* $(i, j) \in \mathcal{V} \times \mathcal{V}$.

In other words, for a connected network of linear systems, the synchronization problem can be seen as a *robust* (or simultaneous) stabilization problem. With "robust" we mean that the stabilization problem must be achieved for any strictly positive eigenvalue of the Laplacian $\lambda_{\mathcal{L}}$, which can be seen as a gain acting on the control term. To fulfill this requirement, a solution is given by employing an *infinite-gain margin* feedback (see e.g., [9, Sect. 3]), that is, a feedback law that achieves stability in the presence of an uncertain factor in front of the gain matrix B, as formalized in the next definition.

Definition 2.2.3 *Given a pair (A, B), we say that K is an infinite-gain margin feedback if the matrix $(A - \kappa BK)$ is Hurwitz for any $\kappa \in [1, \infty)$.*

For linear systems, a general sufficient and necessary condition can be stated as follows.

Lemma 2.2.4 *There exists an infinite-gain margin feedback law for a pair (A, B) if and only if the pair (A, B) is stabilizable. Furthermore, a solution is given by $u = Kx$, with*

$$PA + A^\top P - PBR^{-1}B^\top P = -Q, \qquad K = \tfrac{1}{2}R^{-1}B^\top P, \tag{2.9}$$

for some positive definite symmetric matrices Q, R.

It can be remarked that the solution is based on the algebraic Riccati equation (2.9) that admits a solution if the pair (A, B) is stabilizable. Coming back to our synchronization problem, we may state now the following result.

Lemma 2.2.5 (Synchronization of linear systems) *Consider a network $\mathcal{G} = \{\mathcal{V}, \mathcal{G}, \mathcal{A}\}$ with node dynamics described by* (2.7) *and suppose the pair (A, B) is stabilizable. Then, there exists $\kappa^* > 0$ (which depends on the eigenvalues of the Laplacian matrix L) such that the distributed control law* (2.8)*, with K selected as in* (2.9)*, solves the synchronization Problem 2.2.1 for any $\kappa > \kappa^*$.*

The proof of such a result is shown, for instance, in [5, Sect. II.B]. See also [6, Chap. 5] and references therein for more details. Note that, in general, the value of κ^* depends on the eigenvalue of the Laplacian matrix $\mathcal{L}$ with smaller real part (such an eigenvalue is linked to the connectivity of the graph).

The take-away message we aim to highlight in this section is that, if we aim at developing a theory for general nonlinear dynamics and generic connected networks, we need to be able to solve a robust (incremental) stabilization problem as in Proposition 2.2.2. As a consequence, the key property is given by the extension of the aforementioned infinite-gain margin law in the contraction framework, see, e.g., [7]. This symmetry will be further developed in the next sections.

2.3 Robust Contractive Feedback Design

2.3.1 Riemannian Contraction Conditions for δISS

We begin by considering autonomous nonlinear systems of the form

$$\dot{x} = f(x) \tag{2.10}$$

where $f : \mathbb{R}^n \to \mathbb{R}^n$ is a C^2 vector field. We denote by $X(x_0, t)$ the solution of system (2.10) with initial condition x_0 evaluated at time $t \geq 0$. We assume existence and uniqueness of trajectories. The define the notion of incremental stability according to the following definition.

Definition 2.3.1 *System* (2.10) *is incrementally globally exponentially stable* (*δGES*) *if there exists two strictly positive real numbers* $\lambda, k > 0$ *such that*

$$|X(x_1, t) - X(x_2, t)| \leq k\, |x_1 - x_2|\, e^{-\lambda t} \tag{2.11}$$

for any couple of initial conditions $(x_1, x_2) \in \mathbb{R}^n \times \mathbb{R}^n$ *and for all* $t \geq 0$.

Following the metric approach in [23, 26], a dynamical system of the form (2.10) is δGES if there exists a Riemannian metric for which the mapping $t \mapsto X(x, t)$ is a contracting mapping.

Theorem 2.3.2 *Consider system* (2.10) *and suppose there exist a* C^1 *matrix function* $P : \mathbb{R}^n \to \mathbb{R}^{n\times n}$ *taking symmetric positive definite values and three real numbers* $\overline{p}, \underline{p}, \rho > 0$ *such that the following holds*

$$0 \prec \underline{p} I \preceq P(x) \preceq \overline{p} I\,, \tag{2.12a}$$

$$L_f P(x) \preceq -\rho P(x)\,, \tag{2.12b}$$

for all $x \in \mathbb{R}^n$. *Then the system* (2.10) *is δGES.*

A proof can be found in [23, 25]. A converse theorem can be found in [25, Proposition IV] in the case in which f is a globally Lipschitz vector field. Note that the lower bound in (2.12a) is required to make sure that the whole $\mathbb{R}^n$ space endowed with the Riemannian metric P is complete. Such a condition guarantees that every geodesic (i.e., the shortest curve between (x_1, x_2)) can be maximally extended to $\mathbb{R}$, see, e.g., [28]. By Hopf–Rinow's theorem (see [29, Theorem 1.1]), this implies that the metric is complete and hence that the minimum of the length of any curve γ connecting two point (x_1, x_2) is actually given by the length of the geodesic at any time instant. Moreover, it guarantees that the Lyapunov function defined as $\|\delta_x\|_P^2 := \delta_x^\top P(x)\delta_x$ associated to $\dot{\delta}_x = \frac{\partial f}{\partial x}(x)\delta_x$, is radially unbounded and exponentially decreasing along solutions. The upper bound in (2.12a) is introduced for solutions to be uniformly decreasing with respect to time and to correlate the Riemaniann distance in P to the Euclidean one in (2.11).

We study now the incremental input-to-state (δISS) properties of a system of the form

$$\dot{x} = f(x) + g(x)u \tag{2.13}$$

where $x \in \mathbb{R}^n$ is the state, $u \in \mathbb{R}^m$ is an exogenous signal and $f : \mathbb{R}^n \to \mathbb{R}^n$ and $g : \mathbb{R}^n \to \mathbb{R}^{n\times m}$ are C^2 functions. We denote by $X(x, u, t)$ the solution of system (2.13) starting at initial condition x at time t with input $u = u(t)$ and satisfying the initial value problem

$$\begin{aligned} X(x,u,0) &= x,\\ \frac{\partial X}{\partial t}(x,u,t) &= f(X(x,u,t)) + g(X(x,u,t))u(t). \end{aligned} \tag{2.14}$$

We state the following definition of incremental input-to-state stability (δISS). Note that, in this work, we restrict our attention to δISS properties (2.15) with exponential convergence $e^{-\lambda t}$, but more general definitions considers more general convergences rate characterized by class-$\mathcal{KL}$ functions, see, e.g., [31].

Definition 2.3.3 *System* (2.13) *is incrementally input-to-state stable (δISS) with exponential decay rate, if there exist positive real numbers $k, \lambda, \gamma > 0$ such that*

$$|X(x_1,u_1,t) - X(x_2,u_2,t)| \leq k\,|x_1 - x_2|\,e^{-\lambda t} + \gamma \sup_{s\in[0,t)} |u_1(s) - u_2(s)| \tag{2.15}$$

for all initial conditions $x_1, x_2 \in \mathbb{R}^n$ and for all inputs u_1, u_2 taking values in $\mathcal{U} \subseteq \mathbb{R}^m$, for all $t \geq 0$.

Similar to the result of Theorem 2.3.2, we aim to look for some metric-based sufficient conditions to establish an incremental ISS property. For this, we introduce the notion of Killing vector field.[1]

Definition 2.3.4 *Given a C^1 2-tensor $P : \mathbb{R}^n \to \mathbb{R}^{n\times n}$ and a C^1 matrix function $g : \mathbb{R}^n \to \mathbb{R}^{n\times m}$, we say that g is a Killing vector field with respect to P if*

$$L_{g_i} P(x) = 0, \qquad i = 1,\ldots,m, \qquad \forall x \in \mathbb{R}^n, \tag{2.16}$$

with g_i being the i-th column of g.

The Killing vector property implies that distances between different trajectories generated by the vector field $g(x)$ in the norm $|\cdot|_{P(x)}$ are invariant. Basically, the signals that enter in the directions of the vector field g do not affect the distances, in the sense that different trajectories of the differential equation $\dot{x} = g(x)$ have a distance (associated with the norm provided by P) among them which is constant for any $t \geq 0$. Note that the Killing vector property is always satisfied between two constant matrices P and G.

Based on the previous notion of Killing vector, we have the following result.

Theorem 2.3.5 *Consider system* (2.13) *and suppose that g is a bounded vector field, namely, there exists a real number $\bar{g} > 0$ such that $|g(x)| \leq \bar{g}$ for all $x \in \mathbb{R}^n$. If there exists a C^1 matrix function $P : \mathbb{R}^n \to \mathbb{R}^{n\times n}$ taking symmetric positive definite values and three real numbers $\overline{p}, \underline{p}, \rho > 0$ satisfying*

[1] The name "Killing vector field" takes the name after Wilhelm Killing, a German mathematician.

$$0 \prec \underline{p} I \preceq P(x) \preceq \overline{p} I, \tag{2.17a}$$

$$L_f P(x) \preceq -\rho P(x), \tag{2.17b}$$

$$L_g P(x) = 0, \tag{2.17c}$$

for all $x \in \mathbb{R}^n$, then system (2.13) *is δISS with respect to u.*

A proof of Theorem 2.3.5 can be found in [8].

2.3.2 Infinite-Gain Margin Laws

Consider now a nonlinear system of the form

$$\dot{x} = f(x) + g(x)(u + d) \tag{2.18}$$

with state $x \in \mathbb{R}^n$ control input $u \in \mathbb{R}^m$ and perturbation $d \in \mathbb{R}^m$ satisfying a matching condition [32]. We look for a feedback design satisfying an infinite-gain margin property. We remark that the design of infinite-gain margin laws in the context of input-affine nonlinear systems of the form (2.18) has been investigated in the context of control Lyapunov function [9, Chap. 3] and arise quite naturally in the context of feedback design for passive systems. Here, we study an extension of the linear case in the context of incremental input-to-state stability. The following definition is given.

Definition 2.3.6 *We say that the C^1 function $\psi : \mathbb{R}^n \to \mathbb{R}^m$ is a contractive control law with infinite-gain margin for system* (2.18), *if the system*

$$\dot{x} = f(x) + g(x)\,\kappa\,\psi(x)$$

is δGES for any $\kappa \geq 1$.

Based on the results of Theorem 2.3.5, we have the following "trivial result".

Proposition 2.3.7 *Consider system* (2.18). *Suppose there exists a C^1 function $\psi : \mathbb{R}^p \to \mathbb{R}^m$ and a C^1 matrix function $P : \mathbb{R}^n \to \mathbb{R}^{n\times n}$, taking symmetric positive definite values and three real numbers $\overline{p}$, $\underline{p}$, $\rho > 0$ such that the following holds*

$$0 \preceq \underline{p} I \preceq P(x) \preceq \overline{p} I \tag{2.19a}$$

$$L_{f(x)+\kappa\, g(x)\,\psi(x)} P(x) \preceq -\rho P(x), \tag{2.19b}$$

for all $x \in \mathbb{R}^n$ and all $\kappa \geq 1$. Then, the feedback $u = \psi(x)$ is a contractive control law with infinite-gain margin for system (2.18).

The result of Proposition 2.3.7 being not very constructive, we provide now a set of sufficient conditions which extend *de facto* the linear Algebraic Riccati Inequality (2.9) in our metric context.

Theorem 2.3.8 *Consider system* (2.18), *and suppose there exist a* C^1 *matrix function* $P : \mathbb{R}^n \to \mathbb{R}^n \times \mathbb{R}^n$ *taking positive definite symmetric values, a function* $\alpha : \mathbb{R}^n \to \mathbb{R}^m$ *and real numbers* $\underline{p}, \bar{p}, \rho > 0$ *such that the following hold*

$$0 \preceq \bar{p}I \preceq P(x) \preceq \bar{p}I\,, \tag{2.20a}$$

$$L_f P(x) - P(x)g(x)R^{-1}g^\top(x)P(x) \preceq -\rho P(x)\,, \tag{2.20b}$$

$$L_g P(x) = 0\,, \tag{2.20c}$$

$$\frac{\partial \alpha}{\partial x}^\top(x) = P(x)g(x)\,, \tag{2.20d}$$

for all $x \in \mathbb{R}^n$ *and for some positive definite symmetric matrix* $R \in \mathbb{R}^{m\times m}$. *Then, the function*

$$\psi(x) = -\tfrac{1}{2}R^{-1}\alpha(x) \tag{2.21}$$

is a contractive control law with infinite-gain margin for system (2.18) *according to Definition 2.3.6. Moreover, the closed-loop system is* δ*ISS w.r.t.* d.

A proof of Theorem 2.3.8 can be found in [8]. Note that conditions (2.20) have been also denoted as Control Contraction Metrics in [27, Sect. III.A].

2.3.3 New Relaxed Conditions for Infinite-Gain Margin Laws

The conditions in Theorem 2.3.8 are in general very demanding to be satisfied. For this reason, it is important to study the existence of relaxed conditions. For instance, in the work [21], both integrability and Killing conditions have been assumed to be approximated. Here, we follow a different route based on the next technical result. To this end, we focus here on the case in which there is only one input, namely $m = 1$. Consider a vector field $g : \mathbb{R}^n \mapsto \mathbb{R}$ and a metric P.

Lemma 2.3.9 *Let* $m = 1$ *and suppose that there exist functions* $\alpha : \mathbb{R}^n \mapsto \mathbb{R}$ $\theta : \mathbb{R}^n \mapsto \mathbb{R}_{>0}$ *and* $\gamma : \mathbb{R} \mapsto \mathbb{R}$ *such that*

$$\frac{\partial \alpha}{\partial x}(x)^\top = P(x)g(x)\theta(x), \qquad L_{g\theta}P(x) = \gamma(\alpha(x))\frac{\partial \alpha}{\partial x}(x)^\top\frac{\partial \alpha}{\partial x}(x), \qquad \forall x \in \mathbb{R}^n.$$

Then, there exists $\beta : \mathbb{R} \mapsto \mathbb{R}_{>0}$ *such that* $L_{g\theta\beta}P(x) = 0$ *for all* $x \in \mathbb{R}^n$.

Proof Note that

$$\begin{aligned} L_{g\theta\beta}P(x) = \beta(\alpha(x))L_{g\theta}P(x) + P(x)g(x)\theta(x)\beta'(\alpha(x))\frac{\partial \alpha}{\partial x}(x) \\ + \frac{\partial \alpha}{\partial x}(x)^\top\beta'(\alpha(x))\theta(x)g(x)^\top P(x) \end{aligned}$$

which gives

$$L_{g\theta\beta}P(x) = \left[\beta(\alpha(x))\gamma(\alpha(x)) + 2\beta'(\alpha(x))\right]\frac{\partial\alpha}{\partial x}(x)^\top\frac{\partial\alpha}{\partial x}(x)$$

Hence, if we select

$$\beta(\alpha) = \beta_0 \exp\left(-\frac{1}{2}\int_0^\alpha \gamma(s)ds\right),$$

for any $\beta_0 \in \mathbb{R}$, it yields

$$\beta'(\alpha) = -\frac{1}{2}\gamma(\alpha)\beta(\alpha)$$

and consequently the desired result. □

Based on the previous technical lemma, we have the following theorem ensuring the existence of a contractive infinite-gain margin feedback with relaxed Killing vector conditions. Note that, in this case, we may generically lose the global δ-ISS properties with respect to the disturbance d.

Theorem 2.3.10 *Consider system* (2.18)*, and suppose there exist a* C^1 *matrix function* $P : \mathbb{R}^n \to \mathbb{R}^n \times \mathbb{R}^n$ *taking positive definite symmetric values, functions* $\alpha : \mathbb{R}^n \mapsto \mathbb{R}$, $\theta : \mathbb{R}^n \mapsto \mathbb{R}_{>0}$ *and* $\gamma : \mathbb{R} \mapsto \mathbb{R}$ *such and real numbers* $\underline{p}, \bar{p}, \rho, \underline{\theta}, r, \overline{\gamma} > 0$ *such that the following hold*

$$0 \preceq \underline{p}I \preceq P(x) \preceq \bar{p}I\,, \qquad 0 < \underline{\theta} \leq \theta(x)\,, \qquad \int_0^{\alpha(x)} \gamma(s)ds \leq \overline{\gamma} \tag{2.22a}$$

$$L_f P(x) - \frac{1}{r}P(x)g(x)g^\top(x)P(x) \preceq -\rho P(x)\,, \tag{2.22b}$$

$$L_{g\theta}P(x) = \gamma(\alpha(x))\frac{\partial\alpha}{\partial x}(x)^\top\frac{\partial\alpha}{\partial x}(x)\,, \tag{2.22c}$$

$$\frac{\partial\alpha}{\partial x}^\top(x) = P(x)g(x)\theta(x)\,, \tag{2.22d}$$

for all $x \in \mathbb{R}^n$*. Then, the function* $\psi : \mathbb{R}^n \to \mathbb{R}$ *defined as*

$$\psi(x) = \theta(x)\beta(\alpha(x))\alpha(x)\,, \qquad \beta(\alpha) = \beta_0 \exp\left(-\frac{1}{2}\int_0^\alpha \gamma(s)ds\right), \tag{2.23}$$

with $\beta_0 > 0$ *sufficiently large, is a contractive control law with infinite-gain margin for system* (2.18) *according to Definition 2.3.6.*

Proof Note that we have

$$\begin{aligned} L_{f+\kappa g\psi}P(x) =& L_f P(x) + \kappa\alpha(x)L_{g\theta\beta}P(x) \\ & + \kappa\theta(x)\beta(\alpha(x))\left[P(x)g(x)\frac{\partial\alpha}{\partial x}(x) + \frac{\partial\alpha}{\partial x}(x)^\top g(x)^\top P(x)\right] \end{aligned}$$

Using Lemma 2.3.9 and (2.22b), the previous inequalities gives

$$
\begin{aligned}
L_{f+\kappa g\psi}P(x) &\preceq -\rho P(x) + \left[\frac{1}{r} - 2\kappa\theta(x)^2\beta(\alpha(x))\right] P(x)g(x)g(x)^\top P(x) \\
&\preceq -\rho P(x) + \left[\frac{1}{r} - 2\kappa\beta_0\underline{\theta}^2 e^{-\overline{\gamma}}\right] P(x)g(x)g(x)^\top P(x)
\end{aligned}
$$

Hence, selecting $\beta_0 \geq \frac{1}{2r\underline{\theta}^2 e^{-\overline{\gamma}}}$ yields

$$
L_{f+g\kappa\psi}P(x) \preceq -\rho P(x)
$$

for all x in $\mathbb{R}^n$ and for all $\kappa \geq 1$, concluding the proof. □

We can note that, by taking $\theta = 1$ and $\gamma = 0$ in the previous theorem, the conditions in (2.22) boils down to the conditions (2.20) of Theorem 2.3.8.

2.4 Synchronization of Nonlinear Systems

Based on the previous conditions, we now study the problem of synchronization systems described above in Sect. 2.2.3. First, we recall the results [21, 22] based on the of Theorem 2.3.8 (involving the exact Killing conditions) proved for undirected graphs. Then, we investigate the use of a diffusive coupling based on the conditions of the new Theorem 2.3.10 in the context of synchronization of two agents. We finally conclude this section by conjecturing a general solution for any connected graph.

2.4.1 *Undirected Graphs and Killing Vector Condition*

In this section, we consider undirected graphs and we show that the conditions of Theorem 2.3.8 can be employed to design a distributed diffusive coupling solving the synchronization problem for general nonlinear input-affine dynamics. Note that the network being *undirected*, the communication links are bi-directional (i.e., $e_{ij} = e_{ji}$ for every $i, j = 2, \ldots, N$). We have the following result.

Theorem 2.4.1 *Consider a network* $G = \{\mathcal{V}, \mathcal{G}, \mathcal{A}\}$ *of agents* (2.2)*, and suppose that the communication graph is connected and undirected. Moreover, suppose that the functions* f, g *satisfies the conditions (2.20) of Theorem 2.3.8. Then, the distributed state-feedback control law* (2.4) *with* ψ *selected as in Theorem 2.3.8, namely*

$$
u_i = -\kappa\tfrac{1}{2}R^{-1}\sum_{j=1}^{N} \ell_{ij}\varphi(x_j)
$$

solves the synchronization Problem 2.2.1 for the network of agents given in (2.2) for any $\kappa > \kappa^*$, where κ^* depends on the eigenvalues of the Laplacian matrix L.

The proof of the previous theorem can be found in [21] in the context of leader-connected undirected graphs and in [22] for generic undirected graphs.

2.4.2 Synchronization of Two Agents with Relaxed Killing Conditions

The main limitation of the approach presented in the previous theorem is the complexity of finding a metric P solving all the conditions in (2.20), and in particular satisfying the Killing vector field property in (2.20c) and the integrability condition in (2.20d). As a consequence, building up on the relaxed results of Theorem 2.3.10 we show that synchronization can still be obtained under the approximate integrability condition or approximate Killing vector assumption. We formalize such a result in the context of state synchronization between two agents.

Theorem 2.4.2 *Consider two agents which dynamics is in the form*

$$\begin{aligned} \dot{x}_1 &= f(x_1) + g(x_1)u_1 \,, \\ \dot{x}_2 &= f(x_2) + g(x_2)u_2 \,, \end{aligned} \tag{2.24}$$

and suppose the conditions (2.22) of Theorem 2.3.10 are satisfied. Then, the distributed state-feedback control law

$$\begin{aligned} u_1 &= \kappa\theta(x_1)\beta(\alpha(x_1))\Big[\alpha(x_2) - \alpha(x_1)\Big] \\ u_2 &= \kappa\theta(x_2)\beta(\alpha(x_2))\Big[\alpha(x_1) - \alpha(x_2)\Big] \end{aligned} \tag{2.25}$$

solves the synchronization Problem 2.2.1 when there are only two agents and for any κ sufficiently large.

Proof The main goal is to show that the norm of the difference between agent x_1 and x_2 exponentially decreases to zero. Given (x_1°, x_2°) in $\mathbb{R}^{2n}$, let τ be the time of existence of the solution of the system initialized in (x_1°, x_2°). For t in $[0, \tau)$, let $(X_1(t), X_2(t))$ denote this solution. Consider the function $\Gamma : [0, 1] \times [0, \tau] \mapsto \mathbb{R}^n$, which satisfies $\Gamma(s, t_0) = x_1^\circ + s\,(x_2^\circ - x_1^\circ)$, and where Γ is the solution of the following ordinary differential equation for $0 \leq t < \tau$

$$\frac{\partial \Gamma}{\partial t}(s,t) = f(\Gamma(s,t)) + \kappa\tilde{g}(\Gamma(s,t))\Big[\alpha(X_2(t)) + \alpha(X_1(t)) - 2\alpha(\Gamma(s,t))\Big]\,, \tag{2.26}$$

where we denoted $\tilde{g}(x) = \theta(x)\beta(x)g(x)$. Note that, by uniqueness of the solution, Γ satisfies

$$\Gamma(0,t) = X_1(t), \quad \Gamma(1,t) = X_2(t), \qquad \forall t \in [0,\tau) . \tag{2.27}$$

Consider now the function $V : [0,\tau) \mapsto \mathbb{R}_+$ defined by

$$V(t) = \int_0^1 \frac{\partial \Gamma}{\partial s}(s,t)^\top P(\Gamma(s,t)) \frac{\partial \Gamma}{\partial s}(s,t) ds. \tag{2.28}$$

Note that, for all (k,l) in $\{1,\ldots,n\}^2$, we have

$$\frac{d}{dt}[P(\Gamma(s,t))_{kl}] = \frac{\partial P_{kl}}{\partial x}(\Gamma(s,t)) \frac{\partial \Gamma}{\partial t}(s,t).$$

This implies that for all vector ν in $\mathbb{R}^n$,

$$\begin{aligned} \frac{d}{dt}\left[\nu^\top P(\Gamma(s,t))\nu\right] =& \nu^\top \mathfrak{d}_f P(\zeta_i(s,t)))\nu \\ & -\kappa \nu^\top \mathfrak{d}_{\tilde{g}} P(\Gamma(s,t))\nu \Big[\alpha(X_2(t)) - \alpha(X_1(t)) - 2\alpha(\Gamma(s,t))\Big] . \end{aligned}$$

Hence, we get

$$\begin{aligned} \dot{V}(t) = & \int_0^1 \frac{\partial \Gamma}{\partial s}(s,t)^\top (L_f P(\Gamma(s,t)) + L_{\tilde{g}} P(\Gamma(s,t)) \times \\ & \quad \times (\alpha(X_2(t) - \alpha(X_1(t) - 2\alpha(\Gamma(s,t)))) \frac{\partial \Gamma}{\partial s}(s,t) ds \\ & -4 \int_0^1 \frac{\partial \Gamma}{\partial s}(s,t)^\top P(\Gamma(s,t) \tilde{g}(\Gamma(s,t)) \frac{\partial \alpha}{\partial x}(\Gamma(s,t)) \frac{\partial \Gamma}{\partial s}(s,t). \end{aligned} \tag{2.29}$$

Note that by Lemma 2.3.9, $L_{\tilde{g}} P(x) = 0$. Moreover, with the integrability condition (2.22d), the time derivative of V becomes

$$\begin{aligned} \dot{V}(t) = & \int_0^1 \frac{\partial \Gamma}{\partial s}(s,t)^\top L_f P(\Gamma(s,t)) \frac{\partial \Gamma}{\partial s}(s,t) ds \\ & \qquad -4 \int_0^1 \theta(\Gamma(s,t))^2 \beta(\Gamma(s,t)) \Psi(s,t) \Psi(s,t)^\top ds, \end{aligned} \tag{2.30}$$

where

$$\Psi(s,t) = \frac{\partial \Gamma}{\partial s}(s,t)^\top P(\Gamma(s,t) g(\Gamma(s,t)) . \tag{2.31}$$

From there, with the infinitesimal detectability condition (2.22b), we get

$$\dot{V}(t) \leq -\rho V(t) + \int_0^1 \left[\left(\frac{1}{r} - 2\kappa \underline{\theta}^2 \beta_0 \right) \Psi(s,t) \Psi(s,t)^\top \right] ds . \tag{2.32}$$

Therefore, by selecting κ sufficiently large, we get

$$\dot{V}(t) \leq -\rho V(t).$$

From Gronwall's lemma, this implies $V(t) \leq \exp(-\rho(t))\, V(0)$ for all $t \in [0, \tau)$. Moreover, with (2.27), it yields

$$V(t) \geq \underline{p} \int_0^1 \frac{\partial \Gamma}{\partial s}(s,t)^\top \frac{\partial \Gamma}{\partial s}(s,t) ds \geq \underline{p} \left\| X_1(t) - X_2(t) \right\|^2 . \tag{2.33}$$

Moreover, since

$$V(0) \leq \overline{p} \, \|x_1^\circ - x_2^\circ\|^2, \tag{2.34}$$

it yields for all t in $[0, \tau)$ that

$$\|X_1(t) - X_2(t)\|^2 \leq \exp(-\rho t) \frac{\overline{p}}{\underline{p}} \|x_1^\circ - x_2^\circ\|^2 .$$

Hence, this implies synchronization on the time of existence of the solution and concludes the proof. □

2.4.3 Synchronization with Directed Connected Graphs

We consider in this context a general network of N nonlinear systems described by

$$\dot{x}_i = f(x_i) + g(x_i)u_i, \qquad i = 1, \ldots, N, \tag{2.35}$$

and we suppose that the graph satisfies the following assumption.

Assumption 2.4.3 *The graph $\mathcal{G} = \{\mathcal{V}, \mathcal{E}, \mathcal{A}\}$ is connected.*

We claim that the same feedback law proposed in Theorem 2.4.2 under the previous condition. The main issue in proving such a statement is to find a correct Lyapunov function. As a consequence, we state the following conjecture.

Conjecture 2.4.4 *Consider a network $G = \{\mathcal{V}, \mathcal{G}, \mathcal{A}\}$ of agents* (2.35)*, and suppose Assumption 2.4.3 and the conditions* (2.22) *of Theorem 2.3.10 are satisfied. Then, the distributed state-feedback control law* (2.4) *with ψ given by Theorem 2.3.10, namely*

$$u_i = -\kappa \theta(x_i) \beta(\alpha(x_i)) \sum_{j=1}^{N} \ell_{ij} \alpha(x_j), \tag{2.36}$$

solves the synchronization Problem 2.2.1 for the network of agents given in (2.35) *for any* $\kappa > \kappa^*$, *where* $\kappa^* > 0$ *is a positive real number that depends on the eigenvalues of the Laplacian matrix* L.

We remark that in the recent work [22] we proved such a result for undirected connected graphs with the conditions of Theorem 2.3.8. To show this result, we used a Lyapunov function which is composed by Q error-coordinates (and not $N - 1$), where Q is the number of edges (i.e., of connections between agents). The more general case for directed graph under the relaxed conditions of Theorem 2.3.10 is currently under investigation and it probably involves the construction of a specific Lyapunov function. In the next section, we show that simulations support our claim.

2.5 Illustration

As an illustration, consider a network of Duffing-like equations in which each agent is described by the following dynamics:

$$\dot{x}_i = \begin{pmatrix} x_{i2} \\ -x_{i1} - \frac{3}{\varepsilon}\arctan\left(\varepsilon x_{i1}^3\right) \end{pmatrix} + \begin{pmatrix} 0 \\ b(Cx_i) \end{pmatrix} u_i, \qquad C^\top = \begin{pmatrix} 1 \\ 2 \end{pmatrix}, \tag{2.37}$$

with $x_i = (x_{i1}, x_{i2}) \in \mathbb{R}^2$ and $u \in \mathbb{R}$, where $\varepsilon \in (0, 1]$ and $b : \mathbb{R} \to \mathbb{R}$ is a smooth uniformly bounded function defined as

$$b(s) = (1 + \cos(3s))\sin(s) + 2.$$

Note that the dynamics (2.37) is a modified version of a standard (forced) Duffing oscillator

$$\ddot{y} + y + y^3 = u$$

with $x = (y, \dot{y})$, in which we modified the cubic term with an arctan in order to obtain a globally Lipschitz function, and in which the input u is multiplied by a nonlinear function of the state.

First, we show that we can verify the conditions (2.22) of Theorem 2.3.10. We have

$$\frac{\partial f}{\partial x}(x) = \begin{pmatrix} 0 & 1 \\ -1 - \varphi(x_1) & 0 \end{pmatrix}, \qquad \varphi(s) = \frac{9s^2}{1 + \varepsilon^2 s^6}.$$

It can be verified that φ is uniformly bounded. We select a constant metric $P = P^\top \succ 0$ of the form

$$P = \begin{pmatrix} p_1 & p_2 \\ p_2 & p_3 \end{pmatrix}.$$

Then, condition (2.22d) reads

$$\frac{\partial \alpha}{\partial x}(x)^\top = Pg(x)\theta(x) = \begin{pmatrix} p_2 \\ p_3 \end{pmatrix} b(Cx)\theta$$

which is satisfied by selecting

$$\alpha(x) = \int_0^{Cx} b(s)ds = 2Cx + \frac{1}{2}\cos^2(Cx) - \cos(Cx) - \frac{1}{8}\cos(Cx) + \frac{5}{8}, \tag{2.38}$$

$$\frac{\partial \alpha}{\partial x}(x)^\top = C^\top b(Cx),$$

and imposing $p_2 = \frac{1}{\theta}$, $p_3 = \frac{2}{\theta}$ in P. Then, inequality (2.22b) reads

$$\begin{pmatrix} -\frac{2}{\theta}(1+\varphi(x_1)) & -\frac{2}{\theta}(1+\varphi(x_1)) + p_1 \\ \star & \frac{2}{\theta} \end{pmatrix} - \frac{b(Cx)^2}{r\theta^2}\begin{pmatrix} 1 & 2 \\ 2 & 4 \end{pmatrix} \preceq 0,$$

which is satisfied for any x by selecting any $p_1 > 0$, θ sufficiently large and r sufficiently small because $b(s) > \underline{b} > 0$ for some $\underline{b} > 0$. Finally, the identity (2.22c) reads

$$\begin{aligned} L_{g\theta}P = \theta \frac{\partial}{\partial x}\frac{\partial \alpha}{\partial x}(x)^\top = C^\top C b'(Cx)\theta &= \gamma(\alpha(x))\frac{\partial \alpha}{\partial x}(x)^\top \frac{\partial \alpha}{\partial x}(x) \\ &= \gamma(\alpha(x)) C^\top C b(Cx)^2 \end{aligned}$$

which is satisfied by selecting

$$\gamma(\alpha(s)) = \theta\frac{b'(s)}{b(s)^2}, \qquad \int_0^\alpha \gamma(s)ds = -\left[\frac{1}{b(s)}\right]_0^{\alpha(s)} \leq \bar{\gamma},$$

the latter inequality being satisfied because $b(s)$ is uniformly bounded. Finally, we compute the function β as

$$\beta(\alpha) = \beta_0 \exp\left(-\frac{1}{2}\int_0^\alpha \gamma(s)ds\right) = \beta_0 \exp\left(\frac{\theta}{b(\alpha(Cx))}\right)\exp\left(\frac{\theta}{b(0)}\right) \tag{2.39}$$

The conditions of Theorem 2.3.10 are therefore satisfied. We apply then a distributed feedback law as in Conjecture 2.4.4, namely

$$u_i = -\kappa\theta\beta(x_i)\sum_{j=1}^N \ell_{ij}\alpha(x_j)$$

with the functions α, β selected as above.

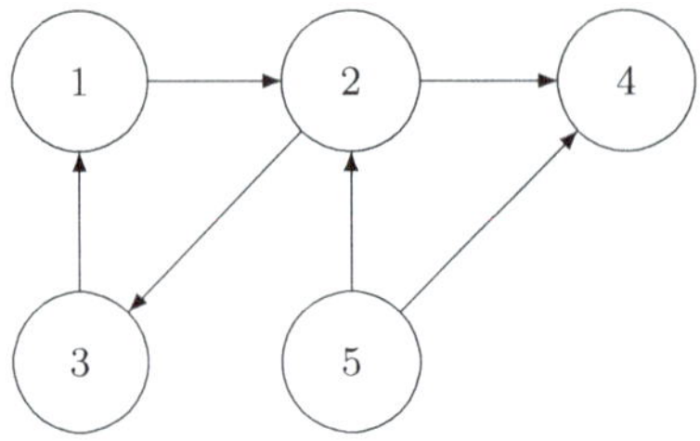

$$L = \begin{pmatrix} 1 & 0 & -1 & 0 & 0 \\ -1 & 2 & 0 & 0 & -1 \\ 0 & -1 & 1 & 0 & 0 \\ 0 & -1 & 0 & 2 & -1 \\ 0 & 0 & 0 & 0 & 0 \end{pmatrix}$$

Fig. 2.1 Direct connected communication graph among 5 agents and associated Laplacian

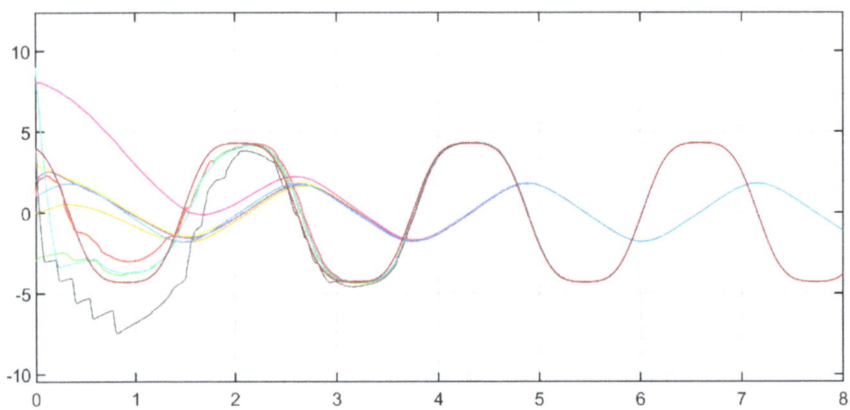

Fig. 2.2 Synchronization of 5 Duffing oscillators

In the simulations, we considered a network of 5 agents connected with the directed graph described in Fig. 2.1. Note that the Laplacian matrix L associated to such a graph is not symmetric. The feedback law is implemented with $\kappa = 1, \theta = 1$ and $\beta_0 = \frac{1}{2}\exp(b(0))$. Figure 2.2 shows convergence of all the agents to the trajectory of the agent 5 for initial conditions randomly chosen but satisfying $|x_i| \leq 10$.

2.6 Conclusions and Perspective

We investigated the problem of multi-agent synchronization described by nonlinear input-affine dynamics. Based on contraction analysis and Riemannian metrics, we provided a new of sufficient conditions for the existence of a contractive infinite-gain margin feedback. We showed that such a feedback can be employed to solve the synchronization problem between two agents and we conjecture that the same type of control law can be employed to solve a synchronization problem between any number of agents exchanging information via a connected directed graphs. Simulations support such a conjecture. The case of Duffing oscillators is carefully studied.

References

1. Mesbahi, M., Egerstedt, M.: Graph Theoretic Methods in Multiagent Networks. Princeton University Press (2010)
2. Simpson-Porco, J.W., Dörfler, F., Bullo, F.: Synchronization and power sharing for droop-controlled inverters in islanded microgrids. Automatica **49**(9), 2603–2611 (2013)
3. Cucuzzella, M., Trip, S., De Persis, C., Cheng, X., Ferrara, A., van der Schaft, A.: A robust consensus algorithm for current sharing and voltage regulation in DC microgrids. IEEE Trans. Control Syst. Technol. **27**(4), 1583–1595 (2018)
4. Scardovi, L., Sepulchrem, R.: Synchronization in networks of identical linear systems. In: 47th IEEE Conference on Decision and Control, pp. 546–551 (2008)
5. Li, Z., Duan, Z., Chen, G., Huang, L.: Consensus of multiagent systems and synchronization of complex networks: a unified viewpoint. IEEE Trans. Circuits Syst. I Regul. Papers **57**(1), 213–224 (2009)
6. Isidori, A.: Lectures in Feedback Design for Multivariable Systems. Springer (2017)
7. Giaccagli, M., Andrieu, V., Tarbouriech, S., Astolfi, D.: Infinite gain margin, contraction and optimality: an LMI-based design. Eur. J. Control **68**, 100685 (2022)
8. Giaccagli, M., Astolfi, D., Andrieu, V.: Further results on incremental input-to-state stability based on contraction-metric analysis. In: 62th IEEE Conference on Decision and Control (2023)
9. Sepulchre, R., Jankovic, M., Kokotovic, P.V.: Constructive Nonlinear Control. Springer (2012)
10. Arcak, M.: Passivity as a design tool for group coordination. IEEE Trans. Autom. Control **52**(8), 1380–1390 (2007)
11. Zhang, F., Trentelman, H.L., Scherpen, J.M.: Fully distributed robust synchronization of networked Lur'e systems with incremental nonlinearities. Automatica **50**(10), 2515–2526 (2014)
12. Isidori, A., Marconi, L., Casadei, G.: Robust output synchronization of a network of heterogeneous nonlinear agents via nonlinear regulation theory. IEEE Trans. Autom. Control **59**(10), 2680–2691 (2014)
13. Arcak, M.: Certifying spatially uniform behavior in reaction-diffusion PDE and compartmental ODE systems. Automatica **47**(6), 1219–1229 (2011)
14. Aminzare, Z., Sontag, E.D.: Synchronization of diffusively-connected nonlinear systems: results based on contractions with respect to general norms. IEEE Trans. Netw. Sci. Eng. **1**(2), 91–106 (2014)
15. di Bernardo, M., Fiore, D., Russo, G., Scafuti, F.: Convergence, consensus and synchronization of complex networks via contraction theory. In: Complex Systems and Networks: Dynamics, Controls and Applications, pp. 313–339 (2016)
16. Coraggio, M., De Lellis, P., di Bernardo, M.: Convergence and synchronization in networks of piecewise-smooth systems via distributed discontinuous coupling. Automatica **129**, 109596 (2021)
17. Pavlov, A., Steur, E., van de Wouw, N.: Nonlinear integral coupling for synchronization in networks of nonlinear systems. Automatica **140**, 110202 (2022)
18. Andreasson, M., Dimarogonas, D.V., Sandberg, H., Johansson, K.H.: Distributed control of networked dynamical systems: static feedback, integral action and consensus. IEEE Trans. Autom. Control **59**(7), 1750–1764 (2014)
19. Panteley, E., Loría, A.: Synchronization and dynamic consensus of heterogeneous networked systems. IEEE Trans. Autom. Control **62**(8), 3758–3773 (2017)
20. Giaccagli, M., Andrieu, V., Astolfi, D., Casadei, G.: Sufficient metric conditions for synchronization of leader-connected homogeneous nonlinear multi-agent systems. IFAC-PapersOnLine **54**(14), 412–417 (2021)
21. Giaccagli, M., Zoboli, S., Astolfi, D., Andrieu, V., Casadei, G.: Synchronization in networks of nonlinear systems: contraction metric analysis and deep-learning for feedback estimation. IEEE Trans. Autom. Control (2024). (under review)
22. Cellier Devaux, A., Astolfi, D., Andrieu, V.: Edges' Riemannian energy analysis for synchronization of multi-agent nonlinear systems over undirected weighted graphs. Automatica (2024). (under review)

23. Lohmiller, W., Slotine, J.J.E.: On contraction analysis for non-linear systems. Automatica **34**(6), 683–696 (1998)
24. Simpson-Porco, J.W., Bullo, F.: Contraction theory on Riemannian manifolds. Syst. Control Lett. **65**, 74–80 (2014)
25. Andrieu, V., Jayawardhana, B., Praly, L.: Transverse exponential stability and applications. IEEE Trans. Autom. Control **61**(11), 3396–3411 (2016)
26. Andrieu, V., Jayawardhana, B., Praly, L.: Characterizations of global transversal exponential stability. IEEE Trans. Autom. Control **66**(8), 3682–3694 (2021)
27. Manchester, I.R., Slotine, J.J.E.: Control contraction metrics: convex and intrinsic criteria for nonlinear feedback design. IEEE Trans. Autom. Control **62**(6), 3046–3053 (2017)
28. Sanfelice, R.G., Praly, L.: Convergence of Nonlinear observers on $\mathbb{R}\mathbb{R}^n$ with a Riemannian Metric (Part I). IEEE Trans. Autom. Control **57**(7), 1709–1722 (2011)
29. Sakai, T.: Riemannian Geometry, vol. 149. American Mathematical Society (1996)
30. Godsil, C., Royle, G.F.: Algebraic Graph Theory. Springer Science & Business Media (2001)
31. Angeli, D.: A Lyapunov approach to incremental stability properties. IEEE Trans. Autom. Control **47**(3), 410–421 (2002)
32. Praly, L., Wang, Y.: Stabilization in spite of matched unmodeled dynamics and an equivalent definition of input-to-state stability. Math. Control, Signals Syst. **9**(1), 1–33 (1996)

Chapter 3
Physics-Based Output-Feedback Consensus-Formation Control of Networked Autonomous Vehicles

Antonio Loría, Emmanuel Nuño, Elena Panteley, and Esteban Restrepo

Abstract We describe a control method for multi-agent vehicles, to make them adopt a formation around a non-pre-specified point on the plane, and with common but non-pre-imposed orientation. The problem may be considered as part of a more complex maneuver in which the robots are summoned to a rendezvous to advance in formation. The novelty and most appealing feature of our control method is that it is physics-based; it relies on the design of distributed dynamic controllers that may be assimilated to second-order mechanical systems. The consensus task is achieved by making the controllers, not the vehicles themselves directly, achieve consensus. Then, the vehicles are steered into a formation by virtue of fictitious mechanical couplings with their respective controllers. We cover different settings of increasing technical difficulty, from consensus formation control of second-order integrators to second-order nonholonomic vehicles and in scenarii including both state- and output-feedback control. In addition, we address the realistic case in which the vehicles communicate over a common WiFi network that introduces time-varying delays. Remarkably, the same physics-based method applies to all the scenarii.

A. Loría
Laboratoire des signaux et systèmes, CNRS, Gif-sur-Yvette, France
e-mail: antonio.loria@cnrs.fr

E. Nuño (✉)
Department of Computer Science at the University of Guadalajara, Guadalajara, Mexico
e-mail: emmanuel.nuno@academicos.udg.mx

E. Panteley
Laboratoire des signaux et systèmes, CNRS, Gif-sur-Yvette, France
e-mail: elena.panteley@cnrs.fr

E. Restrepo
CNRS, University of Rennes, Inria, IRISA, Campus de Beaulieu, 35042 Rennes, France
e-mail: esteban.restrepo@inria.fr

R. Postoyan et al. (eds.), *Hybrid and Networked Dynamical Systems*,
Lecture Notes in Control and Information Sciences 493,
https://doi.org/10.1007/978-3-031-49555-7_3

3.1 Introduction

For first and second-order integrators the leaderless consensus problem, which consists in the state variables of all agents converging to a common value, is well-studied and solved under many different scenarios [52]. However, the solution to this problem is more complex if one considers the agents' dynamics [26, 48], network constraints, such as communication delays [64], unavailability of velocity measurements [41], or nonholonomic constraints that restrict the systems' motion [28].

For autonomous vehicles, which, in contrast to mathematical *models* consisting of first and second-order integrators, do occupy a physical space, the leaderless consensus problem consists in making all robots converge to a *rendezvous* point. That is, the robots are required to coordinate their motions without any pre-established trajectory. Furthermore, because the robots can obviously not occupy the same physical space simultaneously, a formation pattern with an unknown center must be imposed. This is done by specifying for each robot, an offset position from the unknown center [43]. It may be required that positions and orientations converge to a common value [33], or that either only the positions [28] or only the orientations [36] achieve a common equilibrium point.

Rendezvous control is useful in cases where a group of robots must converge to postures that form a desired geometric pattern given any initial configuration in order to subsequently maneuver as a whole [63]. This is a typical two-stage formation problem. In the first, a rendezvous algorithm is required for the stabilization of the agents [11, 21, 55] and in the second a formation-tracking controller is employed [10]–[35].

From a systems viewpoint, rendezvous control of nonholonomic vehicles is inherently a set-point stabilization problem. In that regard, it presents the same technical difficulties as the stabilization of a single robot. In particular, that nonholonomic systems are not stabilizable via time-invariant smooth feedback [6], but either via discontinuous time-invariant control [11] or time-varying smooth feedback [22, 62]. In other words, in contrast to the case of holonomic systems, for systems with nonholonomic constraints stabilization is not a particular case of trajectory tracking, so controllers that solve one problem *generally* cannot solve the other [29]. For multi-agent systems, necessary conditions for rendezvous are laid in [28]. Thus, neither the numerous algorithms for consensus of linear systems nor those for formation-tracking control, notably in a leader-follower configuration, apply to the rendezvous problem for nonholonomic vehicles.

Here we consider a rendezvous problem for second-order (force-controlled) nonholonomic systems interconnected over an undirected static graph and with time-varying measurement delays for which velocity measurements are not available. From a systems viewpoint, this is an output-feedback control problem, with output corresponding to the vehicles' positions and orientations. We emphasize that in spite of the many articles on output-feedback control for the consensus of multi-agent systems—see, e.g., [14, 27, 34, 61], very few address the problem of output-feedback control for nonholonomic vehicles; see for instance, [50] on the leader-follower consensus problem and [18] where a velocity filter has been employed to obviate the need of velocity measurements. In the latter, however, delays are not considered and, more importantly, such problem appears to be unsolvable using the algorithm proposed therein.

In this chapter we explore the consensus-formation control problem for nonholonomic systems under various scenarii. We start by revisiting the consensus control of second-order integrators, and then we show how output-feedback consensus control may be achieved via dynamic feedback, even in the presence of transmission delays. Our controllers are completely distributed because they rely only on the information available to each agent from its neighbors, without requiring any knowledge of the complete network.

Some of the statements presented here are original and others appear in [44] and [45]. In contrast to the latter, however, in this chapter we favor a pedagogical over a technical presentation. Hence, we omit proofs and rather develop in detail the most interesting fact that the stabilization mechanism behind our methods has a clear physical analogy with the stabilization of (under-actuated) flexible-joint robots. In the following section we describe the dynamic model of the nonholonomic agents and we present the formal problem statement. Then, in Sect. 3.3 we revisit the rendezvous problem for linear second-order systems via a state-feedback controller. In Sect. 3.4 we design an output-feedback scheme for nonholonomic vehicles for the undelayed case. In Sect. 3.5 we see how our method applies even in the presence of measurement delays. In Sect. 3.6 we illustrate our findings with a case-study of realistic simulations, for which we used the Gazebo-ROS environment. As customary, we offer some concluding remarks in Sect. 3.7.

3.2 Model and Problem Formulation

3.2.1 *Single-Robot Model*

We consider autonomous vehicles as the one schematically represented in Fig. 3.1. Its position on the plane may be defined as that of its center of mass, with Cartesian coordinates $(x, y) \in \mathbb{R}^2$ and its orientation with respect to the abscissae is denoted

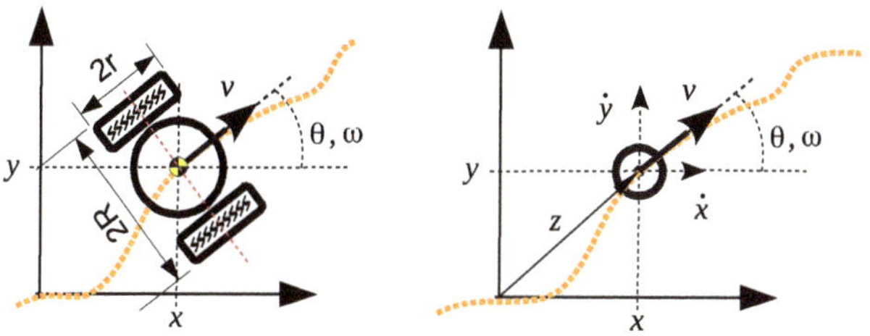

Fig. 3.1 Schematics of a differential-drive mobile robot

by the angle θ. It is assumed that the vehicle may move forward with a velocity $v := [\dot{x}^2 + \dot{y}^2]^{(1/2)}$ and turn with an angular velocity ω. The vehicle, however, cannot move in certain directions (e.g., sideways). This restriction is encoded by the non-integrable velocity constraint

$$\sin(\theta)\dot{x} = \cos(\theta)\dot{y}.$$

From these expressions we obtain the velocity equations

$$\dot{x} = \cos(\theta)v \tag{3.1a}$$
$$\dot{y} = \sin(\theta)v \tag{3.1b}$$
$$\dot{\theta} = \omega, \tag{3.1c}$$

which define a *first-order* model often used in the literature on control of nonholonomic systems—see, e.g., [10, 37, 40, 60] and [35]. In such model the control inputs are the velocities v and ω. Being mechanical systems, however, a more complete model includes a set of Euler–Lagrange equations for the velocity dynamics, i.e.,

$$\dot{v} = F_v(z, \theta, v, \omega) + u_v \tag{3.2a}$$
$$\dot{\omega} = F_\omega(z, \theta, v, \omega) + u_\omega, \tag{3.2b}$$

where F_v and F_ω are smooth functions [39]. Articles on *control* of nonholonomic systems where such *second-order* models are used are considerably scarce in comparison—see, e.g., see [9] and [15] and they are more often found in a single-vehicle setting [13, 19, 24].

Here, we employ a complete second-order model that corresponds to that of so-called differential-drive robots [59]. For the purpose of analysis, only, we assume that the center of mass is aligned with an axis joining the centers of the wheels—see the illustration on the left in Fig. 3.1, so $F_\omega = F_v \equiv 0$, but the model used to test our algorithms in the realistic simulator Gazebo-ROS does not satisfy this assumption.

Thus, the control inputs take the form

$$u_v := \frac{1}{rm}[\tau_1 + \tau_2], \quad u_\omega := \frac{2R}{Ir}[\tau_1 - \tau_2],$$

where m and I are respectively the robot's mass and inertia whereas τ_1 and τ_2 are the torques applied, independently, at each of the wheels.

An essential feature of this model, that is at the basis of the control design, is that Eqs. (3.1)–(3.2) consist of *two* coupled second-order systems driven by independent control inputs. One system determines the linear motion and the other the angular one. To evidence this, we define $z_i := [x_i \ y_i]^\top \in \mathbb{R}^2$, where we introduced the index $i \leq N$ to refer to one among N robots—see the illustration on the right in Fig. 3.1, and rewrite the equations for the ith robot in the form:

$$\text{angular-motion dynamics} \begin{cases} \dot{\theta}_i = \omega_i & (3.3a) \\ \dot{\omega}_i = u_{\omega i}, & (3.3b) \end{cases}$$

$$\text{linear-motion dynamics} \begin{cases} \dot{z}_i = \varphi(\theta_i) v_i, & (3.4a) \\ \dot{v}_i = u_{vi}, & (3.4b) \end{cases}$$

where

$$\varphi(\theta_i) := [\cos(\theta_i) \quad \sin(\theta_i)]^\top. \tag{3.5}$$

This (apparently innocuous) observation is important because the literature is rife with efficient controllers for second-order mechanical systems from which we may draw inspiration for the problem at hand here, even in the context of multi-agent systems [52]. Moreover, even though the subsystems (3.3) and (3.4) are clearly entangled through the function φ they may be dealt with as if decoupled, by replacing θ_i with the trajectory $\theta_i(t)$ since φ is uniformly bounded [30]. Hence, relying on a cascades argument, we may apply a separation principle to design the controllers for the linear and angular-motion subsystems independently. These key features are at the basis of our method to approach the rendezvous problem, which is described next.

3.2.2 The Consensus-Formation Problem

Consider a group of N force-controlled nonholonomic vehicles modeled by (3.3)–(3.5) like the one depicted in Fig. 3.1, each of these robots is assumed to be equipped with positioning sensors that deliver reliable measures of x_i, y_i, and θ_i.

The robots are required to meet in formation at a rendezvous point $z_c := (x_c, y_c)$ and acquire a common orientation θ_c. That is, for each $i \leq N$ the Cartesian positions z_i must converge to $z_c + \delta_i$, where δ_i is a vector that determines the position of the ith vehicle relative to the unknown center of the formation—see Fig. 3.2 for an illustration. More precisely, consider the following problem statement.

Definition 3.2.1 (*Consensus formation*) For each robot (resp. each $i \leq N$), given a vector $\delta_i = [\delta_{xi} \ \delta_{yi}]^\top$, define its translated position $\bar{z}_i := z_i - \delta_i$ (correspondingly, let $\bar{x}_i := x_i - \delta_{xi}$ and $\bar{y}_i := y_i - \delta_{yi}$).

Then, design a distributed controller such that

$$\lim_{t\to\infty} v_i(t) = 0, \quad \lim_{t\to\infty} \bar{z}_i(t) = z_c, \tag{3.6}$$

$$\lim_{t\to\infty} \omega_i(t) = 0, \quad \lim_{t\to\infty} \theta_i(t) = \theta_c \quad \forall i \leq N. \tag{3.7}$$

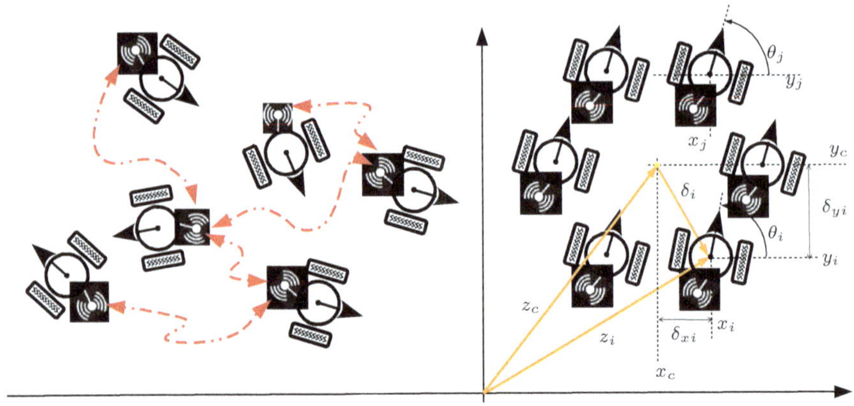

Fig. 3.2 Vehicles initially dispersed communicating over an undirected network (are required to) achieve formation consensus around a rendezvous point

We stress that this is a leaderless consensus problem. That is, neither the coordinates (x_c, y_c) nor the angle θ_c are determined a priori as a reference. They depend on the initial postures, the systems' nonlinear dynamics, and network features (see farther below). In general, this problem, may not be solved using controllers designed to make the vehicles advance in formation while following a leader (virtual or otherwise).

Now, we assume that the vehicles communicate over a WiFi network. The ith robot communicates with a set of neighbors, which we denote by $\mathcal{N}_i$. It is naturally assumed that once a communication is set between two robots i and $j \in \mathcal{N}_i$, the flow of information is bidirectional and is never lost. More precisely, we pose the following.

Assumption 3.2.2 The network may be modeled using an interconnection graph that is undirected, static, and connected.

Remark 3.2.3 In graph theory, a graph is undirected if the nodes exchange information in both directions, it is static if the interconnection is constant, and an undirected graph is connected if any node is reachable from any other node.

Now, because the robots communicate over a WiFi network, the communication between the robots i and j is affected by non-constant time delays. More precisely, we consider the following.

Assumption 3.2.4 The communication from the jth to the ith robot is subject to a variable time delay denoted $T_{ji}(t)$ that is bounded by a known upper-bound $\overline{T}_{ji} \geq 0$ and has bounded time-derivatives, up to the second.

Furthermore, we also assume that the vehicles are equipped only with position and orientation sensors. That is,

Assumption 3.2.5 The velocities v_i and ω_i are not measurable.

Assumptions 3.2.2–3.2.5 coin realistic scenarii of automatic control of multi-agent systems, but all three together have been little addressed in the literature in the context of consensus control of nonholonomic systems [45]. Yet, Assumption 3.2.4 carries certain conservatism in imposing that the delays be differentiable and bounded. Indeed, it must be stressed that, in general, time delays over WiFi networks or the Internet may rather be of a non-smooth nature [1, 3, 33]. Nonetheless, the formal analysis under such condition is considerably intricate and escapes the scope of this document. For a Lyapunov-based analysis of the rendezvous problem under non-differentiable delays, albeit via state feedback, see [32, 33].

3.3 Control Architecture: State-Feedback Case

As previously implied, the controller that we propose relies on the system's structural properties that lead to a separation of the linear- and angular-motion dynamics. For clarity of exposition, we start by revisiting the consensus problem for ordinary second-order systems ("double integrators") via state feedback and without delays. The purpose is to underline the robustness of a commonly used distributed controller, by presenting an original analysis that is helpful to understand the essence of the stabilization mechanism at the basis of our method. In addition, it has the advantage of providing a strict Lyapunov function for consensus of second-order systems.

3.3.1 Consensus Control of Second-Order Systems

The consensus problem for systems with dynamics—*cf.* Eq. (3.3),

$$\ddot{\vartheta}_i = u_i \quad i \le N, \; u_i \in \mathbb{R} \tag{3.8}$$

(that is, steering $\vartheta_i \to \vartheta_c$, $\dot{\vartheta}_i \to 0$, and $\ddot{\vartheta}_i \to 0$ with ϑ_c constant and non-imposed a priori) is now well understood in various settings, e.g., in the case of unmeasurable velocities [2], of measurement delays [17], or with state constraints [51, 53].

For instance, it is well known (see [52]) that if the systems modeled by (3.8) communicate over a network modeled by a directed, static, and connected graph, and the full state is measurable, the distributed control law of proportional-derivative (PD) type

$$u_i = -d_i\dot{\vartheta}_i - p_i \sum_{j\in\mathcal{N}_i} a_{ij}(\vartheta_i - \vartheta_j), \quad d_i,\; p_i > 0, \tag{3.9}$$

where $a_{ij} > 0$ if $j \in \mathcal{N}_i$ and $a_{ij} = 0$ otherwise, solves the consensus problem. More precisely, we have the following.

Proposition 3.3.1 *Consider the system* (3.8) *in closed loop with* (3.9), *that is, the system*

$$\ddot{\vartheta}_i + d_i\dot{\vartheta}_i + p_i \sum_{j\in\mathcal{N}_i} a_{ij}(\vartheta_i - \vartheta_j) = 0. \tag{3.10}$$

Then, the consensus manifold $\{\vartheta_i = \vartheta_j\} \cap \{\dot{\vartheta}_i = 0\}$, *for all* $i,\ j \leq N$ *is globally exponentially stable for any positive values of* d_i *and* p_i.

The proof of global asymptotic stability of the consensus manifold is well reported in the literature [52]. For further development we provide here a simple and original proof of global *exponential* stability based on Lyapunov's direct method.

Let $\vartheta := [\vartheta_1 \ \cdots \ \vartheta_N]^\top$,

$$\tilde{\vartheta} := \vartheta - \frac{1}{N}\mathbf{1}_N\mathbf{1}_N^\top\vartheta, \quad \text{where} \quad \mathbf{1}_N := [1\cdots 1]^\top. \tag{3.11}$$

The ith element of the vector $\tilde{\vartheta}$ denotes the difference between ϑ_i and the average of all the states, i.e., $\vartheta_c := (1/N)\mathbf{1}_N^\top\vartheta$, so consensus is reached if and only if $\tilde{\vartheta} = 0$. In addition, under Assumption 3.2.2, ϑ_c corresponds to the consensus equilibrium point. Now, to abbreviate the notation, we also define

$$R := I - \frac{1}{N}\mathbf{1}_N\mathbf{1}_N^\top.$$

Note that $R = R^\top$ and $\|R\| \leq 1$, where $\|R\|$ corresponds to the induced norm of R, and $\tilde{\vartheta} = R\vartheta$.

Next, we introduce the Laplacian matrix, $L := [\ell_{ij}] \in \mathbb{R}^{N\times N}$, where

$$\ell_{ij} = \begin{cases} \sum_{k\in\mathcal{N}_i} a_{ik} & i = j \\ -a_{ij} & i \neq j. \end{cases} \tag{3.12}$$

By construction, $L\mathbf{1}_N = 0$ and, after Assumption 3.2.2, L is symmetric, it has a unique zero-eigenvalue, and all of its other eigenvalues are strictly positive. Hence, $\text{rank}(L) = N - 1$. Also, the last term on the right-hand side of Eq. (3.9) satisfies

$$\text{col}\Big[\sum_{j\in\mathcal{N}_i} a_{ij}(\vartheta_i - \vartheta_j)\Big] = L\tilde{\vartheta}, \tag{3.13}$$

where $\text{col}[(\cdot)_i]$ denotes a column vector of N elements $(\cdot)_i$. Indeed, by the definition of the Laplacian, we have

$$\text{col}\Big[\sum_{j\in\mathcal{N}_i} a_{ij}(\vartheta_i - \vartheta_j)\Big] = L\Big[\vartheta - \frac{1}{N}\mathbf{1}_N\mathbf{1}_N^\top\vartheta\Big] + \frac{1}{N}L\mathbf{1}_N\mathbf{1}_N^\top\vartheta.$$

However, $L\mathbf{1}_N = 0$, so the right-hand side of the equation above equals to $LR\vartheta$, which corresponds to $L\tilde{\vartheta}$, by definition. These identities are useful to write the closed-loop system (3.8)–(3.9) in the multi-variable form

$$\ddot{\vartheta} = -D\dot{\vartheta} - PL\tilde{\vartheta}, \tag{3.14}$$

where $P := \text{diag}[p_i]$ and $D := \text{diag}[d_i]$. In turn, this serves to notice that the Lyapunov function

$$V_1(\tilde{\vartheta}, \dot{\vartheta}) := \frac{1}{2}\Big[\tilde{\vartheta}^\top L\tilde{\vartheta} + \dot{\vartheta}^\top P^{-1}\dot{\vartheta}\Big] \tag{3.15}$$

is positive definite, even if L is rank deficient. Indeed, the term $\tilde{\vartheta}^\top L\tilde{\vartheta} \geq \lambda_2(L)|\tilde{\vartheta}|^2$, where $\lambda_2(L) > 0$ corresponds to the second eigenvalue of L (that is, the smallest positive eigenvalue). We stress that V_1 is positive, not for *any* $\tilde{\vartheta} \in \mathbb{R}^N\backslash\{0\}$, but for $\tilde{\vartheta}$ as defined in (3.11).

Now, evaluating the total derivative of V_1 along the trajectories of (3.14) and using $L\dot{\tilde{\vartheta}} = L\dot{\vartheta}$ (again, this holds because $L\mathbf{1}_N = 0$) we see that

$$\dot{V}_1(\tilde{\vartheta}, \dot{\vartheta}) = -\dot{\vartheta}^\top P^{-1}D\,\dot{\vartheta}. \tag{3.16}$$

Global asymptotic stability of the consensus manifold $\{(\tilde{\vartheta}, \dot{\vartheta}) = (0, 0)\}$ may be ascertained from (3.16) by invoking Barbashin–Krasovskĭi's theorem [4] (also, but wrongly, known as LaSalle's theorem).

Remark 3.3.2 Note that, in the presence of an additional input α the system (3.14), that is, $\ddot{\vartheta} + D\dot{\vartheta} + PL\tilde{\vartheta} = \alpha$, defines an output strictly passive map $\alpha \mapsto \dot{\vartheta}$. This follows by a direct computation of $\dot{V}_1$ that leads to $\dot{V}_1(\tilde{\vartheta}, \dot{\vartheta}) = -\dot{\vartheta}^\top P^{-1}D\,\dot{\vartheta} + \alpha^\top\dot{\vartheta}$.

As a matter of fact, since the system is linear time-invariant, it is also globally exponentially stable. To see this more clearly, using V_1 it is possible to construct a simple strict Lyapunov function. This is useful to assess the robustness of system (3.8) in closed loop with the consensus control law defined in (3.9) in terms of input-to-state stability.

Let $\varepsilon \in (0, 1)$ and define

$$V_2(\tilde{\vartheta}, \dot{\vartheta}) := V_1(\tilde{\vartheta}, \dot{\vartheta}) + \varepsilon\tilde{\vartheta}^\top P^{-1}\dot{\vartheta}. \tag{3.17}$$

In view of the properties of V_1 it is clear that also V_2 is positive definite and radially unbounded, for sufficiently small values of ε. On the other hand, the total derivative of V_2 along the closed-loop trajectories of (3.14) yields

$$\dot{V}_2(\tilde{\vartheta}, \dot{\vartheta}) = \dot{V}_1 + \varepsilon\Big[\dot{\vartheta}^\top RP^{-1}\dot{\vartheta} - \tilde{\vartheta}^\top P^{-1}D\dot{\vartheta} - \tilde{\vartheta}^\top L\tilde{\vartheta}\Big], \tag{3.18}$$

which, in view of (3.16) and the fact that $\|R\| \leq 1$, implies that

$$\dot{V}_2(\tilde{\vartheta}, \dot{\vartheta}) \leq -c_1|\dot{\vartheta}|^2 - \varepsilon c_2|\tilde{\vartheta}|^2 \tag{3.19}$$

where d_m and p_M are the smallest and largest coefficients of D and P respectively, $c_1 := \frac{d_m}{p_M} - \varepsilon\left[\frac{1}{p_m} + \frac{d_M}{2\lambda p_m}\right]$ and $c_2 := \ell_2 - \frac{\lambda}{2}\frac{d_M}{p_m}$ are positive for appropriate values of λ and $\varepsilon \in (0, 1)$ and any $\ell_2 := \lambda_2(L) > 0$.

The previous analysis is interesting because it leads to the important observation that for perturbed systems with dynamics $\ddot{\vartheta}_i = u_i + \alpha_i$ where α_i is a bounded external disturbance, we have

$$\dot{V}_2(\tilde{\vartheta}, \dot{\vartheta}) \leq -c_1'|\dot{\vartheta}|^2 - \varepsilon c_2'|\tilde{\vartheta}|^2 + c_3|\alpha|^2, \tag{3.20}$$

where $c_1' := c_1 - \frac{\lambda}{2}$, $c_2' := c_2 - \frac{d_m}{2\lambda p_m}$, and $\alpha := [\alpha_1 \cdots \alpha_N]^\top$. So the closed-loop system is input-to-state stable with respect to the input α.

We use the previously established facts in our control-design method. The controllers that we present below, for nonholonomic systems, are devised as second-order mechanical systems in closed loop with a consensus controller, that is, with dynamics reminiscent of (3.14). Then, they are interconnected via virtual springs to the vehicles' dynamics so that, by virtue of reaching consensus among themselves, they stir the vehicles to consensus too.

Thus, following, on one hand, the previous developments for consensus of second-order integrators and, on the other, the fact that the nonholonomic vehicle's dynamics consist mainly in two interconnected second-order mechanical systems, we proceed to design a consensus-formation controller for nonholonomic systems. The input-to-state stability property previously underlined is fundamental to devise the controllers separately, for the linear- and angular-motion dynamics.

3.3.2 State-Feedback Consensus Control of Nonholonomic Vehicles

We start our control design with the angular-motion dynamics (3.3), which we rewrite in the form $\ddot{\theta}_i = u_{\omega i}$ for convenience—*cf.* (3.8). Then, we introduce the control law

$$u_{\omega i} = -d_{\omega i}\dot{\theta}_i - p_i \sum_{j \in \mathcal{N}_i} a_{ij}(\theta_i - \theta_j) \quad d_{\omega i},\ p_{\omega i} > 0, \tag{3.21}$$

which is the same as (3.9), up to an obvious change in the notation. Hence, the closed-loop system (3.3)–(3.21) yields

$$\ddot{\theta}_i + d_{\omega i}\dot{\theta}_i + p_i \sum_{j \in \mathcal{N}_i} a_{ij}(\theta_i - \theta_j) = 0. \tag{3.22}$$

From Proposition 3.3.1 it follows that the consensus manifold $\{\omega_i = 0\} \cap \{\theta_i = \theta_j\}$ is globally exponentially stable for (3.22). Therefore, $\theta_i \to \theta_c$ for all $i \leq N$.

It is left to design a consensus controller of the form (3.9) for the linear-motion dynamics (3.4). In this case, however, the consensus error feedback must take into account the kinematics function φ. That is, we pose the control law—cf. [5]. Let

$$u_{vi} = -d_{vi}v_i - p_{vi}\varphi(\theta_i)^\top \sum_{j\in\mathcal{N}_i} a_{ij}(\bar{z}_i - \bar{z}_j). \tag{3.23}$$

Then, for the purpose of analysis we replace the state variable θ_i with an arbitrary trajectory $\theta_i(t)$. This is possible because $t \mapsto \theta_i(t)$, as a solution of (3.22), exists on $[0, \infty)$ and so does its first derivative, that is, $t \mapsto \omega_i(t)$. Hence, the closed-loop linear-motion dynamics (3.4)–(3.23) may be regarded as a *time-varying* subsystem,[1] decoupled from the angular-motion dynamics. That is,

$$\Sigma_{vi} : \begin{cases} \dot{\bar{z}}_i = \varphi(\theta_i(t))v_i, & \text{(3.24a)}\\ \dot{v}_i = -d_{vi}v_i - p_{vi}\varphi(\theta_i(t))^\top \sum_{j\in\mathcal{N}_i} a_{ij}(\bar{z}_i - \bar{z}_j). & \text{(3.24b)}\end{cases}$$

Then, to analyze the stability of the consensus manifold for (3.24), akin to V_1 in (3.15), we define the Lyapunov function

$$V_3(v, \bar{z}) := \frac{1}{2}\sum_{i\leq N}\left[\frac{1}{p_{vi}}v_i^2 + \frac{1}{2}\sum_{j\in\mathcal{N}_i} a_{ij}|\bar{z}_i - \bar{z}_j|^2\right], \tag{3.25}$$

where $v := [v_1 \cdots v_N]^\top$ and $\bar{z} := [\bar{z}_1 \cdots \bar{z}_N]^\top$—cf. (3.13).

Using the identity

$$\frac{1}{2}\sum_{i\leq N}\sum_{j\in\mathcal{N}_i} a_{ij}(\dot{\bar{z}}_i - \dot{\bar{z}}_j)^\top(\bar{z}_i - \bar{z}_j) = \sum_{i\leq N} a_{ij}\dot{\bar{z}}_i^\top(\bar{z}_i - \bar{z}_j)$$

—see [46] and [8, Lemma 6.1], we compute the total derivative of V_3 along the closed-loop trajectories of (3.24). We obtain

$$\dot{V}_3(v, \bar{z}) = -v^\top D_v P_v^{-1} v, \tag{3.26}$$

where $P_v := \text{diag}[p_{vi}]$ and $D_v := \text{diag}[d_{vi}]$.

Now, the system in (3.24) being non-autonomous, Barbashin–Krasovskĭi's theorem does not apply, but we may use Barbălat's [23] to assess global asymptotic stability. To that end, we first remark that the function V_3 is positive definite and radially unbounded in v_i and $|\bar{z}_i - \bar{z}_j|$ for all $i, j \leq N$. Then, integrating along the trajectories on both sides of the identity (3.26) and of the inequality $\dot{V}_3(v(t), \bar{z}(t)) \leq 0$, we obtain that v_i and $|\bar{z}_i - \bar{z}_j|$ are bounded, i.e., $v_i, |\bar{z}_i - \bar{z}_j| \in \mathcal{L}_\infty$ and $v_i \in \mathcal{L}_2$. In addition, (3.26) implies that the consensus equilibrium defined by $\{v_i = 0,\ \bar{z}_i = \bar{z}_j\}$ is stable.

[1] To prove further see [23, p. 657] and [30].

From (3.24), the boundedness and continuity of $\varphi(\theta_i)$, of $\theta_i(t)$, and of $\omega_i(t) = \dot{\theta}_i(t)$, we see that, also, $\dot{\bar{z}}_i$, $\dot{v}_i$, and, consequently, $\ddot{v}_i \in \mathcal{L}_\infty$. Since $v_i \in \mathcal{L}_2 \cap \mathcal{L}_\infty$ and $v_i \in \mathcal{L}_\infty$ we conclude that $v_i \to 0$. Hence, since

$$\lim_{t\to\infty} \int_0^t \dot{v}_i(s)ds = \lim_{t\to\infty} v_i(t) - v_i(0),$$

we have

$$\lim_{t\to\infty} \int_0^t \dot{v}_i(s)ds = -v_i(0).$$

That is, the limit of $\dot{v}_i$ "exists and is finite" whereas the boundedness of $\ddot{v}_i$ implies that $\dot{v}_i$ is uniformly continuous. Hence, by virtue of Barbălat's lemma, we conclude that $\dot{v}_i \to 0$ as well. In turn, after (3.24) we see that

$$\lim_{t\to\infty} \varphi(\theta_i(t))^\top \sum_{j\in\mathcal{N}_i} a_{ij}\big(\bar{z}_i(t) - \bar{z}_j(t)\big) = 0.$$

This expression, however, does not imply that the consensus objective is reached. Indeed, note that the set of equilibria of the system in (3.24) corresponds to points belonging to the set

$$\mathcal{U} := \Big\{ v_i = 0 \quad \wedge \quad \varphi(\theta_i)^\top \sum_{j\in\mathcal{N}_i} a_{ij}(\bar{z}_i - \bar{z}_j) = 0 \Big\}, \tag{3.27}$$

which admits points such that $\bar{z}_i \neq \bar{z}_j \in \mathbb{R}^2$ because rank $\varphi(\theta_i) = 1$. This means that if orientation consensus is reached and, for instance, $\theta_i(t) \to 0$ then $\bar{x}_i \to x_c$, but $\bar{y}_i \not\to y_c$—see Eq. (3.5).

Remark 3.3.3 This shows that the consensus problem for nonholonomic systems cannot be treated as that for ordinary second-order systems like those discussed in Sect. 3.3.1—cf. [20, 51].

To ensure consensus it is necessary that the set of equilibria correspond to the set $\mathcal{U} \cap \mathcal{U}^\perp$, where

$$\mathcal{U}^\perp := \Big\{ v_i = 0 \quad \wedge \quad \varphi(\theta_i)^{\perp\top} \sum_{j\in\mathcal{N}_i} a_{ij}(\bar{z}_i - \bar{z}_j) = 0 \Big\},$$

and

$$\varphi(\theta_i)^\perp := [-\sin(\theta_i) \quad \cos(\theta_i)]^\top. \tag{3.28}$$

That is, $\varphi(\theta_i)^\perp$ is the annihilator of $\varphi(\theta_i)$ hence, $\varphi(\theta_i)^{\perp\top}\varphi(\theta_i) = \varphi(\theta_i)^\top\varphi(\theta_i)^\perp = 0$.

Roughly speaking, the controller must "pull" out the trajectories that may eventually get "trapped" in the set $\mathcal{U}$, whereas they do not belong to the set $\mathcal{U}^\perp$. To that end, we endow the angular-motion controller with a term that incorporates an external function of time (smooth and bounded) and acts as a perturbation to the angular-motion closed-loop dynamics. This perturbation is designed to *persist* as long as

$$\varphi(\theta_i(t))^{\perp\top} \sum_{j\in\mathcal{N}_i} a_{ij}\big(\bar{z}_i(t) - \bar{z}_j(t)\big) \neq 0. \tag{3.29}$$

More precisely, let ψ_i, $\dot{\psi}_i$, and $\ddot{\psi}_i$ be bounded (belong to $\mathcal{L}_\infty$) let $\dot{\psi}_i$ be persistently exciting [38], that is, let there exist T and $\mu > 0$ such that

$$\int_t^{t+T} \dot{\psi}_i(s)^2 ds \geq \mu \quad \forall t \geq 0. \tag{3.30}$$

Thus, we endow the control law in (3.21) with the additional term

$$\alpha_i(t, \theta_i, \bar{z}_i) := k_{\alpha i}\psi_i(t)\varphi_i(\theta_i)^{\perp\top}(\bar{z}_i - \bar{z}_j), \quad k_{\alpha i} > 0, \tag{3.31}$$

that is, we redefine the control law (3.21) as

$$u_{\omega i} = -d_{\omega i}\dot{\theta}_i - p_i \sum_{j\in\mathcal{N}_i} a_{ij}(\theta_i - \theta_j) + \alpha_i(t, \theta_i, \bar{z}_i). \tag{3.32}$$

To the angular-motion dynamics, α_i acts as a bounded (see Remark 3.3.5 below), hence innocuous, perturbation on the angular-motion closed-loop dynamics

$$\Sigma_{\omega i} \; : \; \ddot{\theta}_i + d_{\omega i}\dot{\theta}_i + p_i \sum_{j\in\mathcal{N}_i} a_{ij}(\theta_i - \theta_j) = \alpha_i(t, \theta_i, \bar{z}_i) \tag{3.33}$$

that impedes θ_i to remain in an unwanted manifold of equilibria, as long as the position consensus has not been achieved, that is, as long as (3.29). That is, through $\theta(t)$ in (3.24), the term α_i injects excitation into the system, as long as the consensus goal remains unachieved.

Now, since the system (3.33) is input-to-state stable with respect to α_i, the resulting closed-loop equations consist in the "cascaded" nonlinear time-varying system formed by Σ_{vi} in (3.24) and (3.33), as illustrated by Fig. 3.3.

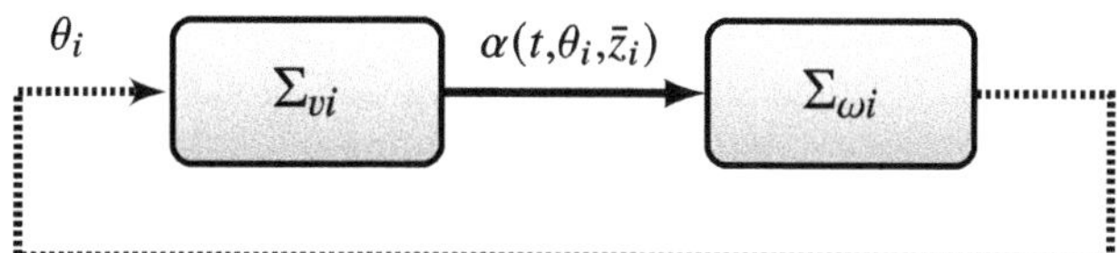

Fig. 3.3 Schematic representation of the closed-loop system, consisting in the error equations for the angular and the linear-motion dynamics. Even though the systems are feedback interconnected, for the purpose of analysis, they may be regarded as in cascade [30], whence the feedback of θ_i is represented by a dashed arrow

Global asymptotic stability of the respective consensus manifolds for Σ_{vi} and $\Sigma_{\omega i}$ with $\alpha_i \equiv 0$, the input-to-state stability of (3.33) and a cascades argument [30] leads to the following statement—*cf.* [5, 43].

Proposition 3.3.4 (State feedback formation consensus) *Consider the system (3.3)–(3.5) in closed loop with* (3.23) *and* (3.32) *with* p_{vi}, d_{vi}, $p_{\omega i}$, *and* $d_{\omega i} > 0$, α_i *as in* (3.31), ψ_i *and* $\dot{\psi}_i$ *bounded, and* ψ_i *persistently exciting. Then, the consensus-formation goal is achieved, that is,* (3.6) *and* (3.7) *hold.*

Remark 3.3.5 (*Sketch of proof of Proposition* 3.3.4) Technically, the proof of this statement follows along the lines discussed above for the case in which $\alpha_i \equiv 0$. In this case, we have a cascade from $\Sigma_{\omega i}$ to Σ_{vi}. If $\alpha_i \not\equiv 0$, the cascade structure is broken, but we can still use the ISS property of (3.33) to *break the loop* [30].

For (3.33) Inequality (3.20) holds (up to obvious changes in the notation) for any continuous α_i. Now, let $t_f \le \infty$ define the maximal length of the interval of existence of solutions and consider (3.20) along the trajectories of (3.33). We have

$$\dot{V}_2(\tilde{\theta}(t), \omega(t)) \le -c_1'|\omega(t)|^2 - \varepsilon c_2'|\tilde{\theta}(t)|^2 + c_3|\alpha(t, \theta(t), \bar{z}(t))|^2 \qquad \forall t \in [0, t_f), \tag{3.34}$$

where $\tilde{\theta} := \theta - \frac{1}{N}\mathbf{1}_N\mathbf{1}_N^\top\theta$, $\theta = \mathrm{col}[\theta_i]$, and $\omega = \dot{\theta}$. On the other hand, on the interval of existence of solutions, (3.26) also holds along the trajectories. Thus, since $\alpha_i(t, \theta_i, \bar{z})$ is linear in $[\bar{z}_i - \bar{z}_j]$, continuous and bounded in t and θ_i, the term $|\alpha(t, \theta(t), \bar{z}(t))|^2$ remains bounded on $t \in [0, t_f)$. Moreover, there exists $c > 0$ such that $|\alpha(t, \theta, \bar{z})|^2 \le cV_3(v, \bar{z})$. It follows from (3.26) and (3.34) that, defining $\nu_4(t) := V_2(\tilde{\theta}(t), \omega(t)) + V_3(v(t), \bar{z}(t))$,

$$\dot{\nu}_4(z(t)) \le c\nu_4(t) \quad \forall t \in [0, t_f). \tag{3.35}$$

Rearranging terms and integrating on both sides of the latter from 0 to t_f we see that

$$\int_{\nu_4(0)}^{\nu_4(t_f)} \frac{1}{c\nu_4} d\nu_4 \le t_f.$$

That is, $\ln\big(\nu_4(t_f)/\nu_4(0)\big) \le t_f$ (without loss of generality we assume that $\nu_4(0) > 0$). Now, since $\nu_4(t) \to \infty$ as $t \to t_f$ it follows that $t_f \not< \infty$—indeed, the postulate that $t_f < \infty$ leads to a contradiction. Therefore, the system is forward complete, (3.26) holds along the trajectories on $[0, \infty)$, $|\bar{z}_i(t) - \bar{z}_j(t)|$ and $v_i(t)$ are bounded on $[0, \infty)$, and so are $\tilde{\theta}(t)$ and $\omega(t)$ in view of the ISS property. On one hand, the presence of α_i prevents the solutions from remaining (close to) the set $\mathcal{U}$ in (3.27) and, on the other, $\alpha_i \to 0$ as the solutions converge to $\mathcal{U} \cap \mathcal{U}^\perp$. The result follows.

Remark 3.3.6 Controllers for nonholonomic systems that make explicit use of persistency of excitation were first used for tracking in [47] and for set-point stabilization in [31], but the underlying ideas are already present in [56, 57]. Nowadays, persistency of excitation is recognized as a fundamental, if not necessary, condition [29], for set-point stabilization of nonholonomic systems via smooth feedback and they are also frequently used in trajectory-tracking scenarii—see, e.g., [12, 25].

3.4 Control Architecture: Output-Feedback Case

In the case that velocities are unmeasurable, the velocity-feedback terms $d_{vi}v_i$ and $d_{\omega i}\omega_i$ cannot be used, so we design dynamic output-feedback controllers for the angular and linear-motion dynamics. These controllers are designed as virtual mass-spring-damper mechanical systems hinged to the vehicles' dynamics. In accordance with the separation-principle approach these controllers are designed independently for the angular and linear-motion dynamics. In addition, as before, we use persistency of excitation to overcome the difficulties imposed by the nonholonomic constraints.

3.4.1 Output-Feedback Orientation Consensus

Expressed as $\ddot{\theta}_i = u_{\omega i}$, the angular-motion dynamics (3.3) corresponds to an elementary Newtonian force-balance equation with unitary inertia. The problem at hand still is to synchronize the angular positions θ_i for N such systems, but since ω_i is not available from measurement, we cannot use the control law in (3.21). Yet, it appears reasonable to conjecture that the objective $\theta_i \to \theta_j$ for all $i, j \leq N$ may be achieved by coupling the subsystems $\ddot{\theta}_i = u_{\omega i}$, via torsional springs, to virtual second-order oscil-

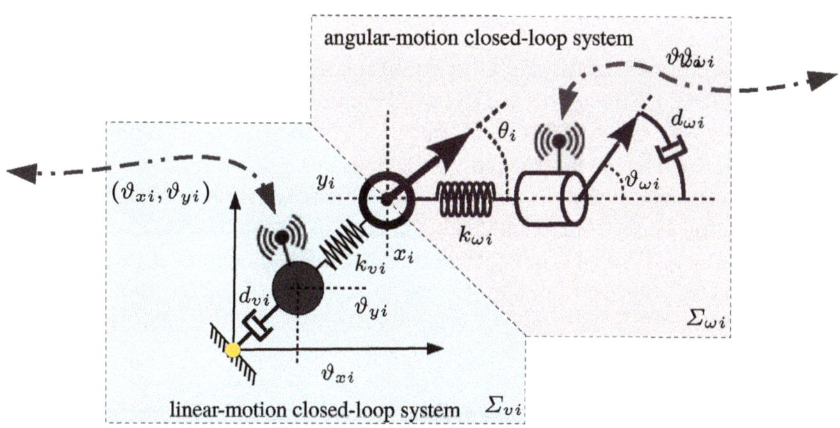

Fig. 3.4 Schematic representation of coupled mass-spring-damper systems: angular motion. It is the controller state variable, $\vartheta_{\omega i}$ that is transmitted to neighboring robots and, correspondingly, $\vartheta_{\omega j}$ is received from $\mathcal{N}_i$ neighbors. The system at the intersection of the two blocks represents the nonholonomic vehicle—*cf.* Fig. 3.1

lators, whose states are available. Then, it is the controllers' variables that are transmitted over the network to achieve consensus among the *controllers*—see Fig. 3.4 for an illustration.

More precisely, consider the dynamic system

$$\ddot{\vartheta}_{\omega i} + d_{\omega i}\dot{\vartheta}_{\omega i} + p_{\omega i} \sum_{j \in \mathcal{N}_i} a_{ij}(\vartheta_{\omega i} - \vartheta_{\omega j}) = \nu_{\omega i}, \tag{3.36}$$

where $\nu_{\omega i}$ is an external input to be defined, the state $\vartheta_{\omega i} \in \mathbb{R}$, and $d_{\omega i}$, $p_{\omega i} > 0$.

As we showed in Sect. 3.3.1, for (3.36) consensus is achieved, that is, there exists a real constant $\vartheta_{\omega c}$, such that $\vartheta_{\omega i} \to \vartheta_{\omega c}$, $\dot{\vartheta}_{\omega i} \to 0$, for all $i \le N$, provided that $d_{\omega i}$, $p_{\omega i} > 0$, and $\nu_{\omega i} = 0$. On the other hand, the system in (3.36) defines a passive map $\nu_{\omega i} \mapsto \dot{\vartheta}_{\omega i}$—*cf.* Remark 3.3.2. Furthermore, the system (3.3b) also defines a passive map, $u_{\omega i} \mapsto \omega_i$. Therefore, it results natural to hinge the systems (3.36) and (3.3) by setting $\nu_{\omega i} := -u_{\omega i}$ and

$$u_{\omega i} := -k_{\omega i}(\theta_i - \vartheta_{\omega i}), \quad k_{\omega i} > 0. \tag{3.37}$$

That is, the coupling $-k_{\omega i}(\theta_i - \vartheta_{\omega i})$ may be interpreted as the force exerted by a torsional spring that hinges the (angular) positions of the two subsystems. The closed-loop system corresponding to the vehicle's angular dynamics is

$$\Sigma_{\omega i} : \begin{cases} \ddot{\theta}_i = -k_{\omega i}(\theta_i - \vartheta_{\omega i}) & \text{(3.38a)} \\ \ddot{\vartheta}_{\omega i} + d_{\omega i}\dot{\vartheta}_{\omega i} + p_{\omega i} \sum_{j \in \mathcal{N}_i} a_{ij}(\vartheta_{\omega i} - \vartheta_{\omega j}) = k_{\omega i}(\theta_i - \vartheta_{\omega i}). & \text{(3.38b)} \end{cases}$$

—see Fig. 3.4.

Note that the right-hand side of Eq. (3.38b) corresponds exactly to that of Eq. (3.10). Therefore, we know that the dynamic controllers (3.36), with $\nu_{\omega i} = 0$ reach position consensus. On the other hand, since each of these systems is passive and so is the map (3.3b) to which it is hinged via the fictitious torsional spring with stiffness $k_{\omega i}$, the vehicles' orientations θ_i are also steered to a consensus manifold $\{\theta_i = \theta_c\}$. This observation stems from an interesting analogy between the control architecture proposed above and (consensus) control of robot manipulators with flexible joints. To better see this, consider the Euler–Lagrange equations for such systems, in closed loop with a proportional-derivative consensus controller like the one defined in (3.9).

We have—see [7, 58],

$$M_i(\theta_i)\ddot{\theta}_i + C_i(\theta_i, \dot{\theta}_i)\dot{\theta}_i + g_i(\theta_i) = -K(\theta_i - \vartheta_i) \tag{3.39a}$$

$$\ddot{\vartheta}_i + d_i\dot{\vartheta}_i + p_i \sum_{j \in \mathcal{N}_i} a_{ij}(\vartheta_i - \vartheta_j) = -K(\vartheta_i - \theta_i) \tag{3.39b}$$

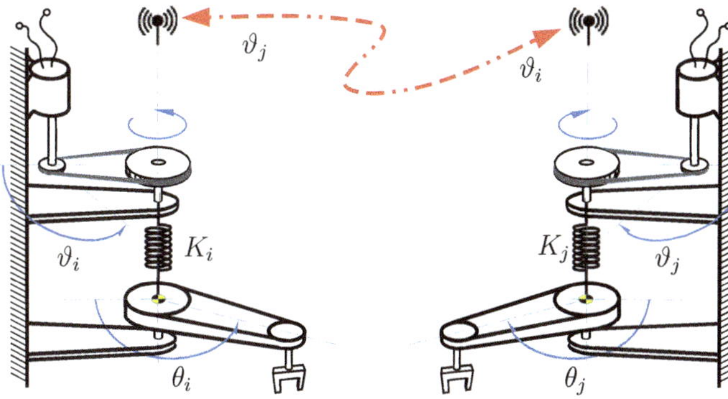

Fig. 3.5 Schematic representation of a pair of flexible-joint SCARA robots, which move on the plane, unaffected by gravity—compare with the mechanism in the pink block in Fig. 3.4. It is assumed that the robots exchange their actuators measurements ϑ_i over an undirected network. As the actuators reach consensus, so do the arms, in view of their respective mechanical couplings arm-actuator

and consensus is reached provided that gravity is cancelled-out and $d_i,\ p_i > 0$; this follows from the main results in [42] (Fig. 3.5).

In (3.39), the generalized coordinate ϑ_i corresponds to the actuators' angular position while θ_i corresponds to the links' position. Hence, in the case of the angular-motion dynamics, the closed-loop equation $\ddot{\theta}_i = -k_{\omega i}(\theta_i - \vartheta_{\omega i})$ may be assimilated to Eq. (3.39a) with unitary inertia $M_i = I$ and null Coriolis and gravitational forces, i.e., $C_i = g_i \equiv 0$. On the other hand, the dynamic controller (3.36) with $\nu_{\omega i} := k_{\omega i}(\theta_i - \vartheta_{\omega i})$ corresponds, up to obvious changes in the notation, to the actuator dynamics in closed loop, that is, Eq. (3.39b).

Proposition 3.4.1 (Output-feedback orientation consensus) *Consider the system (3.3) in closed loop with the dynamic controller defined by* (3.36), (3.37), *and* $\nu_{\omega i} := k_{\omega i}(\theta_i - \vartheta_{\omega i})$. *Let Assumption 3.2.2 hold. Then, there exist constants* θ_c *and* $\vartheta_c \in \mathbb{R}$ *such that, for all i and* $j \leq N$,

$$\lim_{t\to\infty} \theta_i(t) = \lim_{t\to\infty} \theta_j(t) = \theta_c, \qquad \lim_{t\to\infty} \omega_i(t) = 0,$$
$$\lim_{t\to\infty} \vartheta_{\omega i}(t) = \lim_{t\to\infty} \vartheta_{\omega j}(t) = \vartheta_c, \qquad \lim_{t\to\infty} \dot{\vartheta}_j(t) = 0.$$

Proof Consider the function

$$W_3(\dot{\vartheta}_\omega, \vartheta_\omega, \theta, \omega) := W_1(\vartheta_\omega, \dot{\vartheta}_\omega) + W_2(\theta, \omega, \vartheta_\omega), \tag{3.40}$$

where $\vartheta_\omega := [\vartheta_{\omega 1} \cdots \vartheta_{\omega N}]^\top$,

$$W_1(\vartheta_\omega, \dot{\vartheta}_\omega) := \frac{1}{2}\sum_{i\le N}\Big[\frac{\dot{\vartheta}_{\omega i}^2}{p_{\omega i}} + \frac{1}{2}\sum_{j\in\mathcal{N}_i} a_{ij}(\vartheta_{\omega i} - \vartheta_{\omega j})^2\Big], \tag{3.41}$$

$$W_2(\theta, \omega, \vartheta) := \frac{1}{2}\sum_{i\le N}\Big[\frac{\omega_i^2}{p_{\omega i}} + k_{\omega i}(\theta_i - \vartheta_{\omega i})^2\Big]. \tag{3.42}$$

The function W_2 corresponds to the total energy of the mass-spring (closed loop) system $\ddot{\theta}_i = -k_{\omega i}(\theta_i - \vartheta_{\omega i})$; the first term is the kinetic energy and the second the potential energy "stored" in the torsional spring of stiffness $k_{\omega i}$. The function W_1 is reminiscent of V_3 in (3.25). The function W_3 is positive definite in the consensus errors, in ω, ϑ_ω, and $\dot{\vartheta}_\omega$. Moreover, its total derivative along the trajectories of the closed-loop system (3.38)—with $\dot{\theta}_i = \omega_i$— yields

$$\dot{W}_3(\dot{\vartheta}_\omega, \vartheta_\omega, \theta, \omega) = -\frac{1}{2}\sum_{i\le N}\frac{d_{\omega i}}{p_{\omega i}}\dot{\vartheta}_{\omega i}^2. \tag{3.43}$$

Then, the system being autonomous, we may invoke Barbashin–Krasovskĭi's theorem. First, we see that $\dot{W}_3 = 0$ if and only if $\dot{\vartheta}_{\omega i} = 0$. The latter implies that $\ddot{\vartheta}_{\omega i} = 0$ and $\vartheta_{\omega i} =$ const for all $i \le N$. From (3.36) and $\nu_{\omega i} := k_{\omega i}(\theta_i - \vartheta_{\omega i})$ we conclude that $\theta_i =$ const, i.e., $\omega_i = \dot{\omega}_i = 0$. In turn, from $\dot{\omega}_i = -k_{\omega i}(\theta_i - \vartheta_{\omega i}) = -\nu_{\omega i} = 0$ and (3.36) it follows that

$$\sum_{j\in\mathcal{N}_i} a_{ij}(\vartheta_{\omega i} - \vartheta_{\omega j}) = 0 \quad \text{and} \quad \theta_i = \vartheta_{\omega i}\ \forall i, j \le N.$$

Finally, in view of Assumption 3.2.2. The only solution to the equations above is $\theta_i = \vartheta_{\omega i} = \vartheta_c$ for all $i, j \le N$.

3.4.2 Output-Feedback Position Consensus

Akin to the controller for the angular-motion subsystem, to steer the Cartesian positions $\bar{z}_i$ to a consensual point on the plane z_c, we use a second-order dynamic controller system that is reminiscent of the Eq. (3.8) in closed loop with the control (3.9), and an added virtual-spring coupling term, $-k_{vi}(\vartheta_{vi} - \bar{z}_i)$. That is, let

$$\ddot{\vartheta}_{vi} + d_{vi}\dot{\vartheta}_{vi} + p_{vi}\sum_{j\in\mathcal{N}_i} a_{ij}(\vartheta_{vi} - \vartheta_{vj}) = -k_{vi}(\vartheta_{vi} - \bar{z}_i), \tag{3.44}$$

where $\vartheta_{vi} \in \mathbb{R}^2$ and $\dot{\vartheta}_{vi}$ are controller's state variables, and all control gains d_{vi}, p_{vi} and k_{vi} are positive.

Then, the dynamical system (3.44) is coupled to the double (nonholonomic) integrator (3.4)—see Fig. 3.4 for an illustration. In contrast to the case of the angular motion, however, for the linear motion the control input u_{vi} must incorporate the change of coordinates defined by φ. Therefore, we define

$$u_{vi} := -\varphi(\theta_i)^\top k_{vi}(\bar{z}_i - \vartheta_{vi}), \quad k_{vi} > 0 \tag{3.45}$$

—cf. Eq. (3.37). For the angular-motion dynamics, to overcome the effect of the nonholonomic constraints, the angular-motion control law (3.37) is endowed with a perturbation term, that is, we define

$$u_{\omega i} := -k_{\omega i}(\theta_i - \vartheta_{\omega i}) + \alpha_i(t, \theta_i, \vartheta_{vi}, \bar{z}_i), \quad k_{\omega i} > 0. \tag{3.46}$$

The functions α_i are designed to vanish as the control goal is reached. We pose

$$\alpha_i(t, \theta_i, \vartheta_{vi}, \bar{z}_i) := k_{\alpha i}\psi_i(t)\varphi_i(\theta_i)^{\perp\top}(\vartheta_{vi} - \bar{z}_i), \tag{3.47}$$

which vanishes only as the plant and the controller synchronize, that is, if $\varphi_i(\theta_i)^{\perp\top}(\vartheta_{vi} - \bar{z}_i) \equiv 0$.

Thus, the closed-loop system for the linear-motion dynamics becomes

$$\Sigma_{vi}: \begin{cases} \dot{\bar{z}}_i = \varphi(\theta_i(t))v_i, & (3.48a) \\ \dot{v}_i = -\varphi(\theta_i(t))^\top k_{vi}(\bar{z}_i - \vartheta_{vi}), & (3.48b) \\ \ddot{\vartheta}_{vi} + d_{vi}\dot{\vartheta}_{vi} + p_{vi}\sum_{j\in\mathcal{N}_i} a_{ij}(\vartheta_{vi} - \vartheta_{vj}) = -k_{vi}(\vartheta_{vi} - \bar{z}_i); & (3.48c) \end{cases}$$

while the closed-loop equations for the angular-motion dynamics are modified from (3.38) into

$$\Sigma_{\omega i}: \begin{cases} \ddot{\theta}_i = -k_{\omega i}(\theta_i - \vartheta_{\omega i}) + \alpha_i(t, \theta_i, \vartheta_{vi}, \bar{z}_i) & (3.49a) \\ \ddot{\vartheta}_{\omega i} + d_{\omega i}\dot{\vartheta}_{\omega i} + p_{\omega i}\sum_{j\in\mathcal{N}_i} a_{ij}(\vartheta_{\omega i} - \vartheta_{\omega j}) = k_{\omega i}(\theta_i - \vartheta_{\omega i}). & (3.49b) \end{cases}$$

See the blue block in Fig. 3.4 for an illustration.

Note that, as before, in (3.48) we replaced θ_i with $\theta_i(t)$, which is a valid step as a long as the system is forward complete [23, p. 657]. The latter may be established along similar lines as in Remark 3.3.5. Again, the advantage of this is that the overall closed-loop system (3.38)–(3.48) may be regarded as a cascaded nonlinear time-varying system as illustrated in Fig. 3.3 (up to a redefinition of α_i). Thus, we have the following [44].

Proposition 3.4.2 (Output-feedback formation consensus) *Consider the system (3.3)–(3.5) in closed loop with (3.38b),* (3.46), (3.47), (3.45), *and* (3.44), *with* p_{vi}, d_{vi}, $p_{\omega i}$, *and* $d_{\omega i} > 0$, α_i *as in* (3.31), ψ_i *and* $\dot{\psi}_i$ *bounded, and* ψ_i *persistently exciting. Then, the consensus-formation goal is achieved, that is,* (3.6) *and* (3.7) *hold.*

To establish the statement, we use the function

$$\mathcal{V} := \sum_{i \le N} \left[\frac{1}{p_{vi}} \mathcal{E}_{vi} + \frac{1}{4} \sum_{j \in \mathcal{N}_i} a_{ij} |\vartheta_{vi} - \vartheta_{vj}|^2 \right].$$

$$\mathcal{E}_{vi} := \frac{1}{2} \left[v_i^2 + |\dot{\vartheta}_{vi}|^2 + k_{vi} |\vartheta_{vi} - \bar{z}_i|^2 \right].$$

The function $\mathcal{V}$ is positive definite in the consensus errors $(\vartheta_{vi} - \vartheta_{vj})$ and its derivative yields $\dot{\mathcal{V}} \le - \sum_{i \le N} \lambda_{vi} |\dot{\vartheta}_{vi}|^2 \le 0$. Then, based on the statement and proof of Proposition 3.4.1 and arguing as in Remark 3.3.5 one may use Barbălat's lemma to conclude the proof—see [44].

3.5 Output-Feedback Control Under Delays

We showed previously how dynamic output-feedback controllers may be successfully designed based on *physical* considerations. Mostly, by designing the controllers as mechanical systems interconnected, on one hand, to the actual plants and, on the other, among themselves in an undirected network. Besides the neat physical interpretation behind, the control method is versatile, in the sense that one can establish consensus even in the event that the network is affected by time-varying delays.

Technically, the step forward to cover this case only involves using more sophisticated functions. More precisely, we rely on Lyapunov–Krasovskĭi functionals. Remarkably, however, the control architecture remains the same as before, so we continue to rely on the certainty-equivalence principle and we use the controllers' dynamics (3.38b) and (3.44), except that in the present case, the measurement ϑ_j that the ith vehicle receives from its neighbors is affected by a delay. In that regard, the consensus-error correction terms on the left-hand side of (3.38b) and (3.44) depend now on the redefined consensus errors

$$e_{vi} := \sum_{j \in \mathcal{N}_i} a_{ij} \big(\vartheta_{vi} - \vartheta_{vj}(t - T_{ji}(t)) \big), \tag{3.50}$$

whereas for the orientations,

$$e_{\omega i} := \sum_{j \in \mathcal{N}_i} a_{ij} \big(\vartheta_{\omega i} - \vartheta_{\omega j}(t - T_{ji}(t)) \big). \tag{3.51}$$

Note that in both cases, as in previous sections, the errors are defined in the controllers' coordinates and not on robots' measured variables.

Based on (3.45) and (3.44), the certainty-equivalence controller for the linear-motion dynamics, (3.4), is given by

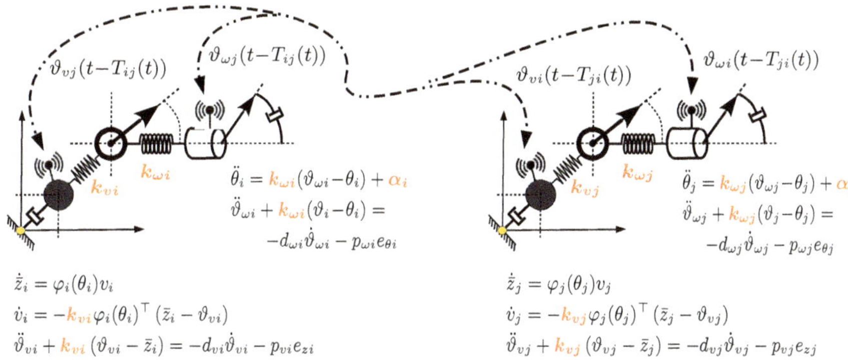

Fig. 3.6 Schematic representation of two vehicles exchanging their measurements, respectively, over a bidirectional link. On each end, we see the vehicles' linear and angular dynamics coupled with their respective controllers and transmitting the states of the latter over the network. Full consensus of the vehicles is achieved due to the mechanical couplings and the underlying spanning tree in the controllers' network

$$u_{vi} = -k_{vi}\varphi(\theta_i)^\top(\bar{z}_i - \vartheta_{vi}), \tag{3.52a}$$
$$\ddot{\vartheta}_{vi} = -d_{vi}\dot{\vartheta}_{vi} - k_{vi}(\vartheta_{vi} - \bar{z}_i) - p_{vi}e_{vi}, \tag{3.52b}$$

whereas, for the angular-motion dynamics, we use

$$u_{\omega i} = -k_{\omega i}(\theta_i - \vartheta_{\omega i}) + \alpha_i(t, \theta_i, \vartheta_{vi}, \bar{z}_i), \tag{3.53a}$$
$$\ddot{\vartheta}_{\omega i} = -d_{\omega i}\dot{\vartheta}_{\omega i} - k_{\omega i}(\vartheta_{\omega i} - \theta_{\omega i}) - p_{\omega i}e_{\omega i}. \tag{3.53b}$$

All constant parameters are defined as above and α_i is defined in (3.47), so the controller *per se* remains the same. That is, α_i in (3.53a) fulfills the same role as explained above. Only the analysis requires more advanced tools. A schematic representation of the networked closed-loop systems is given in Fig. 3.6.

Proposition 3.5.1 (Output-feedback consensus under delay measurements) *Consider the system (3.3)–*(3.5), *under Assumptions 3.2.2–3.2.5, in closed loop with (3.52)–(3.53). Then, the leaderless consensus control goal is achieved, that is,* (3.6) *and* (3.7) *hold provided that*

$$d_{vi} > \frac{1}{2}p_{vi}\sum_{j\in\mathcal{N}_i} a_{ij}\left[\beta_i + \frac{\overline{T}_{ji}^2}{\beta_j}\right] \tag{3.54}$$

$$d_{\omega i} > \frac{1}{2}p_{\omega i}\sum_{j\in\mathcal{N}_i} a_{ij}\left[\varepsilon_i + \frac{\overline{T}_{ji}^2}{\varepsilon_j}\right] \tag{3.55}$$

for some β_i, $\varepsilon_i > 0$, for all $i \leq N$.

The closed-loop equations, according to the logic of separating the linear and the angular-motion dynamics, correspond to

$$\Sigma_{\omega i}:\left\{\begin{aligned}\ddot{\theta}_i &= -k_{\omega i}(\theta_i-\vartheta_{\omega i})+\alpha_i(t,\theta_i,\vartheta_{vi},\bar{z}_i) &&(3.56\text{a})\\ \ddot{\vartheta}_{\omega i} &= -d_{\omega i}\dot{\vartheta}_{\omega i}-k_{\omega i}(\vartheta_{\omega i}-\theta_i)-p_{\omega i}e_{\omega i} &&(3.56\text{b})\end{aligned}\right.$$

$$\Sigma_{vi}:\left\{\begin{aligned}\dot{\bar{z}}_i &= \varphi(\theta_i(t))v_i &&(3.57\text{a})\\ \dot{v}_i &= -k_{vi}\varphi(\theta_i(t))^\top(\bar{z}_i-\vartheta_{vi}) &&(3.57\text{b})\\ \ddot{\vartheta}_{vi} &= -d_{vi}\dot{\vartheta}_{vi}-k_{vi}(\vartheta_{vi}-\bar{z}_i)-p_{vi}e_{vi} \quad . &&(3.57\text{c})\end{aligned}\right.$$

The closed-loop system has the same structure as in previous cases. It is a complex network of feedback-interconnected systems and, yet, we can identify in each node of the network a cascaded system as schematically depicted in Fig. 3.3 (up to a redefinition of the arguments of the "perturbation" α_i). That is, for each robot we may replace the state variables θ_i with fixed, but arbitrary, trajectories $\theta_i(t)$ in (3.57a) and (3.57b), so the dotted feedback line in Fig. 3.3 is disregarded.

Then, the analysis of $\Sigma_{\omega i}$ and Σ_{vi} may again be carried out using a cascade argument. In a nutshell, one needs to establish that:

(*i*) all the trajectories are bounded: for this we employ the Lyapunov–Krasovskĭi functional for the "decoupled" linear-motion dynamics (3.57),

$$\mathcal{V}:=\sum_{i\le N}\left[\frac{1}{p_{vi}}\mathcal{E}_i+\frac{1}{4}\sum_{j\in\mathcal{N}_i}a_{ij}|\vartheta_{vi}-\vartheta_{vj}|^2+\frac{1}{2\beta_i}\Upsilon_i(t,\vartheta_v)\right]$$

$$\mathcal{E}_i := \frac{1}{2}\left[v_i^2+|\dot{\vartheta}_{vi}|^2+k_{vi}|\vartheta_{vi}-\bar{z}_i|^2\right]$$

$$\Upsilon_i(t,\vartheta_v) := \sum_{j\in\mathcal{N}_i}a_{ij}\overline{T}_{ij}\int_{-\overline{T}_{ij}}^{0}\int_{t+\eta}^{t}|\dot{\vartheta}_{vj}(\sigma)|^2 d\sigma d\eta.$$

Using the symmetry of the underlying network's Laplacian one can show that

$$\dot{\mathcal{V}}\le-\sum_{i\le N}\left[\frac{d_{vi}}{p_{vi}}-c(\overline{T}_{ij})\right]|\dot{\vartheta}_{vi}|^2. \tag{3.58}$$

Hence, under the assumption that (3.54) holds, we have $\dot{\mathcal{V}}\le 0$ which implies that the consensus errors $\vartheta_{vi}-\vartheta_{vj}$ as well as v_i and $\dot{\vartheta}_{vi}$ are bounded (at least on the interval of existence of the solutions—see Remark 3.3.5).

On the other hand,

(*ii*) for $\Sigma_{\omega i}$ with $\alpha_i\equiv 0$ we may use the Lyapunov–Krasovskĭi functional

$$\mathcal{W} := \sum_{i\leq N}\left[\frac{1}{p_{\omega i}}\mathcal{H}_i + \frac{1}{4}\sum_{j\in\mathcal{N}_i} a_{ij}(\vartheta_{\omega i} - \vartheta_{\omega j})^2 + \frac{1}{2\varepsilon_i}\Upsilon_i(t, \vartheta_\omega)\right]$$

$$\mathcal{H}_i := \frac{1}{2}\left[\omega_i^2 + \dot\vartheta_{\omega i}^2 + k_{\omega i}(\vartheta_{\omega i} - \theta_i)^2\right]$$

$$\Upsilon_i(t, \vartheta_\omega) := \sum_{j\in\mathcal{N}_i} a_{ij}\overline{T}_{ij}\int_{-\overline{T}_{ij}}^{0}\int_{t+\eta}^{t}|\dot\vartheta_{\omega j}(\sigma)|^2 d\sigma d\eta$$

to obtain that

$$\dot{\mathcal{W}} \leq -\sum_{i\leq N}\left[\frac{d_{\omega i}}{p_{\omega i}} - c(\overline{T}_{ij})\right]\dot\vartheta_{\omega_i}^2. \tag{3.59}$$

Again, under condition (3.55) we have $\dot{\mathcal{W}} \leq 0$, from which one conclude the boundedness of the orientation consensus errors, as well as of the angular velocities ω_i.

From here, one may carry on using recursively Barbălat's lemma and the boundedness of α_i to establish the convergence of multiple signals. Finally,

(*iii*) under the persistently exciting effect of α_i, one ensures that the trajectories can only converge to the consensus manifold, thereby avoiding the undesired equilibria discussed in Sect. 3.3.2.

The detailed proof is available from [45].

3.6 An Illustrative Case-Study

We used the simulator Gazebo-ROS and the Robot Operating System (ROS) interface to evaluate the performance of our controller in a scenario that reproduces as closely as possible that of a laboratory experimental benchmark. Furthermore, for the sake of comparison, we also carried out numerical-integration simulations using Simulink® of MATLAB®.

Gazebo-ROS is an efficient 3D dynamic simulator of robotic systems in indoor and outdoor environments. In contrast to pure numerical-integration-based solvers of differential equations, Gazebo-ROS accurately emulates physical phenomena and dynamics otherwise neglected, such as friction, contact forces, actuator dynamics, slipping, etc. In addition, it offers high-fidelity robot and sensor simulations.

For the test scenario, we employed the model of a PIONEER 3-DX wheeled robot [16], available in Gazebo's library. It must be underlined that for this robot the center of mass is not located on the axis joining the two wheels' centers—cf. Fig. 3.1, as it is assumed at the basis of the developments in the previous sections. More precisely, in Eq. (3.2) the functions F_v and F_ω include Coriolis terms that are quadratic in the velocities, i.e., $\frac{r_i}{3}\omega_i^2$ on the left-hand side of Eq. (3.2a) and $-\frac{r_i m_i}{3I_i}\omega_i v_i$ on the left-hand side of Eq. (3.2b). Akin to an actual experimentation set-up, these constitute dynamic

Table 3.1 Initial conditions and formation offsets

Index	x_i [m]	y_i [m]	θ_i [rad]	δ_{xi} [m]	δ_{yi} [m]
1	8	7	1.57	2	0
2	2	13	0.0	1	2
3	2	9	−0.39	−1	2
4	−2	6	0.39	−2	0
5	1	3	−0.39	−1	−2
6	4	4	−0.39	1	−2

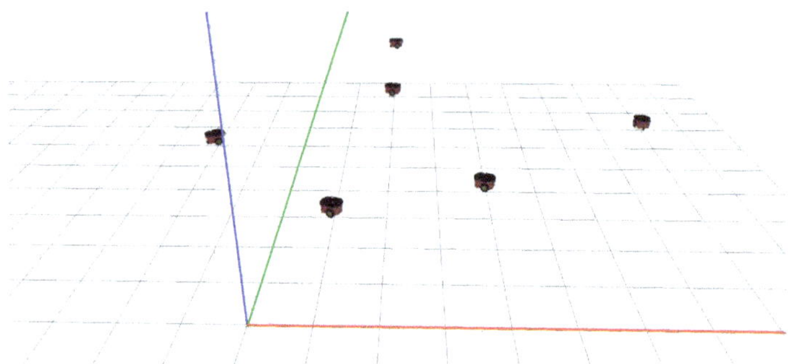

Fig. 3.7 Screenshot of the six PIONEER 3D-X robots' at their initial configuration, in the Gazebo-ROS simulator

effects not considered in the model for which the controller is validated analytically, nor in the simulations carried out with Simulink® of MATLAB®.

Concretely, in the simulation scenario we consider six PIONEER 3D-X robots starting from initial postures as defined in the 2nd–4th columns of Table 3.1.

Also, an illustration of the robots in their initial postures is showed via a screenshot of the Gazebo-ROS simulator's user interface in Fig. 3.7.

It is assumed that the robots communicate over the undirected connected graph like the one illustrated in Fig. 3.8 and with piece-wise constant time-varying delays. For simplicity, all the time delays $T_{ji}(t)$ are taken equal; they are generated randomly following a normal distribution with mean $\mu = 0.3$, variance $\sigma^2 = 0.0003$ and a sample time of 10 ms—see Fig. 3.9. Such time delay (non-smooth but piece-wise continuous) does not satisfy Assumption 3.2.4 since its time-derivative is bounded

Fig. 3.8 Communication topology: undirected connected graph

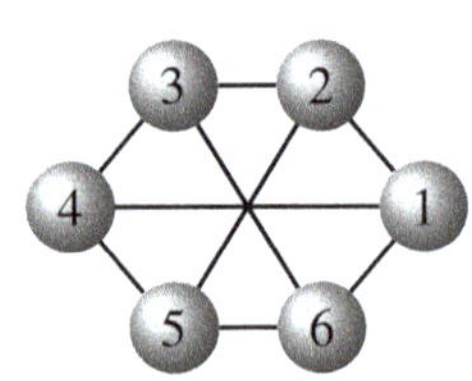

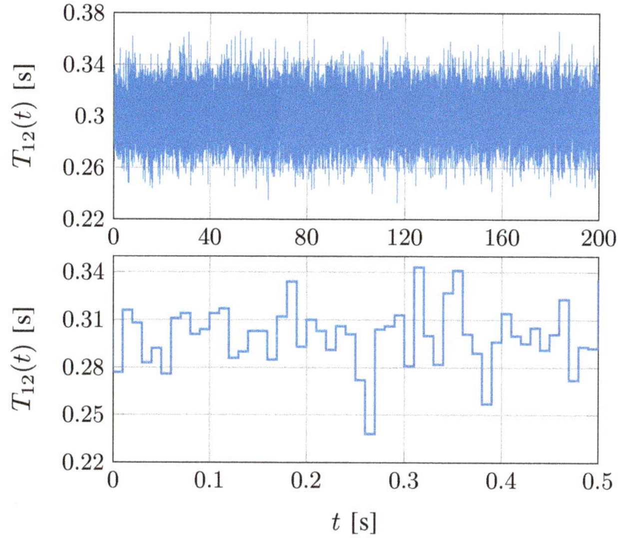

Fig. 3.9 Variable delay between the robot 1 and the received information from neighbor 2

only almost everywhere (that is, except at the points of discontinuity). However, it is considered in the simulations since it is closer to what is encountered in a real-world set-up [1]. Even though the technical Assumption 3.2.4 does not hold, full consensus is achieved (at least practically) in both the MATLAB® and the realistic Gazebo-ROS simulations. This hints at the fact that Assumption 3.2.4 might be relaxed in the analysis, albeit using a different controller—cf. [33].

The desired formation at rendezvous corresponds to a hexagon determined by desired offsets $\delta_i = (\delta_{xi}, \delta_{yi})$ with respect the unknown center of the formation. These constants are presented in the last two columns of Table 3.1—see Fig. 3.2 for an illustration of the target formation.

For a fair and meaningful comparison, the numerical simulations under Simulink® of MATLAB® were performed using the information available on the PIONEER 3D-X robot, from the Gazebo-ROS simulator. For simplicity, it is assumed that all the robots have the same inertial and geometrical parameters given by $m = 5.64$ kg, $I = 3.115$ kg·m^2, $r = 0.09$ m and $R = 0.157$ m.

In both simulators, the control gains were set to $k_{vi} = 1$, $k_{\omega i} = 2$, $d_{vi} = 3$, $p_{vi} = 0.4$, $d_{\omega i} = 2$, $p_{\omega i} = 0.1$, for all $i \in [1, 6]$. These values correspond to magnitudes compatible with the emulated physics of the PIONEER 3D-X robots in Gazebo-ROS and are chosen so that the poles of the second-order system $\ddot{x} = -d_{(\cdot)}\dot{x} - p_{(\cdot)}x$ have negative real parts and the system have an over-damped step-response. The δ-persistently exciting functions α_i, for all $i \in [1, 6]$, were taken as in (3.53) with $k_{\alpha i} = 0.4$ and, for simplicity, (multi)periodic functions

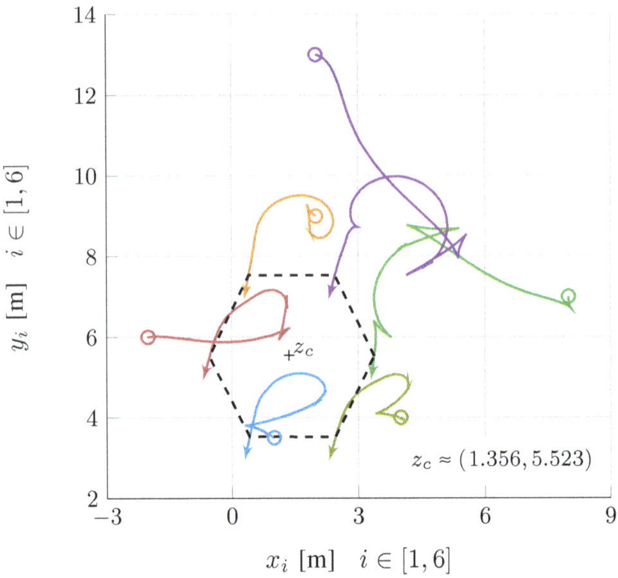

Fig. 3.10 Paths followed by the PIONEER 3D-X robots up to full formation consensus—MATLAB® simulation. A hexagonal formation is achieved with coinciding orientations (illustrated by arrows)

$$\begin{aligned}\psi_i(t) =& 2.5 + \sin(2\pi t) + 0.3\cos(6\pi t) - 0.5\sin(8\pi t) \\ &- 0.1\cos(10\pi t) + \sin(\pi t) \quad \forall i \leq 6. \end{aligned} \tag{3.60}$$

Other parameters, such as the sampling time, were taken equally.

As we mentioned above, however, certain physical phenomena as well as actuator and sensor dynamics, which are hard-coded in Gazebo-ROS, cannot be reproduced in MATLAB®. The consequence of this is clearly appreciated in the figures shown below.

The results obtained with Gazebo-ROS are showed in Figs. 3.11, 3.13, 3.15, and 3.17. Those obtained using Simulink® of MATLAB® are showed in Figs. 3.10, 3.14, 3.16, and 3.18.

In both cases the robots appear to achieve consensus, i.e., to meet at a non-predefined rendezvous point in hexagonal formation and with common non-predefined orientation—see Figs. 3.10 and 3.11, as well as the screenshot of the final postures from Gazebo's graphical interface, Fig. 3.12. Under MATLAB®, the center of the formation is located at (1.356, 5.523), while under Gazebo-ROS it is at (−3.242, −3.597). The consensual orientations are $\theta_c \approx -2.889$ rad under Gazebo-ROS and $\theta_c \approx -1.785$ rad under MATLAB®.

Both simulations illustrate that for networks of nonholonomic vehicles, the initial conditions do not determine the consensus point, as is the case of linear systems interconnected over static undirected connected graphs [52]. Indeed, the consensus

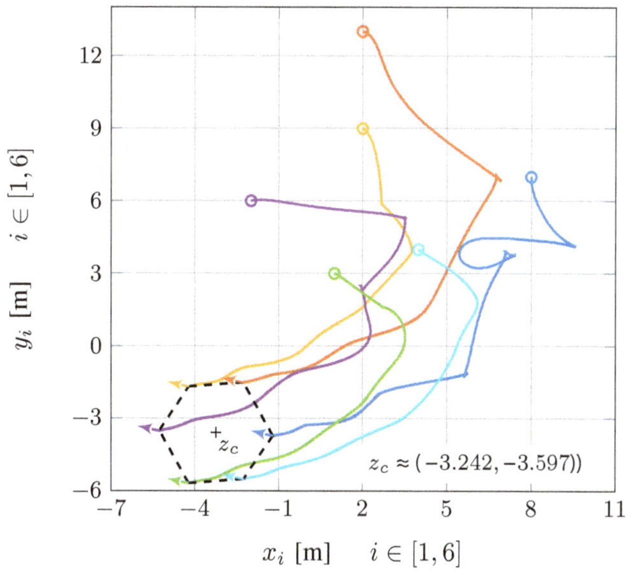

Fig. 3.11 Paths followed by the PIONEER 3D-X robots up to full formation consensus—Gazebo-ROS simulation. A hexagonal formation is achieved with coinciding orientations (illustrated by arrows)

point—in this case the center of the formation and the common orientation—does not correspond to the average of the initial conditions.

The consensus equilibrium heavily depends, as well, on the systems' *nonlinear* dynamics. This is clear both, in Fig. 3.10 which results of a MATLAB® simulation for a network of nonlinear systems modeled as in (3.1)–(3.2) with $F_v = F_\omega = 0$, as well as in Fig. 3.11 which results from a more realistic simulation based on a model that emulates otherwise neglected Coriolis high-order terms, friction, sensor and actuator effects, *etc.* In addition, it appears fitting to recall that the controller, in both cases, is dynamic and time-varying.

Furthermore, it is clear from Figs. 3.10 and 3.11 that the results obtained with either simulator differ considerably in various manners. Obvious discrepancies lay in the position of the center of the consensus formation that is achieved, as well as in the paths followed by the robots.

The differences in the transient behaviors for both simulations are even clearer in the plots of the consensus errors, which, for the purpose of graphic illustration, are defined as the difference between each robot's variables and the corresponding average:

$$e_{zi} := \bar{z}_i - \frac{1}{N} \sum_{j \in \mathcal{N}_i} \bar{z}_j, \quad e_{\theta i} := \theta_i - \frac{1}{N} \sum_{j \in \mathcal{N}_i} \theta_j. \tag{3.61}$$

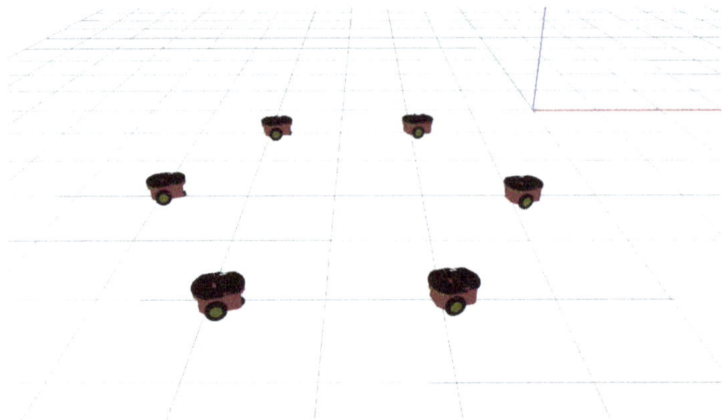

Fig. 3.12 Screenshot of the final configuration in the Gazebo-ROS simulator; the six robots achieving full consensus at the rendezvous point

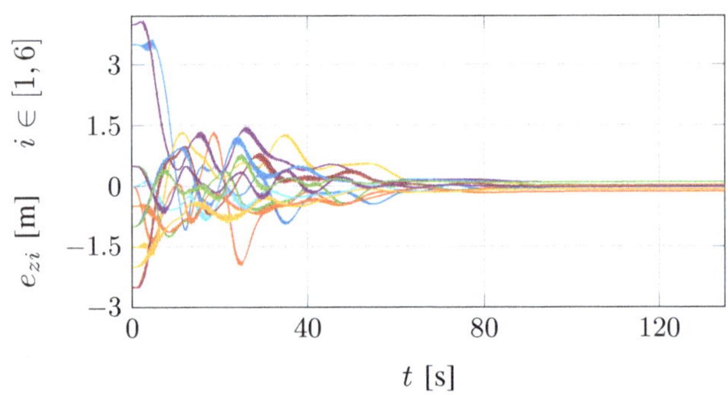

Fig. 3.13 Position consensus errors—Gazebo-ROS simulation

That is, the limits in (3.6) and (3.7) hold if the error trajectories $e_{zi}(t)$ and $e_{\theta i}(t)$ as defined above converge to zero, but the errors in (3.61) do not correspond to variables actually used by the controller nor measured for that matter.

In Fig. 3.13 one can appreciate that such errors do not actually tend to zero, but to a steady-state error—a keen observer will notice that the hexagon in Fig. 3.11 is actually not quite so. In contrast to this, in the simulation obtained using MATLAB®—see Fig. 3.14—the errors converge to zero asymptotically, albeit slowly. The reason is that in the Gazebo-ROS simulation, after a transient, the amplitude of the input torques becomes considerably small in absolute value—see Fig. 3.17.

The presence of a steady-state error and the persistency-of-excitation effect in the controller maintain the input torques oscillating (periodically in this case due to the choice of $\psi_i(t)$ in (3.60)), but, *physically*, they result insufficient to overcome the robots' inertia and friction forces that oppose their forward and angular motions. In contrast to this, in Fig. 3.18 are showed the input torques obtained using Simulink® of MATLAB®. A similar oscillating behavior is observed, but the torques vanish

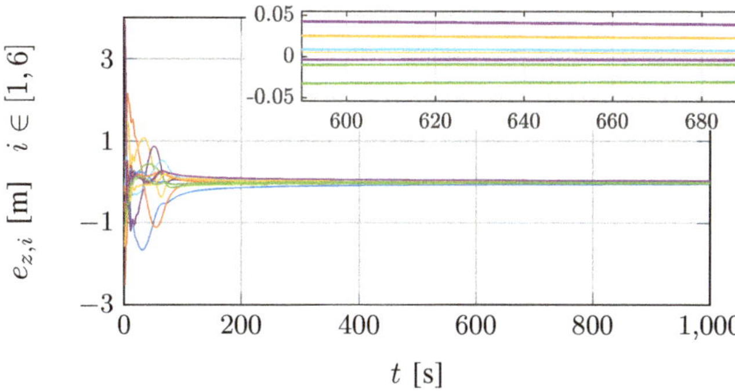

Fig. 3.14 Position consensus errors—MATLAB® simulation

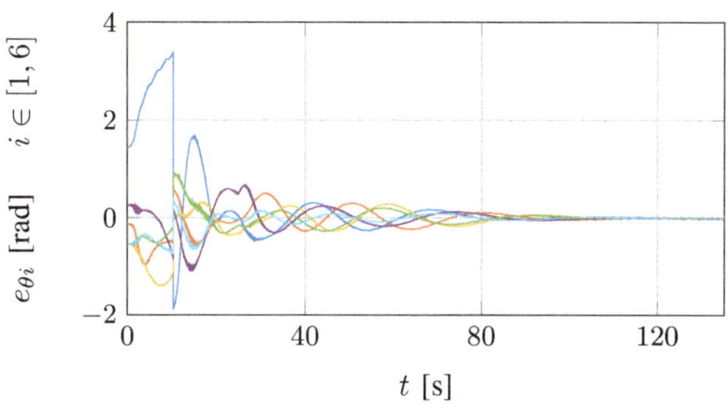

Fig. 3.15 Orientation consensus errors—Gazebo-ROS simulation. The consensus equilibrium $\theta_c \approx -2.889$ rad

asymptotically—notice the order of magnitude in the plots on the right column in Fig. 3.18, in the range of milli-Nm—as the error-dependent persistency of excitation disappears.

It seems fitting to say at this point that the controller gains may be augmented, for instance, to increase the convergence speed, but such values may result incompatible with the robots' and actuators' physical limitations, so it is not done here to conserve a realistic setting.

3.7 Conclusions

We described a simple and appealing control method for formation consensus of differential-drive robots that applies in realistic scenarii including those of vehicles non-equipped with velocity sensors and networks affected by time-varying delays. The controllers that we propose have the neat physical interpretation of a second-

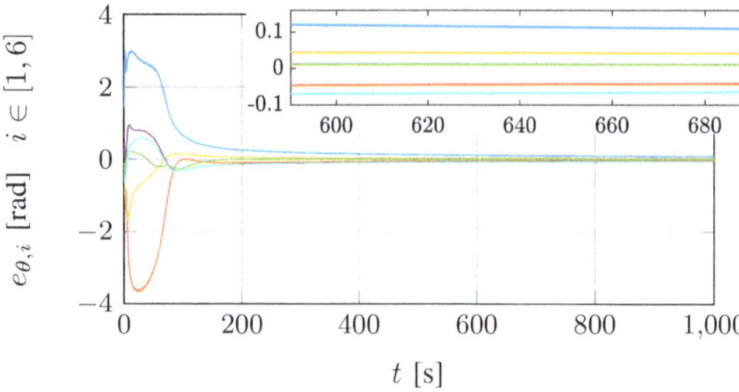

Fig. 3.16 Orientation consensus errors—MATLAB® simulation. The consensus equilibrium $\theta_c \approx -1.785$ rad

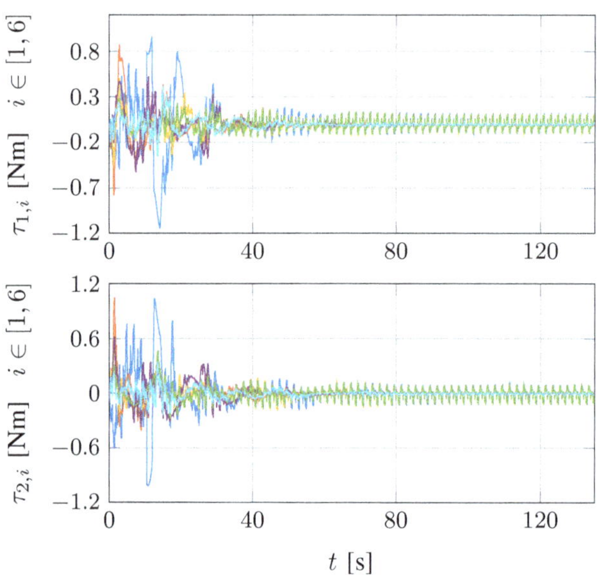

Fig. 3.17 Input torques—Gazebo-ROS simulation

order mechanical system itself, so this technique may be a starting point for observer-less control of multi-agent systems under output feedback. It appears interesting to continue testing the limits of our physics-based method for output-feedback control in a multi-agent systems setting.

Notably, our hypothesis that the graph is undirected and static remains conservative, but the study of multi-agent nonholonomic vehicles with less stringent hypotheses on the topology has been little addressed, even under the assumption that full-state feedback is available. It has certainly been done for second-order integrators, as in [51], for directed-spanning-tree-graph topologies or for high-order systems under

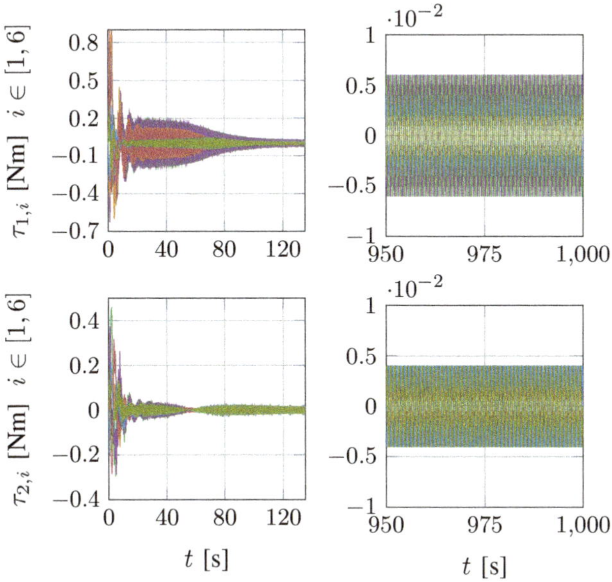

Fig. 3.18 Input torques—MATLAB® simulation

constraints in [54], but the *consensus* control of nonholonomic systems networked over generic directed graphs and under output feedback remains very little explored. A major stumbling block remains the construction of strict Lyapunov functions to establish ISS when the graph is directed and admits cycles (hence, beyond leader-follower configurations). Although the problem was recently solved for linear systems in [49], to the best of our knowledge, it remains largely open for nonholonomic systems.

Another intriguing aspect to investigate further is the influence of the nonlinear dynamics on consensus. Our numerical tests using the Gazebo-ROS simulator clearly show the effects of the nonlinearities in the consensus and the limitations of numerical algorithms bound to solving ordinary differential equations that describe over-simplified models, in which aspects such as unmodelled dynamics, friction and, actuator and sensor dynamics are neglected. It is important to study formally the consensus-formation control problems using models that include the presence of highly nonlinear Lagrangian dynamics—*cf.* [13, 19, 24, 59].

Acknowledgements This work was supported in part by the "Agence Nationale de la Recherche" (ANR) under Grant HANDY ANR-18-CE40-0010.

References

1. Abdessameud, A., Polushin, I.G., Tayebi, A.: Synchronization of Lagrangian systems with irregular communication delays. IEEE Trans. Autom. Control **59**(1), 187–193 (2014)
2. Abdessameud, A., Tayebi, A.: On consensus algorithms for double-integrator dynamics without velocity measurements and with input constraints. Syst. Control Lett. **59**(12), 812–821 (2010)
3. Abdessameud, A., Tayebi, A., Polushin, I.G.: Leader-follower synchronization of Euler–Lagrange systems with time-varying leader trajectory and constrained discrete-time communication. IEEE Trans. Autom. Control **62**(5), 2539–2545 (May2017)
4. Barbashin, E.A., Krasovskĭi, N.N.: Об устойчивости движеняи в целом. Dokl. Akad. Nauk. USSR **86**(3), 453–456 (1952). Commonly (and wrongly) cited in English under: "On the stability of motion in the large"; Correct translation: "On the stability of motion in the *whole*"
5. Bautista-Castillo, A., Lopez-Franco, C., Nuño, E.: Consensus-based formation control for multiple nonholonomic robots. In: 2016 IEEE International Autumn Meeting on Power Electronics and Computing (ROPEC). IEEE (2016)
6. Brockett, R.W.: Asymptotic stability and feedback stabilization. Differ. Geom. Control Theory **27**(1), 181–191 (1983)
7. Burkov, I.V., Zaremba, A.T.: Dynamics of elastic manipulators with electric drives. Izv. Akad. Nauk SSSR Mekh. Tverd. Tela **22**(1), 57–64 (1987). Engl. transl. in Mechanics of Solids, Allerton Press
8. Cao, Y., Ren, W.: Distributed Coordination of Multi-agent Networks: Emergent Problems, Models, and Issues. Springer (2011)
9. Cheng, Y., Jia, R., Du, H., Wen, G., Zhu, W.: Robust finite-time consensus formation control for multiple nonholonomic wheeled mobile robots via output feedback. Int. J. Robust Nonlinear Control **28**(6), 2082–2096 (2018)
10. Consolini, L., Morbidi, F., Prattichizzo, D., Tosques, M.: On a class of hierarchical formations of unicycles and their internal dynamics. IEEE Trans. Autom. Control **57**(4), 845–859 (2012)
11. Dimarogonas, D.V., Kyriakopoulos, K.J.: On the rendezvous problem for multiple nonholonomic agents. IEEE Trans. Autom. Control **52**(5), 916–922 (2007)
12. Dixon, W.E., Dawson, D.M., Zhang, F., Zergeroglu, E.: Global exponential tracking control of a mobile robot system via a PE condition. IEEE Trans. Syst. Man. Cybernet. B **30**(1), 129–142 (2000)
13. Do, K.D., Jiang, Z.-P., Pan, J.: A global output-feedback controller for simultaneous tracking and stabilization of unicycle-type mobile robots. IEEE Trans. Robot. Autom. **20**(3), 589–594 (2004)
14. Dong, W.: Distributed observer-based cooperative control of multiple nonholonomic mobile agents. Int. J. Syst. Sc. **43**(5), 797–808 (2012)
15. El-Hawwary, M.I., Maggiore, M.: Distributed circular formation stabilization for dynamic unicycles. IEEE Trans. Autom. Control **58**(1), 149–162 (2013)
16. Génération Robots. Pioneer P3-DX mobile robot. Accessed May 06, 2021. [Online]. https://www.generationrobots.com/en/402395-robot-mobile-pioneer-3-dx.html
17. González, A., Aragüés, R., López-Nicolás, G., Sagüés, C.: Stability analysis of nonholonomic multiagent coordinate-free formation control subject to communication delays. Int. J. Robust Nonlinear Control **28**(14), 4121–4138 (2018)
18. Hernández, A.T., Loría, A., Nuño, E., Panteley, E.: Consensus-formation control of nonholonomic robots without velocity measurements. In: Proceedings of European Control Conference, pp. 674–679, St. Petersburg, Russia (2020)
19. Huang, J., Wen, C., Wang, W., Jiang, Z.-P.: Adaptive output feedback tracking control of a nonholonomic mobile robot. Automatica **50**, 821–831 (2014)
20. Hui, Q.: Finite-time rendezvous algorithms for mobile autonomous agents. IEEE Trans. Autom. Control **56**(1), 207–211 (2011)
21. Jadbabaie, A., Lin, J., Morse, A.S.: Coordination of groups of mobile autonomous agents using nearest neighbor rules. IEEE Trans. Autom. Control **48**(6), 988–1001 (2003)

22. Jin, J., Gans, N.: Collision-free formation and heading consensus of nonholonomic robots as a pose regulation problem. Robot. Auton. Syst. **95**(9), 25–36 (2017)
23. Khalil, H.: Nonlinear Systems. Macmillan Publishing Co., 2nd ed., New York (1996)
24. Lee, D.: Passive decomposition and control of nonholonomic mechanical systems. IEEE Trans. Robot. **26**(6), 978–992 (2010)
25. Lee, T.C., Song, K.T., Lee, C.H., Teng, C.C.: Tracking control of unicycle-modeled mobile robots using a saturation feedback controller. IEEE Trans. Control Syst. Technol. **9**(2), 305–318 (2001)
26. Li, Z., Ren, W., Liu, X., Fu, M.: Consensus of multi-agent systems with general linear and lipschitz nonlinear dynamics using distributed adaptive protocols. IEEE Trans. Autom. Control **58**(7), 1786–1791 (2013)
27. Liang, X., Wang, H., Liu, Y., Chen, W., Liu, T.: Formation control of nonholonomic mobile robots without position and velocity measurements. IEEE Trans. Robot. **34**(2), 434–446 (2018)
28. Lin, Z., Francis, B., Maggiore, M.: Necessary and sufficient graphical conditions for formation control of unicycles. IEEE Trans. Autom. Control **50**(1), 121–127 (2005)
29. Lizárraga, D.A.: Obstructions to the existence of universal stabilizers for smooth control systems. Math. Control, Sign. Syst. **16**, 255–277 (2004)
30. Loría, A.: From feedback to cascade-interconnected systems: breaking the loop. In: Proceedings of 47th IEEE Conference Decision Control, pp. 4109–4114, Cancun, Mex (2008)
31. Loría, A., Panteley, E., Teel, A.: A new persistency-of-excitation condition for UGAS of NLTV systems: Application to stabilization of nonholonomic systems. In: Proceedings of 5th European Control Conference, pp. 1363–1368, Karlsrühe, Germany (1999)
32. Maghenem, M.: Stability and stabilization of networked time-varying systems. Ph.D. thesis, Univ Paris Saclay, Gif sur Yvette (2017). https://tel.archives-ouvertes.fr/tel-01596158
33. Maghenem, M., Loría, A., Nuño, E., Panteley, E.: Distributed full-consensus control of nonholonomic vehicles under non-differentiable measurement delays. IEEE Control Syst. Lett. **5**(1), 97–102 (2021)
34. Mao, W., Wang, C., Chen, W., Li, X.: Observer-based consensus design for multi-agent systems with unavailable velocities of leader and followers. In: Proceedings of 32nd Chinese Control Conference, pp. 7030–7033 (2013)
35. Marshall, J.A., Broucke, M.E., Francis, B.A.: Formations of vehicles in cyclic pursuit. IEEE Trans. Autom. Control **49**(11), 1963–1974 (2004)
36. Montijano, E., Thunberg, J., Hu, X., Sagüés, C.: Epipolar visual servoing for multirobot distributed consensus. IEEE Trans. Robot. **29**(5), 1212–1225 (2013)
37. Moshtagh, N., Jadbabaie, A.: Distributed geodesic control laws for flocking of nonholonomic agents. IEEE Trans. Autom. Control **52**(4), 681–686 (2007)
38. Narendra, K.S., Annaswamy, A.M.: Stable Adaptive Systems. Prentice-Hall Inc, New Jersey (1989)
39. Neimark, J.I., Fufaev, F.A.: Dynamics of Nonholonomic Systems, vol. 33. A.M.S. Translations of Mathematical Monographs, Providence, RI (1972)
40. Ning, B., Han, Q.: Prescribed finite-time consensus tracking for multiagent systems with nonholonomic chained-form dynamics. IEEE Trans. Autom. Control **64**(4), 1686–1693 (2019)
41. Nuño, E.: Consensus of Euler-Lagrange systems using only position measurements. IEEE Trans. Control Netw. Syst. **5**(1), 489–498 (2018)
42. Nuño, E., Ortega, R.: Achieving consensus of Euler–Lagrange agents with interconnecting delays and without velocity measurements via passivity-based control. IEEE Trans. Control Syst. Technol. **26**(1), 222–232 (2018)
43. Nuño, E., Loría, A., Hernández, A.T., Maghenem, M., Panteley, E.: Distributed consensus-formation of force-controlled nonholonomic robots with time-varying delays. Automatica **120**, 109114 (2020)
44. Nuño, E., Loría, A., Panteley, E.: Leaderless consensus formation control of cooperative multi-agent vehicles without velocity measurements. IEEE Control Syst. Lett. **6**, 902–907 (2022)
45. Nuño, E., Loría, A., Panteley, E., Restrepo-Ochoa, E.: Rendezvous of nonholonomic robots via output-feedback control under time-varying delays. IEEE Trans. Control Syst. Technol. **30**(6), 2707–2716 (2022)

46. Nuño, E., Sarras, I., Loría, A., Maghenem, M., Cruz-Zavala, E., Panteley, E.: Strict Lyapunov-Krasovskĭi functionals for undirected networks of Euler–Lagrange systems with time-varying delays. Syst. Contr. Lett. **135**, 104579 (2020)
47. Panteley, E., Lefeber, E., Loría, A., Nijmeijer, H.: Exponential tracking of a mobile car using a cascaded approach. In: IFAC Workshop on Motion Control, pp. 221–226, Grenoble, France (1998)
48. Panteley, E., Loría, A.: Synchronization and dynamic consensus of heterogeneous networked systems. IEEE Trans. Autom. Control **62**(8), 3758–3773 (2017)
49. Panteley, H., Loría, A., Sukumar, S.: Strict lyapunov functions for consensus under directed connected graphs. In: Proceedings of European Control Conference (ECC), pp. 935–940, St. Petersburg, Russia (2020)
50. Poonawala, H.A., Satici, A.C., Spong, M.W.: Leader-follower formation control of nonholonomic wheeled mobile robots using only position measurements. In: Proceedings of 9th Asian Control Conference, pp. 1–6 (2013)
51. Poonawala, H.A., Spong, M.W.: Preserving strong connectivity in directed proximity graphs. IEEE Trans. Autom. Control **62**(9), 4392–4404 (2017)
52. Ren, W., Beard, R.W.: Distributed Consensus in Multivehicle Cooperative Control. Springer (2005)
53. Restrepo, E., Loría, A., Sarras, I., Marzat, J.: Stability and robustness of edge-agreement-based consensus protocols for undirected proximity graphs. Int. J, Control (2020)
54. Restrepo-Ochoa, E., Loría, A., Sarras, I., Marzat, J.: Robust consensus of high-order systems under output constraints: application to rendezvous of underactuated UAVs. IEEE Trans. Autom. Control **68**(1), 329–342 (2023)
55. Roza, A., Maggiore, M., Scardovi, L.: A Smooth Distributed Feedback for Global Rendezvous of Unicycles. IEEE Trans. Control Netw. Syst. **5**(1), 640–652 (2018)
56. Samson, C.: Time-varying stabilization of a car-like mobile robot. Technical report, INRIA Sophia-Antipolis, 1990. In: Proceedings in Advanced Robot Control, vol. 162. Springer, Berlin (1991)
57. Samson, C.: Control of chained system: application to path following and time-varying point stabilization of mobile robots. IEEE Trans. Autom. Control **40**(1), 64–77 (1995)
58. Spong, M.: Modeling and control of elastic joint robots. ASME J. Dyn. Syst. Meas. Contr. **109**, 310–319 (1987)
59. Tzafestas, S.G.: Introduction to Mobile Robot Control. Elsevier Inc, First ed. (2013)
60. Wang, P., Ding, B.: Distributed RHC for tracking and formation of nonholonomic multi-vehicle systems. IEEE Trans. Autom. Control **59**(6), 1439–1453 (2014)
61. Wang, X., Hong, Y.: Distributed observers for tracking a moving target by cooperative multiple agents with time delays. In: 2009 ICCAS-SICE, pp. 982–987 (2009)
62. Yang, C., Xie, W., Lei, C., Ma, B.: Smooth time-varying formation control of multiple nonholonomic agents. In: Proceedings of Chinese Intelligent of System Conference, pp. 283–291. Springer (2016)
63. Zhao, S.: Affine formation maneuver control of multiagent systems. IEEE Trans. Autom. Control **63**(12), 4140–4155 (2018)
64. Zhao, X., Zheng, X., Ma, C., Li, R.: Distributed consensus of multiple Euler–Lagrange systems networked by sampled-data information with transmission delays and data packet dropouts. IEEE Trans. Autom. Sci. Eng. **14**(3), 1440–1450 (2017)

Chapter 4
Relating the Network Graphs of State-Space Representations to Granger Causality Conditions

Mónika Józsa, Mihály Petreczky, and M. Kanat Camlibel

Abstract In this chapter, we will discuss the problem of estimating the network graphs of state-space representations based on observed data: we observe the output generated by each node of a network of state-space representations, and we would like to reconstruct the communication graph of the network, i.e., we would like to find out which nodes exchange information. The potential exchange of information is not assumed to be observable, i.e., it may take place via hidden internal states. We present an approach based on the notion of Granger causality. The essence of this approach is that there exists a communication link between two nodes, if the outputs generated by the corresponding nodes are related by Granger causality. More precisely, we show an equivalence between the existence of state-space representation in which subsystems corresponding to certain nodes exchange information, and the presence of Granger causality relation between the outputs generated by those subsystems. Since Granger causality can be checked based on observed data, these results open up the possibility of data-driven reverse engineering of the communication graph. We will discuss the case of stochastic linear time-invariant systems, and then the case of stochastic bilinear/LPV/switched systems.

M. Józsa (✉)
Department of Engineering, Control, University of Cambridge, Cambridge CB2 1PZ, UK
e-mail: mj555@eng.cam.ac.uk

M. Petreczky
University of Lille, CNRS, Centrale Lille, UMR 9189 CRIStAL, F-59000 Lille, France
e-mail: mihaly.petreczky@centralelille.fr

M. K. Camlibel
Johann Bernoulli Institute for Mathematics and Computer Science, and Artificial Intelligence, University of Groningen, 9700 AK Groningen, The Netherlands
e-mail: m.k.camlibel@rug.nl

R. Postoyan et al. (eds.), *Hybrid and Networked Dynamical Systems*,
Lecture Notes in Control and Information Sciences 493,
https://doi.org/10.1007/978-3-031-49555-7_4

4.1 Introduction

In this chapter, we are interested in the problem of reverse engineering of the network graph of a stochastic dynamical system. By reverse engineering of the network graph, we mean finding out the network graph of a system based on the observed output of the system. This problem arises in several domains such as systems biology [28, 29, 36, 44], neuroscience [43], smart grids [3, 49], etc. The chapter is based on the thesis [23], and the publications [24–27].

Reverse engineering of the network graph technically means using statistical properties of the observed data for estimating the network graph of the unknown dynamical system whose output is the observed process. To this end, we need to:

- propose a formal mathematical definition of the concept of network graph, and
- relate this definition to statistical properties of the observed data.

Informally, by the *network graph* of a dynamical system, we mean a directed graph, whose nodes correspond to subsystems, such that each subsystem generates a component of the output process. There is an edge from one node to the other, if the subsystem corresponding to the source node sends information to the subsystem corresponding to the target node.

In this chapter, we will consider the following two types of dynamical systems: stochastic *stochastic linear state-space representations* without inputs (*abbreviated by sLTI-SS*) and *stochastic generalized bilinear systems* (*abbreviated by GBSs*).

Informal Definition of sLTI-SS. Informally, a stochastic time-invariant linear state-space representation without input (sLTI-SS for short) is a stochastic dynamical system of the form

$$\begin{aligned} \mathbf{x}(t+1) &= A\mathbf{x}(t) + B\mathbf{v}(t) \\ \mathbf{y}(t) &= C\mathbf{x}(t) + D\mathbf{v}(t) \end{aligned} \tag{4.1}$$

where A, B, C, D are matrices of appropriate sizes, $\mathbf{x}$, $\mathbf{v}$, $\mathbf{y}$ are discrete-time stochastic processes defined on the time axis $\mathbb{Z}$. The processes $\mathbf{x}$, $\mathbf{y}$, and $\mathbf{v}$ are called *state*, *output*, and *noise* process, respectively. Note that the time axis includes negative time instances. Furthermore, the matrix A is assumed to be stable, and there are several assumptions on the statistical properties of the processes involved. A formal definition will be stated in Sect. 4.2.1.

Informal Definition of GBSs. Let $Q = \{1, 2, \ldots, d\}$ for $d > 0$. Consider a collection $\{\boldsymbol{\mu}_q\}_{q\in Q}$ of $\mathbb{R}$-valued stochastic processes, i.e., for each $q \in Q$, $\boldsymbol{\mu}_q$ is a $\mathbb{R}$-valued stochastic process. A *generalized bilinear system* (GBS) with respect to $\{\boldsymbol{\mu}_q\}_{q\in Q}$ is a system of the form

$$\mathcal{G}\begin{cases} \mathbf{x}(t+1) = \sum_{q\in Q}(A_q\mathbf{x}(t) + K_q\mathbf{v}(t))\boldsymbol{\mu}_q(t) \\ \mathbf{y}(t) = C\mathbf{x}(t) + D\mathbf{v}(t). \end{cases} \tag{4.2}$$

where $A_q \in \mathbb{R}^{n\times n}$, $K_q \in \mathbb{R}^{n\times m}$, $C \in \mathbb{R}^{p\times n}$, and $D \in \mathbb{R}^{p\times m}$ are the system matrices, $\mathbf{y}$, $\mathbf{x}$, $\mathbf{v}$ are discrete-time stochastic processes defined on the set of all integers $\mathbb{Z}$ and taking values in $\mathbb{R}^p$, $\mathbb{R}^n$, and $\mathbb{R}^m$, respectively. We call $\mathbf{x}$ the *state process*, $\mathbf{y}$ the *output process*, $\mathbf{v}$ the *noise process*, and $\{\boldsymbol{\mu}_q\}_{q\in Q}$ the collection *input processes* of $\mathcal{G}$.The integer n is called the *dimension* of $\mathcal{G}$. We will impose several additional restrictions on the matrices and the processes of a GBS. These restrictions will be presented in Sect. 4.3.

GBSs is an extension of sLTI-SSs and depending on the choice of $\{\boldsymbol{\mu}_q\}_{q\in Q}$, a GBS may reduce to an sLTI-SS, a jump-linear system with stochastic switching, an autonomous stochastic LPV systems or a stochastic bilinear systems with white noise input. More precisely, if $Q = \{1\}$ and $\boldsymbol{\mu}_1 = 1$, then GBSs are sLTI-SSs. If $\boldsymbol{\mu}_1 = 1$ and for al $\mathbf{p}(t) = \begin{bmatrix}\boldsymbol{\mu}_2(t) & \cdots & \boldsymbol{\mu}_d(t)\end{bmatrix}$ are i.i.d. random variables which are zero mean and whose covariance is the identity matrix, then the GBS $\mathcal{G}$ can be identified either with a stochastic bilinear system with input $\mathbf{p}(t)$, or an LPV system with no input and scheduling signal $\mathbf{p}(t)$. Finally, if $\boldsymbol{\theta}(t)$ are i.i.d. random variables taking values in Q, then with $\boldsymbol{\mu}_q(t) = \begin{cases} 1 & \boldsymbol{\theta}(t) = q \\ 0 & \text{otherwise} \end{cases}$, the GBS $\mathcal{G}$ can be viewed as a stochastic switching system (jump-Markov linear system).

Network Graphs of Dynamical Systems. For sLTI-SS or GBSs, the network graphs are defined as follows. Let $\mathbf{y}$ be an output process of an sLTI-SS or of a GBS and denote the dynamical system whose output is $\mathbf{y}$ by $\mathcal{S}$. Assume that $\mathbf{y}$ is partitioned such that $\mathbf{y} = \begin{bmatrix}\mathbf{y}_1^T, \ldots, \mathbf{y}_n^T\end{bmatrix}^T$ and consider the subsystems $\mathcal{S}_i$, $i = 1, \ldots, n$ of the system $\mathcal{S}$ such that the component $\mathbf{y}_i$ is the output of $\mathcal{S}_i$. Then, the network graph has nodes $\{1, \ldots, n\}$ and there is an edge from node i to node j, if the noise and the state process of $\mathcal{S}_i$ serve as an input to $\mathcal{S}_j$. In fact, for the systems at hand, an edge (i, j) in the network graph corresponds to non-zero blocks in certain matrix parameters of the system. Similarly, the lack of this edge corresponds to zero blocks in those matrices. Intuitively, an edge in the network graph means that information can flow from the subsystem corresponding to a source node to the subsystem corresponding to the target node, but there is no information flowing the other way around. Note that we restrict our attention to transitive and acyclic network graphs. Figure 4.1 illustrates the network graph of an sLTI-SS having the three-node star graph as its network graph. Network graphs offer intuitive and mechanistic explanation of how components of the output process influence each other. However, in general, the same output process can be generated by systems with different network graphs. As a result, interactions between output components depend on the exact dynamical system representing the output process. To overcome this issue, we will use canonical forms of sLTI-SS or GBSs, for which the network graph is well defined.

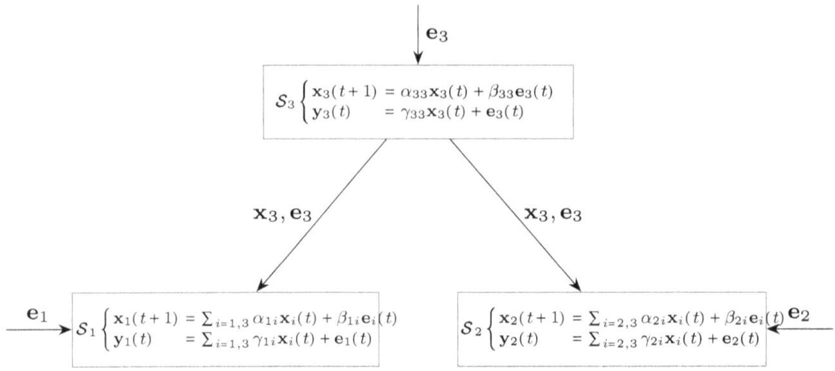

Fig. 4.1 Three-node star graph as a network graph of an sLTI-SS. sLTI-SS representation of a process $\mathbf{y} = [\mathbf{y}_1^T, \mathbf{y}_2^T, \mathbf{y}_3^T]^T$ with the three-node star graph as its network graph: The state and noise process of subsystem $\mathcal{S}_3$ serves as an input to subsystems $\mathcal{S}_1$ and $\mathcal{S}_2$

In order to reverse engineer the network graph of a state-space representation, we need to translate the network graph property of the system to statistical properties of the output of that system. For this purpose, we use Granger causality [18]. Intuitively, $\mathbf{y}_1$ Granger causes $\mathbf{y}_2$, if the best linear predictions of $\mathbf{y}_2$ based on the past values of $\mathbf{y}_1$ and $\mathbf{y}_2$ are better than those only based on the past values of $\mathbf{y}_2$. More generally, consider an output process $\mathbf{y} = \left[\mathbf{y}_1^T, \ldots, \mathbf{y}_n^T\right]^T$ of a dynamical system. Then, the network graph of the system representing $\mathbf{y}$ will be related to the following graph: the nodes correspond to the components $\mathbf{y}_i$ and there is no edge from the node labeled by $\mathbf{y}_i$ to the node labeled by $\mathbf{y}_j$, if $\mathbf{y}_i$ does not Granger cause $\mathbf{y}_j$ in a suitable sense (*conditional Granger causality, GB-Granger causality*, etc.). Note that we described the graph via the lack of edges that correspond to the lack of causality. This is because conventional statistical tests can only be applied to acausality and not to causality.

Conditional Granger non-causality is a general form of Granger non-causality: Informally, $\mathbf{y}_i$ conditionally does not Granger cause $\mathbf{y}_j$ with respect to $\mathbf{z}$ if the knowledge of the past values of $\mathbf{y}_i$, $\mathbf{y}_j$ and $\mathbf{z}$ does not yield a more accurate prediction of the future values of $\mathbf{y}_j$ than the knowledge of the past values of only $\mathbf{y}_j$ and $\mathbf{z}$. In this case, when defining the graph above, we will require that there is no edge from the node $\mathbf{y}_i$ to the node $\mathbf{y}_j$, if $\mathbf{y}_i$ does not conditional Granger cause $\mathbf{y}_j$ with respect to variable $\mathbf{z}$ that consists of suitable output components $\mathbf{y}_k$, $k \neq i, k \neq j$.

The notion of GB-Granger causality will be used only for outputs generated by GBSs driven by some input process $\{\boldsymbol{\mu}_q\}_{q \in Q}$. Intuitively, $\mathbf{y}_1$ does not GB-Granger cause $\mathbf{y}_2$, if the knowledge of the products of past value of $\mathbf{y}_1$ and $\mathbf{y}_2$ with the past vales of $\{\boldsymbol{\mu}_q\}_{q \in Q}$ does not yield a more accurate prediction of future values $\mathbf{y}_2$ than the knowledge of the products of past values of $\mathbf{y}_2$ with the past value of $\{\boldsymbol{\mu}_q\}_{q \in Q}$.

In our first result, that forms the basis of the following results, we show that a process $\mathbf{y} = \left[\mathbf{y}_1^T, \mathbf{y}_2^T\right]^T$ has an sLTI-SS realization in the so-called block triangular form if and only if $\mathbf{y}_1$ does not Granger cause $\mathbf{y}_2$. Informally, an sLTI-SS in block triangular form is a system whose network graph has two nodes, corresponding to

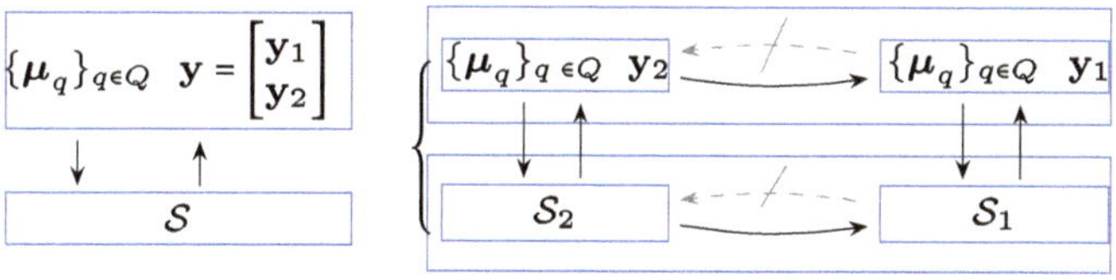

Fig. 4.2 Illustration of the results: Cascade interconnection structure in a GBS S with input $\{\mu_q\}_{q\in Q}$ and output $\mathbf{y}$ decomposed into subsystems S_1 and S_2 in the presence of GB-Granger non-causality from $\mathbf{y}_1$ to $\mathbf{y}_2$ with respect to $\{\mu_q\}_{q\in Q}$

two subsystems generating $\mathbf{y}_1$ and $\mathbf{y}_2$, and an edge from the node associated with $\mathbf{y}_2$ to the node associated with $\mathbf{y}_1$. The results include a realization algorithm on the construction of a minimal sLTI-SS in block triangular form.

This result was extended to more general graphs with three and more nodes [24, 26]. More precisely, let $\mathbf{y} = \left[\mathbf{y}_1^T, \mathbf{y}_2^T, \ldots, \mathbf{y}_n^T\right]^T$ and let us define the graph $G_\mathbf{y}$ associated with $\mathbf{y}$ as follows. We associate each component $\mathbf{y}_i$ of $\mathbf{y}$ with the node i of $G_\mathbf{y}$ and there is no edge from node i to node j, if $\mathbf{y}_i$ does not conditionally Granger cause $\mathbf{y}_j$ with respect to the collection of the components of $\mathbf{y}$ that correspond to the parent nodes of i. We then show that if $G_\mathbf{y}$ is transitive and acyclic, then there exists an sLTI-SS representation of $\mathbf{y}$ whose network graph equals $G_\mathbf{y}$. Conversely, if $\mathbf{y}$ has an sLTI-SS representation in a certain canonical form with network graph $G^{'}$, then the corresponding causality conditions apply and $G^{'}$ coincides with $G_\mathbf{y}$.

The relationship described above for graphs with two nodes was extended to GBSs in [27]. More precisely, it was shown that a process $\mathbf{y} = \left[\mathbf{y}_1^T, \mathbf{y}_2^T\right]^T$ admits a specific GBS realization in block triangular form if and only if $\mathbf{y}_1$ does not GB-Granger cause $\mathbf{y}_2$. The network graph of a GBS realization of $\mathbf{y}$ in block triangular form has two nodes, corresponding to two subsystems generating $\mathbf{y}_1$ and $\mathbf{y}_2$, and an edge from the node associated with $\mathbf{y}_2$ to the node associated with $\mathbf{y}_1$, see Fig. 4.2.

The results of this chapter partially settle a long dispute in neuroscience [14, 16, 43, 44]. There, the purpose is to detect and model interactions between brain regions using e.g., fMRI, EEG, and MEG data. For this purpose, both statistical (model-free) methods [16] and state-space-based methods [14] were used. In the former case, the presence of an interaction was identified with the presence of a statistical relationship (Granger causality) between the outputs associated with various brain regions. In the latter case, the presence of an interaction was interpreted as the presence of an edge in the network graph of a state-space representation, whose parameters were identified from data. However, the formal relationship between these methods was not always clear. This has led to a lively debate regarding the advantages and disadvantages of both methods [11, 42, 44]. The results of this chapter indicate that the two approaches are equivalent under some assumptions.

Besides the concept of network graph introduced in this chapter, there are several other notions for describing the structure of a system or the network of subsystems in a system. Examples of such notions are: feedback modeling [5, 6, 15, 21], dynamical structure function [17, 20, 48], dynamic causal modeling [14, 19, 39], and causality graphs [12, 13]. Also, besides Granger and GB-Granger causality, there are several examples of statistical notions that have an essential role to understand the relation between stochastic processes. We can mention here conditional orthogonality [7, 8], transfer entropy [2] and directional mutual information [33, 35]. For a detailed discussion on the relationship between the results of this chapter and the cited papers see [23]. We would also like to mention [31, 32, 37, 41] on coordinated sLTI-SSs, which served as inspiration for certain classes of sLTI-SSs studied in this chapter.

Finally, there is a large body of literature on identifying transfer functions which are interconnected according to some graph [4, 10, 45–47] and where all the signals exchanged by the transfer functions can be observed. This latter problem is quite different from the problem considered in this chapter.

Outline. In Sect. 4.2, we present an overview of the main results on the relationship between Granger causality and network structure of sLTI-SSs. In Sect. 4.3, we present an overview of the results relating GB-Granger causality with the network structure of GBSs.

4.2 Granger Causality and Network Graph of sLTI-SSs

We will present our results relating Granger causality conditions to the network graph of sLTI-SSs. We start with the case of two processes in Sect. 4.2.2. Then in Sect. 4.2.3, we continue with the case of an arbitrary number of processes whose network graphs are star graphs. The most general case of systems with transitive and acyclic network graphs is treated in Sect. 4.2.4. We conclude by discussing briefly in Sect. 4.2.5 the application of the results to reverse engineering of network graphs. Before presenting the main results, in Sect. 4.2.1, we will introduce the necessary notation and terminology and briefly recall the relevant results from realization theory of sLTI-SS.

4.2.1 Technical Preliminaries: Linear Stochastic Realization Theory

The notation and presentation of this section follows those of [26].

The discrete-time axis is the set of integers $\mathbb{Z}$. The random variable of a process $\mathbf{z}$ at time t is denoted by $\mathbf{z}(t)$. If $\mathbf{z}(t)$ is k-dimensional (for all $t \in \mathbb{Z}$), then we write $\mathbf{z} \in \mathbb{R}^k$ and we call $k = \dim(\mathbf{z})$ the dimension of $\mathbf{z}$. The $n \times n$ identity matrix is denoted by I_n or by I when its dimension is clear from the context. We denote by $\mathcal{H}$ the Hilbert space of zero-mean square-integrable random variables, where the inner product between two random variables y, z is the covariance matrix $E[yz^T]$. The Hilbert space generated by a set $U \subset \mathcal{H}$ is the smallest (w.r.t. set inclusion) closed subspace of $\mathcal{H}$ which contains U. Consider a zero-mean weakly stationary process $\mathbf{z} \in \mathbb{R}^k$. Then $\mathcal{H}^{\mathbf{z}}_{t-}$, $\mathcal{H}^{\mathbf{z}}_{t+}$, $\mathcal{H}^{\mathbf{z}}_{t}$, $t \in \mathbb{Z}$ are the Hilbert spaces generated by the sets $\{\ell^T \mathbf{z}(s) \mid s \in \mathbb{Z}, s < t, \ell \in \mathbb{R}^k\}$, $\{\ell^T \mathbf{z}(s) \mid s \in \mathbb{Z}, s \geq t, \ell \in \mathbb{R}^k\}$, and $\{\ell^T \mathbf{z}(t) | \ell \in \mathbb{R}^k\}$, respectively. If $\mathbf{z}_1, \ldots, \mathbf{z}_n$ are vector valued processes, then $\mathbf{z} = \left[\mathbf{z}_1^T, \ldots, \mathbf{z}_n^T\right]^T$ denotes the process defined by $\mathbf{z}(t) = \left[\mathbf{z}_1^T(t), \ldots, \mathbf{z}_n^T(t)\right]^T, t \in \mathbb{Z}$.

If $\mathbf{z}(t) \in \mathcal{H}$ is a random variable and U is a closed subspace in $\mathcal{H}$, then we denote by $E_l[\mathbf{z}(t) \mid U]$ the orthogonal projection of $\mathbf{z}(t)$ onto U. The orthogonal projection onto U of a random variable $\mathbf{z}(t) = [\mathbf{z}_1(t), \ldots, \mathbf{z}_k(t)]^T$ taking values in $\mathbb{R}^k$ is denoted by $E_l[\mathbf{z}(t)|U]$ and defined element-wise as $E_l[\mathbf{z}(t)|U] := [\hat{\mathbf{z}}_1(t), \ldots, \hat{\mathbf{z}}_k(t)]^T$, where $\hat{\mathbf{z}}_i(t) = E_l[\mathbf{z}_i(t)|U]$, $i = 1, \ldots, k$. That is, $E_l[\mathbf{z}(t)|U]$ is the random variable with values in $\mathbb{R}^k$ obtained by projecting the coordinates of $\mathbf{z}(t)$ onto U. Accordingly, the orthogonality of a multidimensional random variable to a closed subspace in $\mathcal{H}$ is meant element-wise. The orthogonal projection of a closed subspace $U \subseteq \mathcal{H}$ onto a closed subspace $V \subseteq \mathcal{H}$ is written by $E_l[U|V] := \{E_l[u|V], u \in U\}$. Note that, for jointly Gaussian processes $\mathbf{y}$ and $\mathbf{z}$, the orthogonal projection $E_l[\mathbf{y}(t)|\mathcal{H}^{\mathbf{z}}_t]$ is equivalent with the conditional expectation of $\mathbf{y}(t)$ given $\mathbf{z}(t)$.

The processes of sLTI-SS systems belong to the following class: A stochastic process is called *zero-mean square-integrable with rational spectrum (abbreviated by ZMSIR)* if it is weakly stationary, square-integrable, zero-mean, full rank, purely non-deterministic, and its spectral density is a proper rational function. Recall from [34, Definition 9.4.1] that a ZMSIR process $\mathbf{y}$ is said to be *coercive* if its spectrum is strictly positive definite on the unit disk.

For convenience, we now review the formal definition of an sLTI-SS of the form (4.1). An informal definition was presented in the introduction.

Definition 4.2.1 An sLTI-SS is a stochastic dynamical system of the form

$$
\begin{aligned}
\mathbf{x}(t+1) &= A\mathbf{x}(t) + B\mathbf{v}(t) \\
\mathbf{y}(t) &= C\mathbf{x}(t) + D\mathbf{v}(t)
\end{aligned}
$$

where $A \in \mathbb{R}^{n \times n}$, $B \in \mathbb{R}^{n \times m}$, $C \in \mathbb{R}^{p \times n}$, $D \in \mathbb{R}^{p \times m}$ for $n \geq 0$, $m, p > 0$ and $\mathbf{x} \in \mathbb{R}^n$, $\mathbf{y} \in \mathbb{R}^p$, $\mathbf{v} \in \mathbb{R}^m$ are ZMSIR processes. Furthermore, we require that A is stable (all its eigenvalues are inside the open unit circle) and that, for any $t, k \in \mathbb{Z}$, $k \geq 0$, $E[\mathbf{v}(t)\mathbf{v}^T(t-k-1)] = 0$, $E[\mathbf{v}(t)\mathbf{x}^T(t-k)] = 0$, i.e., $\mathbf{v}(t)$ is white noise and uncorrelated with $\mathbf{x}(t-k)$.

Following the classical terminology, we call n, the dimension of the state process, the *dimension* of (4.1). We will say that the sLTI-SS of the form (4.1) is *a realization* of a process $\mathbf{y}$, if the output process of the sLTI-SS equals $\mathbf{y}$. An sLTI-SS realization of $\mathbf{y}$ is called *minimal* realization of $\mathbf{y}$, if it has minimal dimension among all sLTI-SSs which are realizations of $\mathbf{y}$.

Note that the state processes $\mathbf{x}$ and $\mathbf{y}$ are uniquely determined by the noise process $\mathbf{v}$ and the system matrices A, B, C, D, where the convergence of $\mathbf{x}(t) = \sum_{k=0}^{\infty} A^k B\mathbf{v}(t-k)$ is understood in the mean square sense. Hence, an sLTI-SS of the form (4.1) will be identified with the tuple $(A, B, C, D, \mathbf{v})$ or with the tuple

$(A, B, C, D, \mathbf{v}, \mathbf{y})$ to emphasize that it is a realization of $\mathbf{y}$. sLTI-SS realizations of a given process $\mathbf{y}$ are strongly related to deterministic *linear time-invariant state-space (LTI-SS)* realizations of the covariance sequence

$$\{\Lambda_k^{\mathbf{y}} := E[\mathbf{y}(t+k)\mathbf{y}^T(t)]\}_{k=0}^{\infty},$$

see [34, Sect. 6.2]. Below we briefly sketch this relationship, as it plays an important role in deriving the results of the paper. Consider an sLTI-SS $(A, B, C, D, \mathbf{v})$ which is a realization of $\mathbf{y}$. Denote the (time-independent[1]) noise variance matrix by $\Lambda_0^{\mathbf{v}} = E[\mathbf{v}(t)\mathbf{v}^T(t)]$. Then, the variance matrix $\Lambda_0^{\mathbf{x}} = E[\mathbf{x}(t)\mathbf{x}^T(t)]$ of the state process $\mathbf{x}$ of $(A, B, C, D, \mathbf{v})$ is the unique symmetric solution of the Lyapunov equation $\Sigma = A\Sigma A^T + B\Lambda_0^{\mathbf{v}}B^T$ and the covariance $G := E[\mathbf{y}(t)\mathbf{x}^T(t+1)]$ satisfies

$$G = C\Lambda_0^{\mathbf{x}}A^T + D\Lambda_0^{\mathbf{v}}B^T. \tag{4.3}$$

In light of this, the Markov parameters of the input-output map of the LTI-SS $(A, G^T, C, \Lambda_0^{\mathbf{y}})$ are equal to the covariances $\{\Lambda_k^{\mathbf{y}}\}_{k=0}^{\infty}$. More precisely,

$$\Lambda_k^{\mathbf{y}} = CA^{k-1}G^T \qquad k > 0. \tag{4.4}$$

Therefore, sLTI-SS realizations of $\mathbf{y}$ yield LTI-SSs Markov parameters of which are the covariances $\{\Lambda_k^{\mathbf{y}}\}_{k=0}^{\infty}$ of $\mathbf{y}$. Conversely, deterministic LTI-SS systems whose Markov parameters are the covariances $\{\Lambda_k^{\mathbf{y}}\}_{k=0}^{\infty}$ yield an sLTI-SS of $\mathbf{y}$. To this end, we use the following terminology: Recall that $\mathcal{H}_{t-}^{\mathbf{y}}$ denotes the Hilbert space generated by $\mathbf{y}(t-k)$, $k > 0$. We call the process

$$\mathbf{e}(t) := \mathbf{y}(t) - E_l[\mathbf{y}(t)|\mathcal{H}_{t-}^{\mathbf{y}}], \ \forall t \in \mathbb{Z}$$

the *(forward) innovation process* of $\mathbf{y}$. Assume now that $(A, G^T, C, \Lambda_0^{\mathbf{y}})$ is a stable minimal deterministic LTI-SS system whose Markov parameters are the covariances of $\mathbf{y}$, i.e., (4.4) holds. Let $\Sigma_{\mathbf{x}}$ be the minimal symmetric solution[2] of the algebraic Riccati equation

$$\Sigma = A\Sigma A^T + (G^T - A\Sigma C^T)(\Delta(\Sigma))^{-1}(G^T - A\Sigma C^T)^T, \tag{4.5}$$

where $\Delta(\Sigma) = (\Lambda_0^{\mathbf{y}} - C\Sigma C^T)$ and set K as

$$K := (G^T - A\Sigma_{\mathbf{x}}C^T)(\Lambda_0^{\mathbf{y}} - C\Sigma_{\mathbf{x}}C^T)^{-1}. \tag{4.6}$$

[1] Stationarity implies that the (co)variance matrices are time-independent.

[2] For any other symmetric solution $\tilde{\Sigma}$, the matrix $\tilde{\Sigma} - \Sigma$ is positive definite.

Proposition 4.2.2 ([30, Sect. 7.7]) *Let K be as in* (4.6) *and* $\mathbf{e}$ *be the innovation process of* $\mathbf{y}$. *Then the following sLTI-SS realization of* $\mathbf{y}$ *is minimal:*

$$(A, K, C, I, \mathbf{e}) \tag{4.7}$$

Note that, if $\mathbf{x}$ is the state of $(A, K, C, I, \mathbf{e})$, then $\Sigma_{\mathbf{x}} = E[\mathbf{x}(t)\mathbf{x}^T(t)]$, $E[\mathbf{e}(t)\mathbf{e}^T(t)] = (\Lambda_0^{\mathbf{y}} - C\Sigma_{\mathbf{x}}C^T)$ and $K = E[\mathbf{x}(t+1)\mathbf{e}^T(t)]E[\mathbf{e}(t)\mathbf{e}^T(t)]^{-1}$, where is the gain of the steady-state Kalman filter [34, Sect. 6.9]. This motivates the following definition: Let $\mathbf{e}, \mathbf{y} \in \mathbb{R}^p$ be ZMSIR processes and $A \in \mathbb{R}^{n\times n}$, $K \in \mathbb{R}^{n\times p}$, $C \in \mathbb{R}^{p\times n}$, $D \in \mathbb{R}^{p\times p}$. A sLTI-SS $(A, K, C, D, \mathbf{e}, \mathbf{y})$ where $\mathbf{e}$ is the innovation process of $\mathbf{y}$ and $D = I_p$ is called *forward innovation representation of* $\mathbf{y}$ *or a stochastic LTI realization of* $\mathbf{y}$ *in forward innovation form* A forward innovation representation of $\mathbf{y}$ is called *minimal forward innovation representation*, if it is a minimal dimensional stochastic LTI realization of $\mathbf{y}$. The representation in Proposition 4.2.2 is a minimal forward innovation representation, thus we conclude that

Proposition 4.2.3 *Every ZMSIR process* $\mathbf{y}$ *has a minimal forward innovation representation.*

In the sequel, we will use the covariance realization algorithm from Algorithm 4.1 for computing a minimal forward innovation representation of $\mathbf{y}$.

Remark 4.2.4 (Correctness of Algorithm 4.1) Consider a ZMSIR process $\mathbf{y}$ with covariance sequence $\{\Lambda_k^{\mathbf{y}}\}_{k=0}^{\infty}$ and an sLTI-SS $(\tilde{A}, \tilde{B}, \tilde{C}, \tilde{D}, \mathbf{v})$, which is a realization of $\mathbf{y}$. Let $\mathbf{e}$ be the innovation process of $\mathbf{y}$ and N be larger than or equal to the dimension of a minimal stochastic linear system of $\mathbf{y}$. Then it follows from [30, Lemma 7.9, Sect. 7.7] that if $\{A, K, C, \Lambda_0^{\mathbf{e}}\}$ is the output of Algorithm 4.1 with input $\{\Lambda_k^{\mathbf{y}}\}_{k=0}^{2N}$ then $(A, K, C, I, \mathbf{e})$ is a minimal forward innovation representation of $\mathbf{y}$ and $\Lambda_0^{\mathbf{e}} = E[\mathbf{e}(t)\mathbf{e}^T(t)]$.

Minimal forward innovation representations have the following properties.

Proposition 4.2.5 ([34, Proposition 8.6.3]) *An sLTI-SS* $(A, K, C, I, \mathbf{e})$ *realization of* $\mathbf{y}$ *in forward innovation form is minimal if and only if* (A, K) *is controllable and* (A, C) *is observable.*

Proposition 4.2.5 shows that minimality of an sLTI-SS $(A, K, C, I, \mathbf{e})$ realization of $\mathbf{y}$ in forward innovation form can be characterized by minimality of the deterministic system (A, K, C, I). In general, the characterization of minimality in stochastic linear systems is more involved, and it is related to the minimality of the deterministic LTI-SS system $(A, G^T, C, \Lambda_0^{\mathbf{y}})$ associated with the stochastic linear system ([34, Corollary 6.5.5]). The next proposition shows that minimal sLTI-SS realizations of $\mathbf{y}$ in forward innovation form are *isomorphic*, where isomorphism is defined as follows: two sLTI-SS $(A, K, C, I, \mathbf{e})$ and $(\tilde{A}, \tilde{K}, \tilde{C}, I, \mathbf{e})$ are isomorphic if there exists a non-singular matrix T such that $A = T\tilde{A}T^{-1}$, $K = T\tilde{K}$ and $\tilde{C} = CT^{-1}$. Again, in general, the result does not apply to sLTI-SSs which are not in forward innovation form.

Algorithm 4.1 Covariance realization algorithm

Input: $\{\Lambda_k^{\mathbf{y}} := E[\mathbf{y}(t+k)\mathbf{y}^T(t)]\}_{k=0}^{2N}$: Covariance sequence of $\mathbf{y}$

Output: linear system Σ_N

Step 1 Construct the Hankel-matrix

$$H_{N,N+1} = \begin{bmatrix} \Lambda_1^{\mathbf{y}} & \Lambda_2^{\mathbf{y}} & \cdots & \Lambda_{N+1}^{\mathbf{y}} \\ \Lambda_2^{\mathbf{y}} & \Lambda_3^{\mathbf{y}} & \cdots & \Lambda_{N+1}^{\mathbf{y}} \\ \vdots & \vdots & & \vdots \\ \Lambda_N^{\mathbf{y}} & \Lambda_{N+1}^{\mathbf{y}} & \cdots & \Lambda_{2N}^{\mathbf{y}} \end{bmatrix}$$

and compute a decomposition $H_{f,N,N+1} = \mathbf{OR}$, where $\mathbf{O} \in \mathbb{R}^{Np \times n}$ and $\mathbf{R} \in \mathbb{R}^{n \times (N+1)p}$ and rank $\mathbf{R} =$ rank $\mathbf{O} = n$,
Step 2 Consider the decomposition

$$\mathbf{R} = \left[\mathbf{C}_1, \ldots, \mathbf{C}_{N+1}\right],$$

such that $\mathbf{C}_i \in \mathbb{R}^{n \times p}$, $i = 1, 2, \ldots, N+1$, i.e., they are the block columns of $\mathbf{R}$. Define $\overline{\mathbf{R}}, \hat{\mathbf{R}} \in \mathbb{R}^{n \times Np}$ as below

$$\overline{\mathbf{R}} = \left[\mathbf{C}_1, \ldots, \mathbf{C}_N\right],$$
$$\hat{\mathbf{R}} = \left[\mathbf{C}_2, \ldots, \mathbf{C}_{N+1}\right].$$

Step 3 Construct $\Sigma_N = (A, G, C)$ such that

$$G = \text{the first } p \text{ columns of } \mathbf{R}$$
$$C = \text{the first } p \text{ rows of } \mathbf{O}$$
$$A = \hat{\mathbf{R}}\overline{\mathbf{R}}^+,$$

where $\overline{\mathbf{R}}^+$ is the Moore–Penrose pseudo-inverse of $\overline{\mathbf{R}}$.

Step 4 Find the minimal symmetric solution $\Sigma_{\mathbf{x}}$ of the Riccati equation (4.5) ([30, Section 7.4.2]).

Step 5 Set K as in (4.6) and define $\Lambda_0^{\mathbf{e}} := \Lambda_0^{\mathbf{y}} - C\Sigma_{\mathbf{x}}C^T$.

Proposition 4.2.6 ([34, Theorem 6.6.1]) *If* $(A, K, C, I, \mathbf{e})$ *and* $(\tilde{A}, \tilde{K}, \tilde{C}, I, \mathbf{e})$ *are minimal sLTI-SS realizations of* $\mathbf{y}$ *in forward innovation form, then they are isomorphic.*

4.2.2 *Classical Granger Causality and sLTI-SS in Block Triangular Form*

Below we discuss the relationship between Granger causality and existence of an sLTI-SS realization in the so-called block triangular form. Block triangular form defines the class of systems with network graphs that have two nodes and one directed edge (see Definition 4.2.8 below). We will follow the presentation of [26].

We start by defining the notion of Granger causality. Informally, $\mathbf{y}_1$ does not Granger cause $\mathbf{y}_2$, if the best linear predictions of $\mathbf{y}_2$ based on the past values of $\mathbf{y}_2$ are the same as those based only on the past values of $\mathbf{y}$. Recall that $\mathcal{H}_{t-}^{\mathbf{z}}$ denotes the Hilbert space generated by the past $\{\mathbf{z}(t-k)\}_{k=1}^{\infty}$ of $\mathbf{z}$. Then, Granger causality is defined as follows:

Definition 4.2.7 Consider a zero-mean square integrable, wide-sense stationary process $\mathbf{y} = [\mathbf{y}_1^T, \mathbf{y}_2^T]^T$. We say that $\mathbf{y}_1$ *does not Granger cause* $\mathbf{y}_2$ if for all $t, k \in \mathbb{Z}, k \geq 0$ $E_l[\mathbf{y}_2(t+k)|\mathcal{H}_{t-}^{\mathbf{y}_2}] = E_l[\mathbf{y}_2(t+k)|\mathcal{H}_{t-}^{\mathbf{y}}]$. Otherwise, we say that $\mathbf{y}_1$ Granger causes $\mathbf{y}_2$.

The class of sLTI-SSs that Granger causality is associated with later on in this section is defined below.

Definition 4.2.8 An sLTI-SS $(A, K, C, I, [\mathbf{e}_1^T, \mathbf{e}_2^T]^T, [\mathbf{y}_1^T, \mathbf{y}_2^T]^T)$, where $\mathbf{e}_i, \mathbf{y}_i \in \mathbb{R}^{r_i}$, $i = 1, 2$, is said to be in *block triangular form*, if it is in forward innovation form, and

$$A = \begin{bmatrix} A_{11} & A_{12} \\ 0 & A_{22} \end{bmatrix} \; K = \begin{bmatrix} K_{11} & K_{12} \\ 0 & K_{22} \end{bmatrix} \; C = \begin{bmatrix} C_{11} & C_{12} \\ 0 & C_{22} \end{bmatrix}, \tag{4.8}$$

where $A_{ij} \in \mathbb{R}^{p_i \times p_j}$, $K_{ij} \in \mathbb{R}^{p_i \times r_j}$, $C_{ij} \in \mathbb{R}^{r_i \times p_j}$, and $p_i \geq 0$ for $i, j = 1, 2$. If, in addition, $(A_{22}, K_{22}, C_{22}, I_{r_2}, \mathbf{e}_2)$ is a minimal sLTI-SS realization of $\mathbf{y}_2$ in forward innovation form, then $(A, K, C, I, \mathbf{e} = [\mathbf{e}_1^T, \mathbf{e}_2^T]^T, \mathbf{y} = [\mathbf{y}_1^T, \mathbf{y}_2^T]^T)$ is said to be in *causal block triangular form.*

The results that relate Granger causality to sLTI-SS in block triangular form are presented next, see [26, Theorem 1].

Theorem 4.2.9 (Granger causality and network graphs with 2 nodes) *Consider the following statements for a ZMSIR process* $\mathbf{y} = [\mathbf{y}_1^T, \mathbf{y}_2^T]^T$:

1. $\mathbf{y}_1$ *does not Granger cause* $\mathbf{y}_2$*;*
2. *there exists an sLTI-SS realization of* $\mathbf{y}$ *in causal block triangular form;*
3. *there exists a minimal sLTI-SS realization of* $\mathbf{y}$ *in block triangular form;*
4. *there exists an sLTI-SS realization of* $\mathbf{y}$ *in block triangular form.*

Then $1 \iff 2$*. If* $\mathbf{y}$ *is coercive,*[3] *then* $1 \iff 2 \iff 3 \iff 4$.

Algorithm 4.2 below shows how to calculate minimal sLTI-SS in causal block triangular form from the covariance sequence of the output process.

---

[3] See [34, Definition 9.4.1] for the definition of coercivity.

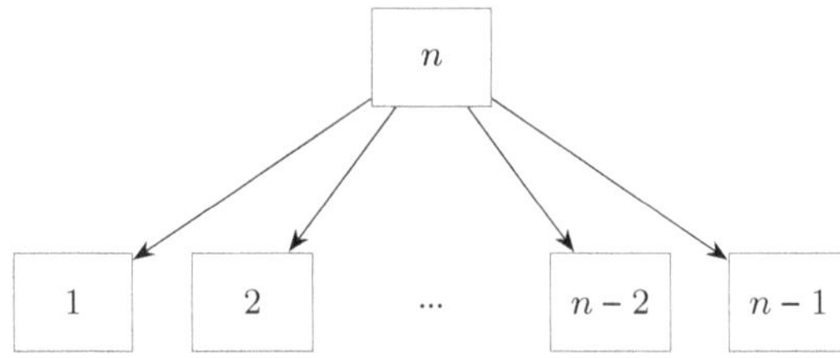

Fig. 4.3 Start-like network graph

Algorithm 4.2 Minimal sLTI-SS in causal block triangular form based on output covariances

Input $\{\Lambda_k^{\mathbf{y}} := E[\mathbf{y}(t+k)\mathbf{y}^T(t)]\}_{k=0}^{2N}$: Covariance sequence of $\mathbf{y} = \left[\mathbf{y}_1^T, \mathbf{y}_2^T\right]^T$
Output $\{A, K, C\}$: System matrices of (4.8)

Step 1 Apply Algorithm 4.1 with input $\{\Lambda_k^{\mathbf{y}}\}_{k=0}^{2N}$ and denote its output by $\{\hat{A}, \hat{K}, \hat{C}\}$.
Step 2 Let $\hat{C} = \left[\hat{C}_1^T \ \hat{C}_2^T\right]^T$ be such that $\hat{C}_i \in \mathbb{R}^{r_i \times n}$. Calculate a non-singular matrix T such that

$$T\hat{A}T^{-1} = \begin{bmatrix} A_{11} & A_{12} \\ 0 & A_{22} \end{bmatrix}, \quad \hat{C}_2 T^{-1} = \left[0 \ C_{22}\right], \tag{4.9}$$

where (A_{22}, C_{22}) is observable.
Step 3 Set $A := T\hat{A}T^{-1}$, $K := T\hat{K}$, $C := \hat{C}T^{-1}$.

Remark 4.2.10 (Correctness of Algorithm 4.2) Consider a ZMSIR process $\mathbf{y} = [\mathbf{y}_1^T, \mathbf{y}_2^T]^T$ with covariance sequence $\{\Lambda_k^{\mathbf{y}}\}_{k=0}^{\infty}$. Let $\mathbf{e}$ be the innovation process of $\mathbf{y}$ and N be any number larger than or equal to the dimension of a minimal sLTI-SS realization of $\mathbf{y}$. If $\mathbf{y}$ satisfies condition 1 of Theorem 4.2.9 and $\{A, K, C\}$ is the output of Algorithm 4.2 with input $\{\Lambda\}_{k=0}^{2N}$, then $(A, K, C, I, \mathbf{e})$ is a minimal sLTI-SS realization of $\mathbf{y}$ in causal block triangular form.

4.2.3 Conditional Granger Causality and sLTI-SS in Coordinated Form

The result above can be generalized to more than two output processes. Let us start by presenting the generalization to the case when the network graph is star like Fig. 4.3.

To this end, in the sequel, we assume that $\mathbf{y} = [\mathbf{y}_1^T, \dots, \mathbf{y}_n^T]^T$ is a ZMSIR process, where $n \geq 2$, $\mathbf{y}_i \in \mathbb{R}^{r_i}$, and $r_i > 0$ for $i = 1, \dots, n$. The following definition describes a subset of sLTI-SSs which has star-like network graph as in Fig. 4.3.

Definition 4.2.11 An sLTI-SS $(A, K, C, I, \mathbf{e} = [\mathbf{e}_1^T, \dots, \mathbf{e}_n^T]^T, \mathbf{y})$, where $\mathbf{e}_i \in \mathbb{R}^{r_i}$, $i = 1, \dots, n$, is *in coordinated form*, if it is in forward innovation form, i.e., $\mathbf{e}$ is the innovation process of $\mathbf{y}$, and

$$A = \left[\begin{array}{cccc|c} A_{11} & 0 & \cdots & 0 & A_{1n} \\ 0 & A_{22} & \ddots & \vdots & A_{2n} \\ \vdots & \ddots & \ddots & 0 & \vdots \\ 0 & \cdots & 0 & \ddots & A_{(n-1)n} \\ \hline 0 & 0 & \cdots & 0 & A_{nn} \end{array}\right], K = \left[\begin{array}{cccc|c} K_{11} & 0 & \cdots & 0 & K_{1n} \\ 0 & K_{22} & \ddots & \vdots & K_{2n} \\ \vdots & \ddots & \ddots & 0 & \vdots \\ 0 & \cdots & 0 & \ddots & K_{(n-1)n} \\ \hline 0 & 0 & \cdots & 0 & K_{nn} \end{array}\right]$$

$$C = \left[\begin{array}{cccc|c} C_{11} & 0 & \cdots & 0 & C_{1n} \\ 0 & C_{22} & \ddots & \vdots & C_{2n} \\ \vdots & \ddots & \ddots & 0 & \vdots \\ 0 & \cdots & 0 & \ddots & C_{(n-1)n} \\ \hline 0 & 0 & \cdots & 0 & C_{nn} \end{array}\right]$$

where $A_{ij} \in \mathbb{R}^{p_i \times p_j}$, $K_{ij} \in \mathbb{R}^{p_i \times r_j}$, $C_{ij} \in \mathbb{R}^{r_i \times p_j}$, and $p_i \geq 0$ for $i, j = 1, \ldots, n$. If, in addition, for each $i = 1, \ldots, n-1$

$$\left(\begin{bmatrix} A_{ii} & A_{in} \\ 0 & A_{nn} \end{bmatrix}, \begin{bmatrix} K_{ii} & K_{in} \\ 0 & K_{nn} \end{bmatrix}, \begin{bmatrix} C_{ii} & C_{in} \\ 0 & C_{nn} \end{bmatrix}, I_{r_i+r_n}, \begin{bmatrix} \mathbf{e}_i \\ \mathbf{e}_n \end{bmatrix}\right) \tag{4.10}$$

is a minimal sLTI-SS realization of $[\mathbf{y}_i^T, \mathbf{y}_n^T]^T$ in causal block triangular form, then $(A, K, C, I, \mathbf{e}, \mathbf{y})$ is called a *sLTI-SS in causal coordinated form.*

If $n = 2$, then Definition 4.2.11 coincides with Definition 4.2.8 of sLTI-SS in block triangular form. Furthermore, if $(A, K, C, I, \mathbf{e})$ is an sLTI-SS in causal coordinated form, then the dimensions of the block matrices A_{ij}, K_{ij}, C_{ij}, $i, j = 1, \ldots, n$ are uniquely determined by $\mathbf{y}$. Definition 4.2.11 is based on the deterministic terminology [31, 41] and on the definition of Gaussian coordinated systems [31, 37]. The term *coordinated* is used because the sLTI-SS at hand can be viewed as consisting of several subsystems; one of which plays the role of a coordinator and the others play the role of agents. More precisely, let $(A, K, C, I, \mathbf{e}, \mathbf{y})$ be an sLTI-SS in coordinated form as in Definition (4.2.11) and let $\mathbf{x} = [\mathbf{x}_1^T, \ldots, \mathbf{x}_n^T]^T$ be its state such that $\mathbf{x}_i \in \mathbb{R}^{p_i}$, $i = 1, \ldots, n$. Then, for $i = 1, \ldots, n-1$,

$$\mathcal{S}_{a_i} \begin{cases} \mathbf{x}_i(t+1) = \sum_{j=\{i,n\}} A_{ij}\mathbf{x}_j(t) + K_{ij}\mathbf{e}_j(t) \\ \mathbf{y}_i(t) = \sum_{j=\{i,n\}} C_{ij}\mathbf{x}_j(t) + \mathbf{e}_i(t) \end{cases} \tag{4.11}$$

$$\mathcal{S}_{c} \begin{cases} \mathbf{x}_n(t+1) = A_{nn}\mathbf{x}_n(t) + K_{nn}\mathbf{e}_n(t) \\ \mathbf{y}_n(t) = C_{nn}\mathbf{x}_i(t) + \mathbf{e}_n(t) \end{cases} . \tag{4.12}$$

Notice that subsystem $\mathcal{S}_{a_i}$ generates $\mathbf{y}_i$ as output, has $\mathbf{x}_i$, $\mathbf{e}_i$ as its state and noise process and takes $\mathbf{x}_n$, $\mathbf{e}_n$ as its inputs, thus takes inputs from subsystem $\mathcal{S}_c$. In contrast, $\mathcal{S}_c$ is autonomous, generating $\mathbf{y}_n$ as output and having $\mathbf{x}_n$, $\mathbf{e}_n$ as its state and noise

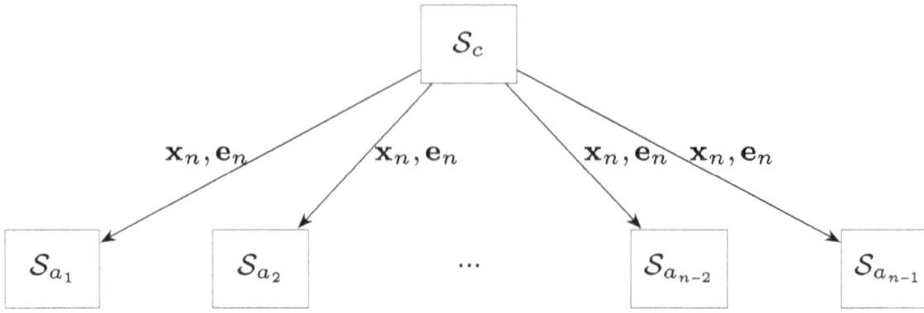

Fig. 4.4 Network graph of an sLTI-SS in coordinated form: $\mathcal{S}_c$ is the coordinator (4.12) and $\mathcal{S}_{a_i}$, $i = 1, \ldots, n-1$ are the agents (4.11)

process but not taking input from subsystems $\mathcal{S}_{a_i}$, $i = 1, \ldots, n-1$ (see Fig. 4.4). We call $\mathcal{S}_c$ the *coordinator* and $\mathcal{S}_{a_i}$ with $i = 1, \ldots, n-1$ the *agents*. Intuitively, the agents do not communicate with each other, only the coordinator sends information ($\mathbf{x}_n$ and $\mathbf{e}_n$) to all agents and does not receive information from them. Next, we define the notion of conditional Granger causality, which will play a central role in characterizing the existence of sLTI-SS realizations in coordinated form.

Definition 4.2.12 Consider a ZMSIR process $\mathbf{y} = [\mathbf{y}_1^T, \mathbf{y}_2^T, \mathbf{y}_3^T]^T$. We say that $\mathbf{y}_1$ *conditionally does not Granger cause* $\mathbf{y}_2$ *with respect to* $\mathbf{y}_3$, if for all $t, k \in \mathbb{Z}$, $k \geq 0$

$$E_l[\mathbf{y}_2(t+k) \mid \mathcal{H}_{t-}^{\mathbf{y}_2,\mathbf{y}_3}] = E_l[\mathbf{y}_2(t+k) \mid \mathcal{H}_{t-}^{\mathbf{y}_1,\mathbf{y}_2,\mathbf{y}_3}].$$

Otherwise, we say that $\mathbf{y}_1$ *conditionally Granger causes* $\mathbf{y}_2$ *with respect to* $\mathbf{y}_3$.

Now we are ready to state the main result relating conditional Granger causality with the existence of sLTI-SSs in coordinated form ([23, 26]).

Theorem 4.2.13 *Consider the following statements for a ZMSIR process* $\mathbf{y} = [\mathbf{y}_1^T, \ldots, \mathbf{y}_n^T]^T$*:*

1. $\mathbf{y}_i$ *does not Granger cause* $\mathbf{y}_n$, $i = 1, \ldots, n-1$*;*
2. $\mathbf{y}_i$ *conditionally does not Granger cause* $\mathbf{y}_j$ *with respect to* $\mathbf{y}_n$, $i, j = 1, \ldots, n-1$, $i \neq j$*;*
3. *there exists a minimal sLTI-SS realization of* $\mathbf{y}$ *in causal coordinated form;*
4. *there exists a sLTI-SS realization of* $\mathbf{y}$ *in causal coordinated form;*
5. *there exists a sLTI-SS realization of* $\mathbf{y}$ *in coordinated form;*

Then, the following hold:

- *1 and 2* $\iff$ *3.*
- *If, in addition,* $\mathbf{y}$ *is coercive, then we have 1 and 2* $\iff$ *4* $\iff$ *5.*

See [23, 26] for an algorithm that computes a minimal sLTI-SS in causal coordinated form from the covariances of $\mathbf{y}$ or any sLTI-SS realization of $\mathbf{y}$. Note that Theorem 4.2.13 and the corresponding algorithm for calculating an sLTI-SS in coordinated form can be used to formulate statistical tests for checking (conditional) Granger causality relationships, see [23, Chap. 7].

4.2.4 Granger Causality Relations and Directed Acyclic Network Graphs

We now present the general case, where the graph induced by Granger causality relations is a directed acyclic graph. To this end, we define the class of transitive acyclic graphs.

Definition 4.2.14 (*TADG*) A directed graph $G = (V, E)$, with set of nodes $V = \{1, \ldots, k\}$ and set of directed edges $E \subseteq V \times V$ is called *acyclic* if there is no cycle i.e., closed directed path. Furthermore, it is *transitive* if for $i, j, l \in V$ the implication $(i, j), (j, l) \in E \implies (i, l) \in E$ holds. The class of transitive acyclic directed graphs is denoted by *TADG*.

For convenience, we make the following assumption that applies for all ZMSIR processes throughout this section. For a process $\mathbf{y} = [\mathbf{y}_1^T, \ldots, \mathbf{y}_n^T]^T$, we assume that none of the components of $\mathbf{y}$ is a white noise process, or equivalently, the dimension of a minimal sLTI-SS realization of $\mathbf{y}_i$ is strictly positive for all $i \in \{1, \ldots, n\}$.

Next, we introduce a *topological ordering* for the nodes $V = \{1, \ldots, n\}$ of a TADG $G = (V, E)$. Throughout this chapter, we use integers to represent nodes of graphs and, without the loss of generality, we assume that if $(i, j) \in E$ then $i > j$.

The class of TADGs will be used to define the class of forward innovation sLTI-SS having TADGs as their network graph. To define this class of sLTI-SSs formally, we need to introduce some new terminology.

Notation for Parent and Non-parent Succeeding Nodes. Let $G = (V = \{1, \ldots, n\}, E)$ be a TADG and consider a node $j \in V$. The set of parent nodes $\{i \in V | (i, j) \in E\}$ of j is denoted by I_j. In addition, the set of non-parent succeeding (with respect to the topological ordering of V) nodes $\{i \in V | i > j, (i, j) \notin E\}$ of j is denoted by $\bar{I}_j$.

The topological ordering on the set of nodes of a TADG graph implies that $I_j, \bar{I}_j \subseteq \{j+1, \ldots, n\}$ for all $j \in \{1, \ldots, n-1\}$. Furthermore, from the definition of $\bar{I}_j$, we have that $I_j \cup \bar{I}_j = \{j+1, \ldots, n\}$. The next notation helps in referring to components of processes beyond the original partitioning of those processes.

Notation for Sub-processes. Consider the finite set $V = \{1, \ldots, n\}$ and a tuple $J = (j_1, \ldots, j_l)$, where $j_1, \ldots, j_l \in V$. Then, for a process $\mathbf{y} = \left[\mathbf{y}_1^T, \ldots, \mathbf{y}_n^T\right]^T$, we denote the sub-process $[\mathbf{y}_{j_1}^T, \ldots, \mathbf{y}_{j_l}^T]^T$ by $\mathbf{y}_{j_1, \ldots, j_l}$ or by $\mathbf{y}_J$. By abuse of terminology, if J is a subset of V and not a tuple, then $\mathbf{y}_J$ will mean process $\mathbf{y}_\alpha$, where α is the tuple obtained by taking the elements of J in increasing order, i.e., if $J = \{j_1, \ldots, j_k\}, j_1 < j_2 < \cdots j_k$, then $\alpha = (j_1, \ldots, j_k)$. However, $\mathbf{y}_{\alpha,\beta}$ always means $[\mathbf{y}_\alpha^T, \mathbf{y}_\beta^T]^T$ regardless the topological order between the elements of α and β.

Next, we introduce what we mean by partition of matrices. Call the set $\{p_i, q_i\}_{i=1}^k$ a *partition* of (p, q), where $p, q > 0$, if $\sum_{i=1}^k p_i = p$ and $\sum_{i=1}^k q_i = q$, where $p_i, q_i > 0$ for $i = 1, \ldots, k$.

Definition 4.2.15 (*Partition of a matrix*) Let $\{p_i, q_i\}_{i=1}^k$ be a partition of (p, q) for some $p, q > 0$. Then *the partition of a matrix* $M \in \mathbb{R}^{p \times q}$ *with respect to* $\{p_i, q_i\}_{i=1}^k$ is a collection of matrices $\{M_{ij} \in \mathbb{R}^{p_i \times q_j}\}_{i,j=1}^k$, such that

$$M = \begin{bmatrix} M_{11} & \cdots & M_{1k} \\ \vdots & \ddots & \vdots \\ M_{k1} & \cdots & M_{kk} \end{bmatrix}.$$

In Definition 4.2.15, the indexing of matrix M refers to the blocks of M and does not refer directly to the elements of M. It is parallel to the component-wise indexing of processes where the components can be multidimensional.

Notation for Sub-matrices. Consider the partition $\{M_{ij} \in \mathbb{R}^{p_i \times q_j}\}_{i,j=1}^k$ of a matrix $M \in \mathbb{R}^{p \times q}$ with respect to the partition $\{p_i, q_i\}_{i=1}^k$ of (p, q). Furthermore, consider the tuples $I = (i_1, \ldots, i_n)$ and $J = (j_1, \ldots, j_m)$, where $i_1, \ldots, i_n, j_1, \ldots, j_m \in \{1, \ldots, k\}$. Then, by the sub-matrix of M indexed by IJ, we mean

$$M_{IJ} := \begin{bmatrix} M_{i_1 j_1} & \cdots & M_{i_1 j_m} \\ \vdots & \ddots & \vdots \\ M_{i_n j_1} & \cdots & M_{i_n j_m} \end{bmatrix}$$

We are now ready to define sLTI-SSs which have a so-called TADG-zero structure:

Definition 4.2.16 Consider a process $\mathbf{y} = \left[\mathbf{y}_1^T, \ldots, \mathbf{y}_n^T\right]^T$ and a TADG $G = (V = \{1, \ldots, n\}, E)$. Let $(A, K, C, I, \mathbf{e})$ be a p-dimensional sLTI-SS realization of $\mathbf{y} \in \mathbb{R}^r$ in forward innovation form. Consider the partition of

- A with respect to $\{p_i, p_i\}_{i=1}^n$,
- K with respect to $\{p_i, r_i\}_{i=1}^n$,
- C with respect to $\{r_i, p_i\}_{i=1}^n$

where $\{p_i, r_i\}_{i=1}^n$ is a partition of (p, r). Then we say that $(A, K, C, I, \mathbf{e})$ has G-*zero structure* if

$$A_{ij} = 0, \;\; K_{ij} = 0, \;\; C_{ij} = 0 \;\; \text{for all } (j, i) \notin E.$$

If, in addition, for all $j \in V$, the tuple $J := (j, \bar{I}_j, I_j)$ defines an sLTI-SS

$$(A_{JJ}, K_{JJ}, C_{JJ}, I, [\mathbf{e}_j^T, \mathbf{e}_{\bar{I}_j}^T, \mathbf{e}_{I_j}^T]^T)$$

which is a realization of $[\mathbf{y}_j^T, \mathbf{y}_{\bar{I}_j}^T, \mathbf{y}_{I_j}^T]^T$ in causal coordinated form (see Definition 4.2.11), then we say that $(A, K, C, I, \mathbf{e})$ has *causal* G*-zero structure*.

Note that, if the graph G in Definition 4.2.16 is $(\{1, 2\}, \{(2, 1)\})$, then Definition 4.2.16 coincides with Definition 4.2.8. In a similar manner, if the graph G

in Definition 4.2.16 is $(\{1, 2, \ldots, n\}, \{(n, 1), (n, 2), \ldots, (n, n-1)\})$ then it coincides with Definition 4.2.11.

If a p-dimensional sLTI-SS $(A, K, C, I, \mathbf{e})$ of $\mathbf{y} \in \mathbb{R}^r$ has causal G-zero structure, where $G = (V, E)$ is a TADG, then the partition $\{p_i, r_i\}_{i=1}^n$ of (p, r) in Definition 4.2.16 is uniquely determined by $\mathbf{y}$.

An sLTI-SS with TADG-zero structure can be viewed as consisting of subsystems where each subsystem generates a component of $\mathbf{y} = [\mathbf{y}_1^T, \ldots, \mathbf{y}_n^T]^T$. More precisely, let $G = (V = \{1, \ldots, n\}, E)$ be a TADG and $(A, K, C, I, \mathbf{e}, \mathbf{y})$ be a p-dimensional sLTI-SS with G-zero structure where A, K and C are partitioned with respect to a partition $\{p_i, q_i\}_{i=1}^k$ of (p, q). Furthermore, let $\mathbf{x} = [\mathbf{x}_1^T, \ldots, \mathbf{x}_n^T]^T$ be its state such that $\mathbf{x}_i \in \mathbb{R}^{p_i}$, $i = 1, \ldots, n$. Then the sLTI-SS with output $\mathbf{y}_j$, $j \in V$ is in the form of

$$\mathcal{S}_j \begin{cases} \mathbf{x}_j(t+1) = A_{jj}\mathbf{x}_j(t) + \left(A_{jI_j}\mathbf{x}_{I_j}(t) + K_{jI_j}\mathbf{e}_{I_j}(t)\right) + K_{jj}\mathbf{e}_j(t) \\ \mathbf{y}_j(t) = C_{jj}\mathbf{x}_j(t) + C_{jI_j}\mathbf{x}_{I_j}(t) + \mathbf{e}_j(t) \end{cases} \tag{4.13}$$

Notice that if $(i, j) \in E$, i.e., i is a parent node of j, then subsystem $\mathcal{S}_j$ takes inputs from subsystem $\mathcal{S}_i$, namely the state and noise processes of $\mathcal{S}_i$. In contrast, if $(j, i) \notin E$, $\mathcal{S}_j$ does not take input from $\mathcal{S}_i$. Intuitively, it means that the subsystems communicate with each other as it is allowed by the directed paths of the graph G. Note that from transitivity, if there is a directed path from $i \in V$ to $j \in V$ then there is also an edge $(i, j) \in E$.

Example 4.2.17 Take the TADG graph $G = (\{1, 2, 3, 4\}, \{(4, 1), (4, 2), (3, 1), (2, 1)\})$ and a process $[\mathbf{y}_1^T, \mathbf{y}_2^T, \mathbf{y}_3^T, \mathbf{y}_4^T]^T$ with innovation process $[\mathbf{e}_1^T, \mathbf{e}_2^T, \mathbf{e}_3^T, \mathbf{e}_4^T]^T$. Then an sLTI-SS with G-zero structure of $[\mathbf{y}_1^T, \mathbf{y}_2^T, \mathbf{y}_3^T, \mathbf{y}_4^T]^T$ is given by

$$\begin{gathered}\begin{bmatrix} \mathbf{x}_1(t+1) \\ \mathbf{x}_2(t+1) \\ \mathbf{x}_3(t+1) \\ \mathbf{x}_4(t+1) \end{bmatrix} = \begin{bmatrix} A_{11} & A_{12} & A_{13} & A_{14} \\ 0 & A_{22} & 0 & A_{24} \\ 0 & 0 & A_{33} & 0 \\ 0 & 0 & 0 & A_{44} \end{bmatrix} \begin{bmatrix} \mathbf{x}_1(t) \\ \mathbf{x}_2(t) \\ \mathbf{x}_3(t) \\ \mathbf{x}_4(t) \end{bmatrix} + \begin{bmatrix} K_{11} & K_{12} & K_{13} & K_{14} \\ 0 & K_{22} & 0 & K_{24} \\ 0 & 0 & K_{33} & 0 \\ 0 & 0 & 0 & K_{44} \end{bmatrix} \begin{bmatrix} \mathbf{e}_1(t) \\ \mathbf{e}_2(t) \\ \mathbf{e}_3(t) \\ \mathbf{e}_4(t) \end{bmatrix} \\ \begin{bmatrix} \mathbf{y}_1(t) \\ \mathbf{y}_2(t) \\ \mathbf{y}_3(t) \\ \mathbf{y}_4(t) \end{bmatrix} = \begin{bmatrix} C_{11} & C_{12} & C_{13} & C_{14} \\ 0 & C_{22} & 0 & C_{24} \\ 0 & 0 & C_{33} & 0 \\ 0 & 0 & 0 & C_{44} \end{bmatrix} \begin{bmatrix} \mathbf{x}_1(t) \\ \mathbf{x}_2(t) \\ \mathbf{x}_3(t) \\ \mathbf{x}_4(t) \end{bmatrix} + \begin{bmatrix} \mathbf{e}_1(t) \\ \mathbf{e}_2(t) \\ \mathbf{e}_3(t) \\ \mathbf{e}_4(t) \end{bmatrix}, \end{gathered} \tag{4.14}$$

where $A_{ij} \in \mathbb{R}^{p_i \times p_j}$, $K_{ij} \in \mathbb{R}^{p_i \times r_j}$, $C_{ij} \in \mathbb{R}^{r_i \times p_j}$ and $\mathbf{y}_i, \mathbf{e}_i \in \mathbb{R}^{r_i}$, $\mathbf{x}_i \in \mathbb{R}^{p_i}$ for some $p_i > 0$, $i, j = 1, 2, 3, 4$. The network graph of this sLTI-SS is the network of the sLTI-SSs $\mathcal{S}_1, \mathcal{S}_2, \mathcal{S}_3, \mathcal{S}_4$ defined in (4.13), generating $\mathbf{y}_1$, $\mathbf{y}_2$, $\mathbf{y}_3$ and $\mathbf{y}_4$, respectively. See Fig. 4.5 for illustration of this network graph.

In order to formulate the main result of this section, we will need to define the notion of G-consistent causality structure for ZMSIR processes.

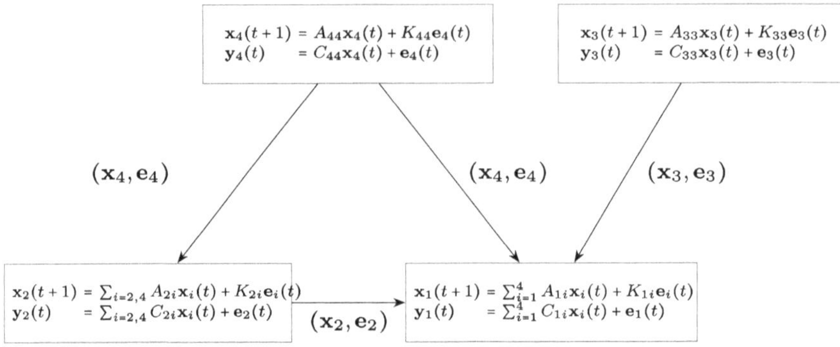

Fig. 4.5 Network graph of the sLTI-SS (4.14) with G-zero structure

Definition 4.2.18 Consider a TADG $G = (V, E)$, where $V = \{1, \ldots, n\}$ and a process $\mathbf{y} = \left[\mathbf{y}_1^T, \ldots, \mathbf{y}_n^T\right]^T$. We say that $\mathbf{y}$ has G-*consistent* causality structure, if $\mathbf{y}_i$ conditionally does not Granger cause $\mathbf{y}_j$ with respect to $\mathbf{y}_{I_j}$ for any $i, j \in V, i \neq j$ such that $(i, j) \notin E$.

Informaly, for a TDAG G, the process $\mathbf{y}$ has G-consistent causality structure, if the edges of G correspond to certain conditional Granger causality relationships among the components of $\mathbf{y}$. If $G = (\{1, 2\}, \{(2, 1)\})$, then Definition 4.2.18 coincides with Definition 4.2.7. Furthermore, if $G = (\{1, 2, 3\}, \{(3, 1), (3, 2), (2, 1)\})$, then Definition 4.2.18 coincides with statements 1 and 2 of Theorem 4.2.13. We are now ready to present our main results on sLTI-SS with TADG network structure [23, 24].

Theorem 4.2.19 *Consider the following statements for a TADG* $G = (V = \{1, \ldots, n\}, E)$ *and a process* $\mathbf{y} = [\mathbf{y}_1^T, \ldots, \mathbf{y}_n^T]^T$*:*

1. $\mathbf{y}$ *has* G*-consistent causality structure;*
2. *there exists an sLTI-SS realization of* $\mathbf{y}$ *with causal* G*-zero structure;*
3. *there exists an sLTI-SS realization of* $\mathbf{y}$ *with* G*-zero structure;*

Then, the following hold:

- $1 \implies 3$;
- $2 \implies 1$.
- *If, in addition,* $\mathbf{y}$ *is coercive, then we have* $1 \iff 2 \iff 3$.

The proof of Theorem 4.2.19 is constructive and it leads to an algorithm for computing an sLTI-SS realization of $\mathbf{y}$ from any sLTI-SS realization of $\mathbf{y}$ or from the covariances of $\mathbf{y}$. See [23, 24] for more details.

4.2.5 Applications of the Theoretical Results

The results described above can be used to develop statistical hypothesis tests for checking various Granger causality based on data. The essence of those methods is to use empirical covariances of the observed processes to calculate an sLTI-SS in causal block triangular form, causal coordinated form, or with causal G-zero structure, respectively. In the ideal case, when true covariances are used, suitable blocks of the matrices of such an sLTI-SS should be zero, if the corresponding Granger causality relationships hold. When approximate covariances are used, the corresponding blocks are not zero, due to approximation error. The statistical hypothesis testing relies on checking if those blocks are significantly different from zero. Details of this approach can be found in [23, Chap. 7].

The results and procedures described in the previous sections can also be viewed as a procedure for reverse engineering of the network graph. Note that the problem of discovering the network graph from data is in principle ill-posed, as systems with different network graphs can generate the same observed behavior. However, if we consider the weaker problem of deciding if there is an underlying system with a particular network graph that generates the observed behavior, then we have a well-posed problem. By using statistical hypothesis testing for checking causal properties of the observed behavior, the theorems on equivalency of causal properties and the network graph of a specific sLTI-SS can be used to solve this weaker problem.

4.3 GB-Granger Causality and Network Graph of GBSs

In this section, we extend our results in Sect. 4.2 to GBSs. The motivation for GBSs is that they include a wide variety of stochastic systems such as sLTI-SSs, subclasses of jump-linear system with stochastic switching, autonomous stochastic LPV systems and stochastic bilinear systems with white noise input.

In the main result of this section, we relate the existence of a GBS realization of $\mathbf{y} = [\mathbf{y}_1^T, \mathbf{y}_2^T]^T$ with a network graph consisting of two nodes and one edge to Granger causality-like properties of $\mathbf{y}$.

We start with formally defining the class of GBSs and introducing necessary terminology and notation. Then, we formulate a Granger causality-like statistical property of $\mathbf{y}$. Finally, we characterize this property with the existence of GBSs whose network graph consists of two nodes and one directed edge.

4.3.1 Technical Preliminaries: GBS Realization Theory

In this section, we define general bilinear state-space representations, adapted from [40], and a realization algorithm for this class of systems. For this, we first introduce the necessary terminology. For the rest of this section, we fix a finite set $Q = \{1, 2, \ldots, d\}$, where d is a positive integer.

In the Introduction, we already described GBSs as a discrete-time stochastic dynamical system of the form (4.2). Equation (4.2) above is incomplete without posing conditions on the input, state, and output processes. More precisely, in order to have a realization theory of representations of the form (4.2), we need to define the class of processes that $\boldsymbol{\mu}$, $\mathbf{x}$, and $\mathbf{y}$ belong to. The following terminology helps us to define these conditions.

Let Q^+ be the set of all finite sequences of elements of Q, i.e., a typical element of Q^+ is a sequence of the form $w = q_1 \cdots q_k$, where $q_1, \ldots, q_k \in Q$. We define the concatenation operation on Q^+ in the standard way: if $w = q_1 \cdots q_k$ and $v = \hat{q}_1 \cdots \hat{q}_l$ where $q_1, \ldots, q_k, \hat{q}_1, ..., \hat{q}_l \in Q$ then the concatenation of w and v, denoted by wv, is defined as the sequence $wv = q_1 \cdots q_k \hat{q}_1 \cdots \hat{q}_l$. In the sequel, it will be convenient to extend Q^+ by adding a formal unit element $\epsilon \notin Q^+$. We denote this set by $Q^* := Q^+ \cup \{\epsilon\}$. The concatenation operation can be extended to Q^* as follows: $\epsilon\epsilon = \epsilon$, and for any $w \in Q^+, \epsilon w = w\epsilon = w$. Let $w = q_1 \cdots q_k \in Q^+$ and $q \in Q$. Then the length of w is defined by $|w| := k$ and the length of ϵ is defined by $|\epsilon| := 0$. Consider a set of matrices $\{M_q\}_{q \in Q}$ where $M_q \in \mathbb{R}^{n \times n}, n \geq 1$ for all $q \in Q$ and let $w = q_1 \cdots q_k \in Q^+$, where $q_1, \ldots, q_k \in Q$. Then, we denote the matrix $M_{q_k} \cdots M_{q_1}$ by M_w and we define $M_\epsilon := I$. In addition, for a set of processes $\{\boldsymbol{\mu}_q\}_{q \in Q}$ and for $w = q_1 \cdots q_k \in Q^+$, where $q_1, \ldots, q_k \in Q$, we denote the process $\boldsymbol{\mu}_{q_k}(t) \cdots \boldsymbol{\mu}_{q_1}(t - |w| + 1)$ by $\boldsymbol{\mu}_w(t)$ and define $\boldsymbol{\mu}_\epsilon(t) :\equiv 1$.

The past of the noise, state, and output processes that are multiplied by the past of the input processes play an important role in defining GBSs. For this reason, we define the following processes:

Definition 4.3.1 Consider a process $\mathbf{r}$ and a set of processes $\{\boldsymbol{\mu}_q\}_{\{}q \in Q\}$. Let $q \in Q$ and $w = q_1 \cdots q_k \in Q^+$, where $q_1, \ldots, q_k \in Q^*$. Then, we define the process

$$\mathbf{z}_w^{\mathbf{r}}(t) := \mathbf{r}(t - |w|)\boldsymbol{\mu}_w(t - 1),$$

which we call the *past* of $\mathbf{r}$ with respect to $\{\boldsymbol{\mu}_q\}_{q \in Q}$.

Definition 4.3.2 Consider a process $\mathbf{r}$ and a set of processes $\{\boldsymbol{\mu}_q\}_{q \in Q}$. Let $q \in Q$ and $w = q_1 \cdots q_k \in Q^+$, where $q_1, \ldots, q_k \in Q^*$. Then, we define the process

$$\mathbf{z}_w^{\mathbf{r}+}(t) := \mathbf{r}(t + |w|)\boldsymbol{\mu}_w(t + |w| - 1),$$

which we call the *future* of $\mathbf{r}$ with respect to $\{\boldsymbol{\mu}_q\}_{q \in Q}$.

Note that for $w = \epsilon$, $\mathbf{z}_\epsilon^{\mathbf{r}}(t) = \mathbf{r}(t)$ and $\mathbf{z}_\epsilon^{\mathbf{r}+}(t) = \mathbf{r}(t)$.

The processes in Definitions 4.3.1 and 4.3.2 slightly differ from the parallel past and future processes of a process used in [40]. We can obtain the processes in Definitions 4.3.1 and 4.3.2 by multiplying the parallel processes (e.g., Eq. (6) in [40]) with a scalar.

Below, we define a set of processes called admissible that will help us in formulating conditions on the input processes of (4.2). For this definition, we first need to recall that all the random variables and stochastic processes are understood with respect to a probability space $(\Omega, \mathcal{F}, P)$, where $\mathcal{F}$ is a σ-algebra over Ω and P is a probability measure on $\mathcal{F}$. Using the standard notation, the conditional expectation of a random variable $\mathbf{z}$ to a σ-algebra $\mathcal{F}^*$ is denoted by $E[\mathbf{z}|\mathcal{F}^*]$. Furthermore, considering a process $\mathbf{z}$ and a time t, the σ-algebra generated by the random variables in the past is denoted by $\mathcal{F}^{\mathbf{z}}_{t-} = \sigma\left(\{\mathbf{z}(k)\}_{k=-\infty}^{t-1}\right)$.

Definition 4.3.3 A set of processes $\{\boldsymbol{\mu}_q\}_{q\in Q}$ is called *admissible* if

- $[\boldsymbol{\mu}_v^T, \boldsymbol{\mu}_w^T]^T$ is weakly stationary for all $v, w \in Q^*$
- there exist real numbers $\{\alpha_q\}_{q\in Q}$ such that for all $t \in \mathbb{Z}$: $\sum_{q\in Q} \alpha_q \boldsymbol{\mu}_q(t) \equiv 1$
- there exist (strictly) positive numbers $\{p_q\}_{q\in Q}$ (where $p_\epsilon := 1$) such that for any $q_1, q_2 \in Q$ and $v_1, v_2 \in Q^*$, where $v_1 v_2 \in Q^+$ the following holds:

$$E[\boldsymbol{\mu}_{v_1 q_1}(t)\boldsymbol{\mu}_{v_2 q_2}(t)| \underset{q\in Q}{\vee} \mathcal{F}^{\boldsymbol{\mu}_q}_{t-}] = \begin{cases} p_{q_1}\boldsymbol{\mu}_{v_1}(t-1)\boldsymbol{\mu}_{v_2}(t-1) & q_1 = q_2 \\ 0 & q_1 \neq q_2 \end{cases}$$

where $\underset{q\in Q}{\vee} \mathcal{F}^{\boldsymbol{\mu}_q}_{t-}$ denotes the smallest sigma algebra containing all $\mathcal{F}^{\boldsymbol{\mu}_q}_{t-}$ for $q \in Q$.

Below, we present some examples of admissible input processes.

- **Zero Mean i.i.d. Input.** Let $Q = \{1, 2, \ldots, d\}$, $\boldsymbol{\mu}_1 = 1$, and assume that for each $i = 2, \ldots, d$, $\boldsymbol{\mu}_i$ is a zero mean and it is an i.i.d. process. and for each $t \in \mathbb{Z}$, $\boldsymbol{\mu}(t)$ is square-integrable and $E[\boldsymbol{\mu}_q^2(t)] = p_q$. Then $\{\boldsymbol{\mu}_q\}_{q\in Q}$ is a collection of admissible input processes.
- **Discrete Valued i.i.d Process.** Assume there exists an i.i.d process $\boldsymbol{\theta}$ which takes its values from a finite set Q. Let $\boldsymbol{\mu}_q(t) = \chi(\boldsymbol{\theta}(t) = q)$ for all $q \in Q, t \in \mathbb{Z}$. $p_q = P(\boldsymbol{\theta}(t) = q)$, $\alpha_q = 1$ for all $q \in Q$. Then $\{\boldsymbol{\mu}_q\}_{q\in Q}$ is a collection of admissible input processes.

Next, we define the class of processes to which the noise, state, and output processes of a GBS belong. The definition involves the concept of conditionally independent σ-algebras: Recall that two σ-algebras $\mathcal{F}_1, \mathcal{F}_2$ are conditionally independent with respect to a third one $\mathcal{F}_3$, if for every event $A_1 \in \mathcal{F}_1$ and $A_2 \in \mathcal{F}_2$ the following holds: $P(A_1 \cap A_2|\mathcal{F}_3) = P(A_1|\mathcal{F}_3)P(A_2|\mathcal{F}_3)$ with probability one [38]. Furthermore, besides the σ-algebra generated by the random variables in the past of $\mathbf{z}$ with respect to a time t, we will work with σ-algebras generated by the random variables in the present and future, denoted by $\mathcal{F}^{\mathbf{z}}_{t} = \sigma(\mathbf{z}(t))$ and $\mathcal{F}^{\mathbf{z}}_{t+} = \sigma\left(\{\mathbf{z}(k)\}_{k=t}^{\infty}\right)$, respectively.

Definition 4.3.4 (*ZMWSSI process*) A stochastic process $\mathbf{r}$ is called *zero-mean weakly stationary* with respect to an admissible set of processes $\{\boldsymbol{\mu}\}_{q\in Q}$ (ZMWSSI) if

- $\mathcal{F}^{\mathbf{r}}_{(t+1)-}$ and $\mathcal{F}^{\boldsymbol{\mu}}_{t+}$ are conditionally independent with respect to $\mathcal{F}^{\boldsymbol{\mu}}_{t-}$
- $[\mathbf{r}^T, (\mathbf{z}^{\mathbf{r}}_v)^T, (\mathbf{z}^{\mathbf{r}}_w)^T]^T$ is zero-mean weakly stationary for all $v, w \in Q^+$.

Note that ZMWSSI abbreviates zero-mean weakly stationary with respect to input. This is because ZMWSSI will describe processes of GBSs where the admissible set of processes will be the set of input processes. We are now ready to define GBSs:

Definition 4.3.5 A system of the form (4.2) is called *generalized stochastic bilinear system* (GBS) w.r.t. $\{\boldsymbol{\mu}_q\}_{q\in Q}$ if the following holds:

- $\{\boldsymbol{\mu}_q\}_{q\in Q}$ is admissible
- $[\mathbf{x}^T, \mathbf{v}^T]$ is ZMWSSI w.r.t. $\{\boldsymbol{\mu}_q\}_{q\in Q}$
- $E[\mathbf{z}^{\mathbf{v}}_w(t)\mathbf{v}^T(t)] = 0$ and $E[\mathbf{z}^{\mathbf{x}}_w(t)\mathbf{v}^T(t)] = 0$ for all $w \in Q^+$
- $E[\mathbf{z}^{\mathbf{x}}_{\hat{q}}(t)\left(\mathbf{z}^{\mathbf{v}}_q(t)\right)^T] = 0$ for all $\hat{q}, q \in Q$
- $\sum_{q\in Q} p_q A_q \otimes A_q$ is stable.

We say that a GBS $\mathcal{G}$ is a *realization* of $\mathbf{y}$, if the output process of $\mathcal{G}$ equals $\mathbf{y}$. Recall that the dimension of a GBS $\mathcal{G}$ of the form (4.2) is the dimension n of its state-space $\mathbb{R}^n$. We say that a GBS $\mathcal{G}$ is a *minimal realization* of $\mathbf{y}$, if the dimension of any GBS realization of $\mathbf{y}$ is not smaller than the dimension of $\mathcal{G}$.

Note that the state and output process of $\mathcal{G}$ are uniquely determined by the noise process and the matrices of $\mathcal{G}$, i.e.,

$$\mathbf{x}(t) = \sum_{q\in Q}\left(K_q\mathbf{z}^{\mathbf{v}}_q(t) + \sum_{k=1}^{\infty}\sum_{w\in Q^+,|w|=k} A_w K_q \mathbf{z}^{\mathbf{v}}_{qw}(t)\right)$$

$$\mathbf{y}(t) = D\mathbf{v}(t) + \sum_{q\in Q}\left(CK_q\mathbf{z}^{\mathbf{v}}_q(t) + \sum_{k=1}^{\infty}\sum_{w\in Q^+,|w|=k} CA_w K_q \mathbf{z}^{\mathbf{v}}_{qw}(t)\right)$$

where the infinite sums involved are convergent in the mean-square sense. See [40]. For this reason, we will identify $\mathcal{G}$ with the tuple

$$\mathcal{G} = (n, p, m, Q, \mathbf{v}, \{\boldsymbol{\mu}_q\}_{q\in Q}, C, D, \{A_q, K_q\}_{q\in Q}) \tag{4.15}$$

We will shortly present a realization algorithm of GBSs that calculates a GBS from the input and output processes $(\{\boldsymbol{\mu}_q\}_{q\in Q}, \mathbf{y})$ in a specific form. This specific form is called GBS in forward innovation form, for which we need to define a so-called GB-innovation process of $\mathbf{y}$ with respect to $(\{\boldsymbol{\mu}_q\}_{q\in Q}$. To this end, we will introduce the following definition. Let $\mathbf{r}$ be a ZMWSSI process w.r.t. a set of admissible processes $\{\boldsymbol{\mu}_q\}_{q\in Q}$. Then, the one-dimensional components of $\mathbf{r}(t)$ and $\mathbf{z}^{\mathbf{r}}_w(t)$ (see Definition 4.3.1) belong to $\mathcal{H}$ for all $t \in \mathbb{Z}$. We denote the Hilbert spaces generated by the one-dimensional components of $\mathbf{r}(t)$ and of $\{\mathbf{z}^{\mathbf{r}}_w(t)\}_{w\in Q^+}$ by $\mathcal{H}^{\mathbf{r}}_t$

and $\mathcal{H}^{\mathbf{z}^{\mathbf{r}}_w}_{t,w\in Q^+}$, respectively. The (orthogonal) linear projection of $\mathbf{r}(t)$ onto a closed subspace $\mathcal{M}$ of $\mathcal{H}$ is meant element-wise and it is denoted by $E_l[\mathbf{r}(t)|\mathcal{M}]$. If all the components of $\mathbf{r}(t)$ are in $\mathcal{M} \subset \mathcal{H}$, then we write $\mathbf{r}(t) \in \mathcal{M}$.

Definition 4.3.6 The *GB-innovation process* of a ZMWSSI process $\mathbf{y}$ with respect to the set of admissible processes $\{\boldsymbol{\mu}_q\}_{q\in Q}$ is defined by

$$\mathbf{e}(t) := \mathbf{y}(t) - E_l[\mathbf{y}(t)|\mathcal{H}^{\mathbf{z}^{\mathbf{y}}_w}_{t,w\in Q^+}].$$

The class of GBSs in forward innovation form is then defined as follows.

Definition 4.3.7 Let $\mathcal{G}$ be a GBS realization of $\mathbf{y}$ of the form (4.2). Then $\mathcal{G}$ is said to be in *in forward innovation form*, if its noise process $\mathbf{v}$ is the GB-innovation process $\mathbf{e}(t) = \mathbf{y}(t) - E_l[\mathbf{y}(t)|\mathcal{H}^{\mathbf{z}^{\mathbf{y}}_w}_{t,w\in Q^+}]$ of $\mathbf{y}$ with respect to the input $\{\boldsymbol{\mu}_q\}_{q\in Q}$ and D is the identity matrix.

In order to give an intuition behind the realization algorithm, by analogy with the sLTI-SS case, we will relate certained covariances of $\mathbf{y}$ with products of the system matrices of GBSs realizations of $\mathbf{y}$. To this end, consider a GBS $\mathcal{G}$ of the form (4.2) which is a realization of $\mathbf{y}$. As in the case for sLTI-SSs, there is a direct relationship between covariances of outputs and products of the system matrices of $\mathcal{G}$. Denote the covariances between $\mathbf{y}$ and its past $\mathbf{z}^{\mathbf{y}}_w$ w.r.t. the input by

$$\begin{aligned} \Lambda^{\mathbf{y}}_w &= E[\mathbf{y}(t)(\mathbf{z}^{\mathbf{y}}_w(t))^T], \quad w \in Q^+ \\ T^{\mathbf{y}}_{q,q} &= E[\mathbf{z}^{\mathbf{y}}_q(t)(\mathbf{z}^{\mathbf{y}}_q(t))^T], \quad q \in Q. \end{aligned}$$

Then, by [40], for all $q \in Q$, $w \in Q^+$,

$$\begin{aligned} \Lambda^{\mathbf{y}}_{qw} &= C A_w G_q p_{qw}, \\ G_q &= A_q P_q C^T + K_q Q_q D^T, \\ T^{\mathbf{y}}_{q,q} &= C P_q C^T + D Q_q D^T. \end{aligned}$$

where $Q_q = E[\mathbf{v}(t)\mathbf{v}^T(t)\boldsymbol{\mu}^2_q(t)]$ and $P_q = E[\mathbf{x}(t)\mathbf{x}^T(t)\boldsymbol{\mu}^2_q(t)]$ and the matrices $\{P_q\}_{q\in Q}$ satisfy

$$P_q = p_q \sum_{q_1\in Q} \left(A_{q_1} P_{q_1} A^T_{q_1} + K_{q_1} Q_{q_1} K^T_{q_1}\right).$$

That is, $\Lambda^{\mathbf{y}}_w$, $w \in Q^+$ can be viewed as generating series of an input-output map generated by a bilinear system [22]. This then suggests that a GBS realization of $\mathbf{y}$ can be computed from a Hankel-matrix constructed from the covariances $\Lambda^{\mathbf{y}}_w$, $w \in Q^+$.

In order to present the corresponding realization algorithm, we need to define Hankel matrices and observability matrices of GBSs. This requires the introduction of a (complete) *lexicographic ordering* ($\prec$) *on* Q^*: $v \prec w$ if either $|v| < |w|$ or if

$v = \nu_1 \dots \nu_k, w = q_1 \dots q_k$ then $\exists l \in \{1, \dots, k\}$ such that $\nu_i = q_i, i < l$ and $\nu_l < q_l$. Let the ordered elements of Q^* be $v_1 = \epsilon, v_2 = q_1, \dots$ and define $M(j) = \frac{d^{j+1}-1}{d-1}$ for $d \geq 2$ and $M(j) = j + 1$ for $d = 1$ as the largest index such that $|v_{M(j)}| \leq j$. Then the matrices that form the block matrices of the Hankel matrix are given by

$$\Psi_w^{\mathbf{y}} := [\Lambda_{1w}^{\mathbf{y}}, \dots, \Lambda_{dw}^{\mathbf{y}}].$$

By using $\Psi_w^{\mathbf{y}}$, we define the (finite) Hankel matrix of $\mathbf{y}$ with respect to $\{\boldsymbol{\mu}_q\}_{q\in Q}$ indexed by i, j as follows

$$H_{i,j}^{\mathbf{y}} := \begin{bmatrix} \Psi_{v_1 v_1}^{\mathbf{y}} & \Psi_{v_2 v_1}^{\mathbf{y}} & \cdots & \Psi_{v_{M(j)} v_1}^{\mathbf{y}} \\ \Psi_{v_1 v_2}^{\mathbf{y}} & \Psi_{v_2 v_2}^{\mathbf{y}} & \cdots & \Psi_{v_{M(j)} v_2}^{\mathbf{y}} \\ \vdots & \vdots & & \vdots \\ \Psi_{v_1 v_{M(i)}}^{\mathbf{y}} & \Psi_{v_2 v_{M(i)}}^{\mathbf{y}} & \cdots & \Psi_{v_{M(j)} v_{M(i)}}^{\mathbf{y}} \end{bmatrix}. \tag{4.16}$$

For presenting the realization algorithm, we also need to introduce observability matrices. The observability matrix $\mathcal{O}_k$ of $(\{A_q\}_{q\in Q}, C)$ up to k is defined by

$$\mathcal{O}_k := \left[(CA_{v_1})^T \ (CA_{v_2})^T \ \cdots \ (CA_{v_k})^T]^T\right].$$

Finally, we will make the technical assumption that the output processes of GBSs are full rank:

Definition 4.3.8 An output process $\mathbf{y}$ of a GBS is called *full rank* if for all $q \in Q$ and $t \in \mathbb{Z}$, the matrix $E[\mathbf{e}(t)\mathbf{e}^T(t)\boldsymbol{\mu}_q^2(t)]$ is strictly positive definite where $\mathbf{e}$ is the GB-innovation process of $\mathbf{y}$ with respect to the input $\{\boldsymbol{\mu}_q\}_{q\in Q}$.

Next, we present Algorithm 4.3, a realization algorithm of GBSs. According to [40], if $N \geq n$, where n is the dimension of a GBS realization of $\mathbf{y}$, or, equivalently, if $N \geq \sup_{i,j\in\mathbb{N}} \text{rank } H_{i,j}^{\mathbf{y}}$, then Algorithm 4.3 returns a minimal GBS realization of $\mathbf{y}$ in forward innovation form. Algorithm 4.3 is equivalent to [40, Algorithm 2], however, there is a nuance between the two algorithms due to that in [40], there is a scalar factor of the processes $\mathbf{z}_w^{\mathbf{y}}(t)$ which factor also effects the formulas for the covariances $\Psi_w^{\mathbf{y}}$ and the Hankel matrix $H_{i,j}^{\mathbf{y}}$.

4.3.2 Extending Granger Causality

We would like to relate the existence of a GBS of $(\{\boldsymbol{\mu}_q\}_{q\in Q}, \mathbf{y})$ with a network graph consisting of two nodes and one edge to Granger causality-like properties of $\mathbf{y}_1$ and $\mathbf{y}_2$. Recall from [9, 30, 34] that sLTI-SSs of $\mathbf{y}$ in forward innovation form can be viewed as optimal predictors which map past values of $\mathbf{y}$ to its best linear prediction. Since the classical definition of Granger causality imposed conditions on such optimal linear predictors, it was possible to translate Granger causality among

Algorithm 4.3 Minimal innovation GBS based on output covariances

Input: $\{\Psi^{\mathbf{y}}_w\}_{\{w\in Q^*,|w|\le N\}}$ and $\{E[\mathbf{z}^{\mathbf{y}}_\sigma(t)(\mathbf{z}^{\mathbf{y}}_q(t))^T]\}_{q\in Q}$: Covariance sequence of $\mathbf{y}$ and its past and variances of $\mathbf{z}^{\mathbf{y}}_q$

Output: GBS $\mathcal{G}=(n,p,m,Q,\mathbf{e},\{\boldsymbol{\mu}_q\}_{q\in Q},C,D,\{A_q,K_q\}_{q\in Q})$ realization of $\mathbf{y}$ in forward innovation form, and the covariance matrix $Q_q=E[\mathbf{e}(t)\mathbf{e}^T(t)\boldsymbol{\mu}^2_q(t)]$ of noise.

Step 1 Form the Hankel matrix $H^{\mathbf{y}}_{N-1,N}$ defined in (4.16)
Step 2 Decompose $H^{\mathbf{y}}_{N-1,N}=\mathbf{OR}$ such that $\mathbf{O}\in\mathbb{R}^{pM(N-1)\times n}$ and $\mathbf{R}\in\mathbb{R}^{n\times pdM(N)}$ have full column and row rank, respectively, and n is the rank of $H^{\mathbf{y}}_{N-1,N}$
Step 3 Take C as the first p rows of $\mathbf{O}$ and $R_{v_i}\in\mathbb{R}^{n\times pd}$ such that $\mathbf{R}=[R_{v_1}\cdots R_{v_{M(N)}}]$
Step 4 Take $G_i\in\mathbb{R}^{n\times p}$ such that $R_{v_1}=[G_1\cdots G_d]$
Step 5 Let A_q be the linear least square solution of

$$A_qR_{v_i}=\frac{1}{p_q}R_{v_iq},\quad i=1,\dots,M(N-1)$$

Step 6 Let $P_q=\lim_{k\to\infty}P^k_q$ where $P^0_q=0$ and P^{i+1}_q, $i=0,1,\dots$ is such that

$$\begin{aligned}Q^i_q&=E[\mathbf{z}^{\mathbf{y}}_q(t)(\mathbf{z}^{\mathbf{y}}_q(t))^T]-CP^i_qC^T\\ K^i_q&=(G_q-A_qP^i_qC^T)(Q^i{}_q)^+\\ P^{i+1}_q&=p_q\sum_{q_1\in Q}\left(A_{q_1}P^i_{q_1}A^T_{q_1}+K^i_{q_1}Q^i_{q_1}(K^i_{q_1})^T\right)\end{aligned}$$

Step 7 Let $K_q=\lim_{i\to\infty}K^i_q$ and $Q_q=\lim_{i\to\infty}Q^i_q$.
return $\mathcal{G}=(n,p,m,Q,\mathbf{e},\{\boldsymbol{\mu}_q\}_{q\in Q},C,D,\{A_q,K_q\}_{q\in Q})$.

components of $\mathbf{y}$ to properties of sLTI-SSs of $\mathbf{y}$. If we extend the same idea to GBSs, then we have to deal with the problem that a GBS in forward innovation form is not a linear predictor based on past values of $\mathbf{y}$, but it is a linear predictor from the past outputs and inputs $\mathbf{z}^{\mathbf{y}}_v(t)$ to the future outputs and inputs $\mathbf{z}^{\mathbf{y}+}_v(t)$. This motivates our extension of Granger causality, where we use the process $\mathbf{z}^{\mathbf{y}+}_w(t)$ rather than $\mathbf{y}(t+|w|)$ and $\mathbf{z}^{\mathbf{y}}_v(t)$ rather than $\mathbf{y}(t-|v|)$, $v,w\in Q^+$:

Definition 4.3.9 (*GB-Granger causality*) We say that $\mathbf{y}_1$ *does not GB-Granger cause* $\mathbf{y}_2$ w.r.t. $\{\boldsymbol{\mu}_q\}_{q\in Q}$ if

$$E_l[\mathbf{z}^{\mathbf{y}_2+}_v(t)|\mathcal{H}^{\mathbf{z}^{\mathbf{y}}_w}_{t,w\in\Sigma^+}]=E_l[\mathbf{z}^{\mathbf{y}_2+}_v(t)|\mathcal{H}^{\mathbf{z}^{\mathbf{y}_2}_w}_{t,w\in\Sigma^+}].\tag{4.17}$$

for all $v\in Q^*$ and $t\in\mathbb{Z}$. Otherwise, $\mathbf{y}_1$ GB-Granger causes $\mathbf{y}_2$ w.r.t. $\{\boldsymbol{\mu}_q\}_{q\in Q}$.

Informally, $\mathbf{y}_1$ *does not GB-Granger cause* $\mathbf{y}_2$, if the best linear predictions of the future of $\mathbf{y}_2$ with respect to $\{\boldsymbol{\mu}_q\}_{q\in Q}$ along $\mathbf{v}$ is the same based on the past of $\mathbf{y}$ or based on the past of $\mathbf{y}_2$ with respect to $\{\boldsymbol{\mu}_q\}_{q\in Q}$ along $\{w\}_{w\in Q^+}$.

Remark 4.3.10 If $\mathbf{y}_1$ does not GB-Granger cause $\mathbf{y}_2$ then it implies that $\mathbf{y}_1$ does not Granger causes $\mathbf{y}_2$. Moreover, if $Q=\{1\}$ and $\boldsymbol{\mu}_1(t)\equiv 1$, then $\mathbf{z}^{\mathbf{y}+}_v(t)=\mathbf{y}(t+|v|)$

and $\mathbf{z}^{\mathbf{y}}_{w}(t) = \mathbf{y}(t - |w|)$ and thus Definitions 4.2.7 and 4.3.9 coincide. The relationship between GB-Granger causality and other concepts of causality, such as conditional independence [1] is more involved and remains a topic of future research.

4.3.3 GB-Granger Causality and Network Graph of GBSs

Next, we present the result on the relationship between GB-Granger causality and network graphs of GBSs. The GBS in question are minimal ones in forward innovation form that can be constructed algorithmically (see Algorithm 4.4 below).

Theorem 4.3.11 *Consider an output* $\mathbf{y} = [\mathbf{y}_1^T, \mathbf{y}_2^T]^T$ *of a GBS where* $\mathbf{y}_i \in \mathbb{R}^{k_i}, i = 1, 2$ *and let* $\mathbf{e} = [\mathbf{e}_1^T, \mathbf{e}_2^T]^T$ *be the innovation process of* $\mathbf{y}$ *w.r.t.* $\{\boldsymbol{\mu}_q\}_{q\in Q}$. *Then,* $\mathbf{y}_1$ *does not GB-Granger cause* $\mathbf{y}_2$ *w.r.t.* $\{\boldsymbol{\mu}_q\}_{q\in Q}$ *if and only if there exists a minimal GBS realization* $\mathcal{G}$ *of* $\mathbf{y}$ *of the form* (4.15) *such that* $\mathcal{G}$ *is in forward innovation form, and its matrices take the following block triangular form:*

$$A_q = \begin{bmatrix} A_{q,11} & A_{q,12} \\ 0 & A_{q,22} \end{bmatrix},\ K_q = \begin{bmatrix} K_{q,11} & K_{q,12} \\ 0 & K_{q,22} \end{bmatrix},\ C = \begin{bmatrix} C_{11} & C_{12} \\ 0 & C_{22} \end{bmatrix}, \tag{4.18}$$

where $A_{q,ij} \in \mathbb{R}^{n_i \times n_j}$, $K_{q,ij} \in \mathbb{R}^{n_i \times k_j}$, $C_{ij} \in \mathbb{R}^{k_i, n_j}$, $q \in Q$, $i, j = 1, 2$ *for some* $n_1 \geq 0$, $n_2 > 0$, *and*

$$\mathcal{G}_2 = (n_2, p, p, Q, \mathbf{e}, \{\boldsymbol{\mu}_q\}_{q\in Q}, C_{22}, I, \{A_{q,22}, K_{q,22}\}_{q\in Q}),$$

is a minimal GBS realization of $\mathbf{y}_2$ *in forward innovation form.*

If $Q = \{1\}$ and $\boldsymbol{\mu}_1(t) \equiv 1$, then GBS reduce to sLTI-SSs and Definitions 4.2.7 and 4.3.9 coincide. As a result, Theorem 4.18 reduces to earlier results on sLTI-SSs and Granger causality (see [26, Theorem 1]). A GBS realization of $\mathbf{y}$ of the form (4.15) in forward innovation form which satisfies (4.18) can be viewed as a cascade interconnection of two subsystems. Define the subsystems

$$\mathcal{S}_1 \begin{cases} \mathbf{x}_1(t+1) = \sum_{q\in Q}(A_{q,11}\mathbf{x}_1(t) + K_{q,11}\mathbf{e}_1(t) + A_{q,12}\mathbf{x}_2(t) + K_{q,12}\mathbf{e}_2(t))\boldsymbol{\mu}_q(t) \\ \quad \mathbf{y}_1(t) = \sum_{i=1}^{2} C_{1i}\mathbf{x}_i(t)) + \mathbf{e}_2(t) \end{cases}$$

$$\mathcal{S}_2 \begin{cases} \mathbf{x}_2(t+1) = (A_{q,22}\mathbf{x}_2(t) + K_{q,22}\mathbf{e}_2(t))\boldsymbol{\mu}_q(t) \\ \quad \mathbf{y}_2(t) = C_{22}\mathbf{x}_2(t) + \mathbf{e}_2(t) \end{cases}$$

Notice that $\mathcal{S}_2$ sends its state $\mathbf{x}_2$ and noise $\mathbf{e}_2$ to $\mathcal{S}_1$ as an external input while $\mathcal{S}_1$ does not send information to $\mathcal{S}_2$. The corresponding network graph is illustrated in Fig. 4.6.

Next, we present a realization algorithm that calculates a minimal innovation GBS in causal block triangular form when $\mathbf{y}_1$ does not GB-Granger cause $\mathbf{y}_2$. Note that this algorithm provides a constructive proof of the necessity part of Theorem 4.3.11.

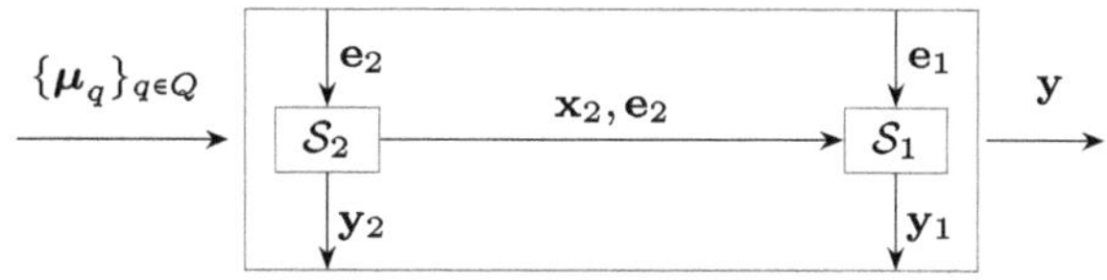

Fig. 4.6 Cascade interconnection of a GBS realization in forward innovation form with system matrices as in (4.18)

Algorithm 4.4 Block triangular minimal GBS

Input $\{\Lambda^{\mathbf{y}}_w := E[\mathbf{y}(t)(\mathbf{z}^{\mathbf{y}}_w(t))^T]\}_{w\in Q^+,|w|\le n}$ and $\{T^{\mathbf{y}}_{q,q} := E[\mathbf{z}^{\mathbf{y}}_q(t)(\mathbf{z}^{\mathbf{y}}_q(t))^T]\}_{q\in Q}$

Output GBS $\mathcal{G} = (n, p, m, Q, \mathbf{e}, \{\mu_q\}_{q\in Q}, C, D, \{A_q, K_q\}_{q\in Q})$ satisfying the conclusion of Theorem 4.3.11

Step 1 Apply Algorithm 4.3 to compute a minimal GBS realization

$$\tilde{\mathcal{G}} = (n, p, m, Q, \mathbf{e}, \{\mu_q\}_{q\in Q}, C, I, \{\tilde{A}_q, \tilde{K}_q\}_{q\in Q})$$

of $\mathbf{y}$ in forward innovation form.
Step 2 Define the sub-matrix consisting of the last k_2 rows of $\tilde{C}$ by $\tilde{C}_2 \in \mathbb{R}^{k_2\times n}$ and define the observability matrix $\tilde{O}_{\mathbf{N}(n)}$

$$\tilde{O}_{\mathbf{N}(n)} = \left[(\tilde{C}_2\tilde{A}_{v_1})^T \cdots (\tilde{C}_2\tilde{A}_{v_k})^T\right]^T,$$

where we used the lexicographic ordering ($\prec$) on Q^*. If $\tilde{O}_{\mathbf{N}(n)}$ is not of full column rank then define the non-singular matrix $T^{-1} = \begin{bmatrix} T_1 & T_2 \end{bmatrix}$ such that the columns of $T_1 \in \mathbb{R}^{n\times n_1}$ is the kernel of $\tilde{O}_{\mathbf{N}(n)}$. If $\tilde{O}_{\mathbf{N}(n)}$ is of full column rank, then set $T = I$.
Define the matrices $A_q = T\tilde{A}_qT^{-1}$, $K_q = T\tilde{K}_q$ for $q \in Q$ and $C = \tilde{C}T^{-1}$.
return $\mathcal{G} = (n, p, m, Q, \mathbf{e}, \{\mu_q\}_{q\in Q}, C, D, \{A_q, K_q\}_{q\in Q})$.

Remark 4.3.12 (Checking GB-Granger causality) Algorithm 4.4 can be used for checking GB-Granger causality as follows. Apply Algorithm 4.4 and check if (i) the returned matrices $\{A_q, K_q\}_{q\in Q}$ and C satisfy

$$A_q = \begin{bmatrix} A_{q,11} & A_{q,12} \\ 0 & A_{q,22} \end{bmatrix} \quad C = \begin{bmatrix} C_{11} & C_{12} \\ 0 & C_{22} \end{bmatrix}, K_q = \begin{bmatrix} K_{q,11} & K_{q,12} \\ 0 & K_{q,22} \end{bmatrix},$$

where $A_{q,ij} \in \mathbb{R}^{n_i\times n_j}$, $C_{ij} \in \mathbb{R}^{k_i\times n_j}$, $K_{i,j} \in \mathbb{R}^{n_i,k_j}$, $i, j = 1, 2$ for some $n_1 \ge 0$, $n_2 > 0$ and if (ii) $\mathcal{S}_2 = (n_2, p, p, Q, \mathbf{e}, \{\mu_q\}_{q\in Q}, C_{22}, I, \{A_{q,22}, K_{q,22}\}_{q\in Q})$ defines a minimal GBS realization of $\mathbf{y}_2$ in forward innovation form. By Theorem 4.3.11, (i) and (ii) are positive if and only if $\mathbf{y}_1$ does not GB-Granger cause $\mathbf{y}_2$. We check whether $\mathcal{S}_2$ is a minimal GBS in forward innovation form as follows. We use Algorithm 4.3 to compute a minimal GBS realization $\bar{\mathcal{S}}_2$ of $\mathbf{y}_2$ in forward innovation form and the covariances $\bar{Q}_q = E[\mathbf{v}(t)\mathbf{v}^T(t)\mu_q^2(t)]$, $q \in Q$ of the innovation process $\mathbf{v}$ of $\mathbf{y}_2$. Then, $\mathcal{S}_2$ is a minimal GBS realization in forward innovation form if and only if $\mathcal{S}_2$ and $\bar{\mathcal{S}}_2$ have the same dimension and the same noise process, i.e., $\mathbf{v} = \mathbf{e}_2$. For checking the latter, we remark that $\mathbf{v}(t) = \mathbf{e}_2(t)$ if and only if for all $i = 1, \ldots, k_2$, $\sum_{q\in Q}\alpha_q^2\bar{Q}_{q,ii} = \sum_{q\in Q}\alpha_q^2 Q_{q,(k_1+i)(k_1+i)}$, where $\{\alpha_q\}_{q\in Q}$ are such that $\sum_{q \in Q} \alpha_q \mu_q = 1$ and $Q_{q,rl}$, $\bar{Q}_{q,kl}$ denotes the $(k, l)th$ entry of the matrix Q_q, $\bar{Q}_q$ respectively.

Note that the covariances used as inputs of Algorithm 4.4 could be estimated from sampled data, the procedure above could be a starting point of a statistical test for checking GB-Granger causality, similar to the one of Granger causality. This remains a topic of future research.

4.4 Conclusions and Future Work

In this chapter, we have presented an overview of results on the relationship between network graphs of state-space representations and Granger causality relations among their outputs. We have discussed two classes of state-space representations: sLTI-SS and GBSs. Similar results can be formulated for the relationship between transfer functions and Granger causality relations [25]. These results could be used for developing statistical hypothesis testing for deciding if the network graph of the system which generates the data has certain interconnections [23].

There are many potential directions for further research. First of all, it would be desirable to extend the results to sLTI-SSs and to stochastic LPV/linear switching systems with control inputs. Reformulations of the results in purely deterministic setting would also be of interest. The latter would require reformulating Granger causality for the deterministic setting. Another direction would be the use of statistical notions which are different from Granger causality. That remains topic of future research as well. Finally, more work needs to be done on statistical hypothesis testing methods for deciding if the data could be generated by a system with a certain network graph. The latter problem is essentially the well-posed version of the generally ill-posed problem of reverse engineering of network graphs. First steps toward this goal were made in [23]. However, a formal theoretical analysis and experimental validation of the methods from [23] is still lacking.

References

1. Amblard, P.-O., Michel, O.J.: On directed information theory and granger causality graphs. J. Comput. Neurosci. **30**(1), 7–16 (2011)
2. Barnett, L., Barrett, A.B., Seth, A.K.: Granger causality and transfer entropy are equivalent for Gaussian variables. Phys. Rev. Lett. **103**, 238701 (2009)
3. Bolognani, S., Bof, N., Michelotti, D., Muraro, R., Schenato, L.: Identification of power distribution network topology via voltage correlation analysis. In: 52nd IEEE Conference on Decision and Control, pp. 1659–1664 (2013)
4. Bombois, X., Colin, K., Van Den Hof, P.M.J., Hjalmarsson, H.: On the informativity of direct identification experiments in dynamical networks. Working paper or preprint, August 2021
5. Caines, P.E.: Weak and strong feedback free processes. IEEE Trans. Autom. Control **21**(5), 737–739 (1976)

6. Caines, P.E., Chan, C.: Feedback between stationary stochastic processes. IEEE Trans. Autom. Control **20**(4), 498–508 (1975)
7. Caines, P.E., Deardon, R., Wynn, H.P.: Conditional orthogonality and conditional stochastic realization. In: Rantzer, A., Byrnes, C.I. (eds.) Directions in Mathematical Systems Theory and Optimization, vol. 286, pp. 71–84. Springer, Berlin, Heidelberg (2003)
8. Caines, P.E., Deardon, R., Wynn, H.P.: Bayes nets of time series: stochastic realizations and projections. In: Pronzato, L., Zhigljavsky, A. (eds.) Optimal Design and Related Areas in Optimization and Statistics. Springer Optimization and its Applications, vol. 28, pp. 155–166. Springer, New York (2009)
9. Caines, P.E.: Linear Stochastic Systems. John Wiley and Sons, New-York (1998)
10. Dankers, A.G.: System identification in dynamic networks. Ph.D. thesis, Delft University of Technology (2014)
11. David, O.: fMRI connectivity, meaning and empiricism: comments on: Roebroeck et al. the identification of interacting networks in the brain using fMRI: model selection, causality and deconvolution. NeuroImage **58**(2), 306–309 (2011)
12. Eichler, M.: Granger causality and path diagrams for multivariate time series. J. Econ. **137**(2), 334–353 (2007)
13. Eichler, M.: Graphical modelling of multivariate time series. Probab. Theory Relat. Fields **153**(1), 233–268 (2012)
14. Friston, K.J., Harrison, L., Penny, W.: Dynamic causal modeling. NeuroImage **19**(4), 1273–1302 (2003)
15. Gevers, M.R., Anderson, B.: On jointly stationary feedback-free stochastic processes. IEEE Trans. Autom. Control **27**(2), 431–436 (1982)
16. Goebel, R., Roebroeck, A., Kim, D.-S., Formisano, E.: Investigating directed cortical interactions in time-resolved fMRI data using vector autoregressive modeling and Granger causality mapping. Magn. Reson. Imaging **21**, 1251–1261 (2003)
17. Gonçalves, J., Howes, R., Warnick, S.: Dynamical structure functions for the reverse engineering of LTI networks. In: 46th IEEE Conference on Decision and Control, pp. 1516–1522 (2007)
18. Granger, C.W.J.: Economic processes involving feedback. Inf. Control **6**(1), 28–48 (1963)
19. Havlicek, M., Roebroeck, A., Friston, K., Gardumi, A., Ivanov, D., Uludag, K.: Physiologically informed dynamic causal modeling of fMRI data. NeuroImage **122**, 355–372 (2015)
20. Howes, R., Eccleston, L., Gonçalves, J., Stan, G.-B., Warnick, S.: Dynamical structure analysis of sparsity and minimality heuristics for reconstruction of biochemical networks. In: 47th IEEE Conference on Decision and Control (2008)
21. Hsiao, C.: Autoregressive modelling and causal ordering of econometric variables. J. Econ. Dyn. Control **4**, 243–259 (1982)
22. Isidori, A.: Direct construction of minimal bilinear realizations from nonlinear input-output maps. IEEE Trans. Autom. Control 626–631 (1973)
23. Jozsa, M.: Relationship between Granger non-causality and network graphs of state-space representations. Ph.D. thesis, University of Groningen (2019)
24. Jozsa, M., Petreczky, M., Camlibel, M.K.: Causality based graph structure of stochastic linear state-space representations. In: 56th IEEE Conference on Decision and Control, pp. 2442–2447 (2017)
25. Jozsa, M., Petreczky, M., Camlibel, M.K.: Relationship between causality of stochastic processes and zero blocks of their joint innovation transfer matrices. In: 20th World Congress of the International Federation of Automatic Control, pp. 4954–4959 (2017)
26. Jozsa, M., Petreczky, M., Camlibel, M.K.: Relationship between Granger non-causality and network graph of state-space representations. IEEE Trans. Autom. Control **64**(8), 912–927 (2019)
27. Jozsa, M., Petreczky, M., Camlibel, M.K.: Causality and network graph in general bilinear state-space representations. IEEE Trans. Autom. Control **65**(8), 3623–3630 (2020)
28. Julius, A.A., Zavlanos, M., Boyd, S., Pappas, G.J.: Genetic network identification using convex programming. IET Syst. Biol. **3**, 155–166 (2009)

29. Kang, T., Moore, R., Li, Y., Sontag, E.D., Bleris, L.: Discriminating direct and indirect connectivities in biological networks. Natl. Acad. Sci. USA **112**, 12893–12898 (2015)
30. Katayama, T.: Subspace Methods for System Identification. Springer-Verlag (2005)
31. Kempker, P.L.: Coordination control of linear systems. Ph.D. thesis, Amsterdam: Vrije Universiteit (2012)
32. Kempker, P.L., Ran, A.C.M., van Schuppen, J.H.: Construction and minimality of coordinated linear systems. Linear Algebr. Appl. **452**, 202–236 (2014)
33. Kramer, G.: Directed information for channels with feedback. Ph.D. thesis, Swiss Federal Institute of Technology Zürich (1998)
34. Lindquist, A., Picci, G.: Linear Stochastic Systems: a Geometric Approach to Modeling. Estimation and Identification, Springe, Berlin (2015)
35. Massey, J.L.: Causality, feedback and directed information. In: International Symposium on Information Theory and its Applications, pp. 27–30 (1990)
36. Nordling, T.E.M., Jacobsen, E.W.: On sparsity as a criterion in reconstructing biochemical networks. In: 18th IFAC World Congress, pp. 11672–11678 (2011)
37. Pambakian, N.: LQG coordination control. Master's thesis, Delft University of Technology (2011)
38. Pearl, J.: Causality: Models, Reasoning and Inference, 1st edn. Cambridge University Press (2000)
39. Penny, W.D., Stephan, K.E., Mechelli, A., Friston, K.J.: Comparing dynamic causal models. NeuroImage **22**(3), 1157–1172 (2004)
40. Petreczky, M., Vidal, R.: Realization theory for a class of stochastic bilinear systems. IEEE Trans. Autom. Control **63**(1), 69–84 (2017)
41. Ran, A.C.M., van Schuppen, J.H.: Coordinated linear systems. In: Coordination Control of Distributed Systems. Lecture Notes in Control and Information Sciences, vol. 456, pp. 113–121 (2014)
42. Roebroeck, A., Formisano, E., Goebel, R.: Reply to Friston and David: after comments on: the identification of interacting networks in the brain using fMRI: model selection, causality and deconvolution. NeuroImage **58**(2), 310–311 (2011)
43. Roebroeck, A., Seth, A.K., Valdes-Sosa, P.A.: Causal time series analysis of functional magnetic resonance imaging data. J. Mach. Learn. Res. Proc. Track **12**, 65–94 (2011)
44. Valdes-Sosa, P.A., Roebroeck, A., Daunizeau, J., Friston, K.J.: Effective connectivity: influence, causality and biophysical modeling. NeuroImage **58**(2), 339–361 (2011)
45. Van den Hof, P.M.J., Dankers, A., Heuberger, P.S.C., Bombois, X.: Identification of dynamic models in complex networks with prediction error methods-basic methods for consistent module estimates. Automatica **49**(10), 2994–3006 (2013)
46. Weerts, H.H.M.: Identifiability and identification methods for dynamic networks. Ph.D. thesis, Eindhoven University of Technology (2018)
47. Weerts, H.H.M., Van den Hof, P.M.J., Dankers, A.G.: Identifiability of linear dynamic networks. Automatica **89**, 247–258 (2018)
48. Yuan, Y., Glover, K., Gonçalves, J.: On minimal realisations of dynamical structure functions. Automatica **55**, 159–164 (2015)
49. Zhang, W., Liu, W., Zang, C., Liu, L.: Multi-agent system based integrated solution for topology identification and state estimation. IEEE Trans. Ind. Inform. **13**(2), 714–724 (2017)

Part II
Hybrid Techniques

Chapter 5
Multi-consensus Problems in Hybrid Multi-agent Systems

Andrea Cristofaro, Francesco D'Orazio, Lorenzo Govoni, and Mattia Mattioni

Abstract In this chapter, we investigate the possible advantages of introducing hybrid behaviors in the control of multi-agent systems. With reference to the consensus problem, two complementary settings are explored: hybrid interaction topology and hybrid linear agent dynamics. The results are then applied and further illustrated in synchronization problems for multi-robot systems equipped with heterogeneous sensors with the aim of opening toward new perspectives.

5.1 Introduction

Networked systems are nowadays well-considered a bridging paradigm among several disciplines spanning, among many others, from physics to engineering, psychology to medicine, biology to computer science. As typical in control theory [9], we refer to a network (or multi-agent) system as composed of several dynamical units (agents) interconnected through a *communication graph*. In this scenario, most control problems are related to driving all systems composing the network toward a *consensus* behavior that might be common to all agents or only to subgroups (e.g., [2, 7, 10, 16, 17]). Most of those works are devoted to either continuous or discrete-time networks that might be heterogeneous in the dynamics of each agent but that are assumed homogeneous in the type of interconnection. Roughly speaking, the

A. Cristofaro (✉) · F. D'Orazio · L. Govoni · M. Mattioni
Dipartimento di Ingegneria Informatica, Automatica e Gestionale A. Ruberti, Sapienza University of Rome, 00185 Rome, Italy
e-mail: cristofaro@diag.uniroma1.it

F. D'Orazio
e-mail: dorazio@diag.uniroma1.it

L. Govoni
e-mail: govoni@diag.uniroma1.it

M. Mattioni
e-mail: mattioni@diag.uniroma1.it

R. Postoyan et al. (eds.), *Hybrid and Networked Dynamical Systems*,
Lecture Notes in Control and Information Sciences 493,
https://doi.org/10.1007/978-3-031-49555-7_5

network is composed of different agents that share, however, the same type of dynamics (i.e., either continuous or discrete) and information that is exchanged.

Starting from our recent works in [5, 6], this chapter is aimed at investigating simple classes of hybrid networks in which we allow each agent to be a hybrid system exchanging information over a hybrid topology. More precisely, with reference to the consensus problem, two complementary settings are explored: hybrid interaction topology and hybrid linear agent dynamics. In the first case, we consider a network of scalar agents communicating according to distinct graphs during flows and jumps. We prove that the number and nature of consensuses depend on the structure of both the union and intersection of the flow and jump graphs, with nodes accordingly clustering and converging to suitable hybrid trajectories. This highlights that the use of a hybrid communication can weaken the demands that are necessary for reaching consensus in terms of connectivity of each independent unity and, consequently, message exchange among all agents. In the second case, we investigate the problem of inducing consensus over a network of hybrid homogenous linear systems via hybrid decentralized control. We show that the consensus can always be enforced provided the existence of a common solution to a continuous-time and a discrete-time algebraic Riccati equation. To conclude, two simulation examples support and corroborate the results and their potentialities, motivating also the generalization to a more general framework. First, we consider the formation control problem of a multi-robot system composed of two heterogeneous subgroups of agents. One of them is characterized by ground robots, e.g., unicycles, whereas the other one is composed of flying robots, e.g., Unmanned Aerial Vehicles (UAVs). Assuming hybrid communication with distinct and disconnected graphs over each subgroup, we show that convergence toward the specified formation is ensured provided that the overall hybrid graph is connected. Then, we address the problem of enforcing rendezvous (in the multi-consensus framework) over a multi-robot system composed of unicycles exchanging heterogeneous information through flow and jump graphs. In detail, the orientation information is exchanged continuously between the agents, whereas the position information is characterized only sporadically in time.

Notation. $\mathbb{C}^+$ and $\mathbb{C}^-$ denote the right- and left-hand sides of the complex plane, respectively. For a given finite set $\mathcal{S}$, $|\mathcal{S}|$ denotes its cardinality. For a closed set $\mathcal{S} \subset \mathbb{R}^n$ and $x \in \mathbb{R}^n$, $\text{dist}(x, \mathcal{S})$ denotes the distance of x from the set $\mathcal{S}$. We denote by 0 either the zero scalar or the zero matrix of suitable dimensions. $\mathbb{1}_c$ denotes the c-dimensional column vector whose elements are all ones while I is the identity matrix of suitable dimensions. Given a matrix $A \in \mathbb{R}^{n\times n}$ then $\sigma\{A\} \subset \mathbb{C}$ denotes its spectrum. A positive definite (semi-definite) matrix $A = A^\top$ is denoted by $A \succ 0$ ($A \succeq 0$); a negative definite (semi-definite) matrix $A = A^\top$ is denoted by $A \prec 0$ ($A \preceq 0$).

5.2 Review on Continuous-Time Systems over Networks

5.2.1 *Directed Graph Laplacians and Almost Equitable Partitions*

Let $\mathcal{G} = (\mathcal{V}, \mathcal{E})$ be a directed graph (or digraph for short) with $|\mathcal{V}| = N, \mathcal{E} \subseteq \mathcal{V} \times \mathcal{V}$. The set of neighbors to a node $\nu \in \mathcal{V}$ is defined as $\mathcal{N}(\nu) = \{\mu \in \mathcal{V} \text{ s.t. } (\mu, \nu) \in \mathcal{E}\}$. For all pairs of distinct nodes $\nu, \mu \in \mathcal{V}$, a directed path from ν to μ is defined as $\nu \rightsquigarrow \mu := \{(\nu_r, \nu_{r+1}) \in \mathcal{E} \text{ s.t. } \cup_{r=0}^{\ell-1} (\nu_r, \nu_{r+1}) \subseteq \mathcal{E} \text{ with } \nu_0 = \nu, \nu_\ell = \mu \text{ and } \ell > 0\}$. The reachable set from a node $\nu \in \mathcal{V}$ is defined as $R(\nu) := \{\nu\} \cup \{\mu \in \mathcal{V} \text{ s.t. } \nu \rightsquigarrow \mu\}$. A set $\mathcal{R}$ is called a reach if it is a maximal reachable set, that is, $\mathcal{R} = R(\nu)$ for some $\nu \in \mathcal{V}$ and there is no $\mu \in \mathcal{V}$ such that $R(\nu) \subset R(\mu)$. Since $\mathcal{G}$ possesses a finite number of vertices, such maximal sets exist and are uniquely determined by the graph itself. Denoting by $\mathcal{R}_i$ for $i = 1, \ldots, \mu$, the reaches of $\mathcal{G}$, the exclusive part of $\mathcal{R}_i$ is defined as $\mathcal{H}_i = \mathcal{R}_i / \cup_{\ell=1, \ell \neq i}^{\mu} \mathcal{R}_\ell$ with cardinality $h_i = |\mathcal{H}_i|$. Finally, the common part of $\mathcal{G}$ is given by $\mathcal{C} = \mathcal{V} / \cup_{i=1}^{\mu} \mathcal{H}_i$ with cardinality $c = |\mathcal{C}|$.

Given two graphs $\mathcal{G}_1 = (\mathcal{V}, \mathcal{E}_1)$ and $\mathcal{G}_2 = (\mathcal{V}, \mathcal{E}_2)$, one defines the *union graph* $\mathcal{G} = \mathcal{G}_1 \cup \mathcal{G}_2 = (\mathcal{V}, \mathcal{E}_1 \cup \mathcal{E}_2)$ and, similarly, the *intersection graph* as $\mathcal{G} = \mathcal{G}_1 \cap \mathcal{G}_2 = (\mathcal{V}, \mathcal{E}_1 \cap \mathcal{E}_2)$.

The Laplacian matrix associated with $\mathcal{G}$ is given by $L = D - A$ with $D \in \mathbb{R}^{N \times N}$ and $A \in \mathbb{R}^{N \times N}$ being, respectively, the in-degree and the adjacency matrices. As proved in [1], L possesses one eigenvalue $\lambda = 0$ with multiplicity coinciding with μ, the number of reaches of $\mathcal{G}$, and the remaining $N - \mu$ with positive real part. Hence, after a suitable re-labeling of nodes, the Laplacian always admits the lower triangular form [3]

$$L = \begin{pmatrix} L_1 & \ldots & 0 & 0 \\ & \ddots & & \vdots \\ 0 & \ldots & L_\mu & 0 \\ M_1 & \ldots & M_\mu & M \end{pmatrix} \tag{5.1}$$

where: $L_i \in \mathbb{R}^{h_i \times h_i}$ $(i = 1, \ldots, \mu)$ is the Laplacian associated with the subgraph $\mathcal{H}_i$ and possessing one eigenvalue in zero with single multiplicity; $M \in \mathbb{R}^{c \times c}$ verifying $\sigma(M) \subset \mathbb{C}^+$ corresponds to the common component $\mathcal{C}$. Thus, the eigenspace associated with $\lambda = 0$ for L is spanned by the right eigenvectors

$$z_1 = \begin{pmatrix} \mathbb{1}_{h_1} \\ \vdots \\ 0 \\ \gamma^1 \end{pmatrix} \quad \ldots \quad z_\mu = \begin{pmatrix} 0 \\ \vdots \\ \mathbb{1}_{h_\mu} \\ \gamma^\mu \end{pmatrix} \tag{5.2}$$

with $\sum_{i=1}^{\mu} \gamma^i = \mathbb{1}_c$ and $M_i \mathbb{1}_{h_i} + M\gamma^i = 0$ for all $i = 1, \ldots, \mu$. In addition, the left eigenvectors associated with the zero eigenvalues are given by

$$\tilde{v}_1^\top = \begin{pmatrix} v_1^\top \dots 0 \; 0 \end{pmatrix} \dots \; \tilde{v}_\mu^\top = \begin{pmatrix} 0 \dots v_\mu^\top \; 0 \end{pmatrix} \tag{5.3}$$

with $v_i^\top = \begin{pmatrix} v_i^1 \dots v_i^{h_i} \end{pmatrix} \in \mathbb{R}^{1 \times h_i}$, $v_i^s > 0$ if the corresponding node is root or zero otherwise. A partition $\pi = \{\rho_1, \dots, \rho_r\}$ of $\mathcal{V}$ is a collection of cells $\rho_i \subseteq \mathcal{V}$ verifying $\rho_i \cap \rho_j = \emptyset$ for all $i \neq j$ and $\cup_{i=1}^r \rho_i = \mathcal{V}$. The characteristic vector of $\rho \subseteq \mathcal{V}$ is given by $p(\rho) = (p_1(\rho) \; \dots p_N(\rho))^\top \in \mathbb{R}^N$ with for $i = 1, \dots, N$

$$p_i(\rho) = \begin{cases} 1 & \text{if } v_i \in \rho \\ 0 & \text{otherwise.} \end{cases}$$

For a partition $\pi = \{\rho_1, \dots, \rho_r\}$ of $\mathcal{V}$, the characteristic matrix of π is $P(\pi) = \begin{pmatrix} p(\rho_1) \dots p(\rho_r) \end{pmatrix}$ with $\mathcal{P} = \text{Im} P(\pi)$ with, by definition of partition, each row of $P(\pi)$ possessing only one element equal to one and all other being zero. Given two partitions π_1 and π_2, π_1 is said to be finer than π_2 ($\pi_1 \preceq \pi_2$) if all cells of π_1 are a subset of some cell of π_2 so implying $\text{Im} P(\pi_2) \subseteq \text{Im} P(\pi_1)$; equivalently, we say that π_2 is coarser than π_1 ($\pi_2 \succeq \pi_1$), with $\text{Im} P(\pi_1) \subseteq \text{Im} P(\pi_2)$. We name $\pi = \mathcal{V}$ the trivial partition as composed of a unique cell with all nodes. Given a cell $\rho \in \mathcal{V}$ and a node $\nu_i \notin \rho$, we denote by $\mathcal{N}(\nu_i, \rho) = \{\nu \in \rho \text{ s.t } (\nu, \nu_i) \in \mathcal{E}\}$ the set of neighbors of ν_i in the cell ρ.

Definition 5.2.1 A partition $\pi^\star = \{\rho_1, \rho_2, \dots, \rho_k\}$ is said to be an *almost equitable partition (AEP)* of $\mathcal{G}$ if, for each $i, j \in \{1, 2, \dots, k\}$, with $i \neq j$, there exists an integer d_{ij} such that $|\mathcal{N}(\nu_i, \rho_j)| = d_{ij}$ for all $\nu_i \in \rho_i$.

In other words, a partition such that each node in ρ_i has the same number of neighbors in ρ_j, for all i, j with $i \neq j$, is an AEP. The property of almost equitability is equivalent to the invariance of the subspaces generated by the characteristic vectors of its cells. In particular, we can give the following equivalent characterization of an AEP $\pi^\star$ [12, 13].

Proposition 5.2.2 *Consider a graph $\mathcal{G}$ and a partition $\pi^\star = \{\rho_1, \rho_2, \dots, \rho_k\}$ with $\mathcal{P}^\star = \text{Im} P(\pi^\star)$. $\pi^\star$ is an* almost equitable partition (AEP) *if and only if $\mathcal{P}^\star$ is L-invariant, that is,*

$$L\mathcal{P}^\star \subseteq \mathcal{P}^\star. \tag{5.4}$$

We say that a non-trivial partition $\pi^\star$ is the coarsest AEP of $\mathcal{G}$ if $\pi^\star \succeq \pi$ for all non-trivial π AEP of $\mathcal{G}$ and, equivalently, $\text{Im} P(\pi^\star) \subseteq \text{Im} P(\pi)$. Algorithms for computing almost equitable partitions are available for arbitrary unweighted digraphs such as, among several others, the one in [13].

5.2.2 Multi-consensus of Continuous-Time Integrators

As proved in [12], when considering multi-agent systems the notion of AEP is linked to the characterization of *multi-consensus*. Roughly speaking, consider a set of $N > 1$ of scalar integrators of the form

$$\dot{x}_i = u_i$$

with $x_i \in \mathbb{R}$ and continuously exchanging information based on a communication graph $\mathcal{G} = (\mathcal{V}, \mathcal{E})$ whose vertices $\nu_i \in \mathcal{V}$ correspond to the ith agent with state x_i $(i = 1, \ldots, N)$. Then, under the coupling rule

$$u_i = -\sum_{\ell:\nu_\ell \in \mathcal{N}(\nu_i)} (x_i - x_\ell)$$

nodes asymptotically cluster into $r = \mu + k$ consensuses for some $k \in \mathbb{N}$ such that $1 \le k \le c$; those clusters are uniquely defined by the coarsest AEP $\pi^\star = \{\rho_1, \ldots, \rho_r\}$ of $\mathcal{G}$: the states of all agents belonging to the same cell of the AEP converge to the same consensus state. More in detail, with a slight abuse of notation and exploiting (5.1), the AEP underlying the multi-consensus of the network is $\pi^\star = \{\mathcal{H}_1, \ldots, \mathcal{H}_\mu, \mathcal{C}_{\mu+1}, \ldots, \mathcal{C}_{\mu+k}\}$ with $\mathcal{C} = \cup_{\ell=1}^{k} \mathcal{C}_{\mu+\ell}$ and all nodes in $\mathcal{C}_{\mu+\ell}$ (of cardinality $c_\ell = |\mathcal{C}_{\mu+\ell}|$) sharing the same components of the vectors γ^i for all $i = 1, \ldots, \mu$. More precisely, setting in (5.2)

$$\gamma^i = \begin{pmatrix} \gamma_1^i \\ \vdots \\ \gamma_c^i \end{pmatrix} \in \mathbb{R}^c, \quad i = 1, \ldots, \mu \tag{5.5}$$

a node $\nu_{N-c+m_1} \in \mathcal{C}$ (with $m_1 \in \{1, \ldots, c\}$) belongs to the cell $\mathcal{C}_{\mu+\ell} \subseteq \mathcal{C}$ if and only if $\gamma_{m_1}^i = \gamma_{m_2}^i$ for all $\nu_{N-c+m_2} \in \mathcal{C}_{\mu+\ell}$, $i = 1, \ldots, \mu$ and some $m_2 \in \{1, \ldots, c\}$. Accordingly, the number of cells partitioning $\mathcal{C}$ (i.e., the integer k) coincides with the number of distinct coefficients of the vector γ^i with $i = 1, \ldots, \mu$.

Consider now the agglomerate network dynamics

$$\dot{x} = -Lx$$

with $x = (x_1 \ \ldots x_N)$, $x_i \in \mathbb{R}$, L the communication graph Laplacian, and the sub-states when $c_0 = h_0 = 0$

$$\begin{aligned} \boldsymbol{x}_i &= \mathrm{col}\{x_{h_1+\cdots+h_{i-1}+1}, \ldots, x_{h_1+\cdots+h_i}\} \in \mathbb{R}^{h_i} \\ \boldsymbol{x}_\ell &= \mathrm{col}\{x_{N-c+c_1+\cdots+c_{\ell-1}}, \ldots, x_{N-c+c_1+\cdots+c_\ell}\} \in \mathbb{R}^{c_\ell} \end{aligned}$$

for $i = 1, \ldots, \mu$ and $\ell = 1, \ldots, k$. One can further rewrite each $\gamma^i \in \mathbb{R}^c$ in (5.2) as

$$\gamma^i = \begin{pmatrix} \gamma_1^i \mathbb{1}_{c_1} \\ \vdots \\ \gamma_k^i \mathbb{1}_{c_k} \end{pmatrix} \in \mathbb{R}^c, \quad \gamma_\ell^i \in \mathbb{R}, \quad \ell = 1, \ldots, k \tag{5.6}$$

with c_ℓ the cardinality of the cell $\mathcal{C}_{\mu+\ell} \subseteq \mathcal{C}$ (i.e., $c_\ell = |\mathcal{C}_{\mu+\ell}|$) so that, as $t \to \infty$, the following holds true:

1. nodes in the same reach $\mathcal{H}_i$ converge to the same consensus value; namely, for $i = 1, \ldots, \mu$ and all $\nu_j \in \mathcal{H}_i$

$$\boldsymbol{x}_i(t) \to x_i^{ss}\mathbb{1}_{h_i}, \quad x_i^{ss} := v_i^\top \boldsymbol{x}_i(0) \in \mathbb{R}$$

   with $v_i^\top \in \mathbb{R}^{h_i}$ being the corresponding component of the left eigenvector (5.3);
2. nodes in $\mathcal{C}$ belonging to the same cell $\mathcal{C}_{\mu+\ell}$ converge to a convex combination of the consensuses induced by the reaches; namely,

$$\boldsymbol{x}_\ell(t) \to x_{c_\ell}^{ss}\mathbb{1}_{c_\ell}, \quad x_{c_\ell}^{ss} := \sum_{i=1}^{\mu} \gamma_\ell^i x_i^{ss}$$

   and $\gamma_\ell^i \in \mathbb{R}$ the distinct components of the vector $\gamma^i \in \mathbb{R}^c$ in (5.6), for $\ell = 1, \ldots, k$.

5.3 Hybrid Multi-agent Systems

We consider a group of $N \in \mathbb{N}$ identical agents, whose state is assumed to be a scalar $x_i \in \mathbb{R}$ for any $i = 1, ..., N$. The evolution of each agent state is assumed to be governed by a hybrid dynamics, i.e., characterized by the interplay of a continuous-time and a discrete-time behavior [8]. The alternate selection of continuous and discrete dynamics can be either driven by specific time patterns or triggered by conditions on the state. In this paper, we consider agents whose dynamics is given by a hybrid integrator with time-driven jumps. In particular, the state of each agent is assumed to obey the following update law:

$$\dot{x}_i(t) = Ax_i(t) + Bu_i(t), \quad t \in \mathbb{R}^+ \setminus \mathcal{J} \tag{5.7a}$$
$$x_i^+(t) = Ex_i(t) + Fv_i(t), \quad t \in \mathcal{J} \tag{5.7b}$$

where $\mathcal{J}$ denotes the sequence of jump instants

$$\mathcal{J} = \{t_j \in \mathbb{R}^+, j = 1, ..., \aleph_{\mathcal{J}} : t_j < t_{j+1}, \ \aleph_{\mathcal{J}} \in \mathbb{N} \cup \infty\}$$

and where $u_i(t), v_i(t) \in \mathbb{R}$ are control inputs and

$$\tau_{\min} < t_{j+1} - t_j < \tau_{\max} \qquad \forall j \in \mathcal{J}. \tag{5.8}$$

The differential equation in (5.7) is referred to as *flow dynamics*, whereas the difference equation corresponds to the *jump dynamics*. To keep track of the jumps, it is convenient to introduce the notion of hybrid time domain as a special case of [8, Definition 2.3].

Definition 5.3.1 A hybrid time domain is a set $\mathcal{T}$ in $[0, \infty) \times \mathbb{N}$ defined as the union of *indexed intervals*

$$\mathcal{T} := \bigcup_{j \in \mathbb{N}} \{ [t_j, t_{j+1}] \times \{j\}\} \tag{5.9}$$

Given a hybrid time domain, its length is defined as $length(\mathcal{T}) = \sup_t \mathcal{T} + \sup_j \mathcal{T}$. A hybrid time domain is said τ-periodic, for some constant $\tau > 0$, if $t_{j+1} - t_j = \tau$ for any $j \in \mathbb{N}$.

In view of the latter definition, we can enhance the notation for the state of the agents as follows:

$$x_i(t, j) \text{ with } (t, j) \in \mathcal{T}, \ i = 1, ..., N.$$

As previously mentioned, the agents are supposed to be connected through a suitable communication graph. However, due to the hybrid nature of the problem, it is reasonable to allow interactions among the agents with different topologies for the flow dynamics and the jump dynamics. To this end, let us consider a flow graph $\mathcal{G}_\mathrm{f} = (\mathcal{V}_\mathrm{f}, \mathcal{E}_\mathrm{f})$ and a jump graph $\mathcal{G}_\mathrm{j} = (\mathcal{V}_\mathrm{j}, \mathcal{E}_\mathrm{j})$, where the sets of vertices satisfy

$$\mathcal{V} := \mathcal{V}_\mathrm{f} = \mathcal{V}_\mathrm{j}$$

as the number of agents remains constant upon jumps. Accordingly, we aim at designing the control inputs in a decentralized way as functions of the states of agent i and its neighbors, i.e.,

$$\begin{aligned} u_i &= f_i(x_i, \{x_k : \nu_k \in \mathcal{N}_{\mathrm{f},i}\}) \\ v_i &= g_i(x_i, \{x_k : \nu_k \in \mathcal{N}_{\mathrm{j},i}\}) \end{aligned}$$

where the set of indexes

$$\mathcal{N}_{\mathrm{f},i} = \{\nu_k \in \mathcal{V} : (\nu_k, \nu_i) \in \mathcal{E}_\mathrm{f}\}, \quad \mathcal{N}_{\mathrm{j},i} = \{\nu_k \in \mathcal{V} : (\nu_k, \nu_i) \in \mathcal{E}_\mathrm{j}\}$$

are, respectively, the flow and the jump neighborhoods of the agent i.

In the following, we will investigate two problems. At first, we address the hybrid network consensus problem that is, when assuming each agent (5.7) as a hybrid integrator

$$\begin{aligned} \dot{x}_i(t) &= u_i(t), & t \in \mathbb{R}^+ \setminus \mathcal{J} \\ x_i^+(t) &= x_i(t) + v_i(t), & t \in \mathcal{J} \end{aligned} \quad i = 1, ..., N \tag{5.10}$$

evolving over a hybrid topology. Then, we consider the hybrid dynamics consensus problem for each general agent (5.7) sharing the same topology during flow periods and jump instants (i.e., $\mathcal{G} = \mathcal{G}_\mathrm{f} = \mathcal{G}_\mathrm{f}$).

5.4 The Hybrid Network Consensus Problem

It is well known that, in classical multi-agent systems, consensus conditions can be attained by implementing simple linear feedback laws, which can be encoded through the Laplacian matrix of the communication graph. We propose a similar approach here, by setting

$$\begin{aligned} u_i(t) &= -\sum\nolimits_{k:\nu_k \in \mathcal{N}_{\mathrm{f},i}} (x_i(t) - x_k(t)) \\ v_i(t) &= -\alpha \sum\nolimits_{k:\nu_k \in \mathcal{N}_{\mathrm{j},i}} (x_i(t) - x_k(t)) \end{aligned} \tag{5.11}$$

where $\alpha > 0$ is a suitable gain to be tuned. Defining the cumulative state $x = [x_1\ x_2\ \cdots\ x_N]^\top$, the dynamics of the multi-agent system driven by (5.11) can be written as

$$\begin{aligned} \dot{x}(t) &= -L_{\mathrm{f}} x(t) & t \in \mathbb{R}^+ \setminus \mathcal{J} \\ x^+(t) &= (I - \alpha L_{\mathrm{j}}) x(t) & t \in \mathcal{J} \end{aligned} \tag{5.12}$$

where L_{f} and L_{j} denote, respectively, the Laplacian matrix of the flow and the jump graph. It is worth noticing that, whilst the negative flow Laplacian $-L_{\mathrm{f}}$ provides a marginally stable continuous dynamics, the jump map has to be made stable by means of a proper tuning of $\alpha > 0$. To this end, we can invoke the following technical result.

Proposition 5.4.1 *Let $\mathcal{G} = (\mathcal{V}, \mathcal{D})$ be a graph with $\bar{N}$ vertices, and let $L \in \mathbb{R}^{\bar{N} \times \bar{N}}$ be its Laplacian matrix. There exists $\alpha > 0$ such that the spectrum of $(I - \alpha L)$ satisfy*

$$\sigma(I - \alpha L) \subset \{z \in \mathbb{C} : |z| < 1\} \cup \{1\} \tag{5.13}$$

$$1 \in \sigma(I - \alpha L) \tag{5.14}$$

Our goal is to establish conditions on the combination of graphs $\mathcal{G}_{\mathrm{f}}, \mathcal{G}_{\mathrm{j}}$ under which a consensus is achieved in the multi-agent system (5.25), and to characterize such consensus.

We now provide the main results to characterize the consensus behaviors of the hybrid network and the corresponding clusters.

5.4.1 *The Multi-consensus Clusters*

In the following result, the multi-consensus clusters arising from (5.25) under the hybrid connection in (5.11) are characterized based on a suitably defined AEP for both the components of the topology.

Theorem 5.4.2 *Consider the hybrid multi-agent system* (5.25) *and let $\mathcal{G}_{\mathrm{f}}$ and $\mathcal{G}_{\mathrm{j}}$ be the flow and jump graphs with Laplacians L_{f} and L_{j}. Then, the multi-consensuses of* (5.25) *are induced by the coarsest almost equitable partition $\pi^\star_{\mathcal{H}}$ for $\mathcal{G}_{\mathrm{f}}$ and $\mathcal{G}_{\mathrm{j}}$ simultaneously.*

Remark 5.4.3 Denote by $\pi_{\mathrm{f}}^{\star}$ and $\pi_{\mathrm{j}}^{\star}$ the coarsest AEPs for, respectively, $\mathcal{G}_{\mathrm{f}}$ and $\mathcal{G}_{\mathrm{j}}$ verifying

$$L_q \mathcal{P}_q^{\star} \subseteq \mathcal{P}_q^{\star}$$

with $\mathcal{P}_q^{\star} = \mathrm{Im} P(\pi_q^{\star}) \equiv \ker L_q$ being the (invariant) eigenspace associated with the non-zero eigenvalues of L_q for $q \in \{\mathrm{f}, \mathrm{j}\}$. Because $\pi_{\mathcal{H}}^{\star}$ is the coarsest AEP for both $\mathcal{G}_{\mathrm{f}}$ and $\mathcal{G}_{\mathrm{j}}$, then the subspace

$$\mathcal{P}_{\mathcal{H}}^{\star} = \mathrm{Im} P(\pi_{\mathcal{H}}^{\star}) \tag{5.15}$$

is L_q-invariant for $q \in \{\mathrm{f}, \mathrm{j}\}$. As a consequence, one necessarily gets $\mathcal{P}_{\mathcal{H}}^{\star} \subseteq \mathcal{P}_{\mathrm{f}}^{\star}$ and $\mathcal{P}_{\mathcal{H}}^{\star} \subseteq \mathcal{P}_{\mathrm{j}}^{\star}$ that implies $\pi_{\mathrm{f}}^{\star} \preceq \pi_{\mathcal{H}}^{\star}$ and $\pi_{\mathrm{j}}^{\star} \preceq \pi_{\mathcal{H}}^{\star}$.

Remark 5.4.4 In the proof of the result above, we have intentionally defined the consensus of (5.25) as a *shared evolution* rather than a set of common asymptotic values the nodes tend to.

At this point, we show that $\pi_{\mathcal{H}}^{\star}$ can be constructed starting from the union and the intersection graphs, respectively, associated with the flow and jump graphs. To this end, denoting by $\mathcal{G}_{\mathrm{un}} = \mathcal{G}_{\mathrm{f}} \cup \mathcal{G}_{\mathrm{j}}$ and $\mathcal{G}_{\mathrm{int}} = \mathcal{G}_{\mathrm{f}} \cap \mathcal{G}_{\mathrm{j}}$, it is always possible [13] to sort nodes of the hybrid network in such a way the Laplacian L_{un} associated with $\mathcal{G}_{\mathrm{un}}$ gets the form

$$L_{\mathrm{un}} = \begin{pmatrix} L_{\mathrm{un},1} & \dots & 0 & 0 \\ \vdots & \ddots & \vdots & \vdots \\ 0 & \dots & L_{\mathrm{un},\mu} & 0 \\ M_{\mathrm{un},1} & \dots & M_{\mathrm{un},\mu} & M_{\mathrm{un}} \end{pmatrix} \tag{5.16}$$

where $L_{\mathrm{un},i}$ are the Laplacians associated with the exclusive reaches $\mathcal{H}_{\mathrm{un},i}$ of $\mathcal{G}_{\mathrm{un}}$ and M_{un} the nonsingular matrix associated with the common $\mathcal{C}_{\mathrm{un}}$ of $\mathcal{G}_{\mathrm{un}}$. As a consequence, exploiting the relation [10]

$$L_{\mathrm{un}} = L_{\mathrm{f}} + L_{\mathrm{j}} - L_{\mathrm{int}} \tag{5.17}$$

one gets for $q \in \{\mathrm{int}, \mathrm{j}, f\}$

$$L_q = \begin{pmatrix} L_{q,1} & \dots & 0 & 0 \\ \vdots & \ddots & \vdots & \vdots \\ 0 & \dots & L_{q,\mu} & 0 \\ M_{q,1} & \dots & M_{q,\mu} & M_q \end{pmatrix}. \tag{5.18}$$

Starting from this, the following result can be proved.

Theorem 5.4.5 *Consider the hybrid multi-agent system* (5.25) *and let* $\mathcal{G}_{\mathrm{f}}$ *and* $\mathcal{G}_{\mathrm{j}}$ *be the flow and jump graphs with Laplacians* L_{f} *and* L_{j}. *Let* $\mathcal{H}_{\mathrm{un},1}, \dots, \mathcal{H}_{\mathrm{un},\mu}$ *and* $\mathcal{C}_{\mathrm{un}}$

be, respectively, the exclusive reaches and the common part of the union graph $\mathcal{G}_{\mathrm{un}} = \mathcal{G}_{\mathrm{f}} \cup \mathcal{G}_{\mathrm{j}}$ *with* $\mathcal{H}_{\mathrm{un},i} \cap \mathcal{C}_{\mathrm{un}} = \emptyset$ *and* $\mathcal{H}_{\mathrm{un},i} \cap \mathcal{H}_{\mathrm{un},j} = \emptyset$ *for all* $i \neq j \in \{1, \ldots, \mu\}$. *Let* $\pi_{\mathcal{C}} = \{\mathcal{C}_{\mu+1}, \ldots, \mathcal{C}_{\mu+k}\}$ *be the coarsest partition of* $\mathcal{C}_{\mathrm{un}}$ *with* $\mathcal{C}_{\mathrm{un}} = \cup_{\ell=1}^{k} \mathcal{C}_{\mu+\ell}$, $c_\ell = |\mathcal{C}_{\mu+\ell}|$ *for* $\ell = 1, \ldots, k$ *and characteristic matrix*

$$\begin{pmatrix} p(\mathcal{C}_{\mu+1}) & \ldots & p(\mathcal{C}_{\mu+k}) \end{pmatrix} = \begin{pmatrix} 0 & \ldots & 0 \\ p_c(\mathcal{C}_{\mu+1}) & \ldots & p_c(\mathcal{C}_{\mu+k}) \end{pmatrix}$$
$$\mathcal{P}_c = Im P_c, \quad P_c = \begin{pmatrix} p_c(\mathcal{C}_{\mu+1}) & \ldots & p_c(\mathcal{C}_{\mu+k}) \end{pmatrix}$$

verifying $M_{\mathrm{int}} \mathcal{P}_c \subseteq \mathcal{P}_c$ *with* $M_{\mathrm{int}} \in \mathbb{R}^{c \times c}$ *as in (5.18). Then, the coarsest almost equitable partition for* $\mathcal{G}_{\mathrm{f}}$ *and* $\mathcal{G}_{\mathrm{j}}$ *is provided by*

$$\pi^\star_{\mathcal{H}} = \pi^\star_{\mathrm{un}} := \{\mathcal{H}_{\mathrm{un},1}, \ldots, \mathcal{H}_{\mathrm{un},\mu}, \mathcal{C}_{\mu+1}, \ldots, \mathcal{C}_{\mu+k}\}. \tag{5.19}$$

Remark 5.4.6 For the hybrid network, multi-consensus is fixed by both the union and intersection graphs $\mathcal{G}_{\mathrm{un}}$ and $\mathcal{G}_{\mathrm{int}}$. As a matter of fact, cells in $\pi^\star_{\mathcal{H}}$ are given by the exclusive reaches of $\mathcal{G}_{\mathrm{un}}$ plus a further partition of nodes in $\mathcal{C}_{\mathrm{un}}$ suitably partitioned so to guarantee M_{int}-invariance of the corresponding characteristic matrix. In other words, once the first μ cells of $\pi^\star_{\mathcal{H}}$ are fixed by the reaches of the union graph, the remaining ones are defined by looking at the subgraph of $\mathcal{G}_{\mathrm{int}}$ arising from M_{int}.

Remark 5.4.7 Whenever $M_{\mathrm{int}} = 0$, multi-consensus is completely fixed by the union graph; namely, cells $\mathcal{C}_{\mu+\ell}$ are such that they provide the coarsest AEP associated with $\mathcal{G}_{\mathrm{un}}$.

Remark 5.4.8 Whenever the union graph possesses no common part (i.e., $\mathcal{C}_{\mathrm{un}} = \emptyset$), the hybrid network exhibits exactly μ consensuses behaviors. As a consequence, if $\mathcal{G}_{\mathrm{un}}$ possesses one consensus only, then the hybrid network converges to a single consensus independently of the clusters of $\mathcal{G}_{\mathrm{f}}$ and $\mathcal{G}_{\mathrm{j}}$.

5.4.2 Convergence Analysis

Let us now discuss the problem of convergence of solutions of (5.25) to the multi-consensus subspace.

First, the following technical result is presented [5].

Proposition 5.4.9 *Let* L_{f}, L_{j} *be the Laplacian matrices defining the hybrid multi-agent dynamics (5.12). Pick* $\tau > 0$ *and define the reverse monodromy matrix*

$$H = e^{-L_{\mathrm{f}}\tau}(I - \alpha L_{\mathrm{j}}) \tag{5.20}$$

where $\alpha > 0$ *is any positive parameter satisfying the conditions:*

$$\alpha < \min_{\lambda \in \sigma(L_{\mathrm{j}}) \setminus \{0\}} \frac{2\mathrm{Re}(\lambda)}{|\lambda|^2} \tag{5.21}$$

$$\alpha < \min_{\lambda \in \sigma(L_{\mathrm{j}}) \setminus \{0\}} \frac{1}{\mathrm{Re}(\lambda)} \tag{5.22}$$

Then the spectrum of H verifies

$$\sigma(H) \subset \mathbb{C}_{\leq 1} := \{z \in \mathbb{C} : |z| < 1\} \cup \{1\}.$$

Let us denote with $\mathcal{K} = \mathcal{P}^{\star}_{\mathcal{H}} \subset \mathbb{R}^N$ the multi-consensus subspace, this being in particular a center invariant subspace [9] for the monodromy matrix, associated with the eigenvalue 1.

Now, by defining the augmented state $\zeta = [\tilde{x}^\top \ j \ s]^\top$ and considering the flow and jump sets

$$\begin{aligned} \mathcal{C} &= \{\zeta = (x, j, s) : x \in \mathbb{R}^N, j \in \mathbb{N}, s \in [0, t_{j+1} - t_j]\} \\ \mathcal{D} &= \{\zeta = (x, j, s) : x \in \mathbb{R}^N, j \in \mathbb{N}, s = t_{j+1} - t_j\} \end{aligned}$$

the hybrid-multi-agent system (5.25) can be rewritten using a hybrid system formalism as

$$\begin{aligned} \dot{\zeta} &= F(\zeta) \quad \zeta \in \mathcal{C} \\ \zeta^+ &= G(\zeta) \; \zeta \in \mathcal{D} \end{aligned} \tag{5.23}$$

where the flow and jump maps are defined, respectively, by

$$F(\zeta) = \begin{bmatrix} -L_{\mathrm{f}} x \\ 0 \\ 1 \end{bmatrix} \quad G(\zeta) = \begin{bmatrix} (I - \alpha L_{\mathrm{j}})x \\ j+1 \\ 0 \end{bmatrix}$$

where we implicitly assume $t_0 = 0$. It is worth stressing that both $\mathcal{C}$ and $\mathcal{D}$ are closed sets thanks to the existence of lower and upper bounds for the dwell time (i.e., $\tau_{\min} < t_{j+1} - t_j < \tau_{\max}$). Now, based on Proposition 5.4.9 and the reformulation (5.23), the following stability result can be proved (see [5]).

Theorem 5.4.10 *The multi-consensus subspace $\mathcal{K}$ is globally asymptotically stable in the hybrid sense [8, Definition 3.6].*

5.5 The Hybrid Dynamics Consensus Problem

We focus now on the complementary setting. At this stage, let us consider a network of LTI agents (5.7) connected through a suitable communication graph $\mathcal{G} = \mathcal{G}_{\mathrm{f}} = \mathcal{G}_{\mathrm{j}} = (\mathcal{V}, \mathcal{E})$ and Laplacian L assumed, with no loss of generality, of the form (5.1). We aim at designing the control inputs in a decentralized way of the form

$$u_i = -\kappa \sum_{k \in \mathcal{N}_i} K(x_i - x_k) \tag{5.24a}$$

$$v_i = -\alpha \sum_{k \in \mathcal{N}_i} H(x_i - x_k) \tag{5.24b}$$

where $\kappa,\ \alpha > 0$ and $K, H \in \mathbb{R}^{m \times n}$ are suitable scalar and matrix gains (referred to as coupling gains and coupling matrices).

Defining the agglomerate states, for $i = 1, \ldots, \mu$, as

$$\begin{aligned}
\boldsymbol{x}_i &= \mathrm{col}\{x_j\}_{j \in \mathcal{H}_i} \in \mathbb{R}^{h_i n},\ \boldsymbol{x}_{\mathcal{C}} = \mathrm{col}\{x_j\}_{j \in \mathcal{C}} \in \mathbb{R}^{cn} \\
\boldsymbol{u}_i &= \mathrm{col}\{u_j\}_{j \in \mathcal{H}_i} \in \mathbb{R}^{h_i m},\ \boldsymbol{u}_{\mathcal{C}} = \mathrm{col}\{u_j\}_{j \in \mathcal{C}} \in \mathbb{R}^{cm} \\
\boldsymbol{v}_i &= \mathrm{col}\{v_j\}_{j \in \mathcal{H}_i} \in \mathbb{R}^{h_i m},\ \boldsymbol{v}_{\mathcal{C}} = \mathrm{col}\{v_j\}_{j \in \mathcal{C}} \in \mathbb{R}^{cm}
\end{aligned}$$

and assuming L of the form (5.1), one can rewrite

$$\begin{aligned}
\boldsymbol{u}_i &= -\kappa(L_i \otimes K)\boldsymbol{x}_i,\ \boldsymbol{v}_i = -\alpha(L_i \otimes H)\boldsymbol{x}_i \\
\boldsymbol{u}_{\mathcal{C}} &= -\kappa \sum_{i=1}^{\mu} (M_i \otimes K)\boldsymbol{x}_i - \kappa(M \otimes K)\boldsymbol{x}_{\mathcal{C}} \\
\boldsymbol{v}_{\mathcal{C}} &= -\alpha \sum_{i=1}^{\mu} (M_i \otimes H)\boldsymbol{x}_i - \alpha(M \otimes H)\boldsymbol{x}_{\mathcal{C}}
\end{aligned}$$

so that the overall network dynamics reads

$$\dot{\boldsymbol{x}}_i = \Big(\big(I_{h_i} \otimes A\big) - \kappa\big(L_i \otimes BK\big)\Big)\boldsymbol{x}_i \tag{5.25a}$$

$$\boldsymbol{x}_i^+ = \Big(\big(I_{h_i} \otimes E\big) - \alpha\big(L_i \otimes FH\big)\Big)\boldsymbol{x}_i \tag{5.25b}$$

$$\dot{\boldsymbol{x}}_{\mathcal{C}} = -\kappa \sum_{i=1}^{\mu} \Big(M_i \otimes BK\Big)\boldsymbol{x}_i + \Big(\big(I_c \otimes A\big) - \kappa\big(M \otimes BK\big)\Big)\boldsymbol{x}_{\mathcal{C}} \tag{5.25c}$$

$$\boldsymbol{x}_{\mathcal{C}}^+ = -\alpha \sum_{i=1}^{\mu} \Big(M_i \otimes FH\Big)\boldsymbol{x}_i + \Big(\big(I_c \otimes E\big) - \alpha\big(M \otimes FH\big)\Big)\boldsymbol{x}_{\mathcal{C}} \tag{5.25d}$$

We investigate the asymptotic cluster properties of the dynamics (5.25) induced by the network interconnection and design the coupling gains and matrices in (5.24) so that nodes asymptotically converge to suitable multi-consensus trajectories.

Remark 5.5.1 In the following, for the sake of simplicity and without loss of generality, we assume $\sigma\{L\} \subset \mathbb{R}_{\geq 0}$. All results to come hold true when considering the real part $\nu \in \mathbb{R}_{\geq 0}$ (flow dynamics) or the modulus $\sqrt{\nu^2 + \omega^2}$ (jump dynamics) of any eigenvalue $\lambda = \nu \pm j\omega \in \sigma\{L\}$ associated to a Jordan block

$$J_\lambda = \begin{pmatrix} \nu & \omega \\ -\omega & \nu \end{pmatrix} \in \mathbb{R}^{2\times 2}.$$

5.5.1 The Hybrid Multi-consensus Dynamics

For the sake of clarity, we first characterize the multi-consensus rising in the hybrid network (5.25) independently on both the sequence of the jump instants and the flow period. Then, we provide a constructive design for the inputs (5.24) making it attractive when assuming, for the sake of simplicity, the case of periodic jumps. In that case, the solution provided for the hybrid network consensus problem coincides with the one we are presenting here.

The next results highlight the consensus dynamics over each exclusive reach via the definition of the so-called mean-field dynamics [11, 15]. In doing so, we extend the results in [9, Chap. 4] for (single) consensus of continuous-time systems to the hybrid multi-consensus case.

Proposition 5.5.2 *Consider the multi-agent system composed of N units evolving as* (5.7) *over communication digraph $\mathcal{G}$ with Laplacian of the form* (5.1). *Let the right and left eigenvectors associated to the zero eigenvalue of L be of the form* (5.2)–(5.3) *with $w_i^\top \mathbb{1}_{h_i} = h_i$ for $i = 1, \ldots, \mu$. Denote by $\sigma\{L_i\} = \{0, \lambda_i^1, \ldots, \lambda_i^{h_i-1}\} \subset \mathbb{C}^+$. Then, all nodes in $\mathcal{H}_i$ $(i = 1, \ldots, \mu)$ converge to a common consensus provided that the hybrid dynamics*

$$\dot{\varepsilon}_{i,\ell} = (A - \kappa\lambda_i^\ell BK)\varepsilon_{i,\ell} \tag{5.26a}$$
$$\varepsilon_{i,\ell}^+ = (E - \alpha\lambda_i^\ell FH)\varepsilon_{i,\ell} \tag{5.26b}$$

are asymptotically stable for all $\lambda_i^\ell \in \sigma\{L_i\} \setminus \{0\}$, $\ell = 1, \ldots, h_i - 1$ and $i = 1, \ldots, \mu$; i.e., $\boldsymbol{x}_i \to \mathbb{1}_{h_i} \otimes x_{s,i}$ with

$$x_{s,i} = \left(w_i^\top \otimes I_n\right)\boldsymbol{x}_i \in \mathbb{R}^n. \tag{5.27}$$

the mean-field unit evolving with mean-field dynamics

$$\dot{x}_{s,i} = Ax_{s,i} \tag{5.28a}$$
$$x_{s,i}^+ = Ex_{s,i}. \tag{5.28b}$$

We can now characterize the consensus arising in the whole network with necessary and sufficient conditions making it asymptotically stable for all nodes. We first show that, as in the continuous-time case [12], nodes in the common regroup into as many clusters as the number (say $p \geq 1$) of the distinct components of the vectors γ^i in (5.5). To this end, let us assume nodes in $\mathcal{C}$ are sorted in such a way that all vectors γ^i in (5.5) get the form (5.6) with $\gamma^i_r \in \mathbb{R}$ for $r = 1, \dots, p$ and verifying $\sum_{r=1}^{p} \gamma^i_r = 1$ and $\sum_{r=1}^{p} c_i = c = |\mathcal{C}|$. In this way, one can denote by $\mathcal{C}_{\mu+r} \subseteq \mathcal{C}$ with $c_p = |\mathcal{C}_{\mu+r}|$ the cell composed of all nodes $j \in \mathcal{C}$ corresponding to the same component γ^i_r of the vector (5.6). Consequently, we denote $\boldsymbol{x}_{\mathcal{C}} = \{\boldsymbol{x}_{\mu+1}, \dots, \boldsymbol{x}_{\mu+p}\}$ with

$$\boldsymbol{x}_{\mu+r} = \text{col}\{x_j\}_{j\in\mathcal{C}_{\mu+r}} \in \mathbb{R}^{c_i n}.$$

Remark 5.5.3 We note that $\mathcal{C} = \cup_{r=1}^{p} \mathcal{C}_{\mu+r}$ with, for all $r_1, r_2 = 1, \dots, p$ and $r_1 \neq r_2$, $\mathcal{C}_{\mu+r_1} \cap \mathcal{C}_{\mu+r_2} = \emptyset$. In addition, as proved in [12], the coarsest nontrivial AEP of a digraph $\mathcal{G}$ is

$$\pi^\star = \{\mathcal{H}_1, \dots, \mathcal{H}_\mu, \mathcal{C}_{\mu+1}, \dots, \mathcal{C}_{\mu+p}\} \tag{5.29}$$

At this point, the following result can be given proving that, under suitable conditions on the coupling gains and matrices, the network induces as many consensuses as the number of cells of the AEP (5.29) associated to $\mathcal{G}$.

Proposition 5.5.4 *Consider the multi-agent system composed of N units evolving as* (5.7) *under the hypotheses of Proposition 5.5.2 and consider $\pi^\star$ as in* (5.29) *be an AEP for $\mathcal{G}$. Let* (5.3) *be the right and left eigenvectors associated to the zero eigenvalue of L verifying $w_i^\top \mathbb{1}_{h_i} = h_i$ with each $\gamma^i \in \mathbb{R}^c$ of the form* (5.6). *Then, the trajectories of the nodes belonging to $\mathcal{C}_{\mu+r} \subseteq \mathcal{C}$ converge to the consensus trajectory*

$$x_{s,\mu+r} = \sum_{i=1}^{\mu} \gamma^i_r x_{s,i} \in \mathbb{R}^n \tag{5.30}$$

for $r = 1, \dots, p$ and $x_{s,i}$ as in (5.27) *if and only if the dynamics*

$$\dot{\varepsilon}_{\mathcal{C},q} =(A - \kappa\lambda^q_M BK)\varepsilon_{\mathcal{C},q} \tag{5.31a}$$

$$\varepsilon^+_{\mathcal{C},q} =(E - \alpha\lambda^q_M FH)\varepsilon_{\mathcal{C},q} \tag{5.31b}$$

are asymptotically stable for all $\lambda^q_M \in \sigma\{M\}$, $q = 1, \dots, c$.

Remark 5.5.5 By the result above, the subspace

$$\mathcal{P}^\star = \text{Im}\{P(\pi^\star) \otimes I_n\} = \ker\{L \otimes I_n\}$$

associated to the AEP (5.29) defines the consensus subspace of the hybrid network (5.25).

Remark 5.5.6 It can be proved that, by Propositions 5.5.2 and 5.5.4, asymptotic stability of the uncontrolled hybrid agents (5.7) ensure convergence to an asymptotically stable consensus trajectory. However, as the intuition suggests, the reverse is not true: when gains are suitably designed, unstable agents converge to a consensus trajectory which is, thus, unbounded.

5.5.2 *The Hybrid Coupling Design*

By Propositions 5.5.2 and 5.5.4, the hybrid network induces $\mu + k$ consensus trajectories: $\mu \geq 1$ independent consensuses arise over each reach $\mathcal{H}_i$; $p \geq 1$ further consensuses are exhibited over the common $\mathcal{C}$. Convergence to such behaviors are guaranteed provided that the coupling gains κ, α and coupling matrices K, H in (5.24) are chosen so to make all dynamics (5.26)–(5.31) asymptotically stable. Those conditions strictly depend on the non-zero eigenvalues of the Laplacian L so that a qualitative design assigning the eigenvalues for all possible $\lambda \in \sigma\{L\}$ might not be feasible in a decentralized way. In the sequel, constructive conditions over (5.24) are given to enforce this property in a robust way. To this end, let us define the quantities

$$\lambda^{\dagger} = \min_{\substack{i=1,\dots,\mu\\ \ell=1,\dots,h_i-1\\ q=1,\dots,c}} \{\lambda_i^{\ell}, \lambda_M^{q}\}$$

$$\lambda^{\circ} = \max_{\substack{i=1,\dots,\mu\\ \ell=1,\dots,h_i-1\\ q=1,\dots,c}} \{\lambda_i^{\ell}, \lambda_M^{q}\}$$

Theorem 5.5.7 *Consider the hybrid system* (5.7) *driven by a control law of the type* (5.24) *under the condition of a bounded dwell time* $\tau_{\min} \leq t_{j+1} - t_j \leq \tau_{\max}$. *Suppose that there exists a common solution* $\hat{Q} = \hat{Q}^{\top} \succ 0$ *to the continuous-time algebraic Riccati inequality*

$$A^{\top}\hat{Q} + \hat{Q}A - \chi_1\hat{Q}BB^{\top}\hat{Q} \prec -\upsilon_1 I \tag{5.32}$$

and to the discrete-time algebraic Riccati inequality

$$E^{\top}\hat{Q}E - \hat{Q} - \chi_2 E^{\top}\hat{Q}F(I + F^{\top}\hat{Q}F)^{-1}F^{\top}\hat{Q}E < -\upsilon_2 I \tag{5.33}$$

for some constants $\upsilon_1, \upsilon_2, \chi_1, > 0$ *and* $\chi_2 \in (0, 1)$ *with*

$$\frac{1 - \sqrt{1 - \chi_2}}{1 + \sqrt{1 - \chi_2}} < \frac{\lambda^{\dagger}}{\lambda^{\circ}} \tag{5.34}$$

Then, if the feedback gains in (5.24) *are selected as*

$$K = B^\top \hat{Q}, \quad H = (I + F^\top \hat{Q} F)^{-1} F^\top \hat{Q} E$$

and

$$\kappa \geq \kappa^\star := \frac{\chi_1}{2\lambda^\dagger}, \ \alpha \in \mathcal{I}_\alpha^\star := \left(\frac{1 - \sqrt{1 - \chi_2}}{\lambda^\dagger}, \frac{1 + \sqrt{1 - \chi_2}}{\lambda^\circ} \right),$$

the hybrid systems in (5.26)–(5.31) *are asymptotically stable for any* i, ℓ, q, *and therefore consensus in* (5.7) *is reached.*

Remark 5.5.8 It is important to stress that the conditions provided by Theorem 5.5.7 are only sufficient and not necessary. On the other hand, they fulfill the nice property of being independent of the particular hybrid time domain.

5.6 Application to Formation Control of a Heterogeneous Multi-robot System

In this section, we address the problem of formation control applied to a multi-robot system composed of two heterogeneous subgroups of $N = 7$ agents each. One of them is characterized by ground robots, e.g., unicycles, whereas the other one is composed of flying robots, e.g., Unmanned Aerial Vehicles (UAVs). Under a preliminary dynamic feedback linearization, each agent is assumed to be modeled as a hybrid double integrator of the form (5.7) characterized by aperiodic jumps with $\tau_{\min} = 0.1$ and $\tau_{\max} = 0.35$ (5.8). The information are exchanged following the topology of the directed graph depicted in Fig. 5.1, where the black arrows represent the edges of $\mathcal{G}_\mathrm{f}$ and the red arrows represent the connections of $\mathcal{G}_\mathrm{j}$. Moreover, if one looks exclusively at the flow graph $\mathcal{G}_\mathrm{f}$, it can be seen that each subgroup has its own leader, coinciding with the unique root of the subgraph, and it is disconnected from the other. The formation control problem aims at making N robots arrange according to a desired shape. In this case, we consider the same shape for both subgroups, where the respective leader must be in the center of the formation. The consensus is achieved only for the planar (x, y) coordinates, with the UAVs keeping a constant height different from zero.

Considering the graph in Fig. 5.1, applying the result in Theorem 5.4.5 and along the lines of Remark 5.4.8, the AEP underlying consensus is given by the trivial one composed of one cell with all nodes included

$$\pi_\mathcal{H}^\star = \{\{\nu_1, \dots, \nu_{14}\}\}$$

As a consequence, although the standalone flow graph induces three consensuses for the UAVs and two consensuses for the unicycles, with partitions

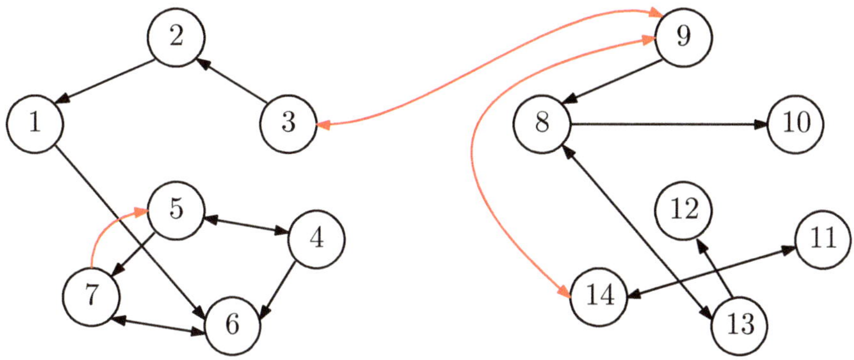

Fig. 5.1 Communication graph used in the application of formation control. Agents $i = 1, \ldots, 7$ are UAVs, while agents $i = 9, \ldots, 14$ are unicycles. The black arrows are the edges of the flow graph, whereas the red arrows represent the connections of the jump graph

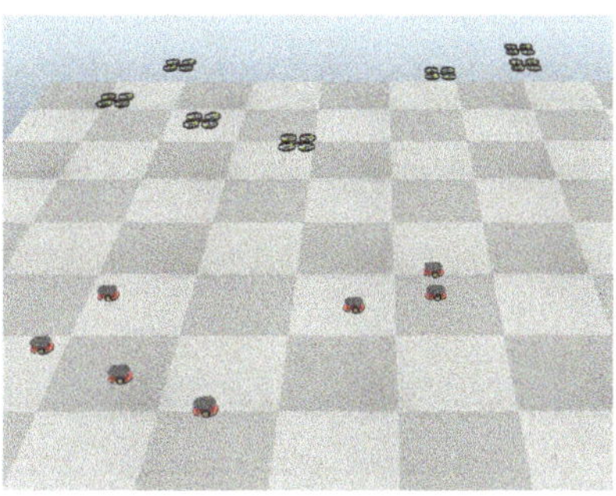

(a) Initial configuration

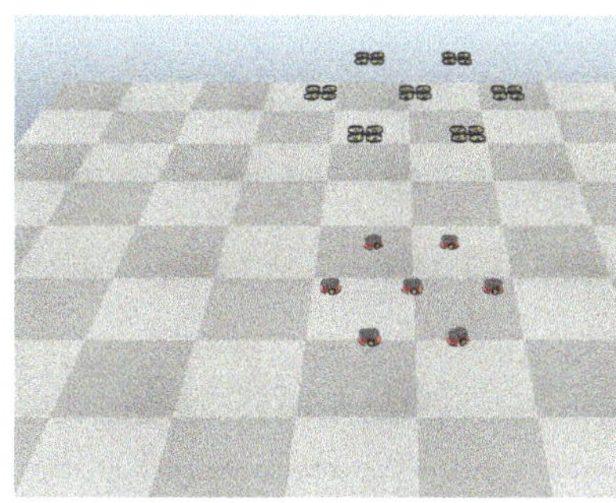

(b) Final configuration

Fig. 5.2 Formation control: CoppeliaSim simulation

$$\begin{aligned}\pi^{\star}_{\text{uav}} &= \{\{\nu_1, \nu_2, \nu_3\}, \{\nu_4, \nu_5\}, \{\nu_6, \nu_7\}\} \\ \pi^{\star}_{\text{uni}} &= \{\{\nu_{11}, \nu_{14}\}, \{\nu_8, \nu_9, \nu_{10}, \nu_{12}, \nu_{13}\}\}\end{aligned} \tag{5.35}$$

the proposed hybrid strategy guarantees the achievement of a unique global consensus among the entire network of robots. In Figs. 5.2 and 5.3, we report the continuous and hybrid evolutions of the heterogeneous multi-robot system. In particular, Fig. 5.3a and b show the multi-consensus achieved under the flow dynamics according to the partitions (5.35), whereas Fig. 5.3c and d report the improvement given by the addition of the jump dynamics leading to a unique consensus with, i.e., the robots placing themselves along the desired pattern. Furthermore, Fig. 5.2 shows the snapshots of the initial and final configurations of a simulation performed in CoppeliaSim, whose complete version can be found at [4].

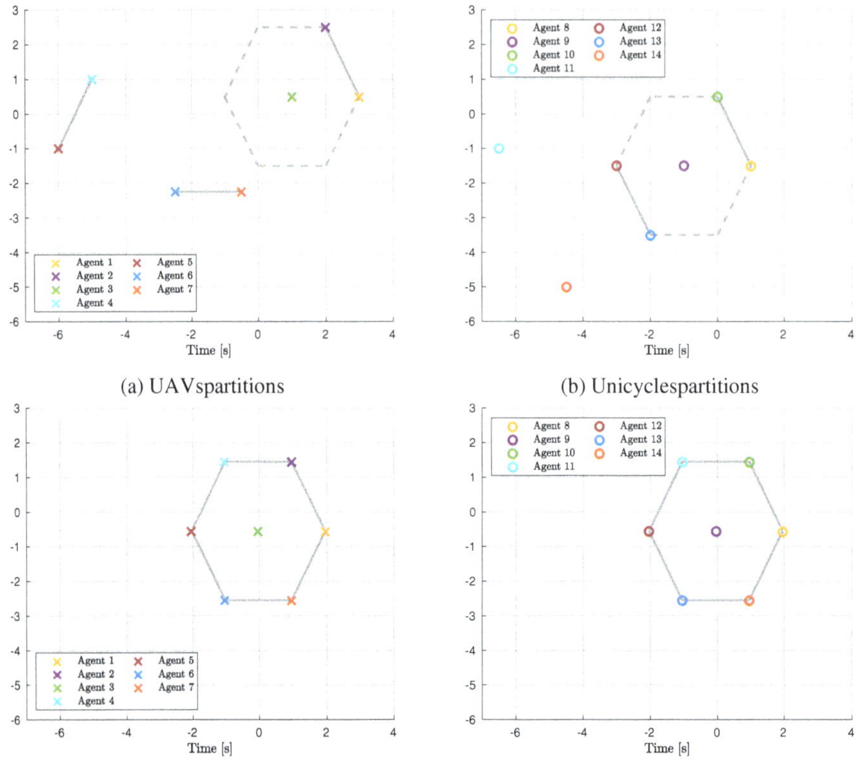

(a) UAVspartitions (b) Unicyclespartitions

(c) UAVshybridconsensus (d) Unicycleshybridconsensus

Fig. 5.3 Formation control. **a** and **b** report the evolution according to the flow dynamics, where the dashed lines represent the desired pattern the robots should draw in a unique consensus situation. **c** and **d** show the unique consensus achieved under the hybrid dynamics

5.7 Toward Future Perspectives: The Example of Rendezvous of Nonholonomic Robots with Heterogeneous Sensors

In this section, we envision an extended version of our framework in which the hybrid multi-consensus is applied to encode the exchange of information of different nature through flow and jump graphs in a multi-robot system. We address the rendezvous problem of nonholonomic robots, e.g., unicycles, in which the agents have to meet at a non-specified point on the (x, y) plane. The orientation information is exchanged continuously between the agents, whereas the position information is impulsively exchanged at aperiodic jumps with $\tau_{\min} = 0.1$ and $\tau_{\max} = 0.35$ (5.8). It is assumed

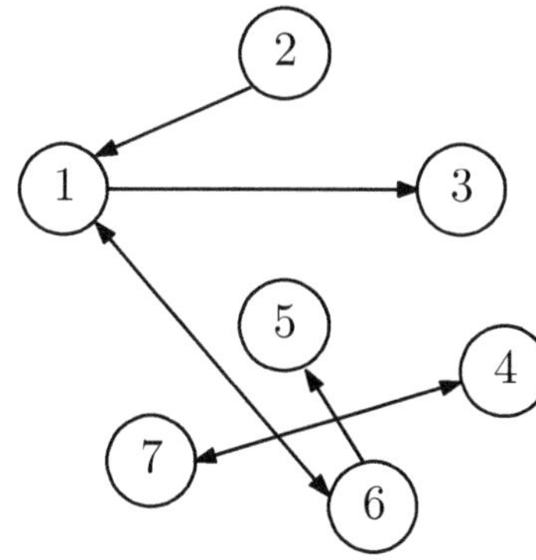

Fig. 5.4 Communication graph used in the application to rendezvous

that $\mathcal{G}_{\mathrm{f}} \equiv \mathcal{G}_{\mathrm{j}}$. We have considered $N = 7$ unicycles that communicate over the graph in Fig. 5.4, where the AEP underlying consensus is given by the following partitions:

$$\pi^{\star}_{\mathcal{H}} = \{\{\nu_4, \nu_7\}, \{\nu_1, \nu_2, \nu_3, \nu_5, \nu_6\}\} \tag{5.36}$$

The unicycles are modeled as a second-order nonholonomic system of the form

$$\begin{cases} \dot{x} = \cos(\theta)v \\ \dot{y} = \sin(\theta)v \\ \dot{\theta} = \omega \\ \dot{v} = u_v \\ \dot{\omega} = u_\omega \end{cases}$$

Defining $z := [x, y]^T$ and $\varphi(\theta) := [\cos(\theta), \sin(\theta)]^T$, we can rewrite the system as two coupled double integrators

$$\text{Driving Dynamics:} \begin{cases} \dot{z} = \varphi(\theta)v \\ \dot{v} = u_v \end{cases} \quad \text{Steering Dynamics:} \begin{cases} \dot{\theta} = \omega \\ \dot{\omega} = u_\omega \end{cases}$$

A PD consensus controller can be used for solving the rendezvous problem, where for each agent $i = 1, \ldots, N$ the control law takes the form

$$u_{v_i} = -d_{v_i} v_i - p_{v_i} \varphi_{v_i}(\theta_i)^T e_{z_i} \qquad \text{with } e_{z_i} = \sum_{j \in \mathcal{N}_i} a_{ij}(z_i - z_j)$$

$$u_{\omega_i} = -d_{\omega_i} \omega_i - p_{\omega_i} e_{\theta_i} \qquad \text{with } e_{\theta_i} = \sum_{j \in \mathcal{N}_i} a_{ij}(\theta_i - \theta_j)$$

where $d_{v_i}, d_{\omega_i}, p_{v_i}, p_{\omega_i} \in \mathbb{R}_{>0}$. In order to prevent θ from getting stuck at undesired equilibria, we add to the angular-motion controller an excitation term [14] of the form

$$\alpha_i(t, \theta_i, e_{zi}) = \psi(t)\varphi_i(\theta_i)^{\perp T} e_{zi}$$

with $\psi(t)$ being persistently exciting and where $\varphi^{\perp}$ is such that $\varphi^T \varphi^{\perp} = 0$. In conclusion, the resulting closed-loop systems are

$$\begin{cases} \dot{z}_i = \varphi(\theta_i) v_i \\ \dot{v}_i = -d_{v_i} v_i - p_{v_i} \varphi_{v_i}(\theta_i)^T e_{z_i} \end{cases}$$

$$\begin{cases} \dot{\theta} = \omega \\ \dot{\omega} = -d_{\omega_i} \omega_i - p_{\omega_i} e_{\theta_i} + \alpha_i(t, \theta_i, e_{z_i}) \end{cases}$$

Figure 5.5 shows the effectiveness of the hybrid multi-consensus in the rendezvous problem when applied to the exchange of different types of information. In particular, Fig. 5.5a reports the clustering of the agents according to the partitions (5.36), which

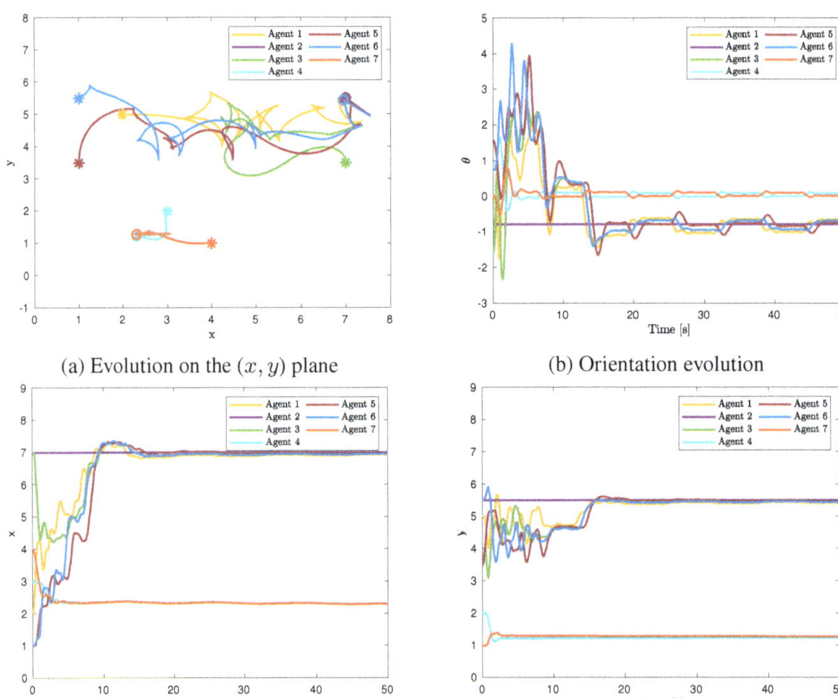

(a) Evolution on the (x, y) plane

(b) Orientation evolution

(c) Evolution along the x-axis

(d) Evolution along the y-axis

Fig. 5.5 Rendezvous of unicycles

can also be observed from the evolution of the unicycles state (θ, x, y) in Fig. 5.5b, c, and d, respectively.

Such results motivate a further investigation into the framework of hybrid consensus, generalizing the model by including the exchange of heterogeneous information during flow and at jumps, possibly with different communication graphs.

5.8 Conclusions

In this chapter, we outline new results on multi-consensus of simple classes of hybrid networks. After characterizing the consensus problem over a class of hybrid integrators with distinct communication topologies at jumps and during flow, the problem has been extended to deal with the dynamical LTI case. Two examples envisage new perspectives devoted to the use of hybrid distributed controllers to deal with multi-robot networks equipped with heterogeneous sensors and exchanging information according to different patterns.

References

1. Agaev, R., Chebotarev, P.: On the spectra of nonsymmetric Laplacian matrices. Linear Algebra Appl. **399**, 157–168 (2005)
2. Cacace, F., Mattioni, M., Monaco, S., Ricciardi Celsi, L.: Topology-induced containment for general linear systems on weakly connected digraphs. Automatica **131**, 109734 (2021)
3. Caughman, J.S., Veerman, J.J.P.: Kernels of directed graph Laplacians. Electron. J. Comb. **13**(1), 39 (2006)
4. Cristofaro, A., D'Orazio, F., Govoni, L., Mattioni, M.: Application to formation control of a heterogeneous multi-robot system (2023)
5. Cristofaro, A., Mattioni, M.: Hybrid consensus for multi-agent systems with time-driven jumps. Nonlinear Anal. Hybrid Syst. **43**, 101113 (2021)
6. Cristofaro, A., Mattioni, M.: Multiconsensus control of homogeneous LTI hybrid systems under time-driven jumps. In: 2022 IEEE 61st Conference on Decision and Control (CDC), pp. 316–321. IEEE (2022)
7. Gambuzza, L.V., Frasca, M.: Distributed control of multiconsensus. IEEE Trans. Autom. Control **66**(5), 2032–2044 (2020)
8. Goebel, R., Sanfelice, R.G., Teel, A.R.: Hybrid Dynamical Systems: Modeling, Stability, and Robustness. Princeton University Press (2012)
9. Isidori, A.: Lectures in Feedback Design for Multivariable Systems. Springer (2017)
10. Jadbabaie, A., Lin, J., Morse, A.S.: Coordination of groups of mobile autonomous agents using nearest neighbor rules. IEEE Trans. Autom. Control **48**(6), 988–1001 (2003)
11. Mattioni, M., Monaco, S.: Cluster partitioning of heterogeneous multi-agent systems. Automatica **138**, 110136 (2022)
12. Monaco, S., Ricciardi Celsi, L.: On multi-consensus and almost equitable graph partitions. Automatica **103**, pp. 53–61 (2019)
13. Monshizadeh, N., Zhang, S., Camlibel, M.K.: Disturbance decoupling problem for multi-agent systems: a graph topological approach. Syst. Control Lett. **76**, 35–41 (2015)

14. Nuño, E., Loría, A., Hernández, T., Maghenem, M., Panteley, E.: Distributed consensus-formation of force-controlled nonholonomic robots with time-varying delays. Automatica **120**, 109114 (2020)
15. Panteley, E., Loría, A.: Synchronization and dynamic consensus of heterogeneous networked systems. IEEE Trans. Autom. Control **62**(8), 3758–3773 (2017)
16. Ren, W., Cao, Y.: Distributed Coordination of Multi-agent Networks: Emergent Problems, Models, and Issues. Springer Science & Business Media (2010)
17. Wang, X., Lu, J.: Collective behaviors through social interactions in bird flocks. IEEE Circuits Syst. Mag. **19**(3), 6–22 (2019)

Chapter 6
Observer Design for Hybrid Systems with Linear Maps and Known Jump Times

Gia Quoc Bao Tran, Pauline Bernard, and Lorenzo Marconi

Abstract This chapter unifies and develops recent developments in observer design for hybrid systems with linear dynamics and output maps, whose jump times are known. We define and analyze the (pre-)asymptotic detectability and uniform complete observability of this class of systems, then present two different routes for observer design. The first one relies on a synchronized Kalman-like observer that gathers observability from both flows and jumps. The second one consists of decomposing the state into parts with different observability properties and coupling observers estimating each of these parts, possibly exploiting an extra fictitious measurement coming from the combination of flows and jumps. These observers are based on a Linear Matrix Inequality (LMI) or the Kazantzis–Kravaris/Luenberger (KKL) paradigm. A comparison of these methods is presented in a table at the end.

6.1 Introduction

Consider a hybrid system with linear maps

$$\mathcal{H}\begin{cases} \dot{x} = Fx + u_c & (x, u_c) \in C \qquad y_c = H_c x \\ x^+ = Jx + u_d & (x, u_d) \in D \qquad y_d = H_d x, \end{cases} \tag{6.1}$$

G. Q. B. Tran (✉) · P. Bernard
Centre Automatique et Systèmes, Mines Paris, Université PSL, 60 boulevard Saint-Michel, 75006 Paris, France
e-mail: gia-quoc-bao.tran@minesparis.psl.eu

P. Bernard
e-mail: pauline.bernard@minesparis.psl.eu

L. Marconi
Department of Electrical, Electronic, and Information Engineering "Guglielmo Marconi", University of Bologna, Bologna, Italy
e-mail: lorenzo.marconi@unibo.it

R. Postoyan et al. (eds.), *Hybrid and Networked Dynamical Systems*,
Lecture Notes in Control and Information Sciences 493,
https://doi.org/10.1007/978-3-031-49555-7_6

where $x \in \mathbb{R}^{n_x}$ is the state, C and D are the flow and jump sets, $y_c \in \mathbb{R}^{n_{y,c}}$ and $y_d \in \mathbb{R}^{n_{y,d}}$ are the outputs known during the flow intervals and at the jump times respectively, $u_c \in \mathbb{R}^{n_x}$ and $u_d \in \mathbb{R}^{n_x}$ are known exogenous input signals, as well as the dynamics matrices F, $J \in \mathbb{R}^{n_x \times n_x}$ and the output matrices $H_c \in \mathbb{R}^{n_{y,c} \times n_x}$, $H_d \in \mathbb{R}^{n_{y,d} \times n_x}$ which are all known and possibly time-varying. Models of the form (6.1) include not only hybrid systems with linear maps described in the setting of [23], but also *switched* and/or *impulsive* systems with linear maps where the active mode is seen as an exogenous signal making (F, J, H_c, H_d) time-varying (see [1, 17, 25] among many other ones), and continuous-time systems with *sporadic* or *multi-rate* sampled outputs where the "jumps" correspond to sampling events, $J = \mathrm{Id}$, $u_d = 0$, $y_c = 0$, and y_d the outputs available at the sampling event [12, 21, 24]. See [23, 28] for some examples of those classes of systems set in the framework of (6.1).

The goal of this chapter is to present in a unified and more complete way recent advances concerning the design of an asymptotic observer for (6.1), assuming that its jump times are known or detected. In practice, we may be interested in estimating only certain trajectories of "physical interest", initialized in some set $\mathcal{X}_0 \subset \mathbb{R}^{n_x}$ and with exogenous terms $(F, J, H_c, H_d, u_c, u_d)$ in some set $\mathcal{U}$ of interest. We then denote $\mathcal{S}_{\mathcal{H}}(\mathcal{X}_0, \mathcal{U})$ as the set of those maximal solutions of interest. Because we look for an *asymptotic* observer, we assume maximal solutions are complete as stated next.

Assumption 6.1.1 Given $\mathcal{X}_0$ and $\mathcal{U}$, each maximal solution in $\mathcal{S}_{\mathcal{H}}(\mathcal{X}_0, \mathcal{U})$ is complete.

Since the jump times of the system $\mathcal{H}$ in (6.1) are known, it is natural to strive for a *synchronized* asymptotic observer of the form

$$\hat{\mathcal{H}} \begin{cases} \dot{\hat{z}} = \Psi_c(\hat{z}, y_c, u_c) & \text{when } \mathcal{H} \text{ flows} \\ \hat{z}^+ = \Psi_d(\hat{z}, y_d, u_d) & \text{when } \mathcal{H} \text{ jumps} \\ \hat{x} = \Upsilon(\hat{z}, y_c, y_d, u_c, u_d), & \end{cases} \tag{6.2}$$

where $\hat{z} \in \mathbb{R}^{n_z}$ is the observer state (with $n_z \geq n_x$ in general), $\Upsilon : \mathbb{R}^{n_z} \times \mathbb{R}^{n_{y,c}} \times \mathbb{R}^{n_{y,d}} \times \mathbb{R}^{n_x} \times \mathbb{R}^{n_x} \to \mathbb{R}^{n_x}$ is the observer output map, $\Psi_c : \mathbb{R}^{n_z} \times \mathbb{R}^{n_{y,c}} \times \mathbb{R}^{n_x} \to \mathbb{R}^{n_z}$ and $\Psi_d : \mathbb{R}^{n_z} \times \mathbb{R}^{n_{y,d}} \times \mathbb{R}^{n_x} \to \mathbb{R}^{n_z}$ are respectively the observer flow and jump maps designed such that each maximal solution $(x, \hat{z})$ to the cascade $\mathcal{H} - \hat{\mathcal{H}}$ initialized in $\mathcal{X}_0 \times \mathbb{R}^{n_z}$ with inputs in $\mathcal{U}$ is complete and verifies

$$\lim_{\substack{t+j \to +\infty \\ (t,j) \in \operatorname{dom} x}} |x(t, j) - \hat{x}(t, j)| = 0. \tag{6.3}$$

The knowledge of the jump times is not only used to trigger the observer jumps at the same time as those of the system, but it can also be used to design the observer maps Ψ_c and Ψ_d. The way this information is exploited varies depending on whether these maps rely on:

- Gains that are computed *offline* (for example via matrix inequalities), according to all possible lengths of flow intervals in between jumps, i.e., they depend on each

individual flow length, not the particular sequence of them. This is the path taken by [1, 6, 12, 18, 22, 24, 29];
- Or, gains that are computed *online* along the time domain of each solution of interest, i.e., they depend on the *sequence* of flow lengths in each particular solution. This is the path taken by all Kalman-like approaches in [17, 26, 28].

In the former case, the design requires some information about the possible duration of flow intervals between successive jumps in each solution of interest (at least after a certain time) as defined next. In the context of switched (resp. sampled) systems, this corresponds to information about the possible switching (resp. sampling) rates.

Definition 6.1.2 (*Set of flow lengths of a hybrid arc*) For a closed subset $\mathcal{I}$ of $[0, +\infty)$ and some $j_m \in \mathbb{N}$, we say that a hybrid arc $(t, j) \mapsto x(t, j)$ *has flow lengths within $\mathcal{I}$ after jump time j_m* if

- $0 \leq t - t_j(x) \leq \sup \mathcal{I}$ for all $(t, j) \in \operatorname{dom} x$;
- $t_{j+1}(x) - t_j(x) \in \mathcal{I}$ holds for all $j \in \mathbb{N}$ with $j \geq j_m$ if $\sup \operatorname{dom}_j x = +\infty$, and for all $j \in \{j_m, j_m + 1, \ldots, \sup \operatorname{dom}_j x - 1\}$ otherwise.

In brief, $\mathcal{I}$ contains all the possible lengths of the flow intervals between successive jumps, at least after some time. The first item is to bound the length of the flow intervals not covered by the second item, namely possibly the first ones before j_m, and the last one, which is $\operatorname{dom}_t x \cap [t_{J(x)}, +\infty)$ where $t_{J(x)}$ is the time when the last jump happens (when defined). If $\mathcal{I}$ is unbounded, the system may admit (eventually) continuous solutions and the observer should correct the estimate at least during flows, while $0 \in \mathcal{I}$ means the hybrid arc can jump more than once at the same time instance or have flow lengths going to zero (including (eventually) discrete and Zeno solutions) and the observer should reduce the estimation error at least at jumps.

From there, one may design either:

- A *flow-based* observer with an innovation term during flows only, exploiting the observability of the full state during flows from y_c when $0 \notin \mathcal{I}$ [1, 6];
- A *jump-based* observer with an innovation term at jumps only, exploiting the detectability of the full state via the combination of flows and jumps from y_d available at the jumps only when $\mathcal{I}$ is bounded [6, 12, 17, 24, 26];
- An observer with innovation terms during both flows and jumps, exploiting the observability from both y_c and y_d and the combination of flows and jumps: this is done via a hybrid Kalman-like approach in [28], or via an observability decomposition in [29], or Lyapunov-based LMIs in [6, 22].

This chapter deals with the third case, where the full state is not necessarily instantaneously observable during flows and not observable from the jump output only. It unifies and extends the work of [28, 29]. More precisely, in Sect. 6.2, we start with an observability analysis allowing us to exhibit necessary conditions and sufficient conditions for observer design: first through hybrid Gramian conditions, and then via an observability decomposition. The latter decomposes the state in two parts: the first one is instantaneously observable through the flow output y_c, while the second one must be detectable from an extended jump output featuring the *available* jump output y_d and an additional *fictitious* one, describing how the non-observable states impact the observable ones at jumps and become *visible* through y_c. While the Gramian-based analysis has led in [28] to a *systematic* hybrid Kalman-like design, we show how the observability decomposition lets us design observers made of:

- A *high-gain* flow-based observer of the state components that are instantaneously observable from y_c;
- A jump-based observer for the remaining components, derived from a *discrete* LMI-based (resp. KKL-based) observer in Sect. 6.3 (resp. Sect. 6.4), thus revisiting and extending [28].

Notation. Let $\mathbb{R}$ (resp. $\mathbb{N}$) denote the set of real numbers (resp. natural numbers, i.e., $\{0, 1, 2, \ldots\}$) and $\mathbb{N}_{>0} = \mathbb{N} \setminus \{0\}$. We denote $\mathbb{R}^{m\times n}$ (resp. $\mathcal{S}^n_{>0}$) as the set of real $(m \times n)$- (resp. symmetric positive definite $(n \times n)$-) dimensional matrices. Given a set S, $\text{int}(S)$ denotes its interior. Let $\Re(z)$ and $\Im(z)$ be the real and imaginary parts of the complex variable z, respectively. Denote Id as the identity matrix of appropriate dimension. Let $|\cdot|$ be the Euclidean norm and $\|\cdot\|$ the induced matrix norm. Let $A^\perp$ be the orthogonal complement of matrix A satisfying $AA^\perp = 0$ and such that $\left(A^\top \; A^\perp\right)$ is invertible, and $A^\dagger$ be the Moore–Penrose inverse of A [20]. For a solution $(t, j) \mapsto x(t, j)$ of a hybrid system, we denote $\text{dom}\, x$ its time domain [13], $\text{dom}_t\, x$ (resp. $\text{dom}_j\, x$) the domain's projection on the ordinary time (resp. jump) component, and for $j \in \text{dom}_j\, x$, $t_j(x)$ the unique time such that $(t_j(x), j) \in \text{dom}\, x$ and $(t_j(x), j-1) \in \text{dom}\, x$, and $\mathcal{T}_j(x) := \{t \in \text{dom}_t\, x : (t, j) \in \text{dom}\, x\}$ (for hybrid systems with inputs, see [23]). The mention of x is omitted when no confusion is possible. A solution x to a hybrid system is *complete* if $\text{dom}\, x$ is unbounded and *Zeno* if it is complete and $\sup \text{dom}_t\, x < +\infty$. Let $\text{diag}(\lambda_1, \lambda_2, \ldots, \lambda_n)$ be the diagonal matrix with entries λ_i. Occasionally, $\star$ denotes the symmetric part, i.e., $\star^\top P = P^\top P$ or sometimes $\star^\top QP = P^\top QP$. Last, let sat_s be a saturation function with level s, i.e., $\text{sat}_s(M) = M$ if $\|M\| \leq s$ and sat_s is bounded otherwise.

6.2 Detectability and Observability Analysis

The existence of an *asymptotic* and *synchronized* observer (6.2) for (6.1) requires (6.1) to be *asymptotically detectable*, in the following sense.

Definition 6.2.1 (*(Pre-)asymptotic detectability with known jump times*) System (6.1) with known jump times, initialized in $\mathcal{X}_0$, and with inputs in $\mathcal{U}$ is *pre-asymptotically detectable* if any complete solutions x_a and x_b in $\mathcal{S}_\mathcal{H}(\mathcal{X}_0, \mathcal{U})$ with the same inputs $(F, J, H_c, H_d, u_c, u_d)$, such that $\text{dom}\, x_a = \text{dom}\, x_b$, and whose flow outputs $y_{a,c}$, $y_{b,c}$ and jump outputs $y_{a,d}$, $y_{b,d}$ satisfy

$$y_{a,c}(t, j) = y_{b,c}(t, j), \qquad \forall t \in \text{int}(\mathcal{T}_j(x_a)), \forall j \in \text{dom}_j\, x_a, \tag{6.4a}$$
$$y_{a,d}(t_j, j-1) = y_{b,d}(t_j, j-1), \qquad \forall j \in \text{dom}_j\, x_a,\ j \geq 1, \tag{6.4b}$$

verify

$$\lim_{\substack{t+j\to+\infty \\ (t,j)\in\text{dom}\, x_a}} |x_a(t, j) - x_b(t, j)| = 0. \tag{6.5}$$

If in addition, all solutions in $\mathcal{S}_{\mathcal{H}}(\mathcal{X}_0, \mathcal{U})$ are complete, then we have *asymptotic detectability*. The set $\mathcal{X}_0$ (resp. $\mathcal{U}$) may be omitted if the property holds for any initial condition in $\mathcal{X}_0 = C \cup D$ (resp. any input).

Remark 6.2.2 In Definition 6.2.1, when all the flow intervals have non-empty interior (i.e., $\text{int}(\mathcal{T}_j(x_a)) \neq \emptyset$ for all $j \in \text{dom}_j\, x_a$), condition (6.4a) is equivalent to

$$y_{a,c}(t, j) = y_{b,c}(t, j), \quad \forall (t, j) \in \text{dom}\, x_a, \tag{6.6}$$

by the continuity of $t \mapsto y_c(t, j)$ during flows for all $j \in \text{dom}_j\, x_a$. On the contrary, when a solution admits consecutive jumps, condition (6.4a) is required only on the flow intervals with a non-empty interior since it holds vacuously on the other ones. In other words, the equality of y_c is only required when the system is *flowing*.

The *necessity* of asymptotic detectability is typically obtained as follows. First, by definition of a synchronized asymptotic observer initialized in $\mathcal{X}_0$ with inputs in $\mathcal{U}$, all solutions in $\mathcal{S}_{\mathcal{H}}(\mathcal{X}_0, \mathcal{U})$ must be complete. Then, pick a pair of solutions (x_a, x_b) as in Definition 6.2.1 verifying (6.4). A solution $\hat{x}$ produced by (6.2) fed with outputs $(y_{a,c}, y_{a,d})$ shares the same time domain and must converge asymptotically to x_a; but, $\hat{x}$ is also a solution to (6.2) fed with outputs $(y_{b,c}, y_{b,d})$ according to (6.4), so that it must also converge asymptotically to x_b. It follows that x_a and x_b necessarily converge asymptotically to each other, thus giving us asymptotic detectability. Note that compared to [5], Definition 6.2.1 is restricted to pairs of complete solutions with the *same* time domain because the knowledge of the jump times is used to trigger the jumps of the observer so that only complete solutions with the same time domain and the same outputs are required to converge asymptotically to each other.

For observer design, one typically requires stronger *observability* assumptions depending on the class of observers and the required observer properties [3]. Observability typically means that the equality of the outputs in (6.4), possibly over a large enough time window, implies that the solutions x_a and x_b are actually the same or said differently, there does not exist any pair of different solutions with the same time domain and the same outputs in the sense of (6.4). Actually, in the context of observer design, a more relevant property is the ability to determine uniquely the current state from the knowledge of the past outputs over a certain time window $\Delta > 0$, which is typically called *backward distinguishability* [27] or *constructibility* [15]. In other words, for all (t, j) in the domain such that $t + j \geq \Delta$, the equality of the outputs (y_c, y_d) along x_a and x_b at all past times (t', j') in the domain such that $0 \leq (t + j) - (t' + j') \leq \Delta$ implies that $x_a(t, j) = x_b(t, j)$ (see later in

Sect. 6.2.1). For continuous systems, this property is equivalent to observability over a time window because of the uniqueness of solutions in forward and backward time. However, they cease to be equivalent in discrete or hybrid systems when the jump maps are not invertible: a system could be constructible without being observable (see [15, Sect. 2.3.3] for a detailed discussion on those notions).

6.2.1 Hybrid Observability Gramian

Consider a pair of solutions x_a and x_b in $\mathcal{S}_{\mathcal{H}}(\mathcal{X}_0, \mathcal{U})$ with the same inputs $(F, J, H_c, H_d, u_c, u_d)$ such that $\operatorname{dom} x_a = \operatorname{dom} x_b := \mathbb{D}$. Then, for all hybrid times $((t', j'), (t, j)) \in \mathbb{D} \times \mathbb{D}$, we have

$$x_a(t, j) - x_b(t, j) = \Phi_{F,J}((t, j), (t', j'))(x_a(t', j') - x_b(t', j')), \tag{6.7}$$

where $\Phi_{F,J}$ is a *hybrid transition matrix* defined as

$$\Phi_{F,J}((t, j), (t', j')) = \phi_F(t, t_{j+1}) \left(\prod_{k=j}^{j'+1} \phi_F(t_{k+1}, t_k) J(t_k, k-1) \right) \phi_F(t_{j'+1}, t'), \tag{6.8}$$

if $t \geq t'$ and $j \geq j'$, and, if the jump matrix J is invertible at the jump times,

$$\Phi_{F,J}((t, j), (t', j')) = \phi_F(t, t_j) \left(\prod_{k=j+1}^{j'} \phi_F(t_{k-1}, t_k) J^{-1}(t_k, k-1) \right) \phi_F(t_{j'}, t'), \tag{6.9}$$

if $t \leq t'$ and $j \leq j'$, with the domain of F and J inherited from $\mathbb{D}$, where ϕ_F denotes the continuous-time transition matrix associated with F, i.e., describing solutions to $\dot{x} = Fx$. By summing and integrating squares, it follows that the equality of the outputs (y_c, y_d) along x_a and x_b between time $(t', j') \in \mathbb{D}$ and a later time $(t, j) \in \mathbb{D}$ is equivalent to

$$(x_a(t', j') - x_b(t', j'))^\top \mathcal{G}_{(F,J,H_c,H_d)}((t', j'), (t, j))(x_a(t', j') - x_b(t', j')) = 0, \tag{6.10}$$

or, assuming the invertibility of J at the jump times,

$$(x_a(t, j) - x_b(t, j))^\top \mathcal{G}^{bw}_{(F,J,H_c,H_d)}((t', j'), (t, j))(x_a(t, j) - x_b(t, j)) = 0, \tag{6.11}$$

where $\mathcal{G}_{(F,J,H_c,H_d)}((t', j'), (t, j))$ (resp. $\mathcal{G}^{bw}_{(F,J,H_c,H_d)}((t', j'), (t, j))$) is the observability Gramian (resp. *backward observability Gramian*) between those times as defined next.

Definition 6.2.3 (*Observability Gramians*) The *observability Gramian* and *backward observability Gramian* of a quadruple (F, J, H_c, H_d) defined on a hybrid time domain $\mathbb{D}$, between time $(t', j') \in \mathbb{D}$ and a later time $(t, j) \in \mathbb{D}$, are defined as

$$\begin{aligned}&\mathcal{G}_{(F,J,H_c,H_d)}((t',j'),(t,j)) = \\ &\int_{t'}^{t_{j'+1}} \star^\top \Psi_c((s,j'),(t',j'))ds + \sum_{k=j'+1}^{j-1} \int_{t_k}^{t_{k+1}} \star^\top \Psi_c((s,k),(t',j'))ds \\ &\qquad + \sum_{k=j'}^{j-1} \star^\top \Psi_d((t_{k+1},k),(t',j')) + \int_{t_j}^{t} \star^\top \Psi_c((s,j),(t',j'))ds,\end{aligned} \tag{6.12}$$

and, when J is invertible at the jump times,

$$\begin{aligned}&\mathcal{G}^{bw}_{(F,J,H_c,H_d)}((t',j'),(t,j)) = \\ &\int_{t'}^{t_{j'+1}} \star^\top \Psi_c((s,j'),(t,j))ds + \sum_{k=j'+1}^{j-1} \int_{t_k}^{t_{k+1}} \star^\top \Psi_c((s,k),(t,j))ds \\ &\qquad + \sum_{k=j'}^{j-1} \star^\top \Psi_d((t_{k+1},k),(t,j)) + \int_{t_j}^{t} \star^\top \Psi_c((s,j),(t,j))ds,\end{aligned} \tag{6.13}$$

respectively, where

$$\Psi_c((s,k),(t,j)) = H_c(s,k)\Phi_{F,J}((s,k),(t,j)), \tag{6.14a}$$

$$\Psi_d((t_{k+1},k),(t,j)) = H_d(t_{k+1},k)\Phi_{F,J}((t_{k+1},k),(t,j)), \tag{6.14b}$$

with all the jump times determined from $\mathbb{D}$.

According to (6.10), we deduce that the observability between times (t', j') and (t, j), namely the ability to reconstruct the initial state $x(t', j')$ from the knowledge of the *future* output until (t, j), is equivalent to the positive definiteness of the observability Gramian over this period. On the other hand, when J is invertible at the jump times, the ability to reconstruct the current state $x(t, j)$ from the knowledge of the *past* output until (t', j'), i.e., the backward distinguishability or constructibility, is characterized by the positive definiteness of the backward observability Gramian over this period according to (6.11). Actually, in that case, both notions are actually equivalent but the backward observability Gramian tends to appear more naturally in the analysis of observers.

Remark 6.2.4 Note that unlike for purely continuous or discrete linear systems, the inputs (u_c, u_d) involved in the hybrid dynamics (6.1) may impact the observability properties since they may change the domain of the solutions and thus the Gramian.

For observer design, the invertibility of the Gramian is typically assumed to be *uniform*, leading to the following hybrid *uniform complete observability*, extending the classical UCO condition of the Kalman and Bucy's filter [16].

Definition 6.2.5 (*Uniform complete observability (UCO)*) The quadruple (F, J, H_c, H_d) defined on a hybrid time domain $\mathbb{D}$ is *uniformly completely observable* with data (Δ, μ) if there exists $\Delta > 0$ and $\mu > 0$ such that for all $((t', j'), (t, j)) \in \mathbb{D} \times \mathbb{D}$ verifying $(t - t') + (j - j') \geq \Delta$, we have

$$\mathcal{G}_{(F,J,H_c,H_d)}((t', j'), (t, j)) \geq \mu\, \mathrm{Id}. \tag{6.15}$$

In [28], this condition is stated with $\mathcal{G}^{bw}$ replacing $\mathcal{G}$ because the former appears directly in the analysis. They are actually equivalent, assuming the uniform invertibility of J at the jump times and the boundedness of F and J. In [28], this UCO condition is exploited to design a *systematic* hybrid Kalman-like observer of the form

$$\begin{cases} \left.\begin{aligned} \dot{\hat{x}} &= F\hat{x} + u_c + PH_c^\top R_c^{-1}(y_c - H_c\hat{x}) \\ \dot{P} &= \lambda P + FP + PF^\top - PH_c^\top R_c^{-1} H_c P \end{aligned}\right\} \text{when } \mathcal{H} \text{ flows} \\ \\ \left.\begin{aligned} \hat{x}^+ &= J\hat{x} + u_d + JK(y_d - H_d\hat{x}) \\ P^+ &= \gamma^{-1} J(I - KH_d)PJ^\top \end{aligned}\right\} \text{when } \mathcal{H} \text{ jumps} \end{cases} \tag{6.16a}$$

with

$$K = PH_d^\top (H_d P H_d^\top + R_d)^{-1}, \tag{6.16b}$$

where $\lambda \geq 0$ and $\gamma \in (0, 1]$ are design parameters, $R_c \in \mathcal{S}_{>0}^{n_{y,c}}$ and $R_d \in \mathcal{S}_{>0}^{n_{y,d}}$ are (possibly time-varying) weighting matrices that are positive definite and are uniformly upper- and lower-bounded. In [28], it is shown that the estimation error:

- Converges asymptotically to zero under UCO and the boundedness of the system matrices *along each of the considered solutions* in $\mathcal{S}_\mathcal{H}(\mathcal{X}_0, \mathcal{U})$;
- Is exponentially stable with an arbitrarily fast rate and robustly stable (as defined in [2] but for hybrid systems) after a certain time, under UCO and the boundedness of the system matrices *uniformly across the considered solutions* in $\mathcal{S}_\mathcal{H}(\mathcal{X}_0, \mathcal{U})$, when additionally λ is sufficiently large and γ is sufficiently small.

Example 6.2.6 Consider a pendulum equipped with an IMU and bouncing on a vertical wall, with angular position $\boldsymbol{\theta} \in \mathbb{R}^3$. The IMU contains a gyroscope measuring its angular velocity $\boldsymbol{\omega} \in \mathbb{R}^3$ and an accelerometer measuring its proper acceleration (linear acceleration minus gravity) $\boldsymbol{y_a} \in \mathbb{R}^3$ in the IMU frame, modulo an unknown constant bias $\boldsymbol{b_a} \in \mathbb{R}^3$. We assume its tilt $\boldsymbol{t} \in \mathbb{R}^3$ is measured at the jump times (when the mass impacts the wall) and that the linear velocity $\boldsymbol{v} \in \mathbb{R}^3$ in the sensor frame can be deduced from the gyroscope measurement via kinematics [31]. We also assume the velocity magnitude is reduced by an unknown constant restitution coefficient $c \in (0, 1]$ at each impact. With these in mind, we model the system in hybrid form with state $x = (\boldsymbol{t}, \boldsymbol{v}, \boldsymbol{b_a}, c)$ as (see [31] for the flow dynamics)

$$\begin{cases} \dot{\boldsymbol{t}} = -[\omega]_\times \boldsymbol{t} \\ \dot{\boldsymbol{v}} = -[\omega]_\times \boldsymbol{v} + \boldsymbol{y_a} - \boldsymbol{b_a} + g\boldsymbol{t} \\ \dot{\boldsymbol{b}}_{\boldsymbol{a}} = 0 \\ \dot{c} = 0 \end{cases} \quad y_c = \boldsymbol{v}, \qquad \begin{cases} \boldsymbol{t}^+ = \boldsymbol{t} \\ \boldsymbol{v}^+ = -c\boldsymbol{v} \\ \boldsymbol{b}_{\boldsymbol{a}}^+ = \boldsymbol{b_a} \\ c^+ = c \end{cases} \quad y_d = (\boldsymbol{v}, \boldsymbol{t}), \tag{6.17}$$

where $[\boldsymbol{\omega}]_\times = \begin{pmatrix} 0 & -\omega_3 & \omega_2 \\ \omega_3 & 0 & -\omega_1 \\ -\omega_2 & \omega_1 & 0 \end{pmatrix}$ and g is the gravitational acceleration, with the flow and jump sets depending on the wall configuration. This system takes the form (6.1) where $F = \begin{pmatrix} -[\omega]_\times & 0 & 0 & 0 \\ g\,\mathrm{Id} & -[\omega]_\times & -\mathrm{Id} & 0 \\ 0 & 0 & 0 & 0 \\ 0 & 0 & 0 & 0 \end{pmatrix}$, $J = \begin{pmatrix} \mathrm{Id} & 0 & 0 & 0 \\ 0 & 0 & 0 & -y_{d,\boldsymbol{v}} \\ 0 & 0 & \mathrm{Id} & 0 \\ 0 & 0 & 0 & 1 \end{pmatrix}$,[1] $u_c = \begin{pmatrix} 0 \\ \boldsymbol{y_a} \\ 0 \\ 0 \end{pmatrix}$, and $u_d = 0$, with $y_{d,\boldsymbol{v}}$ the $\boldsymbol{v}$-component of y_d. We would like to estimate both $\boldsymbol{b_a}$ and c. Assuming the system is persistently not at rest (with external excitation if $c < 1$), it is observable over each period of time containing at least one jump and a flow interval, because

- $\boldsymbol{t}$ is available at jumps and its dynamics are independent of the other state components, making it observable, independently of the input signal $[\boldsymbol{\omega}]_\times$;
- $\boldsymbol{v}$ and $\boldsymbol{b_a}$ are both instantaneously observable during flows from y_c once $\boldsymbol{t}$ is known, also independently of $[\boldsymbol{\omega}]_\times$;
- c is observable at jumps by seeing $\boldsymbol{v}$ as a known input, because $\boldsymbol{v}$ is measured.

It follows that if the pendulum velocity is uniformly lower-bounded (thanks to an appropriate input in the mechanical system, whose effects are in fact contained in $\boldsymbol{y_a}$ in the IMU frame), the observability Gramian computed over a time window Δ larger than the maximal length of flow intervals would be uniformly positive definite. Then, assuming the boundedness of F and J, a Kalman-like observer (6.16) fed with $(y_c, y_d, \boldsymbol{\omega}, \boldsymbol{y_a})$ can be designed for (6.17). Now, consider the case where there is an unknown constant bias $\boldsymbol{b}_g \in \mathbb{R}^3$ in the gyroscope measurement. As a result, $[\boldsymbol{\omega}]_\times$ in (6.17) is replaced by $[\boldsymbol{\omega_m} - \boldsymbol{b}_g]_\times$, where $\boldsymbol{\omega_m}$ is the biased measurement. Assume an estimate $\hat{\boldsymbol{b}}_g$ of $\boldsymbol{b}_g$ is available such that $\hat{\boldsymbol{b}}_g - \boldsymbol{b}_g$ asymptotically vanishes. Then, the dynamics (6.17) may be re-written as

$$\dot{x} = \hat{F}x + (F - \hat{F})x, \tag{6.18}$$

where $[\boldsymbol{\omega_m} - \hat{\boldsymbol{b}}_g]_\times$ replaces $[\boldsymbol{\omega_m} - \boldsymbol{b}_g]_\times$ in $\hat{F}$. Consider the previous Kalman-like observer, but designed with the known $\hat{F}$ instead of F. According to our observability analysis above, the observability of the quadruple (F, J, H_c, H_d) does not depend on F and therefore still holds for $(\hat{F}, J, H_c, H_d)$. The *robust stability* of the Kalman-like observer [28, Theorem 3] then ensures that the error converges asymptotically

[1] Note that J is not invertible at jumps, but it can be made invertible by considering an alternative jump map $\boldsymbol{v}^+ = \boldsymbol{v} + c y_{d,\boldsymbol{v}} - y_{d,\boldsymbol{v}}$ and seeing $-y_{d,\boldsymbol{v}}$ as part of u_d. Simulations have shown that this invertibility may not be necessary for the Kalman-like observer.

to zero because: (i) The UCO condition holds with $\hat{F}$ replacing F, and (ii) The "disturbance" $(F - \hat{F})x$ vanishes asymptotically thanks to $\hat{\boldsymbol{b}}_g$ converging to $\boldsymbol{b}_g$ and the boundedness of x. In other words, we only need to find an asymptotic estimate of $\boldsymbol{b}_g$ and feed it to the hybrid Kalman-like observer of x. For that, notice that the pendulum position $\boldsymbol{\theta}$ verifies the hybrid dynamics

$$\begin{cases} \dot{\boldsymbol{\theta}} = \boldsymbol{\omega_m} - \boldsymbol{b}_g \\ \dot{\boldsymbol{b}}_g = 0 \end{cases} \quad y_c' = 0, \quad \begin{cases} \boldsymbol{\theta}^+ = \boldsymbol{\theta} \\ \boldsymbol{b}_g{}^+ = \boldsymbol{b}_g \end{cases} \quad y_d' = \boldsymbol{\theta}, \tag{6.19}$$

with the flow and jump sets depending on the wall configuration. In other words, $\boldsymbol{\theta}$ has continuous-time dynamics, but with sampled measurement at each impact obtained via the impact condition. Because (6.19) has linear maps and only the jump output, we can design a *jump-based observer* with a constant gain using LMIs on the equivalent *discrete* system sampled at the jumps (see [6, Corollary 5.2]). All in all, still if the pendulum velocity is uniformly lower-bounded away from zero and (F, J) upper-bounded, an observer of x is obtained by the cascade of a jump-based observer of $\boldsymbol{b}_g$ and a Kalman-like one for (6.17) fed with $[\boldsymbol{\omega_m} - \hat{\boldsymbol{b}}_g]_\times$ instead of $[\boldsymbol{\omega}]_\times$.

In what follows, we attempt to analyze more precisely the observability/detectability of the system by decomposing the state according to the different sources of observability. Beyond a finer comprehension, this allows for the design of observers when UCO is not satisfied, for instance under mere detectability properties, or to design observers of smaller dimensions through decoupling (see Table 6.1 for a comparison).

6.2.2 *Observability Decomposition*

In the case where the full state is *instantaneously observable* during flows via the flow output y_c and the system admits an average dwell time, a high-gain flow-based observer (using only y_c) can be designed (see [6, Sect. 4]); and when the full state is observable from the jump output y_d only, a jump-based observer based on an equivalent discrete system can be designed if the jumps are persistent (see [6, Sect. 5]). We are thus interested here in the case where observability rather comes from the combination of flows and jumps and/or the combination of (y_c, y_d). The idea of the decomposition is thus to isolate state components that are instantaneously observable during flows from y_c, from other ones that become visible thanks to y_d or the combination of flows and jumps. It follows that both the flow and jump outputs may need to be fully exploited to reconstruct the state and that neither (eventually) continuous nor discrete/Zeno trajectories are allowed: both flows and jumps need to be persistent at least after a certain time, unlike in Sect. 6.2.1, as assumed next.

Assumption 6.2.7 There exists a $j_m \in \mathbb{N}$ such that solutions have flow lengths within a compact set $\mathcal{I} \subseteq [\tau_m, \tau_M]$ where $\tau_m > 0$ after jump time j_m.

Assumption 6.2.7 means that, for all solutions $x \in \mathcal{S}_{\mathcal{H}}(\mathcal{X}_0, \mathcal{U})$, the hybrid arc $(t, j) \mapsto (x(t, j), t - t_j)$ is solution after some time to the hybrid system

$$\mathcal{H}^\mathcal{T} \begin{cases} \left.\begin{aligned} \dot{x} &= Fx + u_c \\ \dot{\tau} &= 1 \end{aligned}\right\} (x, \tau) \in C^\mathcal{T} & y_c = H_c x \\ \left.\begin{aligned} x^+ &= Jx + u_d \\ \tau^+ &= 0 \end{aligned}\right\} (x, \tau) \in D^\mathcal{T} & y_d = H_d x \end{cases} \tag{6.20a}$$

with the flow and jump sets

$$C^\mathcal{T} = \mathbb{R}^{n_x} \times [0, \tau_M], \qquad D^\mathcal{T} = \mathbb{R}^{n_x} \times \mathcal{I}, \tag{6.20b}$$

where $\tau \in \mathbb{R}$ is a timer keeping track of the time elapsed since the previous jump. Note that $\mathcal{H}^\mathcal{T}$ admits (after the first j_m jumps) a larger set of solutions than $\mathcal{S}_{\mathcal{H}}(\mathcal{X}_0, \mathcal{U})$ since the information of the flow and jump sets are replaced by the knowledge of flow lengths in $\mathcal{I}$ only (as long as the inputs $(F, J, H_c, H_d, u_c, u_d)$ are defined along the time domains of those extra solutions). But, as discussed in Sect. 6.1, when the observer contains gains that are computed offline, based on the knowledge of the possible flow lengths only, it is actually designed for $\mathcal{H}^\mathcal{T}$ instead of $\mathcal{H}$ and it is thus the detectability/observability of $\mathcal{H}^\mathcal{T}$ that is relevant. In that case, the design depends only implicitly on the sets $\mathcal{X}_0$, $\mathcal{U}$, C, and D through the choice of $\mathcal{I}$ satisfying Assumption 6.2.7.

In view of observer design and motivated by [10], we start by proposing a change of variables decomposing the state x of $\mathcal{H}^\mathcal{T}$ into components associated with different types of observability. In order to guarantee the existence of the decomposition, we assume in the next section that the flow pair (F, H_c) is constant. However, the subsequent results of this chapter still hold with (F, H_c) varying, as long as the transformation into the decomposition form exists and is invertible uniformly in time as explained in Remark 6.2.9.

6.2.3 *Observability from y_c During Flows*

Assume (F, H_c) is constant. Let the (flow) observability matrix be

$$\mathcal{O} := \text{row}(H_c, H_c F, \dots, H_c F^{n_x - 1}), \tag{6.21}$$

and assume it is of rank $n_o := \dim \operatorname{Im} \mathcal{O} < n_x$. Consider a basis $(v_i)_{1 \le i \le n_x}$ of $\mathbb{R}^{n_x}$ such that $(v_i)_{1 \le i \le n_o}$ is a basis of the observable subspace and $(v_i)_{n_o + 1 \le i \le n_x}$ is a basis of the non-observable subspace $\ker \mathcal{O}$. Then, we define the invertible matrix $\mathcal{D} := \begin{pmatrix} \mathcal{D}_o & \mathcal{D}_{no} \end{pmatrix}$ where

$$\mathcal{D}_o := \begin{pmatrix} v_1 & \dots & v_{n_o} \end{pmatrix} \in \mathbb{R}^{n_x \times n_o}, \tag{6.22a}$$

$$\mathcal{D}_{no} := \begin{pmatrix} v_{n_o+1} & \dots & v_{n_x} \end{pmatrix} \in \mathbb{R}^{n_x \times n_{no}}, \tag{6.22b}$$

which by definition satisfies for all $\tau \geq 0$,

$$\mathcal{O}\mathcal{D}_{no} = 0, \qquad H_c e^{F\tau} \mathcal{D}_{no} = 0. \tag{6.23}$$

We denote $\mathcal{V} := \mathcal{D}^{-1}$ which we decompose consistently into two parts $\mathcal{V} =: \begin{pmatrix} \mathcal{V}_o \\ \mathcal{V}_{no} \end{pmatrix}$, so that $\mathcal{V}_o x$ represents the part of the state that is instantaneously observable during flows (see [9, Theorem 6.O6]).

A first idea could be to stop the decomposition here and design a sufficiently fast high-gain observer for $\mathcal{V}_o x$ while estimating the rest of the state $\mathcal{V}_{no} x$ through y_d and detectability. However, as noticed in [10, Proposition 6], the fact that $\mathcal{V}_o x$ and $\mathcal{V}_{no} x$ possibly interact with each other during flows prevents achieving stability by further pushing the high gain. The case of such a decomposition where $\mathcal{V}_o x$ and $\mathcal{V}_{no} x$ evolve independently during flows is exploited in a more general context in [30]. Here, because the maps are linear, we can go further and solve this possible coupling by more efficiently decoupling the state components as follows.

Indeed, the estimation of any state that is not instantaneously observable during flows needs to take into account the combination of flows and jumps. That is why it is relevant to exhibit explicitly this combination via the change of coordinates

$$x \mapsto z = \begin{pmatrix} z_o \\ z_{no} \end{pmatrix} = \mathcal{V} e^{-F\tau} x = \begin{pmatrix} \mathcal{V}_o \\ \mathcal{V}_{no} \end{pmatrix} e^{-F\tau} x, \tag{6.24}$$

whose inverse transformation is

$$x = e^{F\tau} \mathcal{D} z = e^{F\tau} (\mathcal{D}_o z_o + \mathcal{D}_{no} z_{no}), \tag{6.25}$$

and which, according to (6.23), transforms $\mathcal{H}^\tau$ into

$$\begin{cases} \dot{z}_o = G_o(\tau) u_c \\ \dot{z}_{no} = G_{no}(\tau) u_c \\ \dot{\tau} = 1 \\ \\ z_o^+ = J_o(\tau) z_o + J_{ono}(\tau) z_{no} + \mathcal{V}_o u_d \\ z_{no}^+ = J_{noo}(\tau) z_o + J_{no}(\tau) z_{no} + \mathcal{V}_{no} u_d \\ \tau^+ = 0, \end{cases} \tag{6.26a}$$

with the flow and jump sets

$$\mathbb{R}^{n_o} \times \mathbb{R}^{n_{no}} \times [0, \tau_M], \qquad \mathbb{R}^{n_o} \times \mathbb{R}^{n_{no}} \times \mathcal{I}, \tag{6.26b}$$

and the measurements

$$y_c = H_{c,o}(\tau)z_o, \qquad y_d = H_{d,o}(\tau)z_o + H_{d,no}(\tau)z_{no}, \tag{6.26c}$$

where $G_o(\tau) = \mathcal{V}_o e^{-F\tau}$, $G_{no}(\tau) = \mathcal{V}_{no} e^{-F\tau}$, $J_o(\tau) = \mathcal{V}_o J e^{F\tau}\mathcal{D}_o$, $J_{ono}(\tau) = \mathcal{V}_o J e^{F\tau}\mathcal{D}_{no}$, $J_{noo}(\tau) = \mathcal{V}_{no} J e^{F\tau}\mathcal{D}_o$, $J_{no}(\tau) = \mathcal{V}_{no} J e^{F\tau}\mathcal{D}_{no}$, $H_{c,o}(\tau) = H_c e^{F\tau}\mathcal{D}_o$, $H_{d,o}(\tau) = H_d e^{F\tau}\mathcal{D}_o$, and $H_{d,no}(\tau) = H_d e^{F\tau}\mathcal{D}_{no}$. This idea of bringing at the jumps the whole combination of flows and jumps is similar to the so-called *equivalent discrete-time system* exhibited in [6] for jump-based observer designs. Notice that by definition and thanks to linearity, the observability decomposition through $\mathcal{V}$ ensures that the flow dynamics of z_o and y_c are totally independent of z_{no}, which only impacts z_o at jumps. In other words, the whole dependence of the observable part on the non-observable part via flows and jumps has been gathered at the jumps. Besides, z_o is by definition instantaneously observable from y_c. More precisely, for any $\delta > 0$, there exists $\alpha > 0$ such that the observability Gramian of the continuous pair $(0, H_{c,o}(\tau))$ satisfies

$$\int_t^{t+\delta} H_{c,o}^\top(s) H_{c,o}(s) ds = \int_t^{t+\delta} \mathcal{D}_o^\top e^{F^\top s} H_c^\top H_c e^{Fs}\mathcal{D}_o ds \geq \alpha \,\mathrm{Id}, \qquad \forall t \geq 0. \tag{6.27}$$

Indeed, this Gramian corresponds to the observability Gramian of the pair (F, H_c) projected onto the observable subspace. This condition is thus related to the *uniform complete observability* of the continuous pair $(0, H_{c,o}(\tau))$ in the Kalman literature [16] (continuous version of the one in Definition 6.2.5), but here with an arbitrarily small window δ. Since z_o is observable via y_c, we propose to estimate z_o sufficiently fast during flows to compensate for the interaction with z_{no} at jumps. Then, intuitively from (6.26), information about z_{no} may be drawn from two sources: the jump output y_d and the part of z_{no} impacting z_o at jumps, namely $J_{ono}(\tau)z_{no}$, which may become "visible" in z_o, via y_c during the following flow interval. This is illustrated in Example 6.2.8 below. Actually, we show next in Sect. 6.2.4 that the detectability of z_{no} comes from these two sources of information only.

Example 6.2.8 Consider a hybrid system of form (6.1) with state $x = (x_1, x_2, x_3, x_4)$, $u_c = 0$, $u_d = 0$, and the matrices

$$F = \begin{pmatrix} 0 & -1 & 0 & 0 \\ 1 & 0 & 0 & 0 \\ 0 & 0 & 0 & -2 \\ 0 & 0 & 2 & 0 \end{pmatrix}, \quad H_c = \begin{pmatrix} 1 & 0 & 0 & 0 \end{pmatrix}, \quad J = \begin{pmatrix} 0 & 0 & 1 & 0 \\ 0 & 0 & 0 & 0 \\ 0 & 0 & 1 & 0 \\ 0 & 0 & 0 & 1 \end{pmatrix}, \quad H_d = \begin{pmatrix} 0 & 0 & 1 & 0 \end{pmatrix}, \tag{6.28}$$

with random flow lengths varying in some compact set $\mathcal{I} \subset \left(0, \frac{\pi}{2}\right)$. It can be seen that only x_1 and x_2 are instantaneously observable during flows from y_c, but x_3 impacts x_1 at jumps (or y_d) and x_4 impacts x_3 during flows. Therefore, we may hope to estimate the full state. In order to decouple those various impacts and analyze detectability more easily, we proceed with the change of variables (6.24). We obtain

$$z_o = \begin{pmatrix} \cos(\tau) & \sin(\tau) & 0 & 0 \\ -\sin(\tau) & \cos(\tau) & 0 & 0 \end{pmatrix} x, \qquad z_{no} = \begin{pmatrix} 0 & 0 & \cos(2\tau) & \sin(2\tau) \\ 0 & 0 & -\sin(2\tau) & \cos(2\tau) \end{pmatrix} x, \tag{6.29}$$

and (6.26) with the matrices $J_o(\tau) = \begin{pmatrix} 0 & 0 \\ 0 & 0 \end{pmatrix}$, $J_{ono}(\tau) = \begin{pmatrix} \cos(2\tau) & -\sin(2\tau) \\ 0 & 0 \end{pmatrix}$, $J_{noo}(\tau) = \begin{pmatrix} 0 & 0 \\ 0 & 0 \end{pmatrix}$, $J_{no}(\tau) = \begin{pmatrix} \cos(2\tau) & -\sin(2\tau) \\ \sin(2\tau) & \cos(2\tau) \end{pmatrix}$, $H_{c,o}(\tau) = \begin{pmatrix} \cos(\tau) & -\sin(\tau) \end{pmatrix}$, $H_{d,o}(\tau) = \begin{pmatrix} 0 & 0 \end{pmatrix}$, and $H_{d,no}(\tau) = \begin{pmatrix} \cos(2\tau) & -\sin(2\tau) \end{pmatrix}$. It can be seen that in this case, the terms $J_{ono}(\tau)z_{no}$ and $H_{d,no}(\tau)z_{no}$ contain the same information on z_{no} and both should be able to let us estimate this part.

Remark 6.2.9 In what follows, a varying pair (F, H_c) can be considered as long as the transformation into the form (6.26) satisfying (6.27) exists and is invertible uniformly in time. This can be done with the transition matrix of F replacing the exponential form, if the observable subspace remains the same at all times. In that case, the jump matrices J_o, J_{ono}, J_{noo}, J_{no}, $H_{c,o}$, $H_{d,o}$, and $H_{d,no}$ are (discrete) known inputs that are no longer functions of τ only, but of the (discrete) jump index, which is not considered in this chapter. Similarly, J could vary at each jump as long as every related condition in the rest of this chapter holds uniformly in u_d.

6.2.4 Detectability Analysis

We first provide a more specific characterization of the (pre-)asymptotic detectability of (6.26) in the case of zero inputs (u_c, u_d). Indeed, we will see in Theorem 6.2.12 that this detectability is relevant to characterize that of the initial system $\mathcal{H}$.

Lemma 6.2.10 *The system (6.26) with known jump times and zero inputs (u_c, u_d) is pre-asymptotically detectable if and only if any of its complete solutions (z, τ) with zero inputs (u_c, u_d) and flow and jump outputs satisfying*

$$y_c(t, j) = 0, \qquad \forall t \in \text{int}(\mathcal{T}_j(z)), \forall j \in \text{dom}_j\, z, \tag{6.30a}$$

$$y_d(t_j, j-1) = 0, \qquad \forall j \in \text{dom}_j\, z, j \geq 1, \tag{6.30b}$$

verifies

$$\lim_{\substack{t+j \to +\infty \\ (t,j) \in \text{dom}\, z}} z(t, j) = 0. \tag{6.31}$$

Proof First, assume (6.26) is pre-asymptotically detectable. Let (z, τ) be a complete solution to (6.26) with zero inputs (u_c, u_d) and with outputs satisfying (6.30). Notice that the hybrid arc (z', τ), with $\text{dom}\, z' = \text{dom}\, z$ and z' constantly zero, is also a solution to (6.26) (thanks to linearity in the maps and the inputs (u_c, u_d) being zero). It can be seen that this solution is complete and also satisfies (6.30). By the pre-asymptotic detectability of (6.26), we have $\lim_{\substack{t+j \to +\infty \\ (t,j) \in \text{dom}\, z}} |z(t, j) - z'(t, j)| = 0$, which implies that $\lim_{\substack{t+j\to+\infty \\ (t,j)\in\operatorname{dom} z}} z(t,j) = 0$. Second, let us prove the converse. For that, assume that any complete solution to (6.26) with zero inputs (u_c, u_d) and with outputs verifying (6.30) is such that z converges to zero. Consider two complete solutions (z_a, τ_a) and (z_b, τ_b) of (6.26) with $\operatorname{dom}(z_a, \tau_a) = \operatorname{dom}(z_b, \tau_b)$, with zero inputs (u_c, u_d), and with outputs satisfying (6.4). By definition of the flow set, z_a and z_b jump at least once. Because the timers are both reset to zero at jumps, we have $\tau_a(t,j) = \tau_b(t,j) := \tau(t,j)$ for all $(t,j) \in \operatorname{dom} z_a$ such that $t \geq t_1$ and $j \geq 1$. By removing $[0, t_1] \times \{0\}$ from the time domain, we see that $(z_a - z_b, \tau)$ is a complete solution to (6.26) with outputs verifying (6.30) (thanks to (6.26) having linear maps in z and the flow and jump sets being independent of z). Therefore, by assumption, we have $\lim_{\substack{t+j\to+\infty \\ (t,j)\in\operatorname{dom} z_a}} (z_a - z_b)(t,j) = 0$, which implies the (pre-)asymptotic detectability of (6.26). □

Note that the equivalence of the *incremental* detectability as in Definition 6.2.1 with the *zero* detectability as in Lemma 6.2.10 is classical for linear continuous or discrete systems, but it is not automatic for hybrid systems with linear maps due to the flow/jump conditions. Here, it holds only because:

- The flow and jump conditions in (6.26) do not depend on z but only on τ;
- τ is determined uniquely after the first jump by the time domain of solutions;
- The inputs (u_c, u_d) are removed, thus avoiding a restriction of solutions of (6.26) due to a mismatch of time domains.

Theorem 6.2.11 *Assume that $0 \notin \mathcal{I}$ and $\mathcal{I}$ is compact. Then, the following three statements are equivalent:*

1. *The hybrid system (6.20) with zero inputs (u_c, u_d) and known jump times is asymptotically detectable;*
2. *The hybrid system (6.26) with zero inputs (u_c, u_d) and known jump times is asymptotically detectable;*
3. *The discrete system defined as*

$$z_{no,k+1} = J_{no}(\tau_k) z_{no,k}, \qquad y_k = H_{d,\text{ext}}(\tau_k) z_{no,k}, \tag{6.32}$$

where $H_{d,\text{ext}}(\tau_k) = \begin{pmatrix} H_{d,no}(\tau_k) \\ J_{ono}(\tau_k) \end{pmatrix}$, with $\tau_k \in \mathcal{I}$ for all $k \in \mathbb{N}$, is asymptotically detectable.

Proof First, notice that all maximal solutions of (6.20) and (6.26) are complete because their dynamics maps are linear and the flow and jump conditions do not depend on the state but only on the timer. Notice also that since $0 \notin \mathcal{I}$, consecutive jumps cannot happen, so that condition (6.4a) is equivalent to (6.6) following Remark 6.2.2. Because (6.20) and (6.26) are the same system modulo a uniformly invertible change of variables, 1. and 2. are equivalent. Then, let us prove that 2. implies 3. So assume 2. holds and consider a solution $(z_{no,k})_{k\in\mathbb{N}}$ of (6.32) with input

$(\tau_k)_{k\in\mathbb{N}}$ in $\mathcal{I}$, such that $y_k = 0$ for all $k \in \mathbb{N}$. We want to show that $(z_{no,k})_{k\in\mathbb{N}}$ asymptotically goes to zero. For that, we build and analyze the complete solution $z = (z_o, z_{no})$ of (6.26) initialized as $z_o(0,0) = 0$, $z_{no}(0,0) = z_{no,0}$, and $\tau(0,0) = 0$ with jumps verifying $\tau(t_j, j-1) = \tau_{j-1} \in \mathcal{I}$ for all $j \geq 1$ and zero inputs (u_c, u_d). It follows[2] from the fact that (i) z_o and z_{no} are constant during flows, (ii) y_c is independent of z_{no}, and (iii) $y_k = 0$ for all $k \in \mathbb{N}$, that for all $j \in \mathbb{N}$,

$$
\begin{aligned}
z_o(t,j) &= 0, && \forall t \in [t_j, t_{j+1}],\\
z_{no}(t,j) &= z_{no,j}, && \forall t \in [t_j, t_{j+1}],\\
y_c(t,j) &= 0, && \forall t \in [t_j, t_{j+1}],\\
z_o(t_{j+1}, j+1) &= J_{ono}(\tau_j) z_{no,j} = 0, &&\\
z_{no}(t_{j+1}, j+1) &= J_{no}(\tau_j) z_{no,j} = z_{no,j+1}, &&\\
y_d(t_{j+1}, j) &= H_{d,no}(\tau_j) z_{no,j} = 0. &&
\end{aligned}
$$

By 2. and Lemma 6.2.10, this implies $\lim_{\substack{t+j\to+\infty \\ (t,j)\in \mathrm{dom}\, z}} z_{no}(t,j) = 0$. Because this solution z_{no} coincides with $(z_{no,k})_{k\in\mathbb{N}}$ at the jumps, we deduce $\lim_{\substack{k\to+\infty \\ k\in\mathbb{N}}} z_{no,k} = 0$, implying 3. Finally, let us prove that 3. implies 2. Consider a complete solution $(z,\tau) = (z_o, z_{no}, \tau)$ of (6.26) with zero inputs (u_c, u_d) and such that (6.30) holds. By definition of the jump set, for all $j \in \mathrm{dom}_j\, z$, $\tau_j := \tau(t_{j+1}, j) \in \mathcal{I}$. Since $0 \notin \mathcal{I}$, the solution admits a dwell time and, because we look for an asymptotic property, we may assume without any loss of generality that the solution starts with a flow, possibly overlooking the first part of the domain (with $j = 0$) in case of a jump at time 0. Since z_o is instantaneously observable during flows according to (6.27), $y_c = 0$ implies that z_o is zero during each flow interval. Next, as $0 \notin \mathcal{I}$, there is no more than one jump at each jump time so that $z_o(t,j) = 0$ for all $(t,j) \in \mathrm{dom}\, z$. Besides, since $\mathcal{I}$ is compact, $\mathrm{dom}_j\, z = \mathbb{N}$ and from (6.26a), we then have $J_{ono}(\tau(t_j, j-1)) z_{no}(t_j, j-1) = 0$ for all $j \in \mathrm{dom}_j\, z$ with $j \geq 1$. Therefore, since z_{no} is constant during flows and $y_d(t_j, j-1) = 0$, for all $j \in \mathrm{dom}_j\, z$, $j \geq 1$, we have for all $j \in \mathbb{N}$,

$$
\begin{aligned}
z_{no}(t_{j+1}, j+1) = J_{no}(\tau(t_{j+1}, j)) z_{no}(t_{j+1}, j) &= J_{no}(\tau_j) z_{no}(t_j, j),\\
H_{d,no}(\tau(t_{j+1}, j)) z_{no}(t_{j+1}, j) &= H_{d,no}(\tau_j) z_{no}(t_j, j) = 0,\\
J_{ono}(\tau(t_{j+1}, j)) z_{no}(t_{j+1}, j) &= J_{ono}(\tau_j) z_{no}(t_j, j) = 0.
\end{aligned}
$$

By considering the sequence $(z_{no}(t_j, j))_{j\in\mathbb{N}}$ solution to (6.32) with input $(\tau_j)_{j\in\mathbb{N}}$ in $\mathcal{I}$ and applying 3., we obtain $\lim_{\substack{j\to+\infty \\ j\in\mathbb{N}}} z_{no}(t_j, j) = 0$. Since z_{no} is constant during flows, we have $\lim_{\substack{t+j\to+\infty \\ (t,j)\in \mathrm{dom}\, z}} = 0$, implying 2. according to Lemma 6.2.10. □

We thus conclude that (at least when the inputs (u_c, u_d) are zero) the asymptotic detectability of (6.20) requires z_{no} to be asymptotically detectable through the output made of the measured output y_d and the *fictitious* one $J_{ono}(\tau) z_{no}$, which describes

[2] Note that $0 \notin \mathcal{I}$ and the compactness of $\mathcal{I}$ are not needed to prove this direction.

how z_{no} impacts z_o at jumps. We insist also that the detectability of z_{no} comes from the combination of flows and jumps and not due to jumps alone since the useful information contained in the flow dynamics and output is gathered at the jumps via the transformation (6.24).

It follows from this analysis that the design of an asymptotic observer for $\mathcal{H}$ with gains computed offline from the knowledge of $\mathcal{I}$ only (without any special consideration of $\mathcal{X}_0, \mathcal{U}, C$, or D), namely an observer for (6.20), without considering the possible restrictions of time domains by the inputs (u_c, u_d), requires the asymptotic detectability of the discrete system (6.32). This is because any time domain with flow lengths in $\mathcal{I}$ is a priori possible. This will be done in Sect. 6.3 with an LMI-based design. On the other hand, if we consider the more precise problem of observer design with time domains restricted to that of solutions in $\mathcal{S}_{\mathcal{H}}(\mathcal{X}_0, \mathcal{U})$, we end up with the following sufficient condition for asymptotic detectability of $\mathcal{H}$.

Theorem 6.2.12 *Suppose Assumptions 6.1.1 and 6.2.7 hold. Then, $\mathcal{H}$ initialized in $\mathcal{X}_0$ with inputs in $\mathcal{U}$ is asymptotically detectable if for each $x \in \mathcal{S}_{\mathcal{H}}(\mathcal{X}_0, \mathcal{U})$, the discrete system (6.32) with input $(\tau_k)_{k\in\mathbb{N}}$ defined as $\tau_k = t_{k+1}(x) - t_k(x)$ for all $k \in \mathbb{N}$ is asymptotically detectable.*

Proof Pick solutions x_a and x_b in $\mathcal{S}_{\mathcal{H}}(\mathcal{X}_0, \mathcal{U})$ with the same inputs $(F, J, H_c, H_d, u_c, u_d)$, such that $\operatorname{dom} x_a = \operatorname{dom} x_b$, and with outputs $y_{a,c}$, $y_{b,c}$, $y_{a,d}$, $y_{b,d}$ satisfying (6.4). By Assumptions 6.1.1 and 6.2.7, these solutions are complete with $\operatorname{dom}_j x_a = \operatorname{dom}_j x_b = \mathbb{N}$. By Assumption 6.2.7, for all $j \geq j_m$, $\tau_j := t_{j+1}(x_a) - t_j(x_a) = t_{j+1}(x_b) - t_j(x_b) \in \mathcal{I}$. Since $0 \notin \mathcal{I}$, the solutions admit a dwell time after the first j_m jumps. Since we look for an asymptotic property, we may discard the first part of the solutions with $j < j_m$, and assume without any loss of generality that they start with a flow interval and have a dwell time. Then, condition (6.4a) is equivalent to (6.6) following Remark 6.2.2. Consider the hybrid signals (z_a, z_b, τ) defined for all $(t, j) \in \operatorname{dom} x_a$ as

$$\tau(t, j) = t - t_j,$$

$$\begin{pmatrix} z_{a,o}(t, j) \\ z_{a,no}(t, j) \end{pmatrix} = \begin{pmatrix} \mathcal{V}_o \\ \mathcal{V}_{no} \end{pmatrix} e^{-F\tau(t,j)} x_a(t, j),$$

$$\begin{pmatrix} z_{b,o}(t, j) \\ z_{b,no}(t, j) \end{pmatrix} = \begin{pmatrix} \mathcal{V}_o \\ \mathcal{V}_{no} \end{pmatrix} e^{-F\tau(t,j)} x_b(t, j).$$

We see that both (z_a, τ) and (z_b, τ) are solutions to (6.26) with the same τ-dependent matrices $(G_o, G_{no}, J_o, J_{ono}, J_{noo}, J_{no}, H_{c,o}, H_{d,o}, H_{d,no})$ and the same inputs (u_c, u_d). Since $z_{a,o}$ and $z_{b,o}$ are instantaneously observable during flows according to (6.27), $y_{a,c} = y_{b,c}$ implies that $z_{a,o} = z_{b,o}$ during each flow interval. Next, as $0 \notin \mathcal{I}$, there is no more than one jump at each jump time so that $z_{a,o}(t, j) = z_{b,o}(t, j)$ for all $(t, j) \in \operatorname{dom} x_a$. Besides, since u_d is the same for both solutions, from (6.26a), we have $J_{ono}(\tau(t_j, j-1))z_{a,no}(t_j, j-1) = J_{ono}(\tau(t_j, j-1))z_{b,no}(t_j, j-1)$ for all $j \geq 1$. Therefore, since $z_{a,no}$ and $z_{b,no}$ evolve in the same way during flows (with the same u_c) and $y_{a,d}(t_j, j-1) = y_{b,d}(t_j, j-1)$, for all

$j \geq 1$, by defining $\tilde{z}_{no} = z_{a,no} - z_{b,no}$, we have that $\tilde{z}_{no}$ is constant during flows and for all $j \in \mathbb{N}$,

$$\begin{aligned} \tilde{z}_{no}(t_{j+1}, j+1) &= J_{no}(\tau(t_{j+1}, j))\tilde{z}_{no}(t_{j+1}, j) = J_{no}(\tau_j)\tilde{z}_{no}(t_j, j), \\ H_{d,no}(\tau(t_{j+1}, j))\tilde{z}_{no}(t_{j+1}, j) &= H_{d,no}(\tau_j)\tilde{z}_{no}(t_j, j) = 0, \\ J_{ono}(\tau(t_{j+1}, j))\tilde{z}_{no}(t_{j+1}, j) &= J_{ono}(\tau_j)\tilde{z}_{no}(t_j, j) = 0. \end{aligned}$$

By considering the sequence $(\tilde{z}_{no}(t_j, j))_{j\in\mathbb{N}}$ solution to (6.32) with input $(\tau_j)_{j\in\mathbb{N}}$ in $\mathcal{I}$ and using the asymptotic detectability of (6.32) for the particular sequence $(\tau_k)_{k\in\mathbb{N}}$ generated by x_a and x_b, we obtain $\lim_{\substack{j\to+\infty \\ j\in\mathbb{N}}} \tilde{z}_{no}(t_j, j) = 0$. Since $\tilde{z}_{no}$ is constant during flows, we have $\lim_{\substack{t+j\to+\infty \\ (t,j)\in \mathrm{dom}\, x_a}} \tilde{z}_{no}(t, j) = 0$, which means that $\lim_{\substack{t+j\to+\infty \\ (t,j)\in \mathrm{dom}\, x_a}} (z_{a,no}(t, j) - z_{b,no}(t, j)) = 0$, implying that $\mathcal{H}$ initialized in $\mathcal{X}_0$ with inputs in $\mathcal{U}$ is asymptotically detectable according to Lemma 6.2.10. □

Unlike Theorems 6.2.11, 6.2.12 does not give a *necessary* condition for detectability (and thus observer design). The reason is that the flow and jump conditions of $\mathcal{H}$ are not taken into account in (6.32). But it still suggests us to build observers for $\mathcal{H}$ under observability/detectability conditions on (6.32) for the flow length sequences $(\tau_k)_{k\in\mathbb{N}}$ appearing in $\mathcal{S}_{\mathcal{H}}(\mathcal{X}_0, \mathcal{U})$. This will be done in Sects. 6.3 and 6.4 through LMI- or KKL-based design.

Remark 6.2.13 Compared to the preliminary work [29], the observer designs in this chapter do not require an extra (constant) transformation decoupling the part of z_{no} detectable from y_d and the part detectable from the fictitious output $J_{ono}(\tau)z_{no}$. We instead consider an extended output in (6.32) and the decomposition (6.26) proposed in this chapter is thus less conservative. For instance, Example 6.2.8 can be cast into the decomposition (6.26), but fails to fall into the scope of the decomposition of [29] unless $\mathcal{I} = \left\{\frac{\pi}{2}\right\}$.

Example 6.2.14 Consider the system in Example 6.2.8. It is possible to check, by computing the (time-varying) observability matrix of the pair $(J_{no}(\tau_k), H_{d,no}(\tau_k))$, that (6.32) is observable for any sequence $(\tau_k)_{k\in\mathbb{N}}$ as long as $\sin(\tau_k + \tau_{k+1}) \neq 0$ at some $k \in \mathbb{N}$, which is the case for $\tau_k \in \mathcal{I}$ since $\mathcal{I} \subset \left(0, \frac{\pi}{2}\right)$. This implies in particular that (6.32) is detectable. Actually, even if $H_d = 0$, i.e., no output is available at jumps, the pair $(J_{no}(\tau_k), J_{ono}(\tau_k))$ is also observable using the same arguments. This means that z_{no} is actually observable through the fictitious measurement of z_o at jumps. We see from this example that thanks to the flow-jump coupling, by using z_o as a fictitious measurement, state components that are not observable during flows from the flow output may become observable via jumps even without any additional measurements at jumps (hidden dynamics), only through the way they impact the observable ones.

6.3 LMI-Based Observer Design from Discrete Quadratic Detectability in Observability Decomposition

Inspired by the detectability analysis of Theorem 6.2.11, we propose a first observer design under the following detectability assumption.

Assumption 6.3.1 Given τ_M and $\mathcal{I}$ defined in Assumption 6.2.7, there exist $Q_{no} \in \mathcal{S}_{>0}^{n_{no}}$, $L_{d,no} : [0, \tau_M] \to \mathbb{R}^{n_{no} \times n_{y,d}}$ bounded on $[0, \tau_M]$ and continuous on $\mathcal{I}$, and $K_{no} \in \mathbb{R}^{n_{no} \times n_o}$ such that

$$\star^\top Q_{no} \left(J_{no}(\tau) - \left(L_{d,no}(\tau) \; K_{no} \right) \begin{pmatrix} H_{d,no}(\tau) \\ J_{ono}(\tau) \end{pmatrix} \right) - Q_{no} < 0, \qquad \forall \tau \in \mathcal{I}. \quad (6.33)$$

We refer the reader to Remark 6.3.3 for constructive methods to solve (6.33), where it is shown that the solvability of (6.33) in Q_{no} and $\left(L_{d,no}(\tau) \; K_{no}\right)$ is equivalent to that of a reduced LMI involving Q_{no} only. Consistently with Theorem 6.2.11, Assumption 6.3.1 requires the detectability of the pair $\left(J_{no}(\tau), \begin{pmatrix} H_{d,no}(\tau) \\ J_{ono}(\tau) \end{pmatrix} \right)$ for each frozen $\tau \in \mathcal{I}$. But it is actually stronger because it further requires Q_{no} and K_{no} to be independent of τ. It corresponds to a stronger version of the *quadratic detectability* of (6.32) defined in [32]. Actually, the detectability of Assumption 6.3.1 allows us to build an observer for any sequence of flow lengths $(\tau_k)_{k \in \mathbb{N}} \in \mathcal{I}$ and thus requires the detectability of the discrete pair for any such sequences, which is still consistent with the result of Theorem 6.2.11. Note that the reason why K_{no} is required to be independent of τ is that it is used to carry out another change of variables in the proof of Theorem 6.3.2 below, allowing us to exhibit the *fictitious* output in the analysis.

6.3.1 LMI-Based Observer Design in the z-Coordinates

Because the flow output matrix $H_{c,o}(\cdot)$ varies and satisfies the observability condition (6.27), we design a flow-based *Kalman-like* observer of z_o during flows using y_c [7]. Its advantage over a Kalman observer is that it admits a *strict* Lyapunov function, allowing for direct robust Lyapunov analysis. Besides, it provides a direct relationship between the Lyapunov matrix and the observability Gramian. Then, as suggested by the detectability analysis, z_{no} should be estimated thanks to both y_d and its interaction with z_o at jumps via a fictitious output. The latter is not available for injection in the observer, but it becomes visible through z_o after the jump, and thus through y_c during flows. This justifies correcting the estimate of z_{no} during flows with y_c, via the gain K_{no}. The dynamics of the observer are then given by

$$
\begin{cases}
\dot{\hat{z}}_o = G_o(\tau)u_c + P^{-1}H_{c,o}^\top(\tau)R^{-1}(\tau)(y_c - H_{c,o}(\tau)\hat{z}_o) \\
\dot{\hat{z}}_{no} = G_{no}(\tau)u_c + K_{no}P^{-1}H_{c,o}^\top(\tau)R^{-1}(\tau)(y_c - H_{c,o}(\tau)\hat{z}_o) \\
\dot{P} = -\lambda P + H_{c,o}^\top(\tau)R^{-1}(\tau)H_{c,o}(\tau) \\
\dot{\tau} = 1 \\
\\
\hat{z}_o^+ = J_o(\tau)\hat{z}_o + J_{ono}(\tau)\hat{z}_{no} + \mathcal{V}_o u_d \\
\hat{z}_{no}^+ = J_{noo}(\tau)\hat{z}_o + J_{no}(\tau)\hat{z}_{no} + \mathcal{V}_{no}u_d \\
\qquad + L_{d,no}(\tau)(y_d - H_{d,o}(\tau)\hat{z}_o - H_{d,no}(\tau)\hat{z}_{no}) \\
P^+ = P_0 \\
\tau^+ = 0,
\end{cases}
\tag{6.34}
$$

with jumps triggered at the same time as $\mathcal{H}$ in the same way as (6.2), $P_0 \in \mathcal{S}_{>0}^{n_o}$, K_{no} and $L_{d,no}$ given by Assumption 6.3.1, and where $\tau \mapsto R(\tau) \in \mathcal{S}_{>0}^{n_{y,c}}$ is a positive definite weighting matrix that is defined and is continuous on $[0, \tau_M]$ to be chosen arbitrarily. The estimate is then recovered by using (6.25) on $\hat{z}$ with the global exponential stability (GES) of the estimation error as stated next.

Theorem 6.3.2 *Under Assumptions 6.1.1, 6.2.7, and 6.3.1, given any $P_0 \in \mathcal{S}_{>0}^{n_o}$, there exists $\lambda^\star > 0$ such that for any $\lambda > \lambda^\star$, there exist $\rho_1 > 0$ and $\lambda_1 > 0$ such that for any solution $x \in \mathcal{S}_\mathcal{H}(\mathcal{X}_0, \mathcal{U})$ and any solution $(\hat{z}, P, \tau)$ of the observer (6.34) with $P(0,0) = P_0$, $\tau(0,0) = 0$, and jumps triggered at the same time as in x, $(\hat{z}, P, \tau)$ is complete and we have*

$$
|x(t,j) - \hat{x}(t,j)| \leq \rho_1 |x(0,0) - \hat{x}(0,0)| e^{-\lambda_1(t+j)}, \qquad \forall (t,j) \in \operatorname{dom} x, \tag{6.35}
$$

with $\hat{x}$ obtained by $\hat{x} = e^{F\tau}\mathcal{D}\hat{z}$ with $\mathcal{D}$ defined in (6.25).

Proof Consider a solution $x \in \mathcal{S}_\mathcal{H}(\mathcal{X}_0, \mathcal{U})$ and a solution $(\hat{z}, P, \tau)$ of the observer (6.34) with $P(0,0) = P_0$ and $\tau(0,0) = 0$, and jumps triggered at the same time as in x. By Assumption 6.1.1, it is complete and so is $(\hat{z}, P, \tau)$. Following (6.24), define

$$
z(t,j) := \mathcal{V}e^{-F\tau(t,j)}x(t,j), \qquad \forall (t,j) \in \operatorname{dom} x, \tag{6.36}
$$

and consider the error $\tilde{z} = (\tilde{z}_o, \tilde{z}_{no}) = (z_o - \hat{z}_o, z_{no} - \hat{z}_{no})$. Because the flow lengths of x are in $\mathcal{I}$ by Assumption 6.2.7 after the first j_m jumps only, the proof consists of two parts: first, we use Lemma 6.6.1 to show the exponential convergence of $\tilde{z}$ starting at hybrid time (t_{j_m}, j_m) by putting the error dynamics into the appropriate form, and then, we analyze the behavior of the error before (t_{j_m}, j_m) using Lemma 6.6.3. Consider first the solution $(\tilde{z}, P, \tau)$ starting from (t_{j_m}, j_m). According to Assumption 6.2.7 and since the observer's jumps are synchronized with those of $\mathcal{H}$, $(\tilde{z}, P, \tau)$ is solution to

$$\begin{cases} \dot{\tilde{z}}_o = & -P^{-1}H_{c,o}^\top(\tau)R^{-1}(\tau)H_{c,o}(\tau)\tilde{z}_o \\ \dot{\tilde{z}}_{no} = & -K_{no}P^{-1}H_{c,o}^\top(\tau)R^{-1}(\tau)H_{c,o}(\tau)\tilde{z}_o \\ \dot{P} = & -\lambda P + H_{c,o}^\top(\tau)R^{-1}(\tau)H_{c,o}(\tau) \\ \dot{\tau} = 1 \\ \\ \tilde{z}_o^+ = J_o(\tau)\tilde{z}_o + J_{ono}(\tau)\tilde{z}_{no} \\ \tilde{z}_{no}^+ = (J_{noo}(\tau) - L_{d,no}(\tau)H_{d,o}(\tau))\tilde{z}_o + (J_{no}(\tau) - L_{d,no}(\tau)H_{d,no}(\tau))\tilde{z}_{no} \\ P^+ = P_0 \\ \tau^+ = 0, \end{cases} \tag{6.37a}$$

with the flow and jump sets

$$\mathbb{R}^{n_o} \times \mathbb{R}^{n_{no}} \times \mathbb{R}^{n_o \times n_o} \times [0, \tau_M], \qquad \mathbb{R}^{n_o} \times \mathbb{R}^{n_{no}} \times \mathbb{R}^{n_o \times n_o} \times \mathcal{I}. \tag{6.37b}$$

We next perform the change of variables

$$\tilde{\eta} = \tilde{z}_{no} - K_{no}\tilde{z}_o, \tag{6.38}$$

which transforms the error system (6.37) into

$$\begin{cases} \dot{\tilde{z}}_o = & -P^{-1}H_{c,o}^\top(\tau)R^{-1}(\tau)H_{c,o}(\tau)\tilde{z}_o \\ \dot{\tilde{\eta}} = 0 \\ \dot{P} = & -\lambda P + H_{c,o}^\top(\tau)R^{-1}(\tau)H_{c,o}(\tau) \\ \dot{\tau} = 1 \\ \\ \tilde{z}_o^+ = \overline{J}_o(\tau)\tilde{z}_o + J_{ono}(\tau)\tilde{\eta} \\ \tilde{\eta}^+ = (\overline{J}_{noo}(\tau) - L_{d,no}(\tau)\overline{H}_{d,o}(\tau))\tilde{z}_o + J_\eta(\tau)\tilde{\eta} \\ P^+ = 0 \\ \tau^+ = 0, \end{cases} \tag{6.39}$$

with the same flow and jump sets where $\overline{J}_o(\tau) = J_o(\tau) + J_{ono}(\tau)K_{no}$, $\overline{J}_{noo}(\tau) = J_{noo}(\tau) + J_{no}(\tau)K_{no} - K_{no}J_o(\tau) - K_{no}J_{ono}(\tau)K_{no}$, $\overline{H}_{d,o}(\tau) = H_{d,o}(\tau) + H_{d,no}(\tau)K_{no}$, and $J_\eta(\tau) = J_{no}(\tau) - L_{d,no}(\tau)H_{d,no}(\tau) - K_{no}J_{ono}(\tau)$, with the flow set $\mathbb{R}^{n_o} \times \mathbb{R}^{n_{no}} \times [0, \tau_M]$ and the jump set $\mathbb{R}^{n_o} \times \mathbb{R}^{n_{no}} \times \mathcal{I}$. From Assumption 6.3.1, $J_\eta(\tau)$ is Schur for all $\tau \in \mathcal{I}$, and more precisely, there exists $Q_\eta \in \mathcal{S}_{>0}^{n_{no}}$ such that

$$J_\eta^\top(\tau)Q_\eta J_\eta(\tau) - Q_\eta < 0, \qquad \forall \tau \in \mathcal{I}. \tag{6.40}$$

Using Lemma 6.6.1, we proceed to prove the GES of the error $(\tilde{z}_o, \tilde{\eta})$ with respect to the value $(\tilde{z}_o, \tilde{\eta})(t_{j_m}, j_m)$. Then the GES with respect to $(\tilde{z}_o, \tilde{\eta})(0, 0)$ is proven using Lemma 6.6.3. Last, because the transformations (6.24) and (6.38) are linear with $\tau \mapsto e^{F\tau}$ bounded with a strictly positive lower bound on the compact set $[0, \tau_M]$, we obtain (6.35) observing that $(\tilde{z}_o, \tilde{\eta}) = \begin{pmatrix} \mathcal{V}_o \\ \mathcal{V}_{no} - K_{no}\mathcal{V}_o \end{pmatrix} e^{-F\tau}(x - \hat{x})$. □

Remark 6.3.3 Applying Schur's lemma and then the elimination lemma [8] to (6.33), we see that $Q_{no} \in \mathcal{S}^{n_{no}}_{>0}$ exists if and only if there exists a solution to the LMI

$$\begin{pmatrix} \begin{pmatrix} H_{d,no}(\tau) \\ J_{ono}(\tau) \end{pmatrix}^{\perp\top} Q_{no} \begin{pmatrix} H_{d,no}(\tau) \\ J_{ono}(\tau) \end{pmatrix}^{\perp} & \star \\ Q_{no} J_{no}(\tau) \begin{pmatrix} H_{d,no}(\tau) \\ J_{ono}(\tau) \end{pmatrix}^{\perp} & Q_{no} \end{pmatrix} > 0, \qquad \forall \tau \in \mathcal{I}. \tag{6.41}$$

If such a Q_{no} is obtained, the gains $L_{d,no}(\cdot)$ and K_{no} are then found by using (6.33) with Q_{no} known. If $\mathcal{I}$ is infinite, then there is an infinite number of LMIs to solve. Actually, it is worth noting that the exponential term $e^{F\tau}$ contained in all the τ-dependent matrices in (6.41) can be expanded using residue matrices [12], as

$$\begin{aligned} e^{F\tau} &= \sum_{i=1}^{\sigma_r} \sum_{j=1}^{m_i^r} R_{ij} e^{\lambda_i \tau} \frac{\tau^{j-1}}{(j-1)!} \\ &+ \sum_{i=1}^{\sigma_c} \sum_{j=1}^{m_i^c} 2 e^{\Re(\lambda_i)\tau} (\Re(R_{ij}) \cos(\Im(\lambda_i)\tau) - \Im(R_{ij}) \sin(\Im(\lambda_i)\tau)) \frac{\tau^{j-1}}{(j-1)!}, \end{aligned} \tag{6.42}$$

where σ_r and σ_c are the numbers of distinct real eigenvalues and complex conjugate eigenvalue pairs; m_i^r and m_i^c are the multiplicity of the real eigenvalue λ_i and of the complex conjugate eigenvalue pair λ_i, λ_i^* in the minimal polynomial of F; $R_{ij} \in \mathbb{R}^{n_x \times n_x}$ are matrices corresponding to the residues associated to the partial fraction expansion of $(sI - F)^{-1}$. This in turn allows $e^{F\tau}$ to be written as a finite sum of matrices affine in N scalar functions $\beta_{ij} = e^{\lambda_i \tau} \tau^{j-1}$, $\gamma_{ij} = e^{\Re(\lambda_i)\tau} \cos(\Im(\lambda_i)\tau)\tau^{j-1}$, and $\gamma_{ij}^* = e^{\Re(\lambda_i)\tau} \sin(\Im(\lambda_i)\tau)\tau^{j-1}$. It then implies that (6.41) can be solved in a polytopic approach, i.e., the LMIs are satisfied for all $\tau \in \mathcal{I}$ compact if they are satisfied at the finite number 2^N of vertices of the polytope formed by these scalar functions when τ varies in $\mathcal{I}$. Alternatively, the LMIs can be solved in a grid-based approach followed by post-analysis of the solution's stability as in [32], possibly with a theoretical proof extended from [24].

6.3.2 LMI-Based Observer Design in the *x*-Coordinates

In this section, we show that the observer can equivalently be implemented directly in the original x-coordinates, with dynamics given by

$$\begin{cases} \dot{\hat{x}} = F\hat{x} + u_c + \mathcal{P}H_c^\top R^{-1}(\tau)(y_c - H_c\hat{x}) \\ \dot{\mathcal{P}} = \lambda\mathcal{P} + F\mathcal{P} + \mathcal{P}F^\top - \mathcal{P}H_c^\top R^{-1}(\tau)H_c\mathcal{P} \\ \dot{\tau} = 1 \\ \\ \hat{x}^+ = J\hat{x} + u_d + \mathcal{D}_{no}L_{d,no}(\tau)(y_d - H_d\hat{x}) \\ \mathcal{P}^+ = (\mathcal{D}_o + \mathcal{D}_{no}K_{no})P_0^{-1}(\mathcal{D}_o + \mathcal{D}_{no}K_{no})^\top \\ \tau^+ = 0, \end{cases} \tag{6.43}$$

with jumps still triggered at the same time as $\mathcal{H}$. The GES of the error is proven in Theorem 6.3.4.

Theorem 6.3.4 *Under Assumptions 6.1.1, 6.2.7, and 6.3.1, given any $P_0 \in \mathcal{S}_{>0}^{n_o}$, there exists $\lambda^\star > 0$ such that for any $\lambda > \lambda^\star$, there exist $\rho_1 > 0$ and $\lambda_1 > 0$ such that for any solution $x \in \mathcal{S}_\mathcal{H}(\mathcal{X}_0, \mathcal{U})$ and for any solution $(\hat{x}, \mathcal{P}, \tau)$ of (6.43) with $\mathcal{P}(0,0) = (\mathcal{D}_o + \mathcal{D}_{no}K_{no})P_0^{-1}(\mathcal{D}_o + \mathcal{D}_{no}K_{no})^\top$, $\tau(0,0) = 0$, and jumps triggered at the same time as in x, $(\hat{x}, \mathcal{P}, \tau)$ is complete and we have*

$$|x(t,j) - \hat{x}(t,j)| \le \rho_1|x(0,0) - \hat{x}(0,0)|e^{-\lambda_1(t+j)}, \qquad \forall (t,j) \in \operatorname{dom} x. \tag{6.44}$$

Proof Pick a solution $x \in \mathcal{S}_\mathcal{H}(\mathcal{X}_0, \mathcal{U})$ and a solution $(\hat{\zeta}, \mathcal{P}, \tau)$ of (6.43) with $\mathcal{P}(0,0) = (\mathcal{D}_o + \mathcal{D}_{no}K_{no})P_0^{-1}(\mathcal{D}_o + \mathcal{D}_{no}K_{no})^\top$, and $\tau(0,0) = 0$, with jumps triggered at the same time as in x. Consider $(\hat{z}, P, \underline{\tau})$ solution to observer (6.34), with $\hat{z}(0,0) = \mathcal{V}\hat{x}(0,0)$, $P(0,0) = P_0$, and $\underline{\tau}(0,0) = 0$, with jumps triggered at the same time as in x. First, notice from its dynamics that $\underline{\tau} = \tau$. Then, applying Theorem 6.3.2, we get

$$|x(t,j) - \underline{\hat{x}}(t,j)| \le \rho_1|x(0,0) - \underline{\hat{x}}(0,0)|e^{-\lambda_1(t+j)}, \qquad \forall (t,j) \in \operatorname{dom} x,$$

where $\underline{\hat{x}} = e^{F\tau}\mathcal{D}\hat{z}$. The proof consists in showing that $\underline{\hat{x}} = \hat{x}$, thus obtaining (6.44). Observe that

$$\dot{\underline{\hat{x}}} = F\underline{\hat{x}} + u_c + L_c(P,\tau)(y_c - H_c\underline{\hat{x}}), \tag{6.45a}$$

during flows and

$$\underline{\hat{x}}^+ = J\underline{\hat{x}} + u_d + L_d(\tau)(y_d - H_d\underline{\hat{x}}), \tag{6.45b}$$

at jumps where

$$\begin{aligned} L_c(P,\tau) &= e^{F\tau}\mathcal{D}\begin{pmatrix} P^{-1}H_{c,o}^\top(\tau)R^{-1}(\tau) \\ K_{no}P^{-1}H_{c,o}^\top(\tau)R^{-1}(\tau) \end{pmatrix} \\ &= e^{F\tau}(\mathcal{D}_o + \mathcal{D}_{no}K_{no})P^{-1}\mathcal{D}_o^\top e^{F^\top\tau}H_c^\top R^{-1}(\tau), \end{aligned} \tag{6.46a}$$

and

$$L_d(\tau) = \mathcal{D}\begin{pmatrix} 0 \\ L_{d,no}(\tau) \end{pmatrix} = \mathcal{D}_{no}L_{d,no}(\tau). \tag{6.46b}$$

From (6.23), $L_c(P,\tau)$ can be re-written as $L_c(P,\tau) = \underline{\mathcal{P}} H_c^\top R^{-1}(\tau)$ where

$$\underline{\mathcal{P}} = e^{F\tau}(\mathcal{D}_o + \mathcal{D}_{no} K_{no}) P^{-1} (\mathcal{D}_o + \mathcal{D}_{no} K_{no})^\top e^{F^\top \tau}. \tag{6.47}$$

Calculating $\dot{\underline{\mathcal{P}}}$ while noting that the time derivative of P^{-1} is $-P^{-1}\dot{P}P^{-1}$, we obtain the same flow/jump dynamics as $\mathcal{P}$ in (6.43). Besides, $\underline{\mathcal{P}}(0,0) = \mathcal{P}(0,0)$, so $\underline{\mathcal{P}} = \mathcal{P}$ thanks to the uniqueness of solutions. We deduce that $\hat{\underline{x}}$ follows the same dynamics as $\hat{x}$, and since

$$\hat{\underline{x}}(0,0) = \mathcal{D}'\hat{z}(0,0) = \hat{x}(0,0),$$

we have $\hat{\underline{x}} = \hat{x}$, which concludes the proof. □

It is interesting to see that the observability provided at jumps by the fictitious output in the non-observable subspace $\mathcal{D}_{no}$ is stored into $\mathcal{P}$ at jumps. This allows the use of y_c to correct the estimate in the non-observable subspace during flows, while the Riccati dynamics of $\mathcal{P}$ instead excites only the observable directions provided by (6.27). In terms of implementation, the observer (6.43) is of a larger dimension than that in (6.34) but it allows us to avoid the online inversion of the change of variables. Actually, the observer (6.43) has the same dimension and the same flow dynamics as the Kalman-like observer proposed in [28]. The difference lies in (i) the jump dynamics, which here contains *a priori* gains K_{no} and $L_{d,no}$ instead of dynamic gains computed online via $\mathcal{P}$, and (ii) the quadratic *detectability* assumption (6.3.1) which replaces the UCO in [28].

Remark 6.3.5 Denote $L_d(\tau) = \mathcal{D}_{no} L_{d,no}(\tau)$ and observe that $L_{d,no}(\tau) = \mathcal{V}_{no} L_d(\tau)$. Similarly, denote $K'_{no} = \mathcal{D}_{no} K_{no}$ and observe that $K_{no} = \mathcal{V}_{no} K'_{no}$. The conditions in Assumption 6.3.1 for the design of $L_{d,no}(\cdot)$ and K_{no} are equivalent to solving for $L_d(\cdot)$ and K'_{no} directly in the x-coordinates and for all $\tau \in \mathcal{I}$,

$$\star^\top \, Q_{no} \mathcal{V}_{no} (J - L_d(\tau) H_d - K'_{no} \mathcal{V}_o J) e^{F\tau} \mathcal{D}_{no} - Q_{no} < 0, \tag{6.48a}$$

$$\mathcal{V}_o L_d(\tau) = 0, \tag{6.48b}$$

$$\mathcal{V}_o K'_{no} = 0. \tag{6.48c}$$

Actually, these are projections of more conservative LMIs (where variables have the full dimensions corresponding to the plant) onto the observable subspaces.

Example 6.3.6 The spiking behavior of a cortical neuron may be modeled with state $\eta = (\eta_1, \eta_2) \in \mathbb{R}^2$ as

$$\begin{cases} \dot{\eta} = \begin{pmatrix} 0.04\eta_1^2 + 5\eta_1 + 140 - \eta_2 + I_{\text{ext}} \\ a(b\eta_1 - \eta_2) \end{pmatrix} & \text{when } \eta_1 \le v_m \\ \eta^+ = \begin{pmatrix} c \\ \eta_2 + d \end{pmatrix} & \text{when } \eta_1 = v_m \end{cases} \tag{6.49}$$

where η_1 is the membrane potential, η_2 is the recovery variable, and I_{ext} represents the (constant) synaptic current or injected DC current [14]. We pick here the parameters as $I_{\text{ext}} = 10$, $a = 0.02$, $b = 0.2$, $c = -55$, $d = 4$, and $v_m = 30$ (all in appropriate units), thus characterizing the neuron type and its firing pattern [14]. The solutions of this system are known to have a dwell time with flow lengths remaining in a compact set $\mathcal{I} = [\tau_m, \tau_M]$ where $\tau_m > 0$, and the jump times can be detected from the discontinuities of the measured output $y_c = \eta_1$. Since $y_c = \eta_1$ is known during flows, we treat $0.04\eta_1^2 + 140 + I_{\text{ext}}$ as a known term that can be compensated using output injection with $u_c = (0.04y_c^2 + 140 + I_{\text{ext}}, 0)$. On the other hand, in the jump map of (6.49), we assume c and d are unknown and include them in the state to be estimated along with (η_1, η_2). We show here that, using the decomposition in this paper, we can design a flow-based observer using the knowledge of $y_c = \eta_1$ during flows only, although the flow dynamics are not observable. In other words, we take $y_d = 0$, which comes back to not using any output injection at jumps. We re-model the system (6.49) extended with (c, d) into the form (6.1) with $x = (x_1, x_2, x_3, x_4) = (\eta_1, \eta_2, c, d) \in \mathbb{R}^4$ and the matrices

$$F = \begin{pmatrix} 5 & -1 & 0 & 0 \\ ab & -a & 0 & 0 \\ 0 & 0 & 0 & 0 \\ 0 & 0 & 0 & 0 \end{pmatrix}, \quad H_c = \begin{pmatrix} 1 & 0 & 0 & 0 \end{pmatrix}, \quad J = \begin{pmatrix} 0 & 0 & 1 & 0 \\ 0 & 1 & 0 & 1 \\ 0 & 0 & 1 & 0 \\ 0 & 0 & 0 & 1 \end{pmatrix}, \quad H_d = \begin{pmatrix} 0 & 0 & 0 & 0 \end{pmatrix}. \tag{6.50}$$

We see that x_1 and x_2 are instantaneously observable from y_c and following (6.24), we get $z_o = \begin{pmatrix} \mu_1(\tau) & \mu_2(\tau) & 0 & 0 \\ \mu_3(\tau) & \mu_4(\tau) & 0 & 0 \end{pmatrix} x \in \mathbb{R}^2$, $z_{no} = \begin{pmatrix} 0 & 0 & 1 & 0 \\ 0 & 0 & 0 & 1 \end{pmatrix} x = (x_3, x_4) \in \mathbb{R}^2$, and the form (6.26) with matrices $J_o(\tau) = \begin{pmatrix} 0 & 0 \\ \mu_5(\tau) & \mu_6(\tau) \end{pmatrix}$, $J_{ono}(\tau) = \begin{pmatrix} 1 & 0 \\ 0 & 1 \end{pmatrix}$, $J_{noo}(\tau) = \begin{pmatrix} 0 & 0 \\ 0 & 0 \end{pmatrix}$, $J_{no}(\tau) = \begin{pmatrix} 1 & 0 \\ 0 & 1 \end{pmatrix}$, $H_{c,o}(\tau) = \begin{pmatrix} \mu_7(\tau) & \mu_8(\tau) \end{pmatrix}$, $H_{d,o}(\tau) = \begin{pmatrix} 0 & 0 \end{pmatrix}$, and $H_{d,no}(\tau) = \begin{pmatrix} 0 & 0 \end{pmatrix}$, where μ_i, $i = 1, 2, \ldots, 8$ are known exponential functions of τ. We see that for any $\tau \in \mathcal{I}$, z_{no} cannot be seen from y_d because $H_{d,no}(\tau) = 0$, but it can be accessed through z_o via $J_{ono}(\tau)$ (*hidden* dynamics). Solving (6.33), we obtain $K_{no} = \begin{pmatrix} 1 & 0 \\ 0 & 1 \end{pmatrix}$ with any $Q_{no} \in \mathcal{S}^1_{>0}$ and any $L_{d,no}(\cdot)$. Let us then take $L_{d,no} = 0$. In this application, we see that (x_3, x_4) is estimated thanks to its interaction with z_o at jumps, the latter being estimated during flows, namely we exploit the *hidden* observability analyzed in Theorem 6.2.11. Then, an LMI-based observer as in (6.34) or (6.43), with the mentioned gains K_{no} and $L_{d,no}$, a weighting matrix taken as $R = \text{Id}$, and a large enough λ, can be designed for this system.

6.4 KKL-Based Observer Design from Discrete Uniform Backward Distinguishability in Observability Decomposition

The idea of this section is to replace the LMI-based design of the observer for the z_{no} part with a *systematic* KKL-based one. To do that, following [27], we exploit a discrete KKL observer design for the discrete linear time-varying system (6.32), for which observability is assumed as suggested by Theorem 6.2.12. The main reasons for relying on this discrete design, as opposed to a discrete Kalman(-like) design, for instance, are two-fold:

- Compared to a Kalman design [11], it admits a *strict* Input-to-State Stable (ISS) Lyapunov function, allowing for an interconnection with the high-gain flow-based observer of z_o;
- Compared to a Kalman-like design [34], its gain with respect to the fictitious output in (6.32) is *constant*, allowing us to re-produce in the analysis a similar change of variables as (6.38) in the LMI-based design of Theorem 6.3.2.

For this method, we make the following assumption.

Assumption 6.4.1 Given a subset $\mathcal{I} \subset [0, +\infty)$, the matrix $J_{no}(\tau)$ is uniformly invertible for all $\tau \in \mathcal{I}$, i.e., there exists $s_J > 0$ such that $\|J_{no}^{-1}(\tau)\| \leq s_J$ for all $\tau \in \mathcal{I}$.

Remark 6.4.2 Contrary to discrete-time systems, the jump map of a hybrid system has little chance of being invertible since it is not a discretization of some continuous-time dynamics. However, here thanks to the transformation (6.24), the flow dynamics are somehow merged with the jumps and thus it is reasonable to expect $J_{no}(\tau)$ to be invertible for all $\tau \in \mathcal{I}$. On the other hand, the ability to deal with the non-invertibility of the dynamics has been studied in [19] (in the linear context). With hybrid systems, it can even be coped with using the non-uniqueness of system representation (see for instance [28, Example 2]). Note that while simulations suggest that the invertibility of the jump dynamics is typically not necessary in the Kalman-like design of [28], and may only be for theoretical analysis, it is needed here to implement the observer (6.60) below, namely to compute T^+ correctly.

6.4.1 *Discrete KKL Observer Design for (6.32)*

Consider the discrete system (6.32) with $\tau_k \in \mathcal{I}$ for all $k \in \mathbb{N}$. Following the KKL spirit, we look for a transformation $(T_k)_{k\in\mathbb{N}}$ such that in the new coordinates $\eta_k := T_k z_{no,k}$, (6.32) follows the dynamics

$$\eta_{k+1} = \gamma A \eta_k + B y_k, \tag{6.51}$$

where $A \in \mathbb{R}^{n_\eta \times n_\eta}$ is Schur, $B \in \mathbb{R}^{n_\eta \times n_{d,\text{ext}}}$ where $n_{d,\text{ext}} := n_{y,d} + n_o$ such that the discrete pair (A, B) is controllable, and $\gamma \in (0, 1]$ is a design parameter. It then follows that the transformation $(T_k)_{k\in\mathbb{N}}$ must be such that for every $(\tau_k)_{k\in\mathbb{N}}$ with $\tau_k \in \mathcal{I}$ for all $k \in \mathbb{N}$,

$$T_{k+1} J_{no}(\tau_k) = \gamma A T_k + B H_{d,\text{ext}}(\tau_k), \tag{6.52}$$

with $H_{d,\text{ext}}(\cdot)$ defined in (6.32) in Theorem 6.2.11. The interest of this form is that an observer of (6.32) in the η-coordinates is a simple filter of the output

$$\hat{\eta}_{k+1} = \gamma A \hat{\eta}_k + B y_k, \tag{6.53}$$

making the error $\tilde{\eta}_k = \eta_k - \hat{\eta}_k$ verify

$$\tilde{\eta}_{k+1} = \gamma A \tilde{\eta}_k, \tag{6.54}$$

and thus is exponentially stable. Then, if $(T_k)_{k\in\mathbb{N}}$ is uniformly left-invertible, the estimate defined by $\hat{z}_{no,k} = T_k^* \hat{\eta}$, where $(T_k^*)_{k\in\mathbb{N}}$ is a bounded sequence of left inverses of $(T_k)_{k\in\mathbb{N}}$ verifying $T_k^* T_k = \text{Id}$ for all $k \geq k^\star$ for some $k^\star \in \mathbb{N}$, is such that for any $(\tau_k)_{k\in\mathbb{N}}$ with $\tau_k \in \mathcal{I}$ for all $k \in \mathbb{N}$, there exist $c_1 > 0$ and $c_2 \in (0, 1)$ such that for any initial conditions $z_{no,0}$ and $\hat{z}_{no,0}$ and for all $k \in \mathbb{N}$,

$$|z_{no,k} - \hat{z}_{no,k}| \leq c_1 c_2^k |z_{no,0} - \hat{z}_{no,0}|. \tag{6.55}$$

From [27, Corollary 1], we know that this is possible under the uniform backward distinguishability of (6.32) as defined next.

Definition 6.4.3 (*Uniform backward distinguishability of* (6.32)) Given $(\tau_k)_{k\in\mathbb{N}}$ with $\tau_k \in \mathcal{I}$ for all $k \in \mathbb{N}$, the system (6.32) is *uniformly backward distinguishable* if for each $i \in \{1, 2, \dots, n_{d,\text{ext}}\}$ its output dimension, there exists $m_i \in \mathbb{N}_{>0}$ such that there exists $\alpha_m > 0$ such that for all $k \geq \max_i m_i$, the backward distinguishability matrix sequence $(\mathbb{O}_k^{bw})_{k\in\mathbb{N}}$ defined as

$$\mathbb{O}_k^{bw} = (\mathbb{O}_{1,k}^{bw}, \mathbb{O}_{2,k}^{bw}, \dots, \mathbb{O}_{n_{d,\text{ext}},k}^{bw}) \in \mathbb{R}^{(\sum_{i=1}^{n_{d,\text{ext}}} m_i) \times n_{no}}, \tag{6.56}$$

where

$$\mathbb{O}_{i,k}^{bw} := \begin{pmatrix} H_{d,\text{ext},i}(\tau_{k-1}) J_{no}^{-1}(\tau_{k-1}) \\ H_{d,\text{ext},i}(\tau_{k-2}) J_{no}^{-1}(\tau_{k-2}) J_{no}^{-1}(\tau_{k-1}) \\ \dots \\ H_{d,\text{ext},i}(\tau_{k-(m_i-1)}) J_{no}^{-1}(\tau_{k-(m_i-1)}) \dots J_{no}^{-1}(\tau_{k-1}) \\ H_{d,\text{ext},i}(\tau_{k-m_i}) J_{no}^{-1}(\tau_{k-m_i}) J_{no}^{-1}(\tau_{k-(m_i-1)}) \dots J_{no}^{-1}(\tau_{k-1}) \end{pmatrix},$$

where $H_{d,\text{ext},i}(\cdot)$ denotes the ith row of the extended output matrix $H_{d,\text{ext}}(\cdot)$ of (6.32), has full rank and satisfies $\mathbb{O}_k^{bw\top} \mathbb{O}_k^{bw} \geq \alpha_m \,\text{Id} > 0$.

Note that the forward version $(\mathbb{O}_k^{fw})_{k\in\mathbb{N}}$ of $(\mathbb{O}_k^{bw})_{k\in\mathbb{N}}$, which is much easier to compute, can be considered as in [27, Remark 3], under the additional assumptions that the m_i are the same for $i \in \{1, 2, \ldots, n_{d,\text{ext}}\}$ and J_{no} is uniformly invertible, namely there exists $c_J > 0$ such that $(J_{no}^{-1}(\tau_k))^\top J_{no}^{-1}(\tau_k) \geq c_J \,\text{Id}$ for all $k \in \mathbb{N}$. Theorem 6.4.4, which is a particular case of [27, Theorems 2 and 3], states the existence and uniform left-invertibility of $(T_k)_{k\in\mathbb{N}}$ solving (6.52).

Theorem 6.4.4 *Consider $n_\eta \in \mathbb{N}_{>0}$, $\gamma > 0$, and a pair $(A, B) \in \mathbb{R}^{n_\eta \times n_\eta} \times \mathbb{R}^{n_\eta \times n_{d,\text{ext}}}$. Under Assumption 6.4.1, for any $(\tau_k)_{k\in\mathbb{N}}$ with $\tau_k \in \mathcal{I}$ for all $k \in \mathbb{N}$ and any $T_0 \in \mathbb{R}^{n_\eta \times n_{no}}$, the transformation $(T_k)_{k\in\mathbb{N}}$ initialized as T_0 and satisfying (6.52) uniquely exists under the following closed form*

$$T_k = (\gamma A)^k T_0 \prod_{p=0}^{k-1} J_{no}^{-1}(\tau_p) + \sum_{p=0}^{k-1} (\gamma A)^{k-p-1} B H_{d,\text{ext}}(\tau_p) \prod_{r=p}^{k-1} J_{no}^{-1}(\tau_r). \tag{6.57}$$

Moreover, for any $(\tau_k)_{k\in\mathbb{N}}$ with $\tau_k \in \mathcal{I}$ for all $k \in \mathbb{N}$ such that (6.32) is uniformly backward distinguishable for some $m_i \in \mathbb{N}_{>0}$, $i \in \{1, 2, \ldots, n_{d,\text{ext}}\}$ and any controllable pairs $(\tilde{A}_i, \tilde{B}_i) \in \mathbb{R}^{m_i \times m_i} \times \mathbb{R}^{m_i}$ with $\tilde{A}_i$ Schur, $i \in \{1, 2, \ldots, n_{d,\text{ext}}\}$, there exists $0 < \gamma^\star \leq 1$ such that for any $0 < \gamma < \gamma^\star$, there exist $k^\star \in \mathbb{N}_{>0}$, $\underline{c}_T > 0$, and $\overline{c}_T > 0$ such that $(T_k)_{k\in\mathbb{N}}$ in (6.57) with

$$A = \text{diag}(\tilde{A}_1, \tilde{A}_2, \ldots, \tilde{A}_{n_{d,\text{ext}}}) \in \mathbb{R}^{n_\eta \times n_\eta}, \tag{6.58a}$$

$$B = \text{diag}(\tilde{B}_1, \tilde{B}_2, \ldots, \tilde{B}_{n_{d,\text{ext}}}) \in \mathbb{R}^{n_\eta \times n_{d,\text{ext}}}, \tag{6.58b}$$

where $n_\eta := \sum_{i=1}^{n_{d,\text{ext}}} m_i$, verifies $T_k^\top T_k \geq \underline{c}_T \,\text{Id}$ for all $k \geq k^\star$ and $\|T_k\| \leq \overline{c}_T$ for all $k \in \mathbb{N}$.

In other words, for γ sufficiently small, $(T_k)_{k\in\mathbb{N}}$ is uniformly left-invertible and upper-bounded after $k^\star$. Note that the dependence of $\gamma^\star$, $\underline{c}_T$, $\overline{c}_T$, and $k^\star$ on $(\tau_k)_{k\in\mathbb{N}}$ and T_0 is only through α_m and the m_i coming from the uniform backward distinguishability and the upper bounds of T_0 and $J_{no}^{-1}(\cdot)$. Note also that the (non-uniform) injectivity of T can be obtained from non-uniform distinguishability conditions, as seen in [27, Example 1], which may suffice in some cases to ensure the convergence of the KKL observer.

Proof These results are the particular case of [27, Theorems 2 and 3]. Note that by continuity, $\tau \mapsto J_{no}(\tau)$ is uniformly invertible on $\mathcal{I}$ because $\mathcal{I}$ is compact, and $\tau \mapsto H_{d,no}(\tau)$ is uniformly bounded on the compact set $[0, \tau_M]$. □

Remark 6.4.5 Interestingly, Definition 6.4.3, whose nonlinear version is defined in [27, Definition 1], coincides with the *uniform complete observability* (UCO) condition required by the discrete Kalman(-like) filter (see [11, Condition (13)], [34, Assumption 3], or [33, Definition 3]), on the pair $\left(J_{no}(\tau), \begin{pmatrix} H_{d,no}(\tau) \\ J_{ono}(\tau) \end{pmatrix}\right)$, which is the discrete version of that in Definition 6.2.5 above, i.e., there exist $m_i \in \mathbb{N}_{>0}$ and $c_o > 0$ such that for all $(\tau_k)_{k\in\mathbb{N}}$ with $\tau_k \in \mathcal{I}$ for all $k \in \mathbb{N}$, we have for all $k \geq \max_i m_i$,

$$\sum_{i=1}^{n_{d,\text{ext}}} \sum_{p=k-m_i}^{k-1} \star^\top H_{d,\text{ext},i}(\tau_p) J_{no}^{-1}(\tau_p) \dots J_{no}^{-1}(\tau_{k-2}) J_{no}^{-1}(\tau_{k-1}) \geq c_o \,\text{Id} > 0. \quad (6.59)$$

It is thus interesting to compare these two discrete observers. In terms of dimensions, the complexity of the Kalman filter is $\frac{n_{no}(n_{no}+1)}{2} + n_{no}$, while that of the KKL observer is $\left(\sum_{i=1}^{n_{d,\text{ext}}} m_i\right) n_{no} + \sum_{i=1}^{n_{d,\text{ext}}} m_i$. Therefore, the Kalman filter is advantageous in dimension compared to the KKL observer. However, the advantage of the latter (besides being applicable in the nonlinear context) is that there exists a strict ISS Lyapunov function of quadratic form that allows us to prove exponential ISS, unlike the discrete Kalman filter [11] whose Lyapunov function is not strict. This advantage is then exploited in the next part, where we design a KKL-based observer to estimate the z_{no} part in the hybrid system (6.26) while the error in z_o is seen as a disturbance. Note that a discrete Kalman-like observer [34] could seem like a possible alternative to the KKL-based one since it also exhibits a strict Lyapunov function. However, the gain multiplied with the fictitious output in the observer must be *constant* during flows for us to perform the analysis (see K_{no} in (6.33)), which is not the case in a Kalman-like observer (unless the pair $(J_{no}(\tau), H_{d,no}(\tau))$ at the jump times is UCO, so without the need for the fictitious output). This is ensured in KKL design since it relies on a transformation into a linear time-invariant form (see below in the proof of Theorem 6.4.7).

Next, in Sect. 6.4.2, we exploit this section's results for the hybrid system (6.26).

6.4.2 KKL-Based Observer Design for (6.26)

The KKL-based observer we propose for (6.26) has the form

$$\left\{\begin{aligned}
\dot{\hat{z}}_o &= G_o(\tau)u_c + P^{-1}H_{c,o}^\top(\tau)R^{-1}(\tau)(y_c - H_{c,o}(\tau)\hat{z}_o)\\
\dot{\hat{\eta}} &= (TG_{no}(\tau) - B_{ono}G_o(\tau))u_c\\
\dot{P} &= -\lambda P + H_{c,o}^\top(\tau)R^{-1}(\tau)H_{c,o}(\tau)\\
\dot{T} &= 0\\
\dot{\tau} &= 1\\
\\
\hat{z}_o^+ &= J_o(\tau)\hat{z}_o + J_{ono}(\tau)\hat{z}_{no} + \mathcal{V}_o u_d\\
\hat{\eta}^+ &= (T^+ J_{noo}(\tau) + \gamma A B_{ono} - B_{ono}J_o(\tau))\hat{z}_o\\
&\quad + \gamma A\hat{\eta} + (T^+\mathcal{V}_{no} - B_{ono}\mathcal{V}_o)u_d + B_{d,no}(y_d - H_{d,o}(\tau)\hat{z}_o)\\
P^+ &= P_0\\
T^+ &= (\gamma AT + B_{d,no}H_{d,no}(\tau) + B_{ono}J_{ono}(\tau))\,\text{sat}_{s_J}(J_{no}^\dagger(\tau))\\
\tau^+ &= 0,
\end{aligned}\right. \quad (6.60a)$$

with

$$\hat{z}_{no} = \text{sat}_{s_T}(T^\dagger)(\hat{\eta} + B_{ono}\hat{z}_o), \tag{6.60b}$$

with jumps triggered at the same time as $\mathcal{H}$ in the same way as (6.2), $P_0 \in \mathcal{S}_{>0}^{n_o}$, $\tau \mapsto R(\tau) \in \mathcal{S}_{>0}^{n_{y,c}}$ is a positive definite weighting matrix that is defined and is continuous on $[0, \tau_M]$ to be chosen only for design purpose, $\gamma \in (0, 1]$, and $(A, B_{d,no}, B_{ono})$ are design parameters to be chosen. Following similar reasoning as in Sect. 6.3.1, those dynamics are picked so that T coincides with the $(T_k)_{k\in\mathbb{N}}$ in Sect. 6.4.1 at jumps and so that the corresponding discrete KKL error dynamics (6.54) appear after a certain change of coordinates, modulo some errors on z_o (see (6.66) below). The difficulty comes here from the fact that the discrete output y_k in the discrete KKL dynamics (6.51) is not fully available at jumps since it contains the fictitious output.

Assumption 6.4.6 Given j_m defined in Assumption 6.2.7, there exist $m_i \in \mathbb{N}_{>0}$ for each $i = 1, 2, \ldots, n_{d,\text{ext}}$ and $\alpha_m > 0$ such that for every complete solution $x \in \mathcal{S}_\mathcal{H}(\mathcal{X}_0, \mathcal{U})$, the sequence of flow lengths $(\tau_j)_{j\in\mathbb{N}, j\geq j_m}$ where $\tau_j = t_{j+1} - t_j$ is such that (6.32), scheduled with that $(\tau_j)_{j\in\mathbb{N}, j\geq j_m}$, is uniformly backward distinguishable with the parameters m_i and α_m following Definition 6.4.3.

Theorem 6.4.7 *Suppose Assumptions 6.1.1, 6.2.7, 6.4.1, and 6.4.6 hold. Define* $n_\eta := \sum_{i=1}^{n_{d,\text{ext}}} m_i$ *and consider for each* $i \in \{1, 2, \ldots, n_{d,\text{ext}}\}$ *a controllable pair* $(\tilde{A}_i, \tilde{B}_i) \in \mathbb{R}^{m_i \times m_i} \times \mathbb{R}^{m_i}$ *with* $\tilde{A}_i$ *Schur. Define then*

$$A = \text{diag}(\tilde{A}_1, \tilde{A}_2, \ldots, \tilde{A}_{n_{d,\text{ext}}}) \in \mathbb{R}^{n_\eta \times n_\eta}, \tag{6.61a}$$

$$B_{d,no} = \text{diag}(\tilde{B}_1, \tilde{B}_2, \ldots, \tilde{B}_{n_{y,d}}) \in \mathbb{R}^{n_\eta \times n_{y,d}}, \tag{6.61b}$$

$$B_{ono} = \text{diag}(\tilde{B}_{n_{y,d}+1}, \tilde{B}_{n_{y,d}+2}, \ldots, \tilde{B}_{n_{d,\text{ext}}}) \in \mathbb{R}^{n_\eta \times n_o}. \tag{6.61c}$$

Given any $\lambda_1 > 0$, *any* $P_0 \in \mathcal{S}_{>0}^{n_o}$, *and any* $T_0 \in \mathbb{R}^{n_\eta \times n_{no}}$, *there exists* $0 < \gamma^\star \leq 1$ *such that there exists* $\lambda^\star > 0$ *such that for any* $0 < \gamma < \gamma^\star$ *and any* $\lambda > \lambda^\star$, *there exist* $\bar{j} \in \mathbb{N}_{>0}$, *saturation levels* $s_T > 0$, $s_J > 0$, *and scalar* $\rho_1 > 0$ *such that for any solution* $x \in \mathcal{S}_\mathcal{H}(\mathcal{X}_0, \mathcal{U})$ *and any solution* $(\hat{z}_o, \hat{\eta}, P, T, \tau)$ *of the observer (6.60) with* $P(0, 0) = P_0$, $T(0, 0) = T_0$, $\tau(0, 0) = 0$, *the chosen* $(A, B_{d,no}, B_{ono})$, sat_{s_T} *at level* s_T, sat_{s_J} *at level* s_J, *and jumps triggered at the same time as in* x, $(\hat{z}_o, \hat{\eta}, P, T, \tau)$ *is complete and we have*

$$|x(t, j) - \hat{x}(t, j)| \leq \rho_1 |x(t_{\bar{j}}, \bar{j}) - \hat{x}(t_{\bar{j}}, \bar{j})| e^{-\lambda_1(t+j)}, \qquad \forall (t, j) \in \text{dom}\, x, j \geq \bar{j}, \tag{6.62}$$

with $\hat{x}$ *obtained by* $\hat{x} = e^{F\tau}\mathcal{D}\hat{z}$ *with* $\mathcal{D}$ *defined in (6.25).*

The parameter $\bar{j}$ is related to j_m in Assumption 6.4.1 and to the number of jumps needed to get the uniform left-injectivity of $(T_k)_{k\in\mathbb{N}}$ in Theorem 6.4.4.

Proof First, according to Assumption 6.2.7, the flow lengths of solutions in $\mathcal{S}_\mathcal{H}(\mathcal{X}_0, \mathcal{U})$ are in the compact set $[0, \tau_M]$, so there exists $\overline{c}_{T,m} > 0$ such that for every solution $x \in \mathcal{S}_\mathcal{H}(\mathcal{X}_0, \mathcal{U})$ and any solution $(\hat{z}_o, \hat{\eta}, P, T, \tau)$ of the observer (6.60) with $P(0,0) = P_0$, $T(0,0) = T_0$, $\tau(0,0) = 0$, and any $\gamma \in (0,1]$, with jumps triggered at the same time as in x, we have $\|T(t,j)\| \leq \overline{c}_{T,m}$ for all $(t,j) \in \operatorname{dom} x$ such that $j \leq j_m$. Then, from Assumptions 6.2.7, 6.4.1, and 6.4.6, and according to Theorem 6.4.4 starting from jump time j_m, there exists $0 < \gamma_0^\star \leq 1$ such that for all $0 < \gamma < \gamma_0^\star$, there exist $j^\star \in \mathbb{N}_{>0}$, $\underline{c}_T > 0$, and $\overline{c}_T > 0$ such that for every solution $x \in \mathcal{S}_\mathcal{H}(\mathcal{X}_0, \mathcal{U})$, the solution $(T_j)_{j\in\mathbb{N}, j\geq j_m}$ to (6.52) with $\tau_j = t_{j+1} - t_j$, initialized at any T_{j_m} verifying $\|T_{j_m}\| \leq \overline{c}_{T,m}$, is uniformly left-invertible for all $j \geq j_m + j^\star$ and uniformly bounded for all $j \geq j_m$, i.e.,

$$T_j^\top T_j \geq \underline{c}_T \operatorname{Id}, \qquad \forall j \geq j_m + j^\star, \qquad \|T_j\| \leq \overline{c}_T, \qquad \forall j \geq j_m.$$

It follows that $(T_j^\dagger)^\top T_j^\dagger \leq \frac{1}{\underline{c}_T} \operatorname{Id}$, for all $j \geq j_m + j^\star$, and there exists a saturation level $s_T > 0$ such that $\operatorname{sat}_{s_T}(T_j^\dagger) = T_j^\dagger$ for all $j \geq j_m + j^\star$. Pick $0 < \gamma < \gamma_0^\star$ and consider a solution $x \in \mathcal{S}_\mathcal{H}(\mathcal{X}_0, \mathcal{U})$ and a solution $(\hat{z}_o, \hat{\eta}, P, T, \tau)$ of the observer (6.60) with $P(0,0) = P_0$, $T(0,0) = T_0$, $\tau(0,0) = 0$, the chosen $(A, B_{d,no}, B_{ono})$, the saturation level s_T, and jumps triggered at the same time as in x. Following (6.24), define

$$z(t,j) := \mathcal{V} e^{-F\tau(t,j)} x(t,j), \qquad \forall (t,j) \in \operatorname{dom} x,$$

and consider the error $\tilde{z} = (\tilde{z}_o, \tilde{z}_{no}) = (z_o - \hat{z}_o, z_{no} - \hat{z}_{no})$. As justified above, $\|T(t_{j_m}, j_m)\| \leq \overline{c}_{T,m}$. Since $\dot{T} = 0$ during flows, the sequence $(T(t_j, j))_{j\in\mathbb{N}, j\geq j_m}$ coincides with the sequence $(T_j)_{j\in\mathbb{N}, j\geq j_m}$ solution to (6.52) with $\tau_j = t_{j+1} - t_j$ for all $j \in \mathbb{N}$ with $j \geq j_m$. Therefore,

$$T^\top(t,j) T(t,j) \geq \underline{c}_T \operatorname{Id}, \qquad \forall (t,j) \in \operatorname{dom} x,\ j \geq j_m + j^\star. \tag{6.63}$$

We now use Corollary 6.6.2 to show exponential convergence of $\tilde{z}$ starting at hybrid time $(t_{j_m+j^\star}, j_m + j^\star)$ by putting the error dynamics into the appropriate form. In order to exploit the KKL design, we define

$$\eta(t,j) = T(t,j) z_{no}(t,j) - B_{ono} z_o(t,j), \qquad \forall (t,j) \in \operatorname{dom} x. \tag{6.64}$$

Notice that η verifies $\dot{\eta} = (T G_{no}(\tau) - B_{ono} G_o(\tau)) u_c$ during flows. From Assumptions 6.2.7 and 6.4.1, after hybrid time (t_{j_m}, j_m) we get $\operatorname{sat}_{s_J}(J_{no}^\dagger(\tau)) = J_{no}^\dagger(\tau)$ and $J_{no}^\dagger(\tau) J_{no}(\tau) = \operatorname{Id}$ so that at jumps,

$$\begin{aligned}\eta^+ = (T^+ J_{noo}(\tau) &+ \gamma A B_{ono} - B_{ono} J_o(\tau)) z_o \\ &+ \gamma A \eta + (T^+ \mathcal{V}_{no} - B_{ono} \mathcal{V}_o) u_d + B_{d,no} H_{d,no}(\tau) z_{no}.\end{aligned} \tag{6.65}$$

Given the dynamics of $\hat{\eta}$ in (6.60), the error $\tilde{\eta} := \eta - \hat{\eta}$ verifies $\dot{\tilde{\eta}} = 0$ during flows and at jumps (after hybrid time (t_{j_m}, j_m)),

$$\tilde{\eta}^+ = (T^+ J_{noo}(\tau) + \gamma A B_{ono} - B_{ono} J_o(\tau) - B_{d,no} H_{d,no}(\tau)) \tilde{z}_o + \gamma A \tilde{\eta}, \tag{6.66}$$

which is a contracting dynamics in $\tilde{\eta}$. After $(t_{j_m+j^\star}, j_m + j^\star)$, we have (i) $T^\dagger T = \text{Id}$ so that $z_{no} = T^\dagger \eta + T^\dagger B_{ono} z_o$, and (ii) $\text{sat}_{s_T}(T^\dagger) = T^\dagger$ so that $\hat{z}_{no} = T^\dagger \hat{\eta} + T^\dagger B_{ono} \hat{z}_o$. Therefore, after $(t_{j_m+j^\star}, j_m + j^\star)$, $(\tilde{z}_o, \tilde{\eta}, P, \tau)$ is solution to

$$\begin{cases} \dot{\tilde{z}}_o = -P^{-1} H_{c,o}^\top(\tau) R^{-1}(\tau) H_{c,o}(\tau) \tilde{z}_o \\ \dot{\tilde{\eta}} = 0 \\ \dot{P} = -\lambda P + H_{c,o}^\top(\tau) R^{-1}(\tau) H_{c,o}(\tau) \\ \dot{\tau} = 1 \\ \\ \tilde{z}_o^+ = \overline{J}_o(\tau) \tilde{z}_o + J_{ono}(\tau) T^\dagger \tilde{\eta} \\ \tilde{\eta}^+ = (T^+ J_{noo}(\tau) + \gamma A B_{ono} - B_{ono} J_o(\tau) - B_{d,no} H_{d,no}(\tau)) \tilde{z}_o + \gamma A \tilde{\eta} \\ P^+ = P_0 \\ \tau^+ = 0, \end{cases} \tag{6.67a}$$

where $\overline{J}_o(\tau) = J_o(\tau) + J_{ono} T^\dagger B_{ono}$, with T^+ seen as a uniformly bounded input, and with the flow and jump sets

$$\mathbb{R}^{n_o} \times \mathbb{R}^{n_\eta} \times \mathbb{R}^{n_o \times n_o} \times [0, \tau_M], \qquad \mathbb{R}^{n_o} \times \mathbb{R}^{n_\eta} \times \mathbb{R}^{n_o \times n_o} \times \mathcal{I}. \tag{6.67b}$$

Since A is Schur, let $Q_\eta \in \mathcal{S}_{>0}^{n_\eta}$ be a solution to the inequality $A^\top Q_\eta A - Q_\eta < 0$. Using Corollary 6.6.2, we prove that there exist $\lambda^\star > 0$ and $0 < \gamma^\star \leq 1$ such that we have the arbitrarily fast GES of the error $(\tilde{z}_o, \tilde{\eta})$ with respect to the value $(\tilde{z}_o, \tilde{\eta})(t_{j_m+j^\star}, j_m + j^\star)$ when $\lambda > \lambda^\star$ and $0 < \gamma < \gamma^\star$. Then the GES in the z-coordinates with respect to $(\tilde{z}_o, \tilde{z}_{no})(t_{j_m+j^\star}, j_m + j^\star)$ is obtained thanks to the uniform left-invertibility of T. Last, the arbitrarily fast GES is recovered in the x-coordinates after hybrid time $(t_{\bar{j}}, \bar{j})$ where $\bar{j} = j_m + j^\star$ by seeing that $\tilde{z} = \mathcal{V} e^{-F\tau} \tilde{x}$ with $\tau \in [0, \tau_M]$. □

Remark 6.4.8 Note that it is the *rate* of convergence that can be arbitrarily fast and not the convergence time since we must anyway wait for j_m jumps for the flow lengths to be in $\mathcal{I}$ and, more importantly, for the $\max_i m_i$ jumps giving us uniform backward distinguishability (see Definition 6.4.3). Furthermore, speeding up the rate may make T poorer conditioned, thus increasing the bound ρ_1, which is the well-known *peaking* phenomenon. This type of result is typical in high-gain KKL designs (see [27]). Note though that this arbitrarily fast convergence rate is an advantage compared to the LMI-based design in Sect. 6.3 where the rate is fixed once the LMI is solved: Corollary 6.6.2 does not apply in that case because the parameters a and Q_η in (6.84) are not independent, i.e., Q_η is not such that (6.84) holds for any $a > 0$.

6.4.3 KKL-Based Observer Design in the (z_o, z_{no})-Coordinates

The rectangular shape of T in observer (6.60) makes the dimension of η larger than that of z_{no}, preventing us from easily writing the observer in the z-coordinates, unlike in Sect. 6.3. Following the spirit of [4], consider a map $\Gamma : \mathbb{R}^{n_\eta \times n_{no}} \to \mathbb{R}^{n_\eta \times (n_\eta - n_{no})}$ such that for any $T \in \mathbb{R}^{n_\eta \times n_{no}}$, $\Gamma(T)$ is a full-rank matrix such that its columns are orthogonal to those of T and $\Gamma^\top(T)\Gamma(T) \geq \text{Id}$. Such a map always exists (see Remark 6.4.9 for an explicit construction method). Then, define $T_e : \mathbb{R}^{n_\eta \times n_{no}} \to \mathbb{R}^{n_\eta \times n_\eta}$ such that

$$T_e(T) = \begin{pmatrix} T & \Gamma(T) \end{pmatrix}, \tag{6.68}$$

which is a square matrix extension of T, that is invertible whenever T is full rank.

Remark 6.4.9 One possible way to construct $\Gamma(T)$ is to exploit a singular value decomposition of T. Indeed, given $T \in \mathbb{R}^{n_\eta \times n_{no}}$ with $n_\eta \geq n_{no}$, consider orthonormal matrices $U(T) \in \mathbb{R}^{n_\eta \times n_\eta}$ and $V(T) \in \mathbb{R}^{n_{no} \times n_{no}}$, as well as a matrix $\Sigma(T) = \begin{pmatrix} \Sigma'(T) \\ 0 \end{pmatrix} \in \mathbb{R}^{n_\eta \times n_{no}}$, with $\Sigma'(T)$ diagonal, such that $T = U(T)\Sigma(T)V^\top(T)$. Let us split $U(T)$ as $U(T) = \begin{pmatrix} U_1(T) & U_2(T) \end{pmatrix}$ where $U_1(T) \in \mathbb{R}^{n_\eta \times n_{no}}$ and $U_2(T) \in \mathbb{R}^{n_\eta \times (n_\eta - n_{no})}$. By taking $\Gamma(T) = U_2(T)$, we see that $\Gamma(T)$ is orthogonal to T and verifies $\Gamma^\top(T)\Gamma(T) = \text{Id}$.

An alternative KKL-based observer to (6.60) can then be implemented in the (z_o, z_{no})-coordinates, as

$$\left\{\begin{aligned}
\dot{\hat{z}}_o &= G_o(\tau)u_c + P^{-1}H_{c,o}^\top(\tau)R^{-1}(\tau)(y_c - H_{c,o}(\tau)\hat{z}_o)\\
(\dot{\hat{z}}_{no}, \dot{\hat{\omega}}) &= (G_{no}(\tau)u_c, 0)\\
&\quad + \text{sat}_{s_{T_e}}((T_e(T))^\dagger)B_{ono}P^{-1}H_{c,o}^\top(\tau)R^{-1}(\tau)(y_c - H_{c,o}(\tau)\hat{z}_o)\\
\dot{P} &= -\lambda P + H_{c,o}^\top(\tau)R^{-1}(\tau)H_{c,o}(\tau)\\
\dot{T} &= 0\\
\dot{\tau} &= 1\\
\\
\hat{z}_o^+ &= J_o(\tau)\hat{z}_o + J_{ono}(\tau)\hat{z}_{no} + \mathcal{V}_o u_d\\
(\hat{z}_{no}^+, \hat{\omega}^+) &= (J_{noo}(\tau)\hat{z}_o + J_{no}(\tau)\hat{z}_{no} + \mathcal{V}_{no}u_d, 0) + \text{sat}_{s_{T_e}}((T_e(T^+))^\dagger)\\
&\quad \times (\gamma A\Gamma(T)\hat{\omega} + B_{d,no}(y_d - H_{d,o}(\tau)\hat{z}_o - H_{d,no}(\tau)\hat{z}_{no}))\\
P^+ &= P_0\\
T^+ &= (\gamma AT + B_{d,no}H_{d,no}(\tau) + B_{ono}J_{ono}(\tau))\,\text{sat}_{s_J}(J_{no}^\dagger(\tau))\\
\tau^+ &= 0,
\end{aligned}\right. \tag{6.69}$$

with jumps still triggered at the same time as $\mathcal{H}$ and initialized as in Theorem 6.4.7 with any $\hat{\omega}(0,0) \in \mathbb{R}^{n_\eta - n_{no}}$. Still in the spirit of [4], $\hat{\omega} \in \mathbb{R}^{n_\eta - n_{no}}$ is an estimate of some fictitious extra state $\omega \in \mathbb{R}^{n_\eta - n_{no}}$ defined on $\text{dom}\, x$ such that $\omega(t, j) = 0$ for all $(t, j) \in \text{dom}\, x$, serving to equalize the dimension in the z_{no} and η coordinates.

Note that along the solutions of (6.69), since T is constant during flows, $T_e(T)$ is also constant during flows and needs to be re-computed only at jumps.

Theorem 6.4.10 *Suppose Assumptions 6.1.1, 6.2.7, 6.4.1, and 6.4.6 hold. Define* n_η *and* $(A, B_{d,no}, B_{ono})$ *as in Theorem 6.4.7. Given any* $\lambda_1 > 0$, *any* $P_0 \in \mathcal{S}^{n_o}_{>0}$, *and any* $T_0 \in \mathbb{R}^{n_\eta \times n_{no}}$, *there exists* $0 < \gamma^\star \leq 1$ *such that there exists* $\lambda^\star > 0$ *such that for any* $0 < \gamma < \gamma^\star$ *and any* $\lambda > \lambda^\star$, *there exist* $\bar{j} \in \mathbb{N}_{>0}$, *saturation levels* $s_{T_e} > 0$, $s_J > 0$, *and scalar* $\rho_1 > 0$ *such that for any solution* $x \in \mathcal{S}_{\mathcal{H}}(\mathcal{X}_0, \mathcal{U})$ *and any solution* $(\hat{z}, \hat{\omega}, P, T, \tau)$ *of the observer (6.69) with* $\hat{\omega}(0, 0) \in \mathbb{R}^{n_\eta - n_{no}}$, $P(0, 0) = P_0$, $T(0, 0) = T_0$, $\tau(0, 0) = 0$, *the chosen* $(A, B_{d,no}, B_{ono})$, $\text{sat}_{s_{T_e}}$ *at level* s_{T_e}, sat_{s_J} *at level* s_J, *and jumps triggered at the same time as in* x, $(\hat{z}, \hat{\omega}, P, T, \tau)$ *is complete and we have*

$$|x(t, j) - \hat{x}(t, j)| \leq \rho_1 |x(t_{\bar{j}}, \bar{j}) - \hat{x}(t_{\bar{j}}, \bar{j})| e^{-\lambda_1 (t+j)}, \qquad \forall (t, j) \in \operatorname{dom} x, j \geq \bar{j}, \tag{6.70}$$

with $\hat{x}$ *obtained by* $\hat{x} = e^{F\tau} \mathcal{D} \hat{z}$ *with* $\mathcal{D}$ *defined in (6.25).*

Proof First, because T in (6.69) verifies the same dynamics as in (6.60), for the same constant $\overline{c}_{T,m}$ as in the proof of Theorem 6.4.7, for every solution $x \in \mathcal{S}_{\mathcal{H}}(\mathcal{X}_0, \mathcal{U})$ and any solution $(\hat{z}, \hat{\omega}, P, T, \tau)$ of the observer (6.69) with $P(0, 0) = P_0$, $T(0, 0) = T_0$, $\tau(0, 0) = 0$, and any $\gamma \in (0, 1]$, with jumps triggered at the same time as in x, we have $\|T(t, j)\| \leq \overline{c}_{T,m}$ for all $(t, j) \in \operatorname{dom} x$ such that $j \leq j_m$. Moreover, considering the same $\gamma^\star$, $j^\star$, $\underline{c}_T$ as in the proof of Theorem 6.4.7, T becomes uniformly left-invertible after hybrid time $(t_{j_m + j^\star}, j_m + j^\star)$, i.e.,

$$T^\top(t, j) T(t, j) \geq \underline{c}_T \operatorname{Id}, \qquad \forall (t, j) \in \operatorname{dom} x, j \geq j_m + j^\star.$$

It follows that $T_e(T)$ also becomes uniformly invertible, since

$$T_e^\top(T) T_e(T) = \begin{pmatrix} T^\top T & 0 \\ 0 & \Gamma^\top(T)\Gamma(T) \end{pmatrix} \geq \min\{\underline{c}_T, 1\} \operatorname{Id}, \quad \forall (t, j) \in \operatorname{dom} x, j \geq j_m + j^\star.$$

After hybrid time $(t_{j_m + j^\star}, j_m + j^\star)$, we thus have

$$((T_e(T(t, j)))^\dagger)^\top (T_e(T(t, j)))^\dagger \leq \max\left\{\frac{1}{\underline{c}_T}, 1\right\} \operatorname{Id}, \tag{6.71}$$

so that there exists a saturation level $s_{T_e} > 0$ such that $\text{sat}_{s_{T_e}}((T_e(T(t, j)))^\dagger) = (T_e(T(t, j)))^\dagger = (T_e(T(t, j)))^{-1}$ for all $(t, j) \in \operatorname{dom} x$ such that $j \geq j_m + j^\star$. Pick $0 < \gamma < \gamma^\star$ and consider a solution $x \in \mathcal{S}_{\mathcal{H}}(\mathcal{X}_0, \mathcal{U})$ and a solution $(\hat{z}, \hat{\omega}, P, T, \tau)$ of the observer (6.69) with $P(0, 0) = P_0$, $T(0, 0) = T_0$, $\tau(0, 0) = 0$, the chosen $(A, B_{d,no}, B_{ono})$, the saturation level s_{T_e}, and jumps triggered at the same time as in x. Define for all $(t, j) \in \operatorname{dom} x$ the change of variables

$$\eta(t, j) = T_e(T(t, j)) \begin{pmatrix} z_{no}(t, j) \\ 0 \end{pmatrix} - B_{ono} z_o(t, j). \tag{6.72}$$

We see that for all $(t, j) \in \operatorname{dom} x$,

$$\eta(t, j) = T(t, j)z_{no}(t, j) - B_{ono}z_o(t, j), \tag{6.73}$$

and hence it verifies the dynamics of η in the proof of Theorem 6.4.7. Let us study the dynamics of the image of $(\hat{z}_o, \hat{z}_{no}, \hat{\omega})$ in the observer (6.69) defined as

$$\hat{\eta}(t, j) = T_e(T(t, j)) \begin{pmatrix} \hat{z}_{no}(t, j) \\ \hat{\omega}(t, j) \end{pmatrix} - B_{ono}\hat{z}_o(t, j). \tag{6.74}$$

We see that after hybrid time $(t_{j_m+j^\star}, j_m + j^\star)$, $\dot{\hat{\eta}} = (TG_{no}(\tau) - B_{ono}G_o(\tau))u_c$ during flows and at jumps,

$$\begin{aligned}
\hat{\eta}^+ &= T_e(T^+) \begin{pmatrix} \hat{z}_{no}^+ \\ \hat{\omega}^+ \end{pmatrix} - B_{ono}\hat{z}_o^+ \\
&= \left(T^+ \; \Gamma^+\right) (J_{noo}(\tau)\hat{z}_o + J_{no}(\tau)\hat{z}_{no}, 0) + \operatorname{sat}_{S_{T_e}}((T_e(T^+))^\dagger)(\gamma A\Gamma(T)\hat{\omega} \\
&\quad + B_{d,no}(y_d - H_{d,o}(\tau)\hat{z}_o - H_{d,no}(\tau)\hat{z}_{no})) - B_{ono}(J_o(\tau)\hat{z}_o + J_{ono}(\tau)\hat{z}_{no}) \\
&= T^+ J_{noo}(\tau)\hat{z}_o + (\gamma AT + B_{d,no}H_{d,no}(\tau) + B_{ono}J_{ono}(\tau))\hat{z}_{no} + \gamma A\Gamma(T)\hat{\omega} \\
&\quad + B_{d,no}(y_d - H_{d,o}(\tau)\hat{z}_o - H_{d,no}(\tau)\hat{z}_{no}) - B_{ono}(J_o(\tau)\hat{z}_o + J_{ono}(\tau)\hat{z}_{no}) \\
&= (T^+ J_{noo}(\tau) - B_{ono}J_o(\tau))\hat{z}_o + \gamma A(T\hat{z}_{no} + \Gamma(T)\hat{\omega}) + B_{d,no}(y_d - H_{d,o}(\tau)\hat{z}_o) \\
&= (T^+ J_{noo}(\tau) + \gamma AB_{ono} - B_{ono}J_o(\tau))\hat{z}_o + \gamma A\hat{\eta} + B_{d,no}(y_d - H_{d,o}(\tau)\hat{z}_o).
\end{aligned} \tag{6.75}$$

Therefore, $\hat{\eta}$ has the same dynamics as in the proof of Theorem 6.4.7 and so by proceeding similarly, we get the results in the (z_o, z_{no}, ω)-coordinates thanks to the uniform invertibility of the transformation. □

6.4.4 KKL-Based Observer Design in the x-Coordinates

In a similar manner as in Sect. 6.3.2, the KKL-based observer can equivalently be implemented directly in the original x-coordinates. Based on the proof of Theorem 6.3.4 by noting that $\dot{T} = 0$ during flows, we can derive the dynamics of the observer in the x-coordinates as

$$\begin{cases} \dot{\hat{\zeta}} = F'\hat{\zeta} + (u_c, 0) + \mathcal{P}H_c'^\top R^{-1}(\tau)(y_c - H_c'\hat{\zeta}) \\ \dot{\mathcal{P}} = \lambda\mathcal{P} + F'\mathcal{P} + \mathcal{P}F'^\top - \mathcal{P}H_c'^\top R^{-1}(\tau)H_c'\mathcal{P} \\ \dot{T} = 0 \\ \dot{\tau} = 1 \\ \\ \hat{\zeta}^+ = J'\hat{\zeta} + (u_d, 0) + \mathcal{D}'_{no}\operatorname{sat}_{s_{T_e}}((T_e(T^+))^\dagger)(B_{d,no}(y_d - H_d\hat{x}) + \gamma A\Gamma(T)\hat{\omega}) \\ \mathcal{P}^+ = \star^\top P_0^{-1}(\mathcal{D}'_o + \mathcal{D}'_{no}\operatorname{sat}_{s_{T_e}}((T_e(T))^\dagger)B_{ono})^\top \\ T^+ = (\gamma AT + B_{d,no}H_d e^{F\tau}\mathcal{D}_{no} + B_{ono}\mathcal{V}_o J e^{F\tau}\mathcal{D}_{no})\operatorname{sat}_{s_J}((\mathcal{V}_{no}Je^{F\tau}\mathcal{D}_{no})^\dagger) \\ \tau^+ = 0, \end{cases} \tag{6.76}$$

where $\hat{\zeta} = (\hat{x}, \hat{\omega})$, with jumps still triggered at the same time as $\mathcal{H}$, where

$$F' = \begin{pmatrix} F & 0 \\ 0 & 0 \end{pmatrix}, \quad H_c' = \begin{pmatrix} H_c & 0 \end{pmatrix}, \quad J' = \begin{pmatrix} J & 0 \\ 0 & 0 \end{pmatrix}, \quad \mathcal{D}'_o = \begin{pmatrix} \mathcal{D}_o \\ 0 \end{pmatrix}, \quad \mathcal{D}'_{no} = \begin{pmatrix} \mathcal{D}_{no} & 0 \\ 0 & \mathrm{Id} \end{pmatrix}.$$

The fact that observability is pumped from discrete time to continuous time via the interaction of $\mathcal{P}$ and T at the jump is recovered, while here it is interesting to see that instead of being reset to a constant as in Sect. 6.3.2, here $\mathcal{P}^+$ depends on T, which adapts to the successive flow lengths.

Theorem 6.4.11 *Suppose Assumptions 6.1.1, 6.2.7, 6.4.1, and 6.4.6 hold. Define* n_η *and* $(A, B_{d,no}, B_{ono})$ *as in Theorem 6.4.7. Given any* $\lambda_1 > 0$*, any* $P_0 \in \mathcal{S}^{n_o}_{>0}$*, and any* $T_0 \in \mathbb{R}^{n_\eta \times n_{no}}$*, there exists* $0 < \gamma^\star \leq 1$ *such that there exists* $\lambda^\star > 0$ *such that for any* $0 < \gamma < \gamma^\star$ *and any* $\lambda > \lambda^\star$*, there exist* $\bar{j} \in \mathbb{N}_{>0}$*, saturation levels* $s_{T_e} > 0$*,* $s_J > 0$*, and scalar* $\rho_1 > 0$ *such that for any solution* $x \in \mathcal{S}_\mathcal{H}(\mathcal{X}_0, \mathcal{U})$ *and any solution* $(\hat{\zeta}, \mathcal{P}, T, \tau)$ *of the observer (6.76) with* $\mathcal{P}(0,0) = \star^\top P_0^{-1}(\mathcal{D}'_o + \mathcal{D}'_{no}\operatorname{sat}_{s_{T_e}}((T_e(T_0)^\dagger)B_{ono})^\top$*,* $T(0,0) = T_0$*,* $\tau(0,0) = 0$*, the chosen* $(A, B_{d,no}, B_{ono})$*,* $\operatorname{sat}_{s_{T_e}}$ *at level* s_{T_e}*,* sat_{s_J} *at level* s_J*, and jumps triggered at the same time as in* x*,* $(\hat{\zeta}, \mathcal{P}, T, \tau)$ *is complete and we have*

$$|x(t,j) - \hat{x}(t,j)| \leq \rho_1 |x(t_{\bar{j}}, \bar{j}) - \hat{x}(t_{\bar{j}}, \bar{j})| e^{-\lambda_1(t+j)}, \qquad \forall (t,j) \in \operatorname{dom} x,\ j \geq \bar{j}. \tag{6.77}$$

Proof Define $\mathcal{D}' = \begin{pmatrix} \mathcal{D}'_o & \mathcal{D}'_{no} \end{pmatrix}$ and its inverse $\mathcal{V}'$. Consider $\lambda^\star$, $\gamma^\star$, $j^\star$, $\underline{c}_T$, and s_{T_e} given by Theorem 6.4.10. Pick a solution $x \in \mathcal{S}_\mathcal{H}(\mathcal{X}_0, \mathcal{U})$ and a solution $(\hat{\zeta}, \mathcal{P}, T, \tau)$ of (6.76) with $\mathcal{P}(0,0) = \star^\top P_0^{-1}(\mathcal{D}'_o + \mathcal{D}'_{no}\operatorname{sat}_{s_{T_e}}((T_e(T_0)^\dagger)B_{ono})^\top$, $T(0,0) = T_0$, $\tau(0,0) = 0$, $0 < \gamma < \gamma^\star$, and $\lambda > \lambda^\star$, with jumps triggered at the same time as in x. Consider $(\hat{\underline{z}}, \hat{\underline{\omega}}, P, \underline{T}, \underline{\tau})$ solution to observer (6.69), with $(\hat{\underline{z}}(0,0), \hat{\underline{\omega}}(0,0)) = \mathcal{V}'\hat{\zeta}(0,0)$, $P(0,0) = P_0$, $\underline{T}(0,0) = T(0,0)$, $\underline{\tau}(0,0) = 0$, and the same parameters, with jumps triggered at the same time as in x. First, notice from their dynamics that $\underline{T} = T$ and $\underline{\tau} = \tau$. Then, applying Theorem 6.4.10, we get

$$|x(t,j) - \hat{\underline{x}}(t,j)| \leq \rho_1 |x(t_{\bar{j}}, \bar{j}) - \hat{\underline{x}}(t_{\bar{j}}, \bar{j})| e^{-\lambda_1(t+j)}, \qquad \forall (t,j) \in \operatorname{dom} x,\ j \geq \bar{j},$$

where $\underline{\hat{x}} = e^{F\tau}\mathcal{D}\hat{z}$. The proof consists in showing that $\underline{\hat{x}} = \hat{x}$ where $\hat{\zeta} = (\hat{x}, \hat{\omega})$, thus obtaining (6.77). Denote $\underline{\hat{\zeta}} = (\underline{\hat{x}}, \underline{\hat{\omega}})$. We start by observing that $\underline{\hat{\zeta}} = e^{F'\tau}\mathcal{D}'(\hat{z}, \underline{\hat{\omega}})$, so that

$$\dot{\underline{\hat{\zeta}}} = F'\underline{\hat{\zeta}} + (u_c, 0) + L'_c(P, T, \tau)(y_c - H'_c\underline{\hat{\zeta}}), \tag{6.78a}$$

during flows and

$$\underline{\hat{\zeta}}^+ = J'\underline{\hat{\zeta}} + (u_d, 0) + L'_{d,x}(T)(y_d - H_d\underline{\hat{x}}) + L'_{d,\omega}(T)\underline{\hat{\omega}}, \tag{6.78b}$$

at jumps where

$$\begin{aligned} L'_c(P, T, \tau) &= e^{F'\tau}\mathcal{D}'\begin{pmatrix} P^{-1}H_{c,o}^\top(\tau)R^{-1}(\tau) \\ \mathrm{sat}_{s_{T_e}}((T_e(T))^\dagger)B_{ono}P^{-1}H_{c,o}^\top(\tau)R^{-1}(\tau)\end{pmatrix} \\ &= e^{F'\tau}(\mathcal{D}'_o + \mathcal{D}'_{no}\,\mathrm{sat}_{s_{T_e}}((T_e(T))^\dagger)B_{ono})P^{-1}H_{c,o}^\top(\tau)R^{-1}(\tau), \end{aligned} \tag{6.79a}$$

and

$$L'_{d,x}(T) = \mathcal{D}'\begin{pmatrix} 0 \\ \mathrm{sat}_{s_{T_e}}((T_e(T^+))^\dagger)B_{d,no}\end{pmatrix} = \mathcal{D}'_{no}\,\mathrm{sat}_{s_{T_e}}((T_e(T^+))^\dagger)B_{d,no}, \tag{6.79b}$$

$$L'_{d,\omega}(T) = \mathcal{D}'\begin{pmatrix} 0 \\ \gamma\,\mathrm{sat}_{s_{T_e}}((T_e(T^+))^\dagger)A\Gamma(T)\end{pmatrix} = \gamma\mathcal{D}'_{no}\,\mathrm{sat}_{s_{T_e}}((T_e(T^+))^\dagger)A\Gamma(T). \tag{6.79c}$$

From (6.23), we actually have $H'_c e^{F'\tau}\mathcal{D}'_{no} = 0$, so $H'_c e^{F'\tau}\mathcal{D}'_{no}\,\mathrm{sat}_{s_{T_e}}((T_e(T))^\dagger)B_{ono} = 0$ and we have $L'_c(P, T, \tau) = \underline{\mathcal{P}}H_c'^\top R^{-1}(\tau)$ where

$$\underline{\mathcal{P}} = \star^\top P^{-1}(\mathcal{D}'_o + \mathcal{D}'_{no}\,\mathrm{sat}_{s_{T_e}}((T_e(T))^\dagger)B_{ono})^\top e^{F'^\top\tau}. \tag{6.80}$$

Calculating $\dot{\underline{\mathcal{P}}}$, we obtain the same flow/jump dynamics as $\mathcal{P}$ in (6.76). Besides, $\underline{\mathcal{P}}(0, 0) = \mathcal{P}(0, 0)$, so $\underline{\mathcal{P}} = \mathcal{P}$ thanks to the uniqueness of solutions. We deduce that $\underline{\hat{\zeta}}$ follows the same dynamics as $\hat{\zeta}$, and since

$$\underline{\hat{\zeta}}(0, 0) = \mathcal{D}'(\hat{z}, \underline{\hat{\omega}})(0, 0) = \hat{\zeta}(0, 0),$$

we have $\underline{\hat{\zeta}} = \hat{\zeta}$, which implies $\underline{\hat{x}} = \hat{x}$ and concludes the proof. □

Example 6.4.12 Consider the same system as in Example 6.3.6. First, with $J_{no}(\tau) = \begin{pmatrix} 1 & 0 \\ 0 & 1 \end{pmatrix}$ for all $\tau \in \mathcal{I}$, Assumption 6.4.1 is satisfied. Since $H_{d,no}(\tau) = 0$ for all $\tau \in \mathcal{I}$, we discard the jump output and only consider the fictitious one described by the matrix $J_{ono}(\tau)$. We then see that with $m_1 = m_2 = 1$, for all sequences of flow lengths $(\tau_j)_{j\in\mathbb{N}}$,

$$\mathbb{O}_j^{bw} = J_{ono}(\tau_j)J_{no}^{-1}(\tau_j) = \begin{pmatrix} 1 & 0 \\ 0 & 1 \end{pmatrix}$$

satisfies $\mathbb{O}_j^{bw\top}\mathbb{O}_j^{bw} = \begin{pmatrix} 1 & 0 \\ 0 & 1 \end{pmatrix} > 0$ for all $j \geq \max\{m_1, m_2\} = 1$. Therefore, Assumption 6.4.6 is satisfied and we can thus design a KKL-based observer where T is of dimension 2×2. Let us take $A = \text{diag}(0.1, 0.2)$, an empty $B_{d,no}$, and $B_{ono} = \begin{pmatrix} 1 & 0 \\ 0 & 1 \end{pmatrix}$. Then, a KKL-based observer as in (6.60), (6.69), or (6.76), with the mentioned $(A, B_{d,no}, B_{ono})$, a weighting matrix taken as $R = \text{Id}$, a large enough λ, and a small enough γ, can be designed for this system.

6.5 Conclusion

This chapter presents and discusses new results on observer design for general hybrid systems with linear maps and known jump times. After defining and discussing the hybrid (pre-)asymptotic detectability and uniform complete observability conditions, we briefly presented again the Kalman-like observer in [28]. We then propose a decomposition of the state of a hybrid system with linear maps and known jump times into a part that is instantaneously observable during flows and a part that is not. A thorough analysis of the asymptotic detectability of the second part is performed, where we show that this part can actually be detectable from an extended output made of the jump output and a fictitious one thanks to the flow-jump combination. A high-gain Kalman-like observer with resets at jumps is proposed to estimate the first part, while two different jump-based algorithms are proposed for the second one. Several examples are provided to illustrate the methods.

A comparison among the mentioned designs, namely the Kalman-like observer [28] and the two designs depending on observability decomposition, is presented in Table 6.1 at the end. While the KKL-based design requires stronger conditions and has a larger dimension than the LMI-based one, it can provide an arbitrarily fast convergence rate of the estimate (achieved after a certain time). Compared to the Kalman-like design in [28], the LMI-based observer has a smaller dimension, whereas the KKL-based one can be bigger or smaller depending on n_o versus n_{no}. Therefore, for a particular application, if n_o is large compared to n_{no}, going through a decomposition is advantageous dimension-wise. Note also that while the Kalman-like observer can easily deal with time-varying matrices in the system dynamics and output, the decomposition method must additionally assume the (uniform) existence of the transformation into (6.26), for example, to solve the LMI (6.33) for the gains $L_{d,no}(\cdot)$ and K_{no}. Furthermore, even though the invertibility of $J_{no}(\tau)$ for all $\tau \in \mathcal{I}$ assumed for the KKL-based method may seem lighter than the invertibility of J at the jump times assumed in the Kalman-like approach [28, Assumption 2], it may turn out stronger since $J_{no}^{\dagger}(\tau)$ is used in the KKL implementation (see (6.60)), while J^{-1} is used only for analysis in the Kalman-like design, not in the implementation (see (6.16)).

Table 6.1 Comparison of observer designs for hybrid systems with linear maps and known jump times. All the conditions here are *sufficient* conditions

	Kalman-like observer (6.16) (Sect. 6.2.1)	LMI-based observer (6.34) (Sect. 6.3)	KKL-based observer (6.60) (Sect. 6.4)
Time domain	No assumption	Persistence of both flows and jumps (Assumption 6.2.7)	
Hybrid model	Invertibility of J at the jump times [28, Assumption 2]	The pair (F, H_c) constant	The pair (F, H_c) constant, invertibility of $J_{no}(\tau)$ for all $\tau \in \mathcal{I}$ (Assumption 6.4.1)
Observability	Uniform complete observability of (6.1) (Definition 6.2.5)	Instantaneous observability of z_o and a form of quadratic detectability of (6.32) for $\tau \in \mathcal{I}$ (Assumption 6.3.1)	Instantaneous observability of z_o and uniform backward distinguishability of (6.32) for the $(\tau_j)_{j\in\mathbb{N}}$ generated by the time domain of x (Assumption 6.4.6)
Dimension	$n_x + \frac{n_x(n_x+1)}{2}$	$n_x + \frac{n_o(n_o+1)}{2} + 1$	$n_o + \left(\sum_{i=1}^{n_{d,\text{ext}}} m_i\right) + \frac{n_o(n_o+1)}{2} + \left(\sum_{i=1}^{n_{d,\text{ext}}} m_i\right) n_{no} + 1$
Convergence rate	Arbitrarily fast, achieved after a certain time [28, Theorem 2]	Determined by the solution of a matrix inequality (Assumption 6.3.1)	Arbitrarily fast, achieved after a certain time (Theorem 6.4.7)
ISS Lyapunov function	Yes [28, Theorem 3]	Yes	Yes

Future work is to study hybrid systems with *nonlinear* maps (and known jump times) by finding transformations into coordinates of possibly higher dimensions and Lyapunov-based sufficient conditions to couple different observers [30], as well as those with *unknown* jump times.

Acknowledgements We thank Florent Di Meglio and Ricardo Sanfelice for their helpful feedback.

6.6 Appendix: Technical Lemmas

6.6.1 Exponential Stability of the Error Dynamics

Consider a hybrid system of the form

$$\begin{cases} \dot{\tilde{z}}_o = -P^{-1}H_{c,o}^\top(\tau)R^{-1}(\tau)H_{c,o}(\tau)\tilde{z}_o \\ \dot{\tilde{\eta}} = 0 \\ \dot{P} = -\lambda P + H_{c,o}^\top(\tau)R^{-1}(\tau)H_{c,o}(\tau) \\ \dot{\tau} = 1 \\ \\ \tilde{z}_o^+ = M_o(u,\tau)\tilde{z}_o + M_{o\eta}(u,\tau)\tilde{\eta} \\ \tilde{\eta}^+ = M_{\eta o}(u,\tau)\tilde{z}_o + M_\eta(\tau)\tilde{\eta} \\ P^+ = P_0 \\ \tau^+ = 0, \end{cases} \tag{6.81a}$$

where $u \in \mathcal{U}$ is the input, with the flow and jump sets

$$\mathbb{R}^{n_o} \times \mathbb{R}^{n_\eta} \times \mathbb{R}^{n_o \times n_o} \times [0, \tau_M], \qquad \mathbb{R}^{n_o} \times \mathbb{R}^{n_\eta} \times \mathbb{R}^{n_o \times n_o} \times \mathcal{I}, \tag{6.81b}$$

where $\mathcal{I}$ is a compact subset of $[\tau_m, \tau_M]$ for some positive τ_m and τ_M.

Lemma 6.6.1 *Assume that:*

1. *$H_{c,o}$ is continuous on $[0, \tau_M]$ and such that the pair $(0, H_{c,o}(\tau))$ satisfies (6.27);*
2. *R is continuous on $[0, \tau_M]$ and $R(\tau) > 0$ for all $\tau \in [0, \tau_M]$;*
3. *M_o, $M_{o\eta}$, and $M_{\eta o}$ are bounded on $\mathcal{U} \times \mathcal{I}$;*
4. *M_η is continuous on $\mathcal{I}$ and there exists $Q_\eta \in \mathcal{S}_{>0}^{n_\eta}$ such that*

$$M_\eta^\top(\tau)Q_\eta M_\eta(\tau) - Q_\eta < 0, \qquad \forall \tau \in \mathcal{I}. \tag{6.82}$$

Then for any $P_0 \in \mathcal{S}_{>0}^{n_o}$, there exists $\lambda^\star > 0$ such that for all $\lambda > \lambda^\star$, there exist $\rho_1 > 0$ and $\lambda_1 > 0$ such that any solution $(\tilde{z}_o, \tilde{\eta}, P, \tau)$ of (6.81) with $P(0,0) = P_0$, $\tau(0,0) = 0$, and $u \in \mathcal{U}$, is complete and verifies

$$|(\tilde{z}_o, \tilde{\eta})(t, j)| \le \rho_1 e^{-\lambda_1(t+j)} |(\tilde{z}_o, \tilde{\eta})(0, 0)|, \qquad \forall (t, j) \in \text{dom}(\tilde{z}_o, \tilde{\eta}, P, \tau). \quad (6.83)$$

Proof First, due to the compactness of $\mathcal{I}$ and Item 4 in Lemma 6.6.1, there exists $a > 0$ such that for all $\tau \in \mathcal{I}$, (6.82) is strengthened into

$$M_\eta^\top(\tau) Q_\eta M_\eta(\tau) - Q_\eta \le -a Q_\eta, \qquad \forall \tau \in \mathcal{I}. \quad (6.84)$$

Since $P(0, 0) = P_0$, $P^+ = P_0$, and $\tau(0, 0) = 0$, the component $(t, j) \mapsto P(t, j)$ of the solution of (6.81) can actually be written as a closed form of the component $(t, j) \mapsto \tau(t, j)$ by defining

$$\mathbb{P}(\tau) = e^{-\lambda\tau} P_0 + \int_0^\tau e^{-\lambda(\tau - s)} H_{c,o}^\top(s) R^{-1}(s) H_{c,o}(s) ds, \quad (6.85)$$

namely $P(t, j) = \mathbb{P}(\tau(t, j))$ for all $(t, j) \in \text{dom}\, x$. Note that since $P_0 > 0$, $\mathbb{P}(\tau)$ is invertible for all $\tau \in [0, \tau_M]$. It follows that the solution is complete and $(\tilde{z}_o, \tilde{\eta}, \tau)$ is the solution to

$$\begin{cases} \dot{\tilde{z}}_o = -\mathbb{P}(\tau)^{-1} H_{c,o}^\top(\tau) R^{-1}(\tau) H_{c,o}(\tau) \tilde{z}_o \\ \dot{\tilde{\eta}} = 0 \\ \dot{\tau} = 1 \\ \\ \tilde{z}_o^+ = M_o(u, \tau)\tilde{z}_o + M_{o\eta}(u, \tau)\tilde{\eta} \\ \tilde{\eta}^+ = M_{\eta o}(u, \tau)\tilde{z}_o + M_\eta(\tau)\tilde{\eta} \\ \tau^+ = 0, \end{cases} \quad (6.86)$$

with the flow set $\mathbb{R}^{n_o} \times \mathbb{R}^{n_\eta} \times [0, \tau_M]$ and the jump set $\mathbb{R}^{n_o} \times \mathbb{R}^{n_\eta} \times \mathcal{I}$. Consider the Lyapunov function

$$V(\tilde{z}_o, \tilde{\eta}, \tau) = e^{\frac{\lambda}{2}\tau} \tilde{z}_o^\top \mathbb{P}(\tau) \tilde{z}_o + k e^{-\epsilon\tau} \tilde{\eta}^\top Q_\eta \tilde{\eta}, \quad (6.87)$$

where $k > 0$ and $\epsilon > 0$. We have for all $\tau \in [\tau_m, \tau_M] \supseteq \mathcal{I}$,

$$\begin{aligned} e^{\frac{\lambda}{2}\tau}\mathbb{P}(\tau) &\ge e^{\frac{\lambda}{2}\tau} \int_{\frac{3}{4}\tau}^{\tau} e^{-\lambda(\tau - s)} H_{c,o}^\top(s) R^{-1}(s) H_{c,o}(s) ds \\ &\ge r_m e^{\frac{\lambda}{4}\tau} \int_{\frac{3}{4}\tau}^{\tau} \mathcal{D}_o^\top e^{F^\top s} H_c^\top H_c e^{Fs} \mathcal{D}_o ds \\ &\ge r_m e^{\frac{\lambda}{4}\tau_m} \int_{\frac{3}{4}\tau}^{\frac{3}{4}\tau + \frac{1}{4}\tau_m} \mathcal{D}_o^\top e^{F^\top s} H_c^\top H_c e^{Fs} \mathcal{D}_o ds \\ &\ge e^{\frac{\lambda}{4}\tau_m} r_m \alpha \,\text{Id} \\ &:= e^{\frac{\lambda}{4}\tau_m} \lambda_m \,\text{Id}, \end{aligned} \quad (6.88)$$

where $r_m > 0$ is a lower bound of the continuous map R on the compact set $[0, \tau_M]$ (thanks to Item 2 in Lemma 6.6.1), $\alpha > 0$ (independent of λ) is obtained by applying

(6.27) with $\delta = \frac{\tau_m}{4}$, and $\lambda_m := r_m\alpha$. On the other hand, from (6.85), for all $\tau \in [0, \tau_m]$, $\mathbb{P}(\tau) \geq e^{-\lambda\tau_m} P_0$, so $e^{\frac{\lambda}{2}\tau}\mathbb{P}(\tau) \geq e^{-\lambda\tau_m} P_0$. Besides, as $\mathbb{P}$ is continuous on the compact set $[0, \tau_M]$, there exists $p_M > 0$ such that $\mathbb{P}(\tau) \leq p_M \,\mathrm{Id}$ for all $\tau \in [0, \tau_M]$. It then follows that there exist $\underline{\rho} > 0$ and $\overline{\rho} > 0$ defined as

$$\underline{\rho} = \min\left\{\mathrm{eig}\left(e^{-\lambda\tau_m} P_0\right), e^{\frac{\lambda}{4}\tau_m}\lambda_m, \mathrm{eig}\left(ke^{-\epsilon\tau_M} Q_\eta\right)\right\}, \tag{6.89a}$$

$$\overline{\rho} = \max\left\{e^{\frac{\lambda}{2}\tau_M} p_M, \mathrm{eig}\left(ke^{-\epsilon\tau_m} Q_\eta\right)\right\}, \tag{6.89b}$$

such that

$$\underline{\rho}|(\tilde{z}_o, \tilde{\eta})|^2 \leq V(\tilde{z}_o, \tilde{\eta}, \tau) \leq \overline{\rho}|(\tilde{z}_o, \tilde{\eta})|^2, \qquad \forall(\tilde{z}_o, \tilde{\eta}) \in \mathbb{R}^n, \forall\tau \in [0, \tau_M]. \tag{6.90}$$

During flows, for all $(\tilde{z}_o, \tilde{\eta}) \in \mathbb{R}^n$ and $\tau \in [0, \tau_M]$,

$$\begin{aligned}
\dot{V} &= e^{\frac{\lambda}{2}\tau}\tilde{z}_o^\top\left(\frac{\lambda}{2}\mathbb{P}(\tau) - 2H_{c,o}^\top(\tau)R^{-1}(\tau)H_{c,o}(\tau) + \dot{\mathbb{P}}(\tau)\right)\tilde{z}_o - \epsilon ke^{-\epsilon\tau}\tilde{\eta}^\top Q_\eta\tilde{\eta} \\
&= e^{\frac{\lambda}{2}\tau}\tilde{z}_o^\top\left(-\frac{\lambda}{2}\mathbb{P}(\tau) - H_{c,o}^\top(\tau)R^{-1}(\tau)H_{c,o}(\tau)\right)\tilde{z}_o - \epsilon ke^{-\epsilon\tau}\tilde{\eta}^\top Q_\eta\tilde{\eta} \\
&\leq -\frac{\lambda}{2}e^{\frac{\lambda}{2}\tau}\tilde{z}_o^\top\mathbb{P}(\tau)\tilde{z}_o - \epsilon ke^{-\epsilon\tau}\tilde{\eta}^\top Q_\eta\tilde{\eta} \\
&\leq -\min\left\{\frac{\lambda}{2}, \epsilon\right\}V.
\end{aligned} \tag{6.91}$$

At jumps, for all $(\tilde{z}_o, \tilde{\eta}) \in \mathbb{R}^n$, $u \in \mathcal{U}$, and $\tau \in \mathcal{I}$,

$$\begin{aligned}
V^+ - V = {}&\star^\top P_0(M_o(u,\tau)\tilde{z}_o + M_{o\eta}(u,\tau)\tilde{\eta}) - e^{\frac{\lambda}{2}\tau}\tilde{z}_o^\top\mathbb{P}(\tau)\tilde{z}_o \\
&+ k\star^\top Q_\eta(M_{\eta o}(u,\tau)\tilde{z}_o + M_\eta(\tau)\tilde{\eta}) - ke^{-\epsilon\tau}\tilde{\eta}^\top Q_\eta\tilde{\eta}.
\end{aligned} \tag{6.92}$$

From Young's inequality, (6.88), (6.84), and Items 3 and 4 in Lemma 6.6.1, there exist non-negative constants c_i, $i = 1, 2, \ldots, 5$ independent of (λ, k, ϵ) such that for any $\kappa > 0$, for all $(\tilde{z}_o, \tilde{\eta}) \in \mathbb{R}^n$, $u \in \mathcal{U}$, and $\tau \in \mathcal{I}$,

$$\begin{aligned}
V^+ - V \leq {}&\left(c_1 + kc_2 + \kappa c_3 - e^{\frac{\lambda}{4}\tau_m}\lambda_m\right)\tilde{z}_o^\top\tilde{z}_o \\
&- \left(k\left(a - \left(1 - e^{-\epsilon\tau_M}\right)\right) - c_4 - \frac{k^2c_5}{\kappa}\right)\tilde{\eta}^\top Q_\eta\tilde{\eta}.
\end{aligned} \tag{6.93}$$

We now show that this quantity can be made negative definite by successively picking the degrees of freedom. For the $\tilde{\eta}$ part, $\exists\epsilon^\star > 0$ such that $0 < \epsilon < \epsilon^\star \implies a - \left(1 - e^{-\epsilon\tau_M}\right) > 0$, $\exists k^\star > 0$ such that $k > k^\star \implies k\left(a - \left(1 - e^{-\epsilon\tau_M}\right)\right) - c_4 > 0$, and $\exists\kappa^\star$ such that $\kappa > \kappa^\star \implies k\left(a - \left(1 - e^{-\epsilon\tau_M}\right)\right) - c_4 - \frac{k^2c_5}{\kappa} > 0$. Then, for the $\tilde{z}_o$

part, $\exists \lambda^\star > 0$ such that $\lambda > \lambda^\star \implies c_1 + kc_2 + \kappa c_3 - e^{\frac{\lambda}{4}\tau_m}\lambda_m < 0$. We deduce that for any $\lambda > \lambda^\star$, there exist $a_c > 0$ and $a_d > 0$ such that for all $(\tilde{z}_o, \tilde{\eta}) \in \mathbb{R}^n$,

$$\dot{V} \leq -a_c V, \qquad \forall \tau \in [0, \tau_M], \tag{6.94a}$$

$$V^+ - V \leq -a_d V, \qquad \forall u \in \mathcal{U}, \forall \tau \in \mathcal{I}. \tag{6.94b}$$

From (6.90) and (6.94), we conclude according to [13, Definition 7.29 and Theorem 7.30] that the set $\mathcal{A} = \{(\tilde{z}_o, \tilde{\eta}, \tau) \in \mathbb{R}^{n_o} \times \mathbb{R}^{n_{no}} \times [0, \tau_M] : \tilde{z}_o = 0, \tilde{\eta} = 0\}$ is GES for the system (6.86). □

Corollary 6.6.2 *Let us now consider (6.81) with $M_\eta(\tau)$ replaced by $\gamma M_\eta(\tau)$ for $\gamma \in (0, \gamma_0^\star]$. Under the same assumptions as in Lemma 6.6.1, for any $\lambda_c > 0$ and any $P_0 \in \mathcal{S}^{n_o}_{>0}$, there exists $\gamma^\star > 0$ such that there exists $\lambda^\star > 0$ such that for any $0 < \gamma < \gamma^\star$ and any $\lambda > \lambda^\star$, there exists $\rho_c > 0$ such that any solution $(\tilde{z}_o, \tilde{\eta}, P, \tau)$ of the new (6.81), with $P(0,0) = P_0$, $\tau(0,0) = 0$, and $u \in \mathcal{U}$, is complete and verifies*

$$|(\tilde{z}_o, \tilde{\eta})(t, j)| \leq \rho_c e^{-\lambda_c(t+j)} |(\tilde{z}_o, \tilde{\eta})(0, 0)|, \qquad \forall (t, j) \in \operatorname{dom}(\tilde{z}_o, \tilde{\eta}, P, \tau). \tag{6.95}$$

Proof This is a modification of the proof of Lemma 6.6.1. Consider the Lyapunov function in (6.87). First, let us show with an appropriate choice of ϵ that for λ sufficiently large and γ sufficiently small, we have for some $a_d > 0$,

$$\dot{V} \leq -2\lambda_c \left(\frac{1}{\tau_m} + 1\right) V, \qquad V^+ < V. \tag{6.96}$$

Following the same analysis as in the proof of Lemma 6.6.1, we obtain that during flows, $\dot{V} \leq -\min\left\{\frac{\lambda}{2}, \epsilon\right\} V$ for all $\tau \in [0, \tau_M]$, and at jumps thanks to (6.82), for all $u \in \mathcal{U}$ and $\tau \in \mathcal{I}$,

$$\begin{aligned} V^+ - V \leq &\left(c_1 + kc_2 + \kappa c_3 - e^{\frac{\lambda}{4}\tau_m}\lambda_m\right) \tilde{z}_o^\top \tilde{z}_o \\ &- \left(k\left(e^{-\epsilon\tau_M} - \gamma^2\right) - c_4 - \frac{k^2 c_5}{\kappa}\right) \tilde{\eta}^\top Q_\eta \tilde{\eta}. \end{aligned} \tag{6.97}$$

Let us pick $\epsilon = 2\lambda_c\left(\frac{1}{\tau_m} + 1\right)$ and define $\lambda_0^\star := 4\lambda_c\left(\frac{1}{\tau_m} + 1\right)$. Then, the first item in (6.96) holds as soon as $\lambda \geq \lambda_0^\star$. Now define $\gamma^\star := \min\left\{\gamma_0^\star, \sqrt{e^{-\epsilon\tau_M}}\right\}$. For any $0 < \gamma < \gamma^\star$, we have $e^{-\epsilon\tau_M} - \gamma^2 > e^{-\epsilon\tau_M} - (\gamma^\star)^2 > 0$, then k, κ, and λ are successively picked (based on $\gamma^\star$) as in the proof of Lemma 6.6.1. The final $\lambda^\star$ is the larger one between this and $\lambda_0^\star$. For any $\lambda > \lambda^\star$ and $0 < \gamma < \gamma^\star$, we obtain (6.96). Second, we deduce (6.95) from (6.96) and the dwell time condition. From (6.96), we get $V(t, j) \leq e^{-2\lambda_c\left(\frac{1}{\tau_m}+1\right)t} V(0, 0)$ for all $(t, j) \in \operatorname{dom}(\tilde{z}_o, \tilde{\eta}, P, \tau)$. Since the flow lengths of (6.81) are at least $\tau_m > 0$ for all $j \geq 1$, we have $j \leq \frac{t}{\tau_m} + 1$ so that $t \geq \frac{t+j-1}{\frac{1}{\tau_m}+1}$, for all $(t, j) \in \operatorname{dom}(\tilde{z}_o, \tilde{\eta}, P, \tau)$. Therefore, $V(t, j) \leq e^{2\lambda_c} e^{-2\lambda_c(t+j)} V(0, 0)$, for all $(t, j) \in \operatorname{dom}(\tilde{z}_o, \tilde{\eta}, P, \tau)$, implying (6.95). □

6.6.2 Boundedness in Finite Time

Lemma 6.6.3 *Consider a hybrid system with state $\eta \in \mathbb{R}^{n_\eta}$ and input $u \in \mathcal{U} \subset \mathbb{R}^{n_u}$:*

$$\begin{cases} \dot{\eta} = M_c(u)\eta & (\eta, u) \in C \\ \eta^+ = M_d(u)\eta & (\eta, u) \in D \end{cases} \tag{6.98}$$

with $M_c, M_d : \mathbb{R}^{n_u} \to \mathbb{R}^{n_\eta \times n_\eta}$. Consider positive scalars m_c, ρ_d, and τ_M, as well as $j_m \in \mathbb{N}$. Then, there exists $\rho > 0$ such that for any solution (η, u) of (6.98) with flow lengths in $[0, \tau_M]$ and such that $\|M_c(u(t, j))\| \leq m_c$ and $\|M_d(u(t, j))\| \leq \rho_d$ for all $u \in \mathcal{U}$ and $(t, j) \in \operatorname{dom} x$, we have $j_m \in \operatorname{dom}_j \eta$ and

$$|\eta(t_{j_m}, j_m)| \leq \rho|\eta(0, 0)|. \tag{6.99}$$

Proof First, the flow lengths of (6.98) are in $[0, \tau_m]$ and solutions are both t- and j-complete. During flows, the evolution of η is characterized by the transition matrix $\Psi_{M_c(u)}$ as

$$\eta(t, j) = \Psi_{M_c(u), u \in \mathcal{U}}(t, t_{j-1})\eta(t_{j-1}, j-1). \tag{6.100}$$

If M_c is uniformly bounded for $u \in \mathcal{U}$, there exists $\rho_c > 0$ such that for all $(t, j) \in \operatorname{dom} \eta$ and all $u \in \mathcal{U}$, $|\eta(t, j)| \leq \rho_c|\eta(t_{j-1}, j-1)|$. Next, we have for all $j \in \operatorname{dom}_j \eta$, $|\eta(t_j, j)| \leq \rho_d|\eta(t_j, j-1)|$. Therefore, for any $j_m \in \mathbb{N}_{>0}$, we have $|\eta(t_{j_m}, j_m)| \leq \rho_c^{j_m} \rho_d^{j_m - 1}|\eta(0, 0)|$, which is (6.99) by seeing that $\rho = \rho_c^{j_m} \rho_d^{j_m - 1}$. □

References

1. Alessandri, A., Coletta, P.: Switching observers for continuous-time and discrete-time linear systems. In: Proceedings of the 2001 American Control Conference, vol. 3, pp. 2516–2521 (2001)
2. Allan, D.A., Rawlings, J., Teel, A.R.: Nonlinear detectability and incremental input/output-to-state stability. SIAM J. Control Optim. **59**(4), 3017–3039 (2021)
3. Bernard, P., Andrieu, V., Astolfi, D.: Observer design for continuous-time dynamical systems. Ann. Rev. Control **53**, 224–248 (2022)
4. Bernard, P., Andrieu, V., Praly, L.: Expressing an observer in preferred coordinates by transforming an injective immersion into a surjective diffeomorphism. SIAM J. Control Optim. **56**(3), 2327–2352 (2018)
5. Bernard, P., Sanfelice, R.: On notions of detectability and observers for hybrid systems. In: IEEE Conference on Decision and Control, pp. 5767–5772 (2020)
6. Bernard, P., Sanfelice, R.G.: Observer design for hybrid dynamical systems with approximately known jump times. Automatica **141**, 110225 (2022)
7. Besancon, G.: Further results on high gain observers for nonlinear systems. In: Proceedings of the 38th IEEE Conference on Decision and Control (Cat. No.99CH36304), vol. 3, pp. 2904–2909 (1999)
8. Boyd, S., El Ghaoui, L., Feron, E., Balakrishnan, V.: Linear matrix inequalities in system and control theory. SIAM (1994)
9. Chen, C.-T.: Linear System Theory and Design. CBS College Publishing (1984)

10. Cox, N., Marconi, L., Teel, A.R.: Isolating invisible dynamics in the design of robust hybrid internal models. Automatica **68**, 56–68 (2016)
11. Deyst, Jr.J., Price, C.: Conditions for asymptotic stability of the discrete minimum-variance linear estimator. IEEE Trans. Autom. Control **13**, 702–705 (1968)
12. Ferrante, F., Gouaisbaut, F., Sanfelice, R., Tarbouriech, S.: State estimation of linear systems in the presence of sporadic measurements. Automatica **73**, 101–109 (2016). November
13. Goebel, R., Sanfelice, R.G., Teel, A.R.: Hybrid Dynamical Systems: Modeling Stability, and Robustness (2012)
14. Izhikevich, E.M.: Simple model of spiking neurons. IEEE Trans. Neural Netw. **14**(6), 1569–1572 (2003)
15. Kailath, T.: Linear Systems, vol. 156. Prentice-Hall, Englewood Cliffs, NJ (1980)
16. Kalman, R.E., Bucy, R.S.: New results in linear filtering and prediction theory. J. Basic Eng. **108**, 83–95 (1961)
17. Medina, E.A., Lawrence, D.A.: State estimation for linear impulsive systems. In: Annual American Control Conference, pp. 1183–1188 (2009)
18. Moarref, M., Rodrigues, L.: Observer design for linear multi-rate sampled-data systems. In: American Control Conference, pp. 5319–5324 (2014)
19. Moore, J.B., Anderson, B.D.O.: Coping with singular transition matrices in estimation and control stability theory. Int. J. Control **31**, 571–586 (1980)
20. Penrose, R.: A generalized inverse for matrices. Math. Proc. Camb. Philos. Soc. **51**(3), 406–413 (1955)
21. Raff, T., Allgöwer, F.: Observers with impulsive dynamical behavior for linear and nonlinear continuous-time systems. In: IEEE Conference on Decision and Control, pp. 4287–4292 (2007)
22. Ríos, H., Dávila, J., Teel, A.R.: State estimation for linear hybrid systems with periodic jumps and unknown inputs. Int. J. Robust Nonlinear Control **30**(15), 5966–5988 (2020)
23. Sanfelice, R.G.: Hybrid Feedback Control. Princeton University Press, Princeton, NJ (2021)
24. Sferlazza, A., Tarbouriech, S., Zaccarian, L.: Time-varying sampled-data observer with asynchronous measurements. IEEE Trans. Autom. Control **64**(2), 869–876 (2019)
25. Tanwani, A., Shim, H., Liberzon, D.: Observability for switched linear systems: characterization and observer design. IEEE Trans. Autom. Control **58**(4), 891–904 (2013)
26. Tanwani, A., Yufereva, O.: Error covariance bounds for suboptimal filters with Lipschitzian drift and Poisson-sampled measurements. Automatica **122**, 109280 (2020)
27. Tran, G.Q.B., Bernard, P.: Arbitrarily fast robust KKL observer for nonlinear time-varying discrete systems. IEEE Trans. Autom. Control (2023). https://ieeexplore.ieee.org/document/10302392
28. Tran, G.Q.B., Bernard, P.: Kalman-like observer for hybrid systems with linear maps and known jump times. In: IEEE Conference on Decision and Control (2023)
29. Tran, G.Q.B., Bernard, P., Di Meglio, F., Marconi, L.: Observer design based on observability decomposition for hybrid systems with linear maps and known jump times. In: 2022 IEEE 61st Conference on Decision and Control (CDC), pp. 1974–1979 (2022)
30. Tran, G.Q.B., Bernard, P., Sanfelice, R.: Coupling flow and jump observers for hybrid systems with known jump times. In: 22nd IFAC World Congress (2023)
31. Vigne, M., Khoury, A.E., Pétriaux, M., Di Meglio, F., Petit, N.: MOVIE: a velocity-aided IMU attitude estimator for observing and controlling multiple deformations on legged robots. IEEE Robot. Autom. Lett. **7**(2), 3969–3976 (2022)
32. Wu, F.: Control of linear parameter varying systems. Ph.D. thesis, University of California at Berkeley (1995)
33. Zhang, Q.: On stability of the Kalman filter for discrete time output error systems. Syst. Control Lett. **107**, 84–91 (2017). July
34. Ţiclea, A., Besançon, G.: Exponential forgetting factor observer in discrete time. Syst. & Control Lett. **62**(9), 756–763 (2013)

Chapter 7
A Joint Spectral Radius for ω-Regular Language-Driven Switched Linear Systems

Georges Aazan, Antoine Girard, Paolo Mason, and Luca Greco

Abstract In this chapter, we introduce some tools to analyze stability properties of discrete-time switched linear systems driven by switching signals belonging to a given ω-regular language. More precisely, we assume that the switching signals are generated by a deterministic Büchi automaton whose alphabet coincides with the set of modes of the switched system. We present the notion of ω-regular joint spectral radius (ω-RJSR), which intuitively describes the contraction of the state when the run associated with the switching signal visits an accepting state of the automaton. Then, we show how this quantity is related to the stability properties of such systems. Specifically, we show how this notion can characterize a class of stabilizing switching signals for a switched system that is unstable for arbitrary switching. Though the introduced quantity is hard to compute, we present some methods to approximate it using Lyapunov and automata-theoretic techniques. More precisely, we show how upper bounds can be computed by solving a convex optimization problem. To validate the results of our work, we finally show a numerical example related to the synchronization of oscillators over a communication network.

G. Aazan (✉)
Université Paris-Saclay, CNRS, ENS Paris-Saclay, Laboratoire Méthodes Formelles, 91190 Gif-sur-Yvette, France
e-mail: georges.aazan@lsv.fr

A. Girard · P. Mason · L. Greco
Université Paris-Saclay, CNRS, CentraleSupélec, Laboratoire des signaux et systèmes, 91190 Gif-sur-Yvette, France
e-mail: antoine.girard@l2s.centralesupelec.fr

P. Mason
e-mail: paolo.mason@l2s.centralesupelec.fr

L. Greco
e-mail: luca.greco@l2s.centralesupelec.fr

R. Postoyan et al. (eds.), *Hybrid and Networked Dynamical Systems*,
Lecture Notes in Control and Information Sciences 493,
https://doi.org/10.1007/978-3-031-49555-7_7

7.1 Introduction

Switched systems are a type of dynamical system that exhibit both continuous and discrete behaviors. They consist of multiple subsystems, each with its own dynamics. At a given time, the active mode is determined by a so-called switching signal. Switched systems have many applications, in particular, they play a crucial role in networked control systems where multiple controllers are connected via a communication network. In this case, switched systems can faithfully model the switching dynamics resulting from communication protocols.

The stability analysis of switched systems is a challenging problem due to the complex dynamics arising from the switching between the modes. For the class of discrete-time switched linear systems with arbitrary switching signals, the notion of joint spectral radius (JSR) [10] is used to analyze their stability. Intuitively, the JSR is a spectral characteristic that characterizes the worst-case asymptotic growth of infinite products of matrices describing the modes. Computing the JSR is, in general, a difficult and challenging task [14], and several methods have been proposed to compute accurate lower and upper bounds [5, 9, 15].

In the past decade, a significant amount of research has been carried out to analyze the stability of switched systems subject to constrained switching. Constraints on the switching signal are usually described by finite-state automata and methods for handling such constraints through Lyapunov functions have been presented in several works [11–13]. An alternative approach to analyze stability with such constrained switching signals is by computing the so-called constrained joint spectral radius (CJSR) [7, 11, 16].

However, there are classes of constrained switching signals that cannot be characterized using classical finite-state automata. For instance, this is the case when switching signals belong to some ω-regular language. An ω-regular language is a type of formal language that consists of infinite sequences of symbols that can be described in terms of a Büchi automaton [3]. A particular example of ω-regular language is that of shuffled switching signals. A switching signal is said to be shuffled if all the modes of the switched systems are activated an infinite number of times. Stability analysis of these constrained systems was studied in [8] where sufficient and necessary conditions for stability were given in terms of multiple Lyapunov function. Further results are found in [1] where a spectral characteristic is defined related to these specific systems; this notion quantifies the contraction rate of the system each time the systems shuffle. The stability of ω-regular language constrained switched systems was studied in [2] where sufficient and necessary conditions were given in terms of existence of Lyapunov functions. However, no approximation of the convergence rate was given in that work. Our chapter extends the results of [1, 2] in order to estimate the convergence rate using a suitable spectral characteristic.

The organization of the chapter is as follows. Section 7.2 presents the necessary background on switched systems and Büchi automata. In Sect. 7.3, the ω-regular joint spectral radius (ω-RJSR) is introduced and several of its properties are established. Sect. 7.4 shows the relationship between the ω-RJSR and stability properties of a switched system driven by switching signals generated by a given Büchi automaton. In Sect. 7.5, we present an approach based on Lyapunov and automata-theoretic techniques to compute an over-approximation of the ω-RJSR. Finally, a numerical example is used in Sect. 7.6 to illustrate the main results of the chapter.

Notation. Given an alphabet Σ, Σ^* denotes the set of all finite words over the alphabet Σ, $\Sigma^+ = \Sigma^* \setminus \{\epsilon\}$ with ϵ the empty word and Σ^ω denotes the set of all infinite words over the alphabet Σ. Given two words $w_1 \in \Sigma^*$, $w_2 \in \Sigma^\omega$, $w_1.w_2$ denotes the element of Σ^ω obtained by concatenation of w_1 and w_2. For a word $w \in \Sigma^+$, w^ω denotes the element of Σ^ω obtained by repeating infinitely w, i.e., $w^\omega = w.w.w\ldots$. $|.|$ denotes the length of a word and is also used to denote the cardinality of a finite set. $\mathbb{R}_0^+$ denotes the set of non-negative real numbers. We use $\|\cdot\|$ to denote an arbitrary norm on $\mathbb{R}^n$ and the associated induced matrix norm defined for $M \in \mathbb{R}^{n\times n}$ by $\|M\| = \sup\limits_{x\neq 0} \frac{\|Mx\|}{\|x\|}$. I_n denotes the $n \times n$ identity matrix. Given a sequence of matrices $(M_k)_{k\in\mathbb{N}}$, with $M_k \in \mathbb{R}^{n\times n}$, we define for $k_1, k_2 \in \mathbb{N}$, with $k_1 \leq k_2$

$$\prod_{k=k_1}^{k_2} M_k = \begin{cases} M_{k_2} \cdots M_{k_1}, & \text{if } k_2 > k_1, \\ M_{k_1}, & \text{if } k_2 = k_1. \end{cases}$$

7.2 Switched Linear Systems Driven by ω-Regular Languages

In this part, we describe the class of systems under consideration and introduce some preliminary results which will be useful for subsequent discussions.

7.2.1 Deterministic Büchi Automaton

A *deterministic Büchi automaton* (DBA) is a tuple $\mathcal{B} = (Q, \Sigma, \delta, q_{init}, F)$ where Q is a finite set of *states*, Σ is the *alphabet*, $\delta : Q \times \Sigma \to Q$ is a *partial transition function*, $q_{init} \in Q$ is the *initial state*, and $F \subseteq Q$ is the set of *accepting states*.

With each finite or infinite word $\sigma = \sigma_0\sigma_1 \cdots \in \Sigma^+ \cup \Sigma^\omega$, we associate, if it exists, a unique *run* $q_0q_1q_2\ldots$ starting from the initial state $q_0 = q_{init}$ and such that $q_{t+1} = \delta(q_t, \sigma_t)$ for all $t = 0, 1, \ldots$. Note that some words in $\Sigma^+ \cup \Sigma^\omega$ may not have associated runs since δ is a partial function. A run $q_0q_1q_2\ldots$ associated with an infinite word $\sigma \in \Sigma^\omega$ is said to be *accepting* if $q_t \in F$ for infinitely many indices $t \in \mathbb{N}$. In other words, a run is accepting if it visits the set F an infinite number of times. The *language* of $\mathcal{B}$, denoted by $Lang(\mathcal{B})$, is the set of all infinite words over the alphabet Σ which have an accepting run in $\mathcal{B}$. It belongs to the class of ω-regular languages [3].

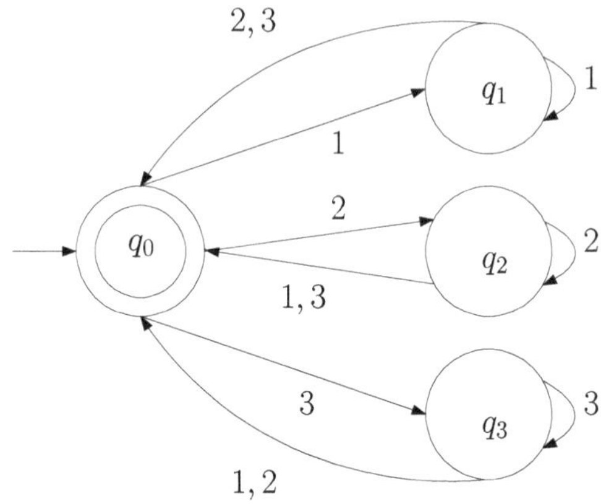

Fig. 7.1 An example of DBA whose language consists of infinite words that do not remain constant after some time. The accepting state is represented with a double circle

Assumption 7.2.1 All the states of the DBA $\mathcal{B}$ are reachable from the initial state, i.e., for all $q \in Q$, there exists a finite run $q_0 q_1 q_2 \dots q_k$ with $q_k = q$; and for any finite run $q_0 q_1 q_2 \dots q_k$, there exists an infinite sequence of states $q_{k+1} q_{k+2} \dots$ such that $q_0 q_1 q_2 \dots q_k q_{k+1} q_{k+2} \dots$ is an accepting run.

Note that this assumption ensures that $Lang(\mathcal{B}) \neq \varnothing$, furthermore there is no loss of generality to suppose that Assumption 7.2.1 holds true since it can be shown easily that for any DBA, there exists a DBA with the same language and satisfying Assumption 7.2.1. Hence, in the rest of the chapter, Assumption 1 is supposed as always satisfied.

An example of a DBA is shown in Fig. 7.1. In this example, the set of states is $\{q_0, q_1, q_2, q_3\}$ and the alphabet is $\Sigma = \{1, 2, 3\}$. The initial state is $q_{init} = q_0$ and the set of accepting states is $F = \{q_0\}$. The transition function δ is represented by the arrows in Fig. 7.1. The language of this DBA consists of all infinite words $\sigma = \sigma_0 \sigma_1 \cdots \in \Sigma^\omega$ that do not remain constant after some time:

$$\sigma \in Lang(\mathcal{B}) \iff \forall t \in \mathbb{N}, \exists t' \geq t,\ \sigma_{t'} \neq \sigma_t.$$

Now given a DBA $\mathcal{B}$, and an infinite word $\sigma \in Lang(\mathcal{B})$, let us denote $q_0 q_1 \dots$ the associated run in $\mathcal{B}$ and define the following quantities:

- The sequence of *accepting instants* $(\tau_k^\sigma)_{k \in \mathbb{N}}$ defined by $\tau_0^\sigma = 0$, and for all $k \in \mathbb{N}$,

$$\tau_{k+1}^\sigma = \min \left\{ t > \tau_k^\sigma \ | q_t \in F \right\}.$$

- The *accepting index* $\kappa^\sigma : \mathbb{N} \to \mathbb{N}$ given by

$$\kappa^\sigma(t) = \max\{k \in \mathbb{N} |\ \tau_k^\sigma \leq t\}.$$

- The *accepting rate* γ^σ as

$$\gamma^\sigma = \liminf_{t\to+\infty} \frac{\kappa^\sigma(t)}{t}.$$

Intuitively, τ_k^σ is the instant of the kth visit of the run to the set F and the accepting index $\kappa^\sigma(t)$ counts the number of times the run has visited the set F up to time t. It is worth noting that since $\sigma \in Lang(\mathcal{B})$, we have $\lim_{t\to\infty} \kappa^\sigma(t) = \infty$. The accepting rate γ^σ characterizes the frequency of visit to the set F over all time.

In the next lemma, we establish a tight upper bound on the accepting rate of all switching sequences belonging to a language of a given DBA $\mathcal{B}$. Before introducing this result, let us define the notion of cycle in $\mathcal{B}$. A *cycle* is a sequence of states $q_1 \cdots q_n$ such that for all $i \neq j$, $q_i \neq q_j$; for all $i = 1, \ldots, n-1$, there exists $\sigma_i \in \Sigma$ such that $q_{i+1} = \delta(q_i, \sigma_i)$; and there exists $\sigma_n \in \Sigma$ such that $q_1 = \delta(q_n, \sigma_n)$. We denote the set of cycles in $\mathcal{B}$ by $C_\mathcal{B}$. For a cycle $c \in C_\mathcal{B}$, we denote by F_c the set of accepting states appearing in c. Formally, $F_c = \{q_i \in F : c = q_1 \cdots q_n\}$.

Lemma 7.2.2 *Given a DBA $\mathcal{B}$, let $M = \max_{c\in C_\mathcal{B}} \frac{|F_c|}{|c|}$. Then,*

$$\forall \sigma \in Lang(\mathcal{B}),\ \forall t \in \mathbb{N},\ \kappa^\sigma(t) \leq |Q| + Mt. \tag{7.1}$$

Moreover, $\max_{\sigma\in Lang(\mathcal{B})} \gamma^\sigma = M.$

Proof Let $\sigma \in Lang(\mathcal{B})$, let $q_0q_1\ldots$ be the associated run, and let $t \in \mathbb{N}$. The accepting index $\kappa^\sigma(t)$ is given by the number of occurrences of elements of F in the sequence $r_0 = q_1 \ldots q_t$. Since $|r_0| = t$, if $t \leq |Q|$, it is clear that $\kappa^\sigma(t) \leq |Q|$ and (7.1) holds. If $t > |Q|$, then there exists at least one state appearing twice in r_0. Let $1 \leq i < j \leq t$ be the indices of the first state appearing twice in r_0. Then, there exists a cycle $c_1 = q_i \ldots q_{j-1} \in C_\mathcal{B}$. Moreover, the number of occurrences of elements of F in r_0 is given by the sum of the number of occurrences of elements of F in the cycle c_1 and in the sequence $r_1 = q_1 \ldots q_{i-1}q_j \ldots q_t$ (or in the sequence $r_1 = q_j \ldots q_t$ if $i = 1$). By repeating this reasoning, we get that the number of occurrences of elements of F in r_0 is bounded by the sum of the number of occurrences of elements of F in a collection of cycles $c_1, \ldots, c_k$ and in a sequence of states r_k such that $|r_k| \leq |Q|$. The number of occurrences of elements of F in r_k is bounded by $|r_k| \leq |Q|$. The total number of occurrences of elements of F in the cycles $c_1, \ldots, c_k$ is bounded by $M(|c_1| + \cdots + |c_k|) \leq Mt$. Hence, it follows that (7.1) holds and we directly obtain that

$$\sup_{\sigma\in Lang(\mathcal{B})} \gamma^\sigma \leq M.$$

To show that this bound is tight, let us consider $c \in C_\mathcal{B}$ such that $\frac{|F_c|}{|c|} = M$. From Assumption 7.2.1, it follows that there exists a run of $\mathcal{B}$ of the form $r = w.c^\omega$. Let σ be an associated element of Σ^ω. Then, it is easy to see that $\gamma^\sigma = M$. □

As an illustration, consider the DBA $\mathcal{B}$ shown in Fig. 7.1. This DBA contains six cycles $q_1, q_2, q_3, q_0q_1, q_0q_2$, and q_0q_3. Then, the maximal accepting rate for infinite words $\sigma \in Lang(\mathcal{B})$ is $M = \frac{1}{2}$.

7.2.2 Switched Linear Systems

Now let us consider a discrete-time switched linear system in which the switching sequences are elements of the language of a given DBA. More precisely, given a Büchi automaton $\mathcal{B} = (Q, \Sigma, \delta, q_{init}, F)$ with alphabet $\Sigma = \{1, \dots, m\}$, and a finite set of matrices $\mathcal{A} = \{A_1, \dots, A_m\}$ with $A_i \in \mathbb{R}^{n \times n}$, $i \in \Sigma$, the discrete-time switched linear system $(\mathcal{A}, \mathcal{B})$ is described by the following equation:

$$x(t+1) = A_{\theta(t)} x(t), \tag{7.2}$$

where $t \in \mathbb{N}$, $x(t) \in \mathbb{R}^n$ is the state and $\theta : \mathbb{N} \to \Sigma$ is the switching signal with $\theta \in Lang(\mathcal{B})$.[1] Given an initial condition $x_0 \in \mathbb{R}^n$, and a switching signal $\theta \in Lang(\mathcal{B})$, the trajectory with $x(0) = x_0$ is unique, denoted by $\mathbf{x}(\cdot, x_0, \theta)$ and given by

$$\forall t \in \mathbb{N}, \ \mathbf{x}(t, x_0, \theta) = \mathbb{A}_{\theta,t} x_0,$$

where $\mathbb{A}_{\theta,0} = I_n$, and

$$\mathbb{A}_{\theta,t} = \prod_{s=0}^{t-1} A_{\theta(s)}, \ \forall t \geq 1.$$

In the context of switched linear systems with constrained switching signals, let us recall an important quantity, the constrained joint spectral radius (CJSR), see e.g., [13]. This notion characterizes the asymptotic rate of growth of constrained products of matrices. Given a finite set of matrices $\mathcal{A}$ and a Büchi automaton $\mathcal{B}$, let us denote by $\Theta_{\mathcal{B}}$ the set of switching sequences θ such that there exists a sequence of states $p_0 p_1 \dots$ such that $p_{t+1} = \delta(p_t, \theta(t))$ for all $t \in \mathbb{N}$. Let us remark that all elements of $Lang(\mathcal{B})$ belong to $\Theta_{\mathcal{B}}$. Then, the CJSR of $(\mathcal{A}, \mathcal{B})$ is defined by

$$\rho(\mathcal{A}, \mathcal{B}) = \lim_{t \to +\infty} \max \left\{ \left\| \mathbb{A}_{\theta,t} \right\|^{\frac{1}{t}} : \theta \in \Theta_{\mathcal{B}} \right\}.$$

The following lemma relates this notion to trajectories of (7.2).

Lemma 7.2.3 *For all $\rho > \rho(\mathcal{A}, \mathcal{B})$, there exists a $C \geq 1$ such that*

$$\|\mathbf{x}(t, x_0, \theta)\| \leq C \rho^t \|x_0\|, \qquad \forall x_0 \in \mathbb{R}^n, \theta \in \Theta_{\mathcal{B}}, t \in \mathbb{N}.$$

Proof This is a direct consequence of [13, Theorem 1.1] after scaling the matrices in $\mathcal{A}$ by $\frac{1}{\rho}$. □

The latter inequality happens to be conservative in the case of switching sequences generated by a Büchi automaton $\mathcal{B}$. Therefore, to better estimate the convergence rate of such systems, we introduce in the next section a new spectral characteristic specific to switched linear systems driven by ω-regular switching sequences.

[1] We identify a switching signal $\theta : \mathbb{N} \to \Sigma$ with the infinite word $\theta(0)\theta(1)\dots$ and use the notation $\theta \in Lang(\mathcal{B})$ for $\theta(0)\theta(1)\cdots \in Lang(\mathcal{B})$

7.3 ω-Regular Joint Spectral Radius

In this section, we introduce the ω-Regular Joint Spectral Radius and establish several of its properties.

Definition 7.3.1 Let $\rho > \boldsymbol{\rho}(\mathcal{A}, \mathcal{B})$, the ω-Regular Joint Spectral Radius relative to $(\mathcal{A}, \mathcal{B}, \rho)$ (ρ-ω-RJSR for short) is defined as

$$\boldsymbol{\lambda}(\mathcal{A}, \mathcal{B}, \rho) = \limsup_{k \to +\infty} \left(\sup_{\theta \in Lang(\mathcal{B})} \left(\frac{\left\| \mathbb{A}_{\theta, \tau_k^\theta} \right\|}{\rho^{\tau_k^\theta}} \right)^{1/k} \right). \tag{7.3}$$

To better understand the characteristics of the ρ-ω-RJSR, we are going to analyze some properties of the function $\boldsymbol{\lambda}(\mathcal{A}, \mathcal{B}, \cdot)$.

Proposition 7.3.2 *For all $\rho > \boldsymbol{\rho}(\mathcal{A}, \mathcal{B})$, the value $\boldsymbol{\lambda}(\mathcal{A}, \mathcal{B}, \rho)$ does not change if we replace the induced matrix norm in* (7.3) *by any other matrix norm.*

Proof This is straightforward from the equivalence of norms in a finite-dimensional subspace. □

Proposition 7.3.3 *The function $\rho \mapsto \boldsymbol{\lambda}(\mathcal{A}, \mathcal{B}, \rho)$ is non-increasing, takes values in* $[0, 1)$ *and, for all $\boldsymbol{\rho}(\mathcal{A}, \mathcal{B}) < \rho_1 \leq \rho_2$,*

$$\boldsymbol{\lambda}(\mathcal{A}, \mathcal{B}, \rho_2) \leq \left(\frac{\rho_1}{\rho_2} \right)^{\frac{1}{M}} \boldsymbol{\lambda}(\mathcal{A}, \mathcal{B}, \rho_1) \tag{7.4}$$

where M is the same as in Lemma 7.2.2.

Proof Let $\rho > \boldsymbol{\rho}(\mathcal{A}, \mathcal{B})$, from Lemma 7.2.3, we get that there exists $C_1 \geq 1$ such that

$$0 \leq \frac{\left\| \mathbb{A}_{\theta, t} \right\|}{\rho^t} \leq C_1, \quad \forall \theta \in \Theta_{\mathcal{B}}, \forall t \in \mathbb{N}. \tag{7.5}$$

Using the fact that $Lang(\mathcal{B}) \subseteq \Theta_{\mathcal{B}}$, we get that

$$0 \leq \frac{\left\| \mathbb{A}_{\theta, t} \right\|}{\rho^t} \leq C_1, \quad \forall \theta \in Lang(\mathcal{B}), \forall t \in \mathbb{N}. \tag{7.6}$$

By considering $t = \tau_k^\theta$, raising to the power $1/k$ and taking the supremum over all switching signals in $Lang(\mathcal{B})$, we find

$$0 \leq \sup_{\theta \in Lang(\mathcal{B})} \left(\frac{\left\| \mathbb{A}_{\theta, \tau_k^\theta} \right\|}{\rho^{\tau_k^\theta}} \right)^{1/k} \leq C_1^{1/k}.$$

Taking the lim sup of all terms yields $\boldsymbol{\lambda}(\mathcal{A}, \mathcal{B}, \rho) \in [0, 1]$.

Now, let us consider $\boldsymbol{\rho}(\mathcal{A}, \mathcal{B}) < \rho_1 \leq \rho_2$. From (7.1), it follows that $\tau_k^\theta \geq \frac{k-|Q|}{M}$. Then, we have for all $k \geq 1$ and for all $\theta \in Lang(\mathcal{B})$

$$\frac{\|\mathbb{A}_{\theta,\tau_k^\theta}\|}{\rho_2^{\tau_k^\theta}} = \frac{\|\mathbb{A}_{\theta,\tau_k^\theta}\|}{\rho_1^{\tau_k^\theta}} \left(\frac{\rho_1}{\rho_2}\right)^{\tau_k^\theta} \leq \frac{\|\mathbb{A}_{\theta,\tau_k^\theta}\|}{\rho_1^{\tau_k^\theta}} \left(\frac{\rho_1}{\rho_2}\right)^{\frac{k-|Q|}{M}}.$$

Raising to the power $1/k$ and taking the supremum over all switching signals in $Lang(\mathcal{B})$ yields

$$\sup_{\theta \in Lang(\mathcal{B})} \left(\frac{\|\mathbb{A}_{\theta,\tau_k^\theta}\|}{\rho_2^{\tau_k^\theta}}\right)^{1/k} \leq \sup_{\theta \in Lang(\mathcal{B})} \left(\frac{\|\mathbb{A}_{\theta,\tau_k^\theta}\|}{\rho_1^{\tau_k^\theta}}\right)^{1/k} \left(\frac{\rho_1}{\rho_2}\right)^{\frac{k-|Q|}{kM}}.$$

Now we take the lim sup of both terms and we get (7.4), which implies that $\rho \mapsto \boldsymbol{\lambda}(\mathcal{A}, \mathcal{B}, \rho)$ is non-increasing. Now, let us assume that there exists $\rho_2 > \boldsymbol{\rho}(\mathcal{A}, \mathcal{B})$ such that $\boldsymbol{\lambda}(\mathcal{A}, \mathcal{B}, \rho_2) = 1$. It follows from (7.4) that for all $\rho_1 \in (\boldsymbol{\rho}(\mathcal{A}, \mathcal{B}), \rho_2)$, $\boldsymbol{\lambda}(\mathcal{A}, \mathcal{B}, \rho_2) < \boldsymbol{\lambda}(\mathcal{A}, \mathcal{B}, \rho_1)$, which contradicts the fact that $\boldsymbol{\lambda}(\mathcal{A}, \mathcal{B}, \rho_1) \in [0, 1]$. Hence, $\boldsymbol{\lambda}(\mathcal{A}, \mathcal{B}, \rho) \in [0, 1)$ for all $\rho > \boldsymbol{\rho}(\mathcal{A}, \mathcal{B})$. □

In order to get rid of the dependence of ρ in the ρ-ω-RJSR, it is natural to introduce the following definition.

Definition 7.3.4 The ω-Regular Joint Spectral Radius (ω-RJSR) of $(\mathcal{A}, \mathcal{B})$ is defined as

$$\boldsymbol{\lambda}(\mathcal{A}, \mathcal{B}) = \lim_{\rho \to \boldsymbol{\rho}(\mathcal{A},\mathcal{B})^+} \boldsymbol{\lambda}(\mathcal{A}, \mathcal{B}, \rho). \tag{7.7}$$

Since by Proposition 7.3.3, $\boldsymbol{\lambda}(\mathcal{A}, \mathcal{B}, \cdot)$ is bounded and non-increasing in ρ, the right limit at $\boldsymbol{\rho}(\mathcal{A}, \mathcal{B})$ in (7.7) exists and the ω-RJSR is well-defined. We can now show some of its properties.

Proposition 7.3.5 *The ω-RJSR enjoys the following properties:*

(i) $\boldsymbol{\lambda}(\mathcal{A}, \mathcal{B})$ *belongs to* $[0, 1]$ *and is independent of the choice of the norm;*
(ii) For all $K \in \mathbb{R}$, $K \neq 0$, *we have* $\boldsymbol{\lambda}(K\mathcal{A}, \mathcal{B}) = \boldsymbol{\lambda}(\mathcal{A}, \mathcal{B})$.

Proof The first statement follows from (7.7) and by the properties of $\boldsymbol{\lambda}(\mathcal{A}, \mathcal{B}, \cdot)$ proved in Propositions 7.3.2 and 7.3.3. Concerning the second item, let $K \in \mathbb{R}$, $K \neq 0$, we have by (7.7)

$$\boldsymbol{\lambda}(K\mathcal{A}, \mathcal{B}) = \lim_{\rho \to \boldsymbol{\rho}(K\mathcal{A},\mathcal{B})^+} \boldsymbol{\lambda}(K\mathcal{A}, \mathcal{B}, \rho) = \lim_{\rho \to \boldsymbol{\rho}(\mathcal{A},\mathcal{B})^+} \boldsymbol{\lambda}(K\mathcal{A}, \mathcal{B}, |K|\rho)$$

where the second equality comes from the property of the CJSR, $\boldsymbol{\rho}(K\mathcal{A}, \mathcal{B}) = |K|\boldsymbol{\rho}(\mathcal{A}, \mathcal{B})$, see e.g., [13]. Furthermore, from (7.3), one can deduce that for all $\rho > \boldsymbol{\rho}(\mathcal{A}, \mathcal{B})$, $\boldsymbol{\lambda}(K\mathcal{A}, \mathcal{B}, |K|\rho) = \boldsymbol{\lambda}(\mathcal{A}, \mathcal{B}, \rho)$, therefore $\boldsymbol{\lambda}(K\mathcal{A}, \mathcal{B}) = \boldsymbol{\lambda}(\mathcal{A}, \mathcal{B})$. □

Let us remark that, while the CJSR $\boldsymbol{\rho}(\mathcal{A}, \mathcal{B})$ belongs to $\mathbb{R}_0^+$, the ω-RJSR $\boldsymbol{\lambda}(\mathcal{A}, \mathcal{B})$ is always in $[0, 1]$. Intuitively, while the CJSR provides an estimate of the contraction rate (when $\boldsymbol{\rho}(\mathcal{A}, \mathcal{B}) < 1$) or of the expansion rate (when $\boldsymbol{\rho}(\mathcal{A}, \mathcal{B}) > 1$) of the system state at each time step for constrained switching signals, the ω-RJSR measures how much additional contraction is obtained each time the set of accepting states F is visited. Theorem 7.4.1 in the next section provides theoretical ground to this interpretation.

7.4 ω-Regular Language-Driven Switched Systems and ρ-ω-RJSR

In this section, we show how the ρ-ω-RJSR relates to stability properties of constrained switched linear systems. In particular, we show how the ρ-ω-RJSR allows us to compute bounds on the accepting rate ensuring stability. The next result clarifies the relationship between the ρ-ω-RJSR and the behavior of the trajectories of (7.2).

Theorem 7.4.1 *For all $\rho > \boldsymbol{\rho}(\mathcal{A}, \mathcal{B})$, for all $\lambda \in (\boldsymbol{\lambda}(\mathcal{A}, \mathcal{B}, \rho), 1]$, there exists $C \geq 1$ such that*

$$\|\mathbf{x}(t, x_0, \theta)\| \leq C\rho^t \lambda^{\kappa^\theta(t)} \|x_0\|, \ \forall x_0 \in \mathbb{R}^n, \theta \in Lang(\mathcal{B}), t \in \mathbb{N}. \tag{7.8}$$

Conversely, if the matrices in $\mathcal{A}$ are invertible, if there exists $C \geq 1$, $\rho \geq 0$ and $\lambda \in [0, 1]$ such that (7.8) holds, then either $\rho > \boldsymbol{\rho}(\mathcal{A}, \mathcal{B})$ and $\lambda \geq \boldsymbol{\lambda}(\mathcal{A}, \mathcal{B}, \rho)$, or $\rho = \boldsymbol{\rho}(\mathcal{A}, \mathcal{B})$ and $\lambda \geq \boldsymbol{\lambda}(\mathcal{A}, \mathcal{B})$.

Proof We start by proving the direct result. Let $\rho > \boldsymbol{\rho}(\mathcal{A}, \mathcal{B})$ and $\lambda \in (\boldsymbol{\lambda}(\mathcal{A}, \mathcal{B}, \rho), 1]$. By definition of $\boldsymbol{\lambda}(\mathcal{A}, \mathcal{B}, \rho)$, there exists $k_0 \geq 1$ such that

$$\sup_{\theta \in Lang(\mathcal{B})} \left(\frac{\|\mathbb{A}_{\theta, \tau_k^\theta}\|}{\rho^{\tau_k^\theta}} \right)^{1/k} \leq \lambda, \qquad \forall k \geq k_0.$$

It follows that

$$\left\|\mathbb{A}_{\theta, \tau_k^\theta}\right\| \leq \rho^{\tau_k^\theta} \lambda^k, \quad \forall \theta \in Lang(\mathcal{B}), \forall k \geq k_0. \tag{7.9}$$

Then, let $C_1 \geq 1$ be such that (7.6) holds. In particular, for $t = \tau_k^\theta$ and for $\theta \in Lang(\mathcal{B})$, we obtain from (7.6) that

$$\left\|\mathbb{A}_{\theta, \tau_k^\theta}\right\| \leq C_1 \rho^{\tau_k^\theta}, \quad \forall \theta \in Lang(\mathcal{B}), \forall k \in \mathbb{N}. \tag{7.10}$$

Then, let $C_2 = C_1 \lambda^{-k_0}$, then $C_2 \geq 1$ and it follows from (7.9) and (7.10) that

$$\left\|\mathbb{A}_{\theta, \tau_k^\theta}\right\| \leq C_2 \rho^{\tau_k^\theta} \lambda^k, \quad \forall \theta \in Lang(\mathcal{B}), \forall k \in \mathbb{N}. \tag{7.11}$$

Let $\theta \in Lang(\mathcal{B}), t \in \mathbb{N}$, and $k = \kappa^{\theta}(t)$, we have $\mathbb{A}_{\theta,t} = \mathbb{A}_{\theta',t-\tau_k^{\theta}}\mathbb{A}_{\theta,\tau_k^{\theta}}$ where $\theta' \in \Theta_{\mathcal{B}}$ is given by $\theta'(s) = \theta(\tau_k^{\theta} + s)$, for all $s \in \mathbb{N}$. By (7.5), we get that

$$\left\|\mathbb{A}_{\theta',t-\tau_k^{\theta}}\right\| \leq C_1\rho^{t-\tau_k^{\theta}}. \tag{7.12}$$

Then, let $C = C_1C_2$, by (7.11) and (7.12), we get

$$\left\|\mathbb{A}_{\theta,t}\right\| \leq \left\|\mathbb{A}_{\theta',t-\tau_k^{\theta}}\right\|\left\|\mathbb{A}_{\theta,\tau_k^{\theta}}\right\| \leq C\lambda^{\kappa^{\theta}(t)}\rho^t.$$

Hence, (7.8) holds.

We now prove the converse result. By definition of induced matrix norm, (7.8) is equivalent to the following:

$$\|\mathbb{A}_{\theta,t}\| \leq C\rho^t\lambda^{\kappa^{\theta}(t)}, \quad \forall\theta \in Lang(\mathcal{B}),\ \forall t \in \mathbb{N}. \tag{7.13}$$

Since $\lambda \in [0, 1]$ and $\kappa^{\theta}(t) \in \mathbb{N}$ we also have

$$\|\mathbb{A}_{\theta,t}\| \leq C\rho^t, \quad \forall\theta \in Lang(\mathcal{B}),\ \forall t \in \mathbb{N}. \tag{7.14}$$

Now, let $\theta \in \Theta_{\mathcal{B}}$ and $t \in \mathbb{N}$. From Assumption 7.2.1, there exists $\theta' \in Lang(\mathcal{B})$ and $t_0 \leq |Q|$ such that $\theta(s) = \theta'(t_0 + s)$, for all $s = 0, \ldots, t$. Then, since all matrices in $\mathcal{A}$ are invertible, we get

$$\mathbb{A}_{\theta,t} = \mathbb{A}_{\theta',t+t_0}\mathbb{A}_{\theta',t_0}^{-1}.$$

Let us denote $D = \max_{A\in\mathcal{A}} \|A^{-1}\|$, then

$$\|\mathbb{A}_{\theta,t}\| \leq \|\mathbb{A}_{\theta',t+t_0}\|\ \|\mathbb{A}_{\theta',t_0}^{-1}\| \leq \|\mathbb{A}_{\theta',t+t_0}\|\ D^{t_0}.$$

From (7.14), we then get that

$$\|\mathbb{A}_{\theta,t}\| \leq C\rho^{t+t_0}D^{t_0} \leq C'\rho^t$$

with $C' = C\max(1, \rho D)^{|Q|}$. Hence, for all $t \in \mathbb{N}$,

$$\sup_{\theta\in\Theta}\left\|\mathbb{A}_{\theta,t}\right\| \leq C'\rho^t.$$

Raising both side of the inequality to the power $\frac{1}{t}$ and taking the limit, one gets $\rho(\mathcal{A}, \mathcal{B}) \leq \rho$. Considering (7.13) and fixing $t = \tau_k^{\theta}$, we have

$$\|\mathbb{A}_{\theta,\tau_k^{\theta}}\| \leq C\rho^{\tau_k^{\theta}}\lambda^k, \quad \forall\theta \in Lang(\mathcal{B}),\ \forall k \in \mathbb{N}$$

which implies that

$$\left(\frac{\|\mathbb{A}_{\theta,\tau_k^\theta}\|}{\rho^{\tau_k^\theta}}\right)^{1/k} \leq C^{1/k}\lambda, \quad \forall \theta \in Lang(\mathcal{B}), \ \forall k \geq 1. \tag{7.15}$$

If $\rho > \boldsymbol{\rho}(\mathcal{A}, \mathcal{B})$, taking the supremum over all switching signals in $Lang(\mathcal{B})$ and the lim sup as k goes to infinity yields $\boldsymbol{\lambda}(\mathcal{A}, \mathcal{B}, \rho) \leq \lambda$. If $\rho = \boldsymbol{\rho}(\mathcal{A}, \mathcal{B})$, then, for all $\rho' > \boldsymbol{\rho}(\mathcal{A}, \mathcal{B})$, (7.15) gives

$$\left(\frac{\|\mathbb{A}_{\theta,\tau_k^\theta}\|}{\rho'^{\tau_k^\theta}}\right)^{1/k} \leq C^{1/k}\lambda, \quad \forall \theta \in Lang(\mathcal{B}), \ \forall k \geq 1.$$

Then, it follows that for all $\rho' > \boldsymbol{\rho}(\mathcal{A}, \mathcal{B})$, $\boldsymbol{\lambda}(\mathcal{A}, \mathcal{B}, \rho') \leq \lambda$ and hence by taking the limit we get $\boldsymbol{\lambda}(\mathcal{A}, \mathcal{B}) \leq \lambda$. □

The previous theorem provides a bound on the growth of the state and can be used to derive conditions for stabilization of system (7.2) using ω-regular switching sequences with a minimal accepting rate as shown in the following result.

Corollary 7.4.2 *Let $\theta \in Lang(\mathcal{B})$, if there exists $\rho > \boldsymbol{\rho}(\mathcal{A}, \mathcal{B})$ such that $\gamma^\theta > -\frac{\ln(\rho)}{\ln(\boldsymbol{\lambda}(\mathcal{A},\mathcal{B},\rho))}$, then*

$$\lim_{t\to+\infty} \|\mathbf{x}(t, x_0, \theta)\| = 0, \quad \forall x_0 \in \mathbb{R}^n. \tag{7.16}$$

Proof If $\gamma^\theta > -\frac{\ln(\rho)}{\ln(\boldsymbol{\lambda}(\mathcal{A},\mathcal{B},\rho))}$, then there exists $\lambda \in (\boldsymbol{\lambda}(\mathcal{A}, \mathcal{B}, \rho), 1)$ such that $\gamma^\theta > -\frac{\ln(\rho)}{\ln(\lambda)}$ and $\epsilon > 0$ such that $\epsilon < \gamma^\theta + \frac{\ln(\rho)}{\ln(\lambda)}$. Then, from the definition of accepting rate, there exists $t_0 \in \mathbb{N}$ such that

$$\frac{\kappa^\theta(t)}{t} > \gamma^\theta - \frac{\epsilon}{2}, \qquad \forall t \geq t_0.$$

From Theorem 7.4.1, there exists $C \geq 1$ such that for all $x_0 \in \mathbb{R}^n$, for all $t \geq t_0$.

$$\begin{aligned}
\|\mathbf{x}(t, x_0, \theta)\| &\leq C\rho^t \lambda^{\kappa^\theta(t)} \|x_0\| \\
&\leq C\rho^t \lambda^{(\gamma^\theta - \frac{\epsilon}{2})t} \|x_0\| \\
&\leq C\rho^t \lambda^{(-\frac{\ln(\rho)}{\ln(\lambda)} + \frac{\epsilon}{2})t} \|x_0\| \\
&= C\lambda^{\frac{\epsilon}{2}t} \|x_0\|,
\end{aligned}$$

from which (7.16) follows. □

As a result of the previous corollary, even if the switched system is unstable for constrained switching (i.e., if $\boldsymbol{\rho}(\mathcal{A}, \mathcal{B}) > 1$), it can still be stabilized by visiting the set F frequently enough. From Lemma 7.2.2, we know that the accepting rate always satisfies $\gamma^\theta \leq M$. Consequently, stabilization in this scenario is only possible if there exists $\rho > \boldsymbol{\rho}(\mathcal{A}, \mathcal{B})$ such that $-\frac{\ln(\rho)}{\ln(\boldsymbol{\lambda}(\mathcal{A},\mathcal{B},\rho))} < M$.

7.5 Computing Upper Bounds of the ρ-ω-RJSR

In this section, we present an approach for computing upper bounds of the ρ-ω-RJSR using a combination of Lyapunov and automata-theoretic techniques. Our approach is mainly based on the following result:

Theorem 7.5.1 *If there exist* $V : Q \times \mathbb{R}^n \to \mathbb{R}_0^+$, $\alpha_1, \alpha_2, \rho > 0$ *and* $\lambda \in [0, 1]$ *such that the following inequalities hold true for every* $x \in \mathbb{R}^n$

$$\alpha_1 \|x\| \leq V(q, x) \leq \alpha_2 \|x\|, \qquad q \in Q \tag{7.17}$$

$$V(q', A_i x) \leq \rho V(q, x), \qquad q \in Q, i \in \Sigma, \delta(q, i) = q' \notin F \tag{7.18}$$

$$V(q', A_i x) \leq \rho \lambda V(q, x), \qquad q \in Q, i \in \Sigma, \delta(q, i) = q' \in F \tag{7.19}$$

then the bound (7.8) *holds. Conversely, if the matrices in* $\mathcal{A}$ *are invertible, and the bound* (7.8) *holds for some* $\rho > 0$, $\lambda \in [0, 1]$ *and* $C \geq 1$, *then there exists a function* $V : Q \times \mathbb{R}^n \to \mathbb{R}_0^+$ *such that the inequalities* (7.17)–(7.19) *are satisfied.*

Proof Let us consider an initial condition $x_0 \in \mathbb{R}^n$ and a switching signal $\theta \in Lang(\mathcal{B})$, let $q_0 q_1 q_2 \ldots$ be the accepting run associated with θ. We denote $x(\cdot) = \mathbf{x}(\cdot, x_0, \theta)$ and we define the function $W : \mathbb{N} \to \mathbb{R}_0^+$ by $W(t) = \frac{V(q_t, x(t))}{\rho^t}$ for all $t \in \mathbb{N}$. It follows from (7.18) and (7.19) that $W(t+1) \leq W(t)$ for all $t \in \mathbb{N}$. From (7.19), we get that

$$\forall k \geq 1, \; W(\tau_k^\theta) \leq \lambda W(\tau_k^\theta - 1).$$

From the monotonicity of W, we deduce that

$$\forall k \geq 1, \; W(\tau_k^\theta) \leq \lambda W(\tau_{k-1}^\theta).$$

By induction on k, we get that

$$\forall k \in \mathbb{N}, \; W(\tau_k^\theta) \leq \lambda^k W(0).$$

Now let $t \in \mathbb{N}$, and let $k \in \mathbb{N}$ such that $t \in [\tau_k^\theta, \tau_{k+1}^\theta)$, then the accepting index is $\kappa^\theta(t) = k$. We get from the monotonicity of W that $W(t) \leq \lambda^{\kappa^\theta(t)} W(0)$. Finally from (7.17), we get, for all $t \in \mathbb{N}$, $x_0 \in \mathbb{R}^n$ that

$$\|x(t)\| \leq \frac{\rho^t}{\alpha_1} W(t) \leq \frac{\rho^t}{\alpha_1} \lambda^{\kappa^\theta(t)} W(0) \leq \frac{\alpha_2}{\alpha_1} \rho^t \lambda^{\kappa^\theta(t)} \|x_0\|.$$

Hence, the bound (7.8) holds with $C = \frac{\alpha_2}{\alpha_1}$.

The proof of the converse result follows the same proof procedure as in [2, Theorem 3] with a slight modification, that is the scale of the matrices in $\mathcal{A}$ by $\rho > \boldsymbol{\rho}(\mathcal{A}, \mathcal{B})$. □

As a direct consequence of the previous result and Theorem 7.4.1, we have the following corollary.

Corollary 7.5.2 *Let us assume the matrices in* $\mathcal{A}$ *are invertible. Let* $\rho > 0$ *and* $\lambda \in [0, 1]$ *such that there exists a function* V *satisfying the conditions of Theorem 7.5.1, then either* $\rho > \boldsymbol{\rho}(\mathcal{A}, \mathcal{B})$ *and* $\lambda \geq \boldsymbol{\lambda}(\mathcal{A}, \mathcal{B}, \rho)$*, or* $\rho = \boldsymbol{\rho}(\mathcal{A}, \mathcal{B})$ *and* $\lambda \geq \boldsymbol{\lambda}(\mathcal{A}, \mathcal{B})$*. Conversely, for all* $\rho > \boldsymbol{\rho}(\mathcal{A}, \mathcal{B})$*, for all* $\lambda \in (\boldsymbol{\lambda}(\mathcal{A}, \mathcal{B}, \rho), 1]$*, there exists a function* V *satisfying the conditions of Theorem 7.5.1.*

Corollary 7.5.2 shows that tight upper bounds of the CJSR and of the ω-ρ-RJSR can be obtained by computing Lyapunov functions satisfying the conditions in Theorem 7.5.1. Limiting the search to quadratic Lyapunov functions, the conditions (7.17)–(7.19) can straightforwardly be translated into linear matrix inequalities (LMIs) for which efficient solvers exist. However, the tightness of the conditions in Theorem 7.5.1 is lost when constraining the Lyapunov functions to be quadratic.

7.6 Numerical Example

We consider a multi-agent system consisting of m discrete-time oscillators with identical dynamics given by

$$z_i(t+1) = Rz_i(t) + u_i(t), \;\; i = 1, \ldots, m$$

where $z_i(t) \in \mathbb{R}^2$, $u_i(t) \in \mathbb{R}^2$ and $R = \mu \begin{pmatrix} \cos(\phi) & -\sin(\phi) \\ \sin(\phi) & \cos(\phi) \end{pmatrix}$ with $\phi = \frac{\pi}{6}$, $\mu = 1.02$. The input $u_i(t)$ is used for synchronization purpose and is based on the available information at time t.

To exchange information, the agents are communicating over a network with ring topology. More precisely, there exist m communication channels in the network: one channel between agent i and agent $i+1$, for $i = 1, \ldots, m-1$, and one channel between agent m and agent 1. Hence, the vertices of the undirected communication graph are given by the agents of the network $V = \{1, \ldots, m\}$ and the set of edges are given by the communication channels $E = \{e_1, \ldots, e_m\} \subseteq V \times V$, where

$$e_i = \begin{cases} (i, i+1) & \text{if } i = 1, \ldots m-1 \\ (m, 1) & \text{if } i = m \end{cases}$$

At each instant, only one of these channels is active and the active channel is selected by a switching signal $\theta : \mathbb{N} \to \Sigma = \{1, \ldots, m\}$. More precisely, at time $t \in \mathbb{N}$, the active channel is $e_{\theta(t)}$.

Then, the input value is given as follows:

$$u_1(t) = \begin{cases} k(z_m(t) - z_1(t)), & \text{if } \theta(t) = m \\ k(z_2(t) - z_1(t)), & \text{if } \theta(t) = 1 \\ 0, & \text{otherwise}, \end{cases}$$

$$u_i(t) = \begin{cases} k(z_{i-1}(t) - z_i(t)), & \text{if } \theta(t) = i - 1 \\ k(z_{i+1}(t) - z_i(t)), & \text{if } \theta(t) = i \\ 0 & \text{otherwise} \end{cases}$$

for $i = 2, \ldots, m - 1$,

$$u_m(t) = \begin{cases} k(z_{m-1}(t) - z_m(t)), & \text{if } \theta(t) = m - 1 \\ k(z_1(t) - z_m(t)), & \text{if } \theta(t) = m \\ 0, & \text{otherwise} \end{cases}$$

where k is a control gain. Denoting the vector of synchronization errors as $x(t) = (x_1(t)^\top, \ldots, x_{m-1}(t)^\top)^\top$ with $x_i(t) = z_{i+1}(t) - z_i(t)$, the error dynamics is described by a $2(m-1)$-dimensional switched linear system of the form:

$$x(t+1) = A_{\theta(t)} x(t) \tag{7.20}$$

where the matrices A_k for $k = 1, \ldots, m$ can be easily obtained from the multi-agent dynamics and the input values given above.

Similar to consensus problems [4], we expect the stability of (7.20) to be related to the connectivity, for all $t \in \mathbb{N}$, of the graph $G_t = (V, E_t)$, where $E_t = \bigcup_{s \geq t} \{e_{\theta(s)}\}$.

Let us consider the DBA $\mathcal{B} = (Q, \Sigma, \delta, q_{init}, F)$, where the set of states is $Q = 2^E \setminus E$ with 2^E the set of subsets of E; the alphabet is $\Sigma = \{1, \ldots, m\}$; the initial state is $q_{init} = \varnothing$ and the set of accepting states is $F = \{\varnothing\}$. The transition function δ is given as follows:

$$\delta(q, i) = \begin{cases} \varnothing & \text{if } (V, q \cup \{e_i\}) \text{ is connected} \\ q \cup \{e_i\} & \text{otherwise.} \end{cases}$$

For $m = 3$, the automaton $\mathcal{B}$ corresponds to the DBA represented in Fig. 7.1. For all switching signal $\theta \in Lang(\mathcal{B})$, we get by construction of $\mathcal{B}$ that for all $k \in \mathbb{N}$, the graph $G_k = (V, E_k)$ is connected where

$$E_k = \bigcup_{\tau_k^\theta \leq s \leq \tau_{k+1}^\theta - 1} \{e_{\theta(s)}\}.$$

Then, it follows that $\theta \in Lang(\mathcal{B})$ if and only if $G = (V, E_t)$ is connected for all $t \in \mathbb{N}$. Moreover, by analyzing the cycles of $\mathcal{B}$, we can check that the maximal achievable accepting rate established in Lemma 7.2.2 is given by $M = \frac{1}{m-1}$.

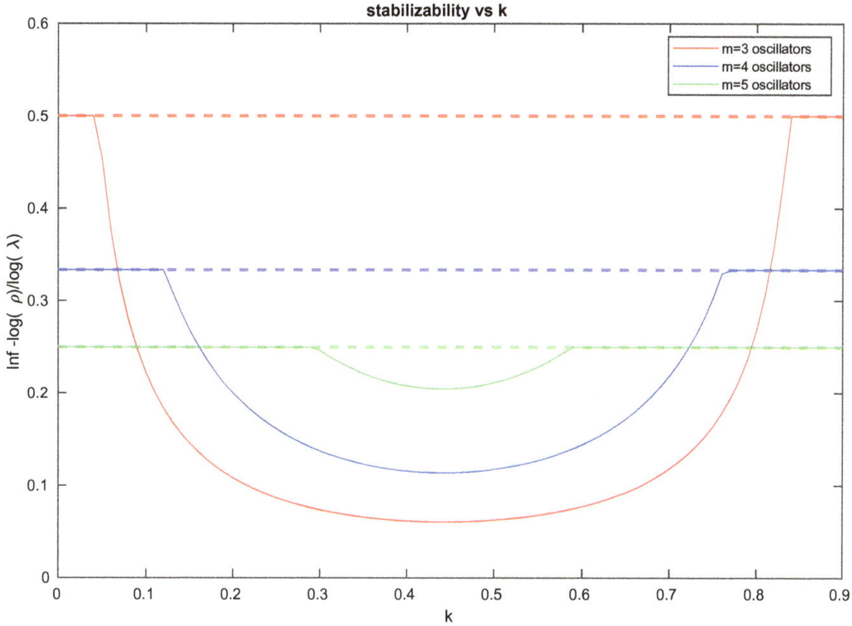

Fig. 7.2 Minimal accepting rate γ^* as a function of the control gain k and of the number of oscillators m

Let us remark that for the DBA $\mathcal{B}$ defined above, any arbitrary switching signal belongs to $\Theta_{\mathcal{B}}$. Hence, the CJSR of $(\mathcal{A}, \mathcal{B})$ coincides with the JSR of the set $\mathcal{A}$. We can then compute $\boldsymbol{\rho}(\mathcal{A}, \mathcal{B})$ using the JSR toolbox [15].

We then use Theorem 7.5.1 to compute an upper bound $\overline{\boldsymbol{\lambda}}(\rho)$ on the ρ-ω-RJSR by solving the LMIs associated with (7.17)–(7.19). Note that

$$\gamma^* := \inf_{\rho > \boldsymbol{\rho}(\mathcal{A},\mathcal{B})} -\frac{\ln(\rho)}{\ln(\overline{\boldsymbol{\lambda}}(\rho))} \geq \inf_{\rho > \boldsymbol{\rho}(\mathcal{A},\mathcal{B})} -\frac{\ln(\rho)}{\ln(\boldsymbol{\lambda}(\mathcal{A}, \mathcal{B}, \rho))}.$$

In this example, we observe numerically that the infimum is reached for $\rho = \boldsymbol{\rho}(\mathcal{A}, \mathcal{B})$ if it satisfies $\gamma^* < M = \frac{1}{m-1}$ (in which case the system is stabilizable, see the discussion after Corollary 7.4.2), and at infinity otherwise, with $\gamma^* = \frac{1}{m-1}$.

We aim at finding the minimal accepting rate γ^* required to stabilize the system with an arbitrary number of oscillators, therefore we computed γ^* for several values of $k \in (0, 1)$ and $m \in \{3, 4, 5\}$ and we show the result in Fig. 7.2. In this figure, the horizontal lines correspond to the maximal achievable accepting rate $\frac{1}{m-1}$. From Corollary 7.4.2, we know that the switched system can be stabilized if we use switching signals belonging to the language of $\mathcal{B}$, whose accepting rate satisfies $\gamma^\theta \in (\gamma^*, \frac{1}{m-1}]$. Hence, in order to stabilize the system, we should carefully select the control gain k such that $\gamma^* < \frac{1}{m-1}$.

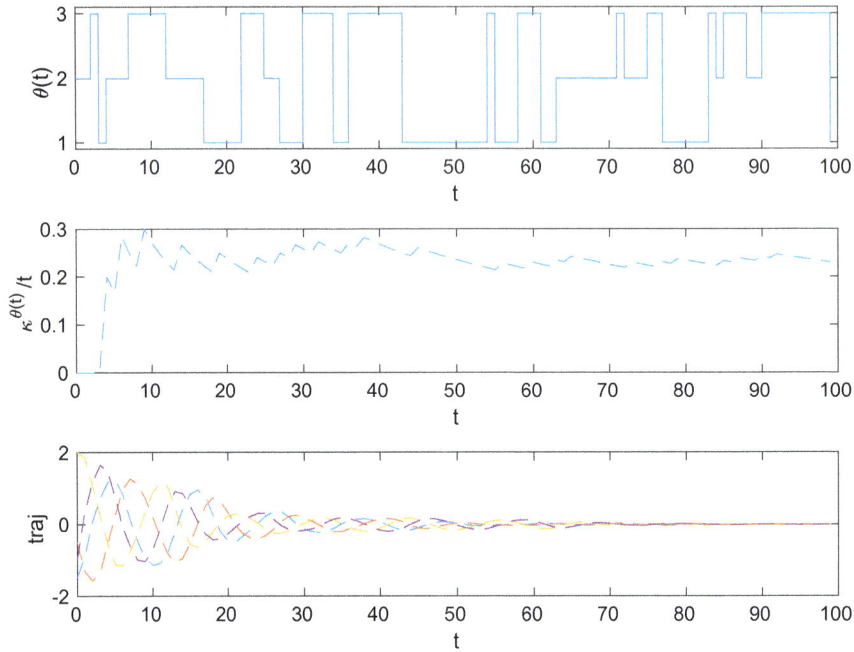

Fig. 7.3 Time evolution of the synchronization error $x(t)$ and the switching signal $\theta(t)$ for an accepting rate higher than $\gamma^* = 0.22$

We also verified that for $m = 6$ oscillators, $\gamma^* > \frac{1}{m-1}$ for all control gain $k \in (0, 1)$. Therefore, for 6 oscillators, there is no suitable choice of control gain k such that the system can be stabilized using switching signals with a sufficiently high accepting rate.[2]

We now proceed with some illustrative numerical simulations. Let us consider a system composed of $m = 3$ oscillators with control gain $k = 0.1$, then from Fig. 7.2, the corresponding lower bound on the accepting rate is $\gamma^* = 0.22$. Let us consider the initial synchronization error $x(0) = (-1.5\ -0.5\ 2\ -1)^\top$. We consider random switching signals generated by a discrete-time Markov chain with 3 states such that $\mathbb{P}(\theta(t+1) = j|\theta(t) = j) = p$, and $\mathbb{P}(\theta(t+1) = j|\theta(t) = i) = \frac{1-p}{2}$ where $p \in (0, 1)$ and $i, j \in \{1, 2, 3\}$, $i \neq j$, $t \geq 0$. From the definition of the DBA $\mathcal{B}$, it is easy to see that the smaller p, the higher the accepting rate.

We first consider $p = 0.8$. Figure 7.3 shows the switching signal $\theta(t)$, the evolution of $\frac{\kappa^\theta(t)}{t}$, and the synchronization errors $x_1(t), x_2(t)$. It is interesting to note that $\gamma^\theta \geq 0.23 > 0.22$ and that the system stabilizes as expected. Then, we consider a switching signal with $p = 0.98$. The result of the simulation is shown in Fig. 7.4. We can check that $\gamma^\theta < 0.22$ and that the system does not stabilize.

[2] The MATLAB® scripts of this numerical example are available at the following repository: https://github.com/georgesaazan/w-regular-oscillators.

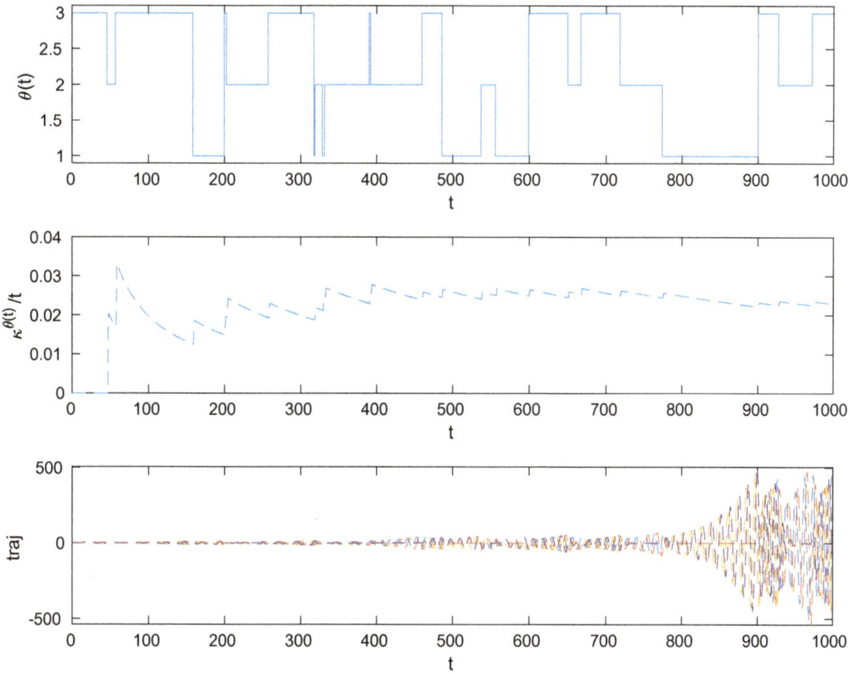

Fig. 7.4 Time evolution of the synchronization error $x(t)$ and the switching signal $\theta(t)$ for an accepting rate smaller than $\gamma^* = 0.22$

7.7 Conclusion

In this chapter, we introduced the ρ-ω-RJSR, a special notion of joint spectral radius for discrete-time switched linear systems driven by infinite sequences generated by a given DBA. We have also shown its relation to stability properties of switched systems. We proposed a method based on Lyapunov functions and automata-theoretic techniques to compute upper bounds on this quantity. Finally, we illustrated our results by an application of the ρ-ω-RJSR to the synchronization of unstable oscillators.

This work can open several research perspectives for the future. Although we presented a method for computing upper bounds on the ρ-ω-RJSR, we did not provide a method for computing lower bounds. In view of the results obtained in this chapter, lower bounds will play a crucial role for determining the stabilizability of such systems. On the other hand, one may ask if these results can be adapted to the case of switching sequences belonging to a certain quantitative language, since the latter can be seen as a generalization of the ω-regular language [6].

Acknowledgements The authors would like to thank Laurent Fribourg for useful discussions. This work was supported in part by the "Agence Nationale de la Recherche" (ANR) under Grant HANDY ANR-18-CE40-0010.

References

1. Aazan, G., Girard, A., Greco, L., Mason, P.: Stability of shuffled switched linear systems: a joint spectral radius approach. Automatica **143**, 110434 (2022)
2. Aazan, G., Girard, A., Mason, P., Greco, L.: Stability of discrete-time switched linear systems with ω-regular switching sequences. In: ACM International Conference on Hybrid Systems: Computation and Control, pp. 1–7 (2022)
3. Baier, C., Katoen, J.-P.: Principles of Model Checking. MIT Press (2008)
4. Blondel, V.D., Hendrickx, J.M., Olshevsky, A., Tsitsiklis, J.N.: Convergence in multiagent coordination, consensus, and flocking. In: IEEE Conference on Decision and Control, pp. 2996–3000 (2005)
5. Blondel, V.D., Nesterov, Y.: Computationally efficient approximations of the joint spectral radius. SIAM J. Matrix Anal. Appl. **27**(1), 256–272 (2005)
6. Chatterjee, K., Doyen, L., Henzinger, T.A.: Quantitative languages. ACM Trans. Comput. Logic **11**(4), 1–38 (2010)
7. Dai, X.: A Gel'fand-type spectral radius formula and stability of linear constrained switching systems. Linear Algebra Appl. **436**(5), 1099–1113 (2012)
8. Girard, A., Mason, P.: Lyapunov functions for shuffle asymptotic stability of discrete-time switched systems. IEEE Control Syst. Lett. **3**(3), 499–504 (2019)
9. Gripenberg, G.: Computing the joint spectral radius. Linear Algebra Appl. **234**, 43–60 (1996)
10. Jungers, R.: The Joint Spectral Radius: Theory and Applications, vol. 385. Springer Science & Business Media (2009)
11. Kozyakin, V.: The Berger-Wang formula for the Markovian joint spectral radius. Linear Algebra Appl. **448**, 315–328 (2014)
12. Lee, J.-W., Dullerud, G.E.: Uniformly stabilizing sets of switching sequences for switched linear systems. IEEE Trans. Autom. Control **52**(5), 868–874 (2007)
13. Philippe, M., Essick, R., Dullerud, G.E., Jungers, R.M.: Stability of discrete-time switching systems with constrained switching sequences. Automatica **72**, 242–250 (2016)
14. Tsitsiklis, J.N., Blondel, V.D.: The Lyapunov exponent and joint spectral radius of pairs of matrices are hard - when not impossible - to compute and to approximate. Math. Control Signals Syst. **10**(1), 31–40 (1997)
15. Vankeerberghen, G., Hendrickx, J., Jungers, R.M.: JSR: a toolbox to compute the joint spectral radius. In: International conference on Hybrid Systems: Computation and Control, pp. 151–156 (2014)
16. Xu, X., Acikmese, B.: Approximation of the constrained joint spectral radius via algebraic lifting. IEEE Trans. Autom. Control (2020)

Chapter 8
Control of Uncertain Nonlinear Fully Linearizable Systems

Sophie Tarbouriech, Christophe Prieur, Isabelle Queinnec, Luca Zaccarian, and Germain Garcia

Abstract This chapter proposes a hybrid control scheme for fully linearizable nonlinear systems, subject to uncertainty. Adopting a hybrid dynamic framework allows providing LMI-based tools for designing a sampled-data feedback controller whose internal state comprises the held value of the plant input. This controller state is discretely updated at exactly the classical linearizing control law, which is held constant during the continuous evolution of the closed loop. The updates happen at suitably triggered jumps, whose triggering rules stem from two different Lyapunov-based sets of inequalities, the first one ensuring robust in-the-small stability properties and the second one ensuring more desirable robustness in-the-large. Simulation results illustrate the effectiveness of the proposed hybrid control scheme.

S. Tarbouriech (✉) · I. Queinnec
LAAS-CNRS, University of Toulouse, CNRS, Toulouse, France
e-mail: tarbour@laas.fr

I. Queinnec
e-mail: queinnec@laas.fr

L. Zaccarian
MAC, LAAS-CNRS, University of Toulouse, CNRS, Toulouse, France
e-mail: zaccarian@laas.fr

Dipartimento di Ingegneria Industriale, University of Trento, Trento, Italy

G. Garcia
LAAS-CNRS, University of Toulouse, INSA, Toulouse, France
e-mail: garcia@laas.fr

C. Prieur
University Grenoble Alpes, CNRS, Grenoble-INP, GIPSA-lab, F-38000, Grenoble, France
e-mail: christophe.prieur@gipsa-lab.fr

R. Postoyan et al. (eds.), *Hybrid and Networked Dynamical Systems*,
Lecture Notes in Control and Information Sciences 493,
https://doi.org/10.1007/978-3-031-49555-7_8

8.1 Introduction

In this chapter, we focus on fully linearizable nonlinear systems, which correspond to a subclass of the more common class of linearizable systems [10]. Even if this class is restrictive, it may represent several kinds of systems, in particular in robotics and aeronautical systems [11, 14, 15, 17, 20]. To design controllers for this kind of systems, which are affine in the input, Nonlinear Dynamic Inversion (NDI) reveals to be a popular approach, corresponding to a specific type of feedback linearization scheme. Indeed, based on a plant inversion, NDI techniques provide nonlinear controllers enjoying desirable properties around the operating point of a given domain. However, NDI techniques suffer from the lack of robustness in particular due to the non-perfect knowledge of the system dynamics (and potentially inaccurate sensing of the output signals). Some directions to robustify the NDI techniques have been proposed, for example, by over-bounding the uncertain terms and using H_{infinity} techniques. See the works [1–3, 5, 6, 12] where both uncertainties affecting the dynamical equations are taken into account, in addition to possible additional dynamics neglected in the modeling phase. Besides robustness aspects, existing works also address performance metrics in the NDI context: see [4, 9] in the anti-windup loop design and $\mathcal{L}_2$-gain contexts. In the more general nonlinear context, see also the studies in [19, 21, 22].

This chapter proposes a new scheme for robustified NDI techniques for fully linearizable nonlinear systems, subject to uncertainties. We consider uncertainties affecting the nonlinear functions involved in the entry channel of the control input and satisfying suitable norm-related bounds. Our hybrid dynamic feedback can be tuned by solving suitable linear matrix inequalities (LMIs) providing a computationally attractive mean for the selection of the parameters of our hybrid controller. The resulting controller is dynamic in the sense that its internal state corresponds to a sample-and-hold version of the classical linearizing feedback. The updates of the linearizing feedback are triggered by suitable Lyapunov-based conditions that may be selected by following two strategies: the first one is simpler and leads to in-the-small robustness guarantees, while the second one is more involved and provides stronger guarantees covering a potentially large set of uncertainties whose size is well characterized by design. Numerical results illustrate the effectiveness of the proposed feedback solutions. The implementation of the hybrid controller is also discussed.

The chapter is organized as follows. The system considered together with the problem that we intend to solve are formally stated in Sect. 8.2. In Sect. 8.3, the results ensuring robustness in-the-small are developed and commented. Section 8.4 revisits these first results and then proposes a more sophisticated selection of the feedback gain and of the triggering function, which enables obtaining a guaranteed robustness in-the-large. In Sect. 8.5, the way to implement the hybrid controller is discussed. Furthermore, a simulation example is discussed in order to illustrate the features of the proposed Hybrid NDI technique.

Notation. I and 0 denote the identity matrix and the null matrix of appropriate dimensions, respectively. The Euclidean norm of a vector is denoted by $|\cdot|$. For any $A \in \mathbb{R}^{m\times n}$, $A^\top$ denotes the transpose of A. For two symmetric matrices, A and B, $A > B$ means that $A - B$ is positive definite.

8.2 The Proposed Hybrid Control Scheme

Consider a nonlinear plant with a continuous-time state-space model given by

$$\dot{x} = Ax + B\gamma(x)(u - \alpha(x)), \tag{8.1}$$

where A in $\mathbb{R}^{n\times n}$, B in $\mathbb{R}^{n\times m}$, x in $\mathbb{R}^n$ is the state and u in $\mathbb{R}^m$ is the control signal.

The robust control design problem addressed in this chapter stems from assuming that the nonlinear functions $\gamma : \mathbb{R}^n \to \mathbb{R}^{m\times m}$ and $\alpha : \mathbb{R}^n \to \mathbb{R}^m$ in (8.1) can be decomposed as

$$\gamma(x) = \gamma_n(x) + \tilde{\gamma}(x), \quad \forall x \in \mathbb{R}^n \tag{8.2a}$$

$$\alpha(x) = \alpha_n(x) + \tilde{\alpha}(x), \quad \forall x \in \mathbb{R}^n, \tag{8.2b}$$

where $\gamma_n : \mathbb{R}^n \to \mathbb{R}^{m\times m}$ and $\alpha_n : \mathbb{R}^n \to \mathbb{R}^m$ (seen in this work as the nominal functions) are known continuously differentiable functions such that $\gamma_n(x)$ is invertible for all x in $\mathbb{R}^n$, and the mismatch functions $x \mapsto \tilde{\gamma}(x)$ and $x \mapsto \tilde{\alpha}(x)$ represent all the uncertainties satisfying suitable properties to be characterized next. We assume the following regularity conditions on γ, α and their decompositions. Note that the assumption $\alpha_n(0) = 0$ is without loss of generality if one performs an input transformation.

Assumption 8.2.1 Functions γ_n and α_n are continuous differentiable, and the functions $\tilde{\gamma}$ and $\tilde{\alpha}$ are locally Lipschitz. Moreover, it holds $\alpha_n(0) = 0$. Finally, for all x in $\mathbb{R}^n$, the matrix $\gamma_n(x)$ is an invertible matrix of $\mathbb{R}^{m\times m}$, and the function $x \mapsto \gamma_n(x)^{-1}$, for all $\mathbb{R}^n$, is continuous.

Based on the decompositions in (8.2), it is natural to introduce the known function $\phi : \mathbb{R}^n \to \mathbb{R}^m$ defined, for all $x \in \mathbb{R}^n$, as

$$\phi(x) := \gamma_n(x)^{-1}Kx + \alpha_n(x). \tag{8.3}$$

The rationale behind the control function ϕ is that the system (8.1) with $\gamma = \gamma_n$ and $\alpha = \alpha_n$ in closed loop with $u = \phi$ reduces to the linear system $\dot{x} = (A + BK)x$. Note that, due to the properties of γ_n and α_n, we have that ϕ is continuous and satisfies $\phi(0) = 0$.

The goal of this chapter is to provide LMI-based tools for the design of a hybrid dynamic controller whose internal state is exactly the signal u in (8.1), which is held

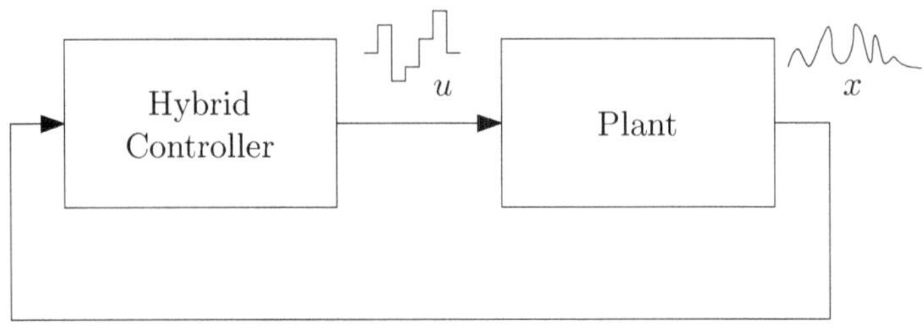

Fig. 8.1 The proposed hybrid Lyapunov-induced event-triggered stabilizer

constant during the continuous evolution of the closed loop and is discretely updated at suitably triggered jumps, whose triggering rules stem from a Lyapunov-based set of inequalities.

Figure 8.1 shows a pictorial representation of the proposed architecture, where the control input u obeys the following hybrid dynamics

$$\dot{u} = 0, \qquad (x, u) \in \mathcal{C} \tag{8.4a}$$
$$u^+ = \phi(x), \qquad (x, u) \in \mathcal{D}. \tag{8.4b}$$

The sets $\mathcal{C}$ and $\mathcal{D}$ are the so-called "flow" and "jump" sets, where continuous and discrete evolution is allowed, respectively. These sets are selected by the following event-triggering rule

$$\mathcal{C} = \{(x, u) \in \mathbb{R}^{n+m} : \psi(x, u) \leq 0\} \tag{8.5a}$$
$$\mathcal{D} = \{(x, u) \in \mathbb{R}^{n+m} : \psi(x, u) \geq 0\}, \tag{8.5b}$$

where function ψ has to be designed in such a way that robust closed-loop asymptotic stability is guaranteed in light of specific classes of allowed perturbations $\tilde{\gamma}$ and $\tilde{\alpha}$ in (8.2).

The feedback interconnection between the hybrid controller (8.4), (8.5) and plant (8.1), corresponding to the block diagram in Fig. 8.1, can be described by the following hybrid closed-loop system:

$$\begin{cases} \dot{x} = Ax + B\gamma(x)(u - \alpha(x)) \\ \dot{u} = 0, \end{cases} \qquad (x, u) \in \mathcal{C} \tag{8.6a}$$
$$\begin{cases} x^+ = x \\ u^+ = \phi(x), \end{cases} \qquad (x, u) \in \mathcal{D}. \tag{8.6b}$$

In the next sections we provide LMI-based selections of ψ and K in (8.5) and (8.3) that induce different levels of robustness. All of our selections are inspired by the Lyapunov-based construction in [16]. In particular, we will rely on the following Lyapunov function candidate for the closed loop defined by, for all x in $\mathbb{R}^n$ and for all u in $\mathbb{R}^m$,

$$V(x,u) = x^\top Px + (u-\phi(x))^\top M(u-\phi(x)), \tag{8.7}$$

where function $\phi : \mathbb{R}^n \to \mathbb{R}^m$, as defined in (8.3), is continuously differentiable due to the properties of γ_n and α_n, and $P \in \mathbb{R}^{n\times n}$, $M \in \mathbb{R}^{m\times m}$ are two symmetric positive definite matrices to be designed. We emphasize that the overall closed-loop state is (x, u) due to the fact that controller (8.4) is a dynamical system having state u.

The first design, given in Sect. 8.3, is the most intuitive application of the techniques of [16] to the context of (8.4), (8.5), which ensures robustness in-the-small, that is, for sufficiently small uncertainties of $\tilde{\alpha}$ and $\tilde{\gamma}$. The second design, given in Sect. 8.4, is more sophisticated and provides a guaranteed robustness margin for $\tilde{\alpha}$ and $\tilde{\gamma}$.

8.3 Hybrid Loop Design for Robustness in the Small

For our first design, inspired by [16, Eq. (9)], we select function ψ in (8.5) as, for all (x, u) in $\mathbb{R}^{n+m}$,

$$\psi(x,u) = \frac{\partial V(x,u)}{\partial x}\big(Ax + B\gamma_n(x)(u-\alpha_n(x)\big) + \mu V(x,u), \tag{8.8}$$

where we denoted

$$\frac{\partial V(x,u)}{\partial x} = 2\left(x^\top P - (u-\phi(x))^\top M\frac{\partial \phi(x)}{\partial x}\right), \tag{8.9}$$

and where μ is a positive scalar. With selection (8.8), following the same rationale as in [16, Eq. (9)], the flow and jump sets in (8.5) become

$$\mathcal{C} = \{(x,u) \in \mathbb{R}^{n+m} : \dot{V}_n(x,u) \le -\mu V(x,u)\} \tag{8.10a}$$
$$\mathcal{D} = \{(x,u) \in \mathbb{R}^{n+m} : \dot{V}_n(x,u) \ge -\mu V(x,u)\}, \tag{8.10b}$$

where $\mu \in \mathbb{R}$ is a tunable positive scalar indirectly associated with the dwell time between consecutive jumps and

$$\dot{V}_n(x,u) := \frac{\partial V(x,u)}{\partial x}\big(Ax + B\gamma_n(x)(u-\alpha_n(x)\big) \tag{8.11}$$

is the directional derivative of V in (8.7) along the nominal plant dynamics (8.1) (note that $\dot{u} = 0$ allows to disregard the direction u in the derivative of V).

For this selection to be effective, we assume the following properties for the matrices M, P and K of our design.

Property 8.3.1 Matrix P in $\mathbb{R}^{n\times n}$ is symmetric positive definite and, for a positive value μ, and a matrix K in $\mathbb{R}^{m\times n}$, the following holds:

$$(A+BK)^\top P + P(A+BK) \leq -2\mu P. \tag{8.12}$$

It is well known that Property 8.3.1 is satisfied for some (small enough) selection of $\mu > 0$ and a suitable gain K if and only if the pair (A, B) is stabilizable in the linear sense, therefore we regard this property as mild. Property 8.3.1 imposes the nonlinear constraint (8.12) on the parameters P, K and μ to be selected, but it is possible to transform the corresponding design problem into a quasi-convex problem (that can be solved as a generalized eigenvalue problem), as clarified in the next lemma.

Lemma 8.3.2 *For any solution of the following quasi-convex problem in the decision variables $Q = Q^\top > 0$, Q in $\mathbb{R}^{n\times n}$, X in $\mathbb{R}^{m\times n}$ and $\mu > 0$*

$$AQ + BX + QA^\top + X^\top B^\top \leq -2\mu Q, \tag{8.13}$$

the selection $P = Q^{-1}$ and $K = XQ^{-1}$ satisfies (8.12).

Proof Left- and right- multiplying (8.12) by matrix $Q = P^{-1}$ and exploiting the fact that $X = KP^{-1} = KQ$, it is immediate to see that (8.13) is equivalent to (8.12). □

Once again, note that the quasi-convex problem (8.13) admits a solution if and only if the pair (A, B) is stabilizable. With selection (8.8) and parameters P, K and μ satisfying Property 8.3.1, we can ensure global asymptotic stability of the origin for any uncertainty $\tilde{\alpha}$ and $\tilde{\gamma}$ satisfying the following assumption.

Assumption 8.3.3 The mismatch functions $x \mapsto \tilde{\alpha}(x)$ and $x \mapsto \tilde{\gamma}(x)$ in (8.2) are locally Lipschitz and such that, for all (x, u) in $\mathcal{C}$,

$$\frac{\partial V(x,u)}{\partial x} B\Big(\tilde{\gamma}(x)(u - \alpha_n(x)) - \tilde{\alpha}(x)(\gamma_n(x) + \tilde{\gamma}(x))\Big) \leq \frac{\mu}{2} V(x,u). \tag{8.14}$$

Remark 8.3.4 (On the limitation of the robustness-in-the-small property) Due to the positive definiteness and continuity of function V in (8.7), Assumption 8.3.3 always holds for small enough (locally Lipschitz) functions $\tilde{\alpha}$ and $\tilde{\gamma}$ that are zero at zero. Nevertheless, due to the fact that the mismatch $\tilde{\alpha}$ and $\tilde{\gamma}$ is not taken into account in the stabilizer design and only considered a posteriori, the robustness margin might be small. This shortcoming of our first design is addressed in our second design of Sect. 8.4, where the stabilizing gain K and the functions ψ of our hybrid feedback law are tuned based on a suitable characterization of the uncertainties.

The main result of this section is stated in the theorem below and can be viewed as an immediate application of the construction in [16, Theorem 1, item 2.] to the particular case of the closed-loop system (8.6) with (8.10).

Theorem 8.3.5 *Under Assumption 8.2.1 and Assumption 8.3.3, for any selection of $M = M^\top > 0$ and any selection of P, K and μ satisfying Property 8.3.1, the origin of the hybrid closed-loop system* (8.6) *with* (8.10) *is globally asymptotically stable.*

Remark 8.3.6 (Rationale behind the control design) Intuitively speaking, the hybrid feedback controller (8.4), (8.5) with selection (8.8) and P, K, μ satisfying Assumption 8.3.3 stems from wanting to exploit the intrinsic robustness-in-the-small (as established, e.g., in [7, Chap. 7]) of a nominal solution initially disregarding the presence of the mismatch functions $\tilde{\gamma}$ and $\tilde{\alpha}$, and emulating an ideal continuous nominal feedback of the type $u = \phi(x)$, which would ensure $\dot{x} = (A + BK)x$. For this closed loop, via (8.12), Property 8.3.1 ensures that the Lyapunov function $x^\top Px$ decreases with rate at least -2μ. The emulation-based feedback in (8.4), (8.10) then exploits the fact that immediately after a jump, one has $\dot{V}_n(x, u) = \dot{V}_n(x, \phi(x)) \leq -2\mu V(x, u)$, so that the solution is either at zero or in the interior of the flow set. As a consequence, some flow will occur (and a corresponding decrease of the Lyapunov function V will be experienced) until half of the decrease $-2\mu V(x, u)$ remains in place. At some point, possibly, one will get $\dot{V}_n(x, u) \geq -\mu V(x, u)$ and a new sample (or jump) will be triggered.

The robustness ensured by Assumption 8.3.3 within this setting is the mere robustness that one can obtain through the (possibly small) margin available in the Lyapunov decrease bound. Such a margin cannot be explicitly quantified, if not through the convoluted expressions in (8.14) and should be seen as a "robustness in the small" result. More clever generalizations of the sets $\mathcal{C}$ and $\mathcal{D}$ and the ensuing design of K, μ and P are discussed in the next section.

Let us now prove Theorem 8.3.5 by following an approach similar in nature to [16, Theorem 1, item 2].

Proof To prove Theorem 8.3.5, we exploit function V in (8.7), which is positive definite and radially unbounded with respect to both state variables (x, u), due to the positive definiteness of P and M.

Before studying the properties of V, let us first characterize the flow dynamics of the state x in (8.6) as follows:

$$\begin{aligned}\dot{x} &= Ax + B\gamma(x)(u - \alpha(x)) \\ &= Ax + B\gamma_n(x)(u - \alpha_n(x)) + B\tilde{\gamma}(x)(u - \alpha_n(x)) - B\gamma(x)\tilde{\alpha}(x). \end{aligned} \tag{8.15}$$

Then, we may compute the directional derivative of the function V defined by (8.7) along this dynamics. Using the expression of $\mathcal{C}$ in (8.10) and the selection of ψ in (8.8), we have, for all (x, u) in $\mathcal{C}$,

$$\begin{aligned}\dot{V} &= \frac{\partial V(x,u)}{\partial x}(Ax + B\gamma(x)(u-\alpha(x)))\\ &= \dot{V}_n(x,u) + \frac{\partial V(x,u)}{\partial x}B\Big(\tilde{\gamma}(x)(u-\alpha_n(x)) - (\gamma_n(x)+\tilde{\gamma}(x))\tilde{\alpha}(x)\Big)\\ &\leq -\mu V(x,u) + \frac{\mu}{2}V(x,u) \leq -\frac{\mu}{2}V(x,u),\end{aligned} \tag{8.16}$$

where we used (8.14) in the last line.

Consider now the variation of V across jumps from $\mathcal{D}$ in (8.10) and let us first show that

$$(x,u) \in \mathcal{D}\setminus\{(0,0)\} \quad \Rightarrow \quad u \neq \phi(x). \tag{8.17}$$

To prove (8.17), two cases need to be inspected:

- Consider first the case $x = 0$, which, to the end of proving (8.17), implies $u \neq 0$. Since by Assumption 8.2.1 we have $\alpha_n(0) = 0$, then $\phi(x) = \phi(0) = 0 \neq u$, as required in (8.17). This concludes the proof of (8.17) in the first case.
- Let us now consider the case $x \neq 0$ and assume by contradiction that $u = \phi(x)$. Then, using the definition of $\dot{V}_n(x,u)$ in (8.11), and the expression of $\frac{\partial V(x,u)}{\partial x}$ in (8.9), we have

$$\begin{aligned}\dot{V}_n(x,u) &= 2x^\top P(Ax + B\gamma_n(x)(\phi(x)-\alpha_n(x)) = 2x^\top P(A+BK)x\\ &\leq -2\mu x^\top Px = -2\mu V(x,u),\end{aligned}$$

where we used in the second line inequality (8.12) in Property 8.3.1. Since $x \neq 0$, then $V(x,u) > 0$ and $\dot{V}_n(x,u) \leq -2\mu V(x,u) < -\mu V(x,u)$, then from (8.10) we have $(x,u) \notin \mathcal{D}$, which is a contradiction with the assumption in the left part of (8.17). Thus the implication (8.17) is proven in the second case.

Therefore the proof of (8.17) is complete.

Using (8.17) and positive definiteness of M, we have that $(x,u) \in \mathcal{D}\setminus\{(0,0)\}$ implies $(u-\phi(x))^\top M(u-\phi(x)) > 0$ and since $x^+ = \phi(x)$, then we have

$$\begin{aligned}\Delta V(x) &= V(x^+) - V(x)\\ &= x^\top Px - x^\top Px - (u-\phi(x))^\top M(u-\phi(x)) < 0, \quad \forall (x,u) \in \mathcal{D}\setminus\{(0,0)\}.\end{aligned} \tag{8.18}$$

Based on the fact that the origin is a compact set, that the hybrid closed loop satisfies the hybrid basic conditions ($\mathcal{C}$ and $\mathcal{D}$ are closed and the flow and jump maps are both continuous), and that $(x,u) = (0,0)$ implies $(x^+,u^+) = (0,0)$, the negative definiteness of $\dot{V}$ and ΔV established in (8.16) and (8.18), respectively, allow concluding global asymptotic stability of the origin by applying [8, Theorem 20]. □

8.4 Hybrid Loop Design for Robustness in the Large

Our second design corresponds to a more sophisticated selection of the feedback gain K and of the triggering function ψ, which enables obtaining a guaranteed robustness in the large. In particular, we start by stating the main assumption on the perturbation functions $\tilde{\gamma}$ and $\tilde{\alpha}$, which is made here more explicit, as compared to Assumption 8.3.3, and corresponds to potentially large perturbations.

Assumption 8.4.1 For functions γ_n, $\tilde{\gamma}$ and $\tilde{\alpha}$, there exist symmetric positive definite matrices Q_γ and Q_α in $\mathbb{R}^{m\times m}$ such that

$$\tilde{\gamma}(x)^\top Q_\gamma \tilde{\gamma}(x) \leq \gamma_n(x)^\top \gamma_n(x), \qquad \forall x \in \mathbb{R}^n, \tag{8.19a}$$

$$(\gamma(x)\tilde{\alpha}(x))^\top Q_\alpha \gamma(x)\tilde{\alpha}(x) \leq x^\top x, \qquad \forall x \in \mathbb{R}^n. \tag{8.19b}$$

As compared to the parallel conditions (8.14) required in Assumption 8.3.3, condition (8.19) is clearly more explicit and potentially less restrictive, because it directly focuses on the uncertain functions, without any immediate link with the controller state u, or the matrix P of the Lyapunov function. More specifically, (8.19a) establishes a proportionality bound between the size of the uncertainty $\tilde{\gamma}$ and the size of the nominal function γ_n, scaled via matrix Q_γ, while (8.19b) requires that the product between γ and $\tilde{\alpha}$ remains within a ball, suitably scaled by matrix Q_α.

Our second design is based on generalizing Property 8.3.1 to an intrinsically robust formulation enjoyed by parameters P and K in view of the characterization of the uncertainties $\tilde{\alpha}$ and $\tilde{\gamma}$ via matrices Q_α and Q_γ, as per Assumption 8.4.1. This formulation includes two new positive scalar parameters η_γ and η_α, which provide useful degrees of freedom for the convex LMI-based tuning of the control gains, clarified in the follow-up lemma. In our second design we assume the following for scalars η_α, η_γ, μ and matrices M, P and K.

Property 8.4.2 Matrix $P \in \mathbb{R}^{n\times n}$ is symmetric positive definite and, for positive scalars μ, η_γ and η_α, and a matrix K in $\mathbb{R}^{m\times n}$, the following holds:

$$\begin{aligned}&(A+BK)^\top P + P(A+BK) + 2\mu P \\ &\quad + \eta_\gamma P B Q_\gamma^{-1} B^\top P + \eta_\gamma^{-1} K^\top K + \eta_\alpha P B Q_\alpha^{-1} B^\top P + \eta_\alpha^{-1} I \leq 0.\end{aligned} \tag{8.20}$$

Similar to Property 8.3.1, also the constraint (8.20) imposed in Property 8.4.2 is a nonlinear function of the tuning parameters P, K, μ, η_α and η_γ, but it is possible to transform the corresponding design problem into a quasi-convex problem, as clarified next.

Lemma 8.4.3 *Given symmetric positive definite matrices Q_γ and Q_α in $\mathbb{R}^{m\times m}$ as in Assumption 8.4.1, for any solution of the following quasi-convex optimization in the decision variables $Q = Q^\top > 0$, $Q \in \mathbb{R}^{n\times n}$, $X \in \mathbb{R}^{m\times n}$, η_α, η_γ, and $\mu > 0$*

$$\begin{bmatrix} AQ + BX + QA^\top + X^\top B^\top + \eta_\gamma B Q_\gamma^{-1} B^\top + \eta_\alpha B Q_\alpha^{-1} B^\top + 2\mu Q & X^\top & Q \\ X & -\eta_\gamma I & 0 \\ Q & 0 & -\eta_\alpha I \end{bmatrix} < 0 \tag{8.21}$$

the selection $P = Q^{-1}$ *and* $K = XQ^{-1}$ *satisfies* (8.20).

Proof Applying a Schur complement (8.21) is equivalent to

$$\begin{bmatrix} AQ + BX + QA^\top + X^\top B^\top + \eta_\gamma B Q_\gamma^{-1} B^\top + \eta_\alpha B Q_\alpha^{-1} B^\top + 2\mu Q + \eta_\alpha^{-1} Q^2 & X^\top \\ X & -\eta_\gamma I \end{bmatrix} < 0,$$

where we can apply a second Schur complement to obtain

$$AQ + BX + QA^\top + X^\top B^\top + \eta_\gamma B Q_\gamma^{-1} B^\top + \eta_\alpha B Q_\alpha^{-1} B^\top + 2\mu Q + \eta_\alpha^{-1} Q^2 + \eta_\gamma^{-1} X^\top X < 0$$

and after replacing $X = KQ$ the proof follows the same steps as that of Lemma 8.3.2 by pre- and post-multiplying by $P = Q^{-1}$. □

We are now ready to introduce the robust generalization of the intuitive selection in (8.8) of the triggering function ψ, which now includes the following additive terms

$$\psi_\alpha(x, u) := \frac{\eta_\alpha}{4} \frac{\partial V(x, u)}{\partial x} B Q_\alpha^{-1} B^\top \left(\frac{\partial V(x, u)}{\partial x}\right)^\top + \eta_\alpha^{-1} x^\top x \tag{8.22}$$

$$\psi_\gamma(x, u) := \frac{\eta_\gamma}{4} \frac{\partial V(x, u)}{\partial x} B Q_\gamma^{-1} B^\top \left(\frac{\partial V(x, u)}{\partial x}\right)^\top + \eta_\gamma^{-1} |\gamma_n(x)(u - \alpha_n(x))|^2, \tag{8.23}$$

related to the required robustness margin. Function ψ is then selected as

$$\psi(x, u) = \frac{\partial V(x, u)}{\partial x}\big(Ax + B\gamma_n(x)(u - \alpha_n(x))\big) + \mu V(x, u) + \psi_\alpha(x, u) + \psi_\gamma(x, u), \tag{8.24}$$

where once again $\frac{\partial V(x,u)}{\partial x}$ is defined as in (8.9). With selection (8.24), the flow and jump sets in (8.5) correspond to the following robust generalization of the sets in (8.10):

$$\mathcal{C} = \{(x, u) \in \mathbb{R}^{n+m} : \dot{V}_n(x, u) + \psi_\alpha(x, u) + \psi_\gamma(x, u) \le -\mu V(x, u)\} \tag{8.25a}$$

$$\mathcal{D} = \{(x, u) \in \mathbb{R}^{n+m} : \dot{V}_n(x, u) + \psi_\alpha(x, u) + \psi_\gamma(x, u) \ge -\mu V(x, u)\}, \tag{8.25b}$$

where $\dot{V}_n(x, u)$ is defined in (8.11).

The main result of this section is stated in the theorem below, which can be viewed as a robustification of the result of Theorem 8.3.5 where we explicitly take into account the uncertainties.

Theorem 8.4.4 *Under Assumption 8.2.1, for any selection of $M = M^\top > 0$ and any selection of P, K, η_α, η_γ and μ satisfying Property 8.4.2, the origin of the hybrid closed loop* (8.25), (8.6) *is globally asymptotically stable for any functions γ_n, α_n, $\tilde{\gamma}$ and $\tilde{\alpha}$ satisfying Assumption 8.4.1.*

Proof The proof follows similar steps as those of the proof of Theorem 8.3.5, even though the additional terms appearing in the control law have to be taken into consideration in the derivations. Once again, we rely on the Lyapunov function V in (8.7), which is positive definite and radially unbounded with respect to (x, u), due to positive definiteness of both P and M.

To suitably characterize the directional derivative of the function V in (8.7) along the flow dynamics, we start from the characterization (8.15) already given in the proof of Theorem 8.3.5. By regarding the perturbation terms $B\tilde{\gamma}(x)(u - \alpha_n(x))$ and $B\gamma(x)\tilde{\alpha}(x)$ in (8.15) as undesirable ones, we prove next that they satisfy the following bounds for all (x, u):

$$\left| \frac{\partial V(x,u)}{\partial x} B\tilde{\gamma}(x)(u - \alpha_n(x)) \right| \le \psi_\gamma(x, u) \tag{8.26}$$

$$\left| \frac{\partial V(x,u)}{\partial x} B\gamma(x)\tilde{\alpha}(x) \right| \le \psi_\alpha(x, u). \tag{8.27}$$

Proof of (8.26) Let us start by manipulating (8.19a) via a Schur complement to obtain

$$Q_\gamma^{-1} - \tilde{\gamma}(x)(\gamma_n(x)^\top \gamma_n(x))^{-1}\tilde{\gamma}(x)^\top \ge 0, \tag{8.28}$$

where matrix $\gamma_n(x)^\top \gamma_n(x)$ is invertible due to Assumption 8.2.1. Recall now that, by the Young inequality, for any vectors v, w in $\mathbb{R}^m$, and any symmetric positive definite and invertible matrix S in $\mathbb{R}^{m\times m}$, we have $2|v^\top w| \le v^\top S^{-1} v + w^\top S w$. Then we obtain, by selecting $v = \tilde{\gamma}(x)^\top B^\top \frac{\partial V(x,u)}{\partial x}^\top$, $w = u - \alpha_n(x)$ and $S = \frac{2}{\eta_\gamma}\gamma_n(x)^\top \gamma_n(x)$ in the first step, and using (8.28) in the second step

$$\begin{aligned}
&\left| \frac{\partial V(x,u)}{\partial x} B\tilde{\gamma}(x)(u - \alpha_n(x)) \right| \\
&\le \frac{1}{2}\frac{\partial V(x,u)}{\partial x} B\tilde{\gamma}(x)\frac{\eta_\gamma}{2}(\gamma_n(x)^\top \gamma_n(x))^{-1}\tilde{\gamma}(x)^\top B^\top \frac{\partial V(x,u)}{\partial x}^\top \\
&\qquad + \frac{1}{2}(u - \alpha_n(x))^\top 2\eta_\gamma^{-1}\gamma_n(x)^\top \gamma_n(x)(u - \alpha_n(x)) \\
&\le \frac{\eta_\gamma}{4}\frac{\partial V(x,u)}{\partial x} B Q_\gamma^{-1} B^\top \frac{\partial V(x,u)}{\partial x}^\top + \eta_\gamma^{-1}\Big(\gamma_n(x)(u - \alpha_n(x))\Big)^\top \gamma_n(x)(u - \alpha_n(x)) \\
&= \psi_\gamma(x, u),
\end{aligned}$$

thus proving (8.26).

Proof of (8.27) Using again the Young inequality $2|v^\top w| \leq v^\top S^{-1} v + w^\top S w$ with $v = B^\top \frac{\partial V(x,u)}{\partial x}^\top$, $w = \gamma(x)\tilde{\alpha}(x)$ and $S = 2\eta_\alpha^{-1} Q_\alpha$ in the first step, and then applying (8.19b) in the second step, we obtain

$$\begin{aligned}
&\left|\frac{\partial V(x,u)}{\partial x} B\gamma(x)\tilde{\alpha}(x)\right| \\
&\leq \frac{1}{2}\frac{\partial V(x,u)}{\partial x} B\frac{\eta_\alpha}{2} Q_\alpha^{-1} B^\top \frac{\partial V(x,u)}{\partial x}^\top + \frac{1}{2}\Big(\gamma(x)\tilde{\alpha}(x)\Big)^\top 2\eta_\alpha^{-1} Q_\alpha \gamma(x)\tilde{\alpha}(x) \\
&\leq \frac{\eta_\alpha}{4}\frac{\partial V(x,u)}{\partial x} B Q_\alpha^{-1} B^\top \frac{\partial V(x,u)}{\partial x}^\top + \eta_\alpha^{-1} x^\top x \\
&= \psi_\alpha(x,u),
\end{aligned}$$

thus proving (8.27).

Now that we have proven the bounds (8.26), (8.27), we are ready to characterize the directional derivative of V along the flow dynamics (8.15). Using the expression of $\mathcal{C}$ in (8.25) and the selection of ψ in (8.24), proceeding in a similar way to (8.16), and exploiting (8.26), (8.27), we have, for all (x, u) in $\mathcal{C}$,

$$\begin{aligned}
\dot{V} &= \frac{\partial V(x,u)}{\partial x}(Ax + B\gamma(x)(u - \alpha(x))) \\
&= \dot{V}_n(x,u) + \frac{\partial V(x,u)}{\partial x} B\tilde{\gamma}(x)(u - \alpha_n(x)) - \frac{\partial V(x,u)}{\partial x} B\gamma(x)\tilde{\alpha}(x) \\
&\leq -\psi_\alpha(x,u) - \psi_\gamma(x,u) - \mu V(x,u) + \psi_\alpha(x,u) + \psi_\gamma(x,u) \\
&\leq -\mu V(x,u).
\end{aligned} \tag{8.29}$$

To characterize the variation of V across jumps from $\mathcal{D}$ in (8.25) we proceed as in the proof of Theorem 8.3.5, and first show that implication (8.17) (reported here again for the convenience of the reader) holds

$$(x,u) \in \mathcal{D} \setminus \{(0,0)\} \quad \Rightarrow \quad u \neq \phi(x). \tag{8.30}$$

To prove (8.30), two cases need to be inspected:

- We first consider the case $x = 0$, which, to the end of proving (8.30), implies $u \neq 0$. Since by Assumption 8.2.1 we have $\alpha_n(0) = 0$, then $\phi(x) = \phi(0) = 0 \neq u$, as required in (8.30). This concludes the proof of (8.30) in the first case.
- For the case $x \neq 0$, mimicking the proof of Theorem 8.3.5, let us assume by contradiction that $u = \phi(x)$.

Let us note that, due to (8.7), it holds that

$$V(x, \phi(x)) = x^\top P x \tag{8.31}$$

and, due to (8.9),

$$\frac{\partial V(x, \phi(x))}{\partial x} = 2x^\top P. \tag{8.32}$$

From (8.22), (8.23) and (8.32), we get

$$\psi_\alpha(x, \phi(x)) = \eta_\alpha x^\top P B Q_\alpha^{-1} B^\top P x + \eta_\alpha^{-1} x^\top x, \tag{8.33}$$
$$\psi_\gamma(x, \phi(x)) = \eta_\gamma x^\top P B Q_\gamma^{-1} B^\top P x + \eta_\gamma^{-1} |\gamma_n(x)(\phi(x) - \alpha_n(x))|^2. \tag{8.34}$$

Now, using the definition of $\dot{V}_n(x, u)$ in (8.11), and the expression of $\frac{\partial V(x,u)}{\partial x}$ in (8.9), we may exploit the inequality (8.20) in Property 8.4.2 to obtain

$$\begin{aligned}
\dot{V}_n(x, u) &= 2x^\top P(Ax + B\gamma_n(x)(\phi(x) - \alpha_n(x)) = 2x^\top P(A + BK)x \\
&\le -2\mu x^\top P x - \eta_\gamma x^\top P B Q_\gamma^{-1} B^\top P x - \eta_\gamma^{-1} x^\top K^\top K x \\
&\qquad\qquad - \eta_\alpha x^\top P B Q_\alpha^{-1} B^\top P x - \eta_\alpha^{-1} x^\top x \\
&= -2\mu V(x, \phi(x)) - \psi_\gamma(x, \phi(x)) - \psi_\alpha(x, \phi(x)) \\
&= -2\mu V(x, u) - \psi_\gamma(x, u) - \psi_\alpha(x, u),
\end{aligned}$$

where we used in the next to last line the fact that $\gamma_n(x)(\phi(x) - \alpha_n(x)) = Kx$ (see the discussion after (8.3)) with (8.31), (8.33) and (8.34). Since $x \neq 0$, then $V(x, u) > 0$ and $\dot{V}_n(x, u) \le -2\mu V(x, u) - \psi_\gamma(x, u) - \psi_\alpha(x, u) < -\mu V(x, u) - \psi_\gamma(x, u) - \psi_\alpha(x, u)$, and then from (8.25) we have $(x, u) \notin \mathcal{D}$, which is a contradiction with the assumption in the left part of (8.30). Thus the implication (8.30) is proven in second case.

Therefore the proof of (8.30) is complete.

The rest of the proof is the same as that of Theorem 8.3.5, but is reported here for completeness. Using (8.30) and positive definiteness of M, we have that $(x, u) \in \mathcal{D} \setminus \{(0, 0)\}$ implies $(u - \phi(x))^\top M(u - \phi(x)) > 0$ and since $x^+ = \phi(x)$, then we have across jumps

$$\begin{aligned}
\Delta V(x) &= V(x^+) - V(x) \\
&= x^\top P x - x^\top P x - (u - \phi(x))^\top M(u - \phi(x)) < 0, \quad \forall (x, u) \in \mathcal{D} \setminus \{(0, 0)\}.
\end{aligned} \tag{8.35}$$

Based on the fact that the origin is a compact set, that the hybrid closed loop satisfies the hybrid basic conditions ($\mathcal{C}$ and $\mathcal{D}$ are closed and the flow and jump maps are both continuous), and that $(x, u) = (0, 0)$ implies $(x^+, u^+) = (0, 0)$, the negative definiteness of $\dot{V}$ and ΔV established in (8.16) and (8.18), respectively, allow concluding global asymptotic stability of the origin by applying [8, Theorem 20]. □

8.5 Numerical Illustration

8.5.1 *Implementation*

To implement (8.8) and determine whether the state (x, u) belongs to the jump and flow sets, using (8.9), we compute an explicit expression for $\frac{\partial \phi(x)}{\partial x} \in \mathbb{R}^{m \times n}$. To this end, due to the invertibility of $\gamma_n(x)$, and exploiting the fact that, for each $i = 1, \ldots, n$, it holds that $\frac{\partial(\gamma_n^{-1}(x))}{\partial x_i}\gamma_n(x) + \gamma_n^{-1}(x)\frac{\partial \gamma_n(x)}{\partial x_i} = 0$, we obtain, with (8.3),

$$
\begin{aligned}
&\frac{\partial \phi(x)}{\partial x} \\
&= \frac{\partial \alpha_n(x)}{\partial x} + \gamma_n^{-1}(x)K - \left[\gamma_n^{-1}(x)\frac{\partial \gamma_n(x)}{\partial x_1}\gamma_n^{-1}(x)Kx, \ldots, \gamma_n^{-1}(x)\frac{\partial \gamma_n(x)}{\partial x_n}\gamma_n^{-1}(x)Kx\right] \\
&= \frac{\partial \alpha_n(x)}{\partial x} + \gamma_n^{-1}(x)\left(K - \left[\frac{\partial \gamma_n(x)}{\partial x_1}\gamma_n^{-1}(x)Kx, \ldots, \frac{\partial \gamma_n(x)}{\partial x_n}\gamma_n^{-1}(x)Kx\right]\right),
\end{aligned}
$$

where x_i denotes the ith component of vector x.

For the selection of the feedback gain K we may follow the LMI-based design approach proposed in Lemmas 8.3.2 and 8.4.3 but it is appropriate to also provide some optimality criterion, because Lemmas 8.3.2 and 8.4.3 only present feasibility conditions. To obtain a numerically sensible solution, we consider an optimization-based formulation that minimizes the norm of the control gain K, by first noticing that both (8.13) and (8.21) are homogeneous in the decision variables, so that imposing $Q > I_n$ does not reduce the feasibility set. In particular, note that scaling any feasible solution by a positive scalar preserves feasibility and does not change the resulting gain $K = XQ^{-1}$. More specifically, after fixing a desired convergence rate μ, we solve the following quasi-convex optimization problem

$$
\begin{aligned}
&\min_{Q,X,\bar{k}} \bar{k} \\
&\text{subject to} \begin{bmatrix} \bar{k}I_n & X^\top \\ X & \bar{k}I_n \end{bmatrix} \geq 0\,; \, Q \geq I_n\,; \, (13)
\end{aligned}
\tag{8.36}
$$

when designing a robust in-the-small selection, and the following quasi-convex optimization problem

$$
\begin{aligned}
&\min_{Q,X,\bar{k},\eta_\alpha,\eta_\gamma} \bar{k} \\
&\text{subject to} \begin{bmatrix} \bar{k}I_n & X^\top \\ X & \bar{k}I_n \end{bmatrix} \geq 0\,; \; Q \geq I_n\,; \; (21)
\end{aligned}
\tag{8.37}
$$

when designing a robust in-the-large selection. In both cases, with the change of variable $K = XQ^{-1}$, the design ensures that $K^\top K \leq \bar{k}^2$.

8.5.2 Example

In order to illustrate the results of the chapter, consider (8.1) with the following data inspired by [12]:

$$
\begin{aligned}
A &= \begin{bmatrix} -1 & 0.3 \\ 0 & 0.5 \end{bmatrix}, \quad B = \begin{bmatrix} 0.5 \\ -10 \end{bmatrix}, \\
\gamma_n(x) &= 2.14 + 0.72 \sin(x_2), \qquad \forall (x_1, x_2) \in \mathbb{R}^2, \\
\alpha_n(x) &= x_1(x_2 + 0.5), \qquad \forall (x_1, x_2) \in \mathbb{R}^2.
\end{aligned} \tag{8.38}
$$

Consider potential uncertainty $\tilde{\gamma}$ and $\tilde{\alpha}$ satisfying $|\tilde{\gamma}(x)| \leq \sqrt{0.1}\gamma_n(x)$, and $|\tilde{\alpha}(x)(x)| \leq \sqrt{0.1}|\alpha_n(x)|$, for all (x_1, x_2) in $\mathbb{R}^2$, respectively, which corresponds to select $Q_\gamma = 10$ and $Q_\alpha = 10$ in Assumption 8.4.1.

Let us select for the numerical evaluation the convergence rate $\mu = 1.1$. The above-described optimization problems (8.36) and (8.37) are implemented with Yalmip [13] and the MATLAB® Lmilab solver. When solving problem (8.36) we obtain the robust-in-the-small controller:

$$K_{\text{small}} = [0.026987 \;\; 0.17146].$$

Similarly, we solve problem (8.37) by selecting the uncertainty parameters $Q_\gamma = Q_\alpha = 10$, and we obtain the robust-in-the-large controller:

$$K_{\text{large}} = [0.89763 \;\; 1.3531].$$

These controllers are then used to simulate the hybrid closed-loop nonlinear system, implemented in MATLAB®/Simulink® with the Hybrid Systems Simulation Toolbox (HyEQ) [18] in various operating conditions discussed below, with the initial condition

$$x^0 = [x_1^0 \;\; x_2^0]^\top = [1 \;\; 1.5]^\top, \quad u^0 = 0.$$

Figure 8.2 illustrates the influence on the closed-loop response of the choice of matrix $M = M^\top > 0$, which appears in the Lyapunov function V in (8.7). While any positive definite M is allowed, it is clear from the expression of V that large selections of M penalizes the difference between the (held and) applied control input u and the emulated feedback $\phi(x)$, being quite directly related to the triggering of the jumps, especially with large penalties. More specifically, the larger is M, the smaller is the allowed distance between u and $\phi(x)$ and the larger is the number of overall expected jumps. This phenomenon is well illustrated by the simulation results reported in Table 8.1.

For the selection $M = 5$, Fig. 8.3 illustrates the influence of the uncertainty $\tilde{\alpha}$ and $\tilde{\gamma}$ on the controlled systems in closed loop with both the robust in-the-large and the robust in-the-small controllers. Three different selections of the uncertainty are used

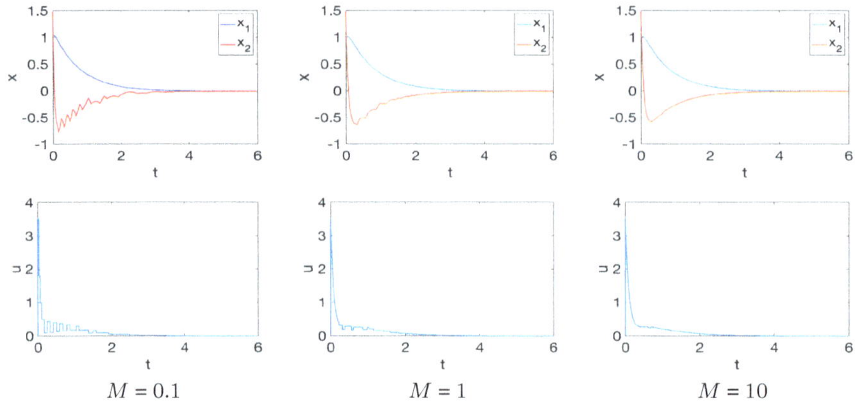

Fig. 8.2 Influence of M. Control gain K_{large}, nominal case. Evolution of the plant state x (top row) and of the plant input (bottom row)

Table 8.1 Influence of M on the number of jumps ($t_{\text{final}} = 6s$). Control gain K_{large}, nominal case (no uncertainty)

M	0.1	1	10
Jumps	29	57	324

and reported at the bottom of the left, middle, and right plots. One can check that stability is preserved when closing the loop with the robust-in-the-large controller, for various uncertainties all satisfying the conditions (8.19) of Assumption 8.4.1. On the other hand, the closed-loop system with the robust-in-the-small controller already exhibits non-converging responses for uncertainties $\tilde{\gamma}$ and $\tilde{\alpha}$ as large as 5% of γ_n and α_n, respectively.

The increased robustness of the robust in-the-large stabilizer is also apparent from Table 8.2, which reports the number of jumps performed by the solution over a horizon $[0, t_{\text{final}}]$ with $t_{\text{final}} = 6s$ for various selections of the uncertainties and when using the stabilizer K_{small} (top row) or K_{large} (bottom row). The different selections of the perturbations $\tilde{\gamma}$ and $\tilde{\alpha}$ are parametrized by a scalar δ as follows: $\tilde{\gamma} = \delta\gamma_n$ and $\tilde{\alpha} = \delta\alpha_n$, so that the first column ($\delta = -0.1$) corresponds to a reduced $\gamma = 0.9\gamma_n$, $\alpha = 0.9\gamma_n$ while the remaining columns describe an increased selection of γ and α. The jumps are only reported in Table 8.2 when obtaining converging solutions, whereas with both controllers, a stability limit is attained when increasing δ too much. When the stability limit is reached, the simulation shows persistent oscillations of the type shown at the middle top plot of Fig. 8.3. The increased robustness margin of controller K_{large} is apparent by the fact that such a stability limit is attained with $\delta = 0.05$ for K_{small} (a 5% maximum variation) and with $\delta = 0.5$ for K_{large} (a 50% maximum variation, which is one order of magnitude larger). For larger selections of δ as compared to such stability limit, the closed-loop responses present exponentially diverging branches of the type shown in the right top plot of Fig. 8.3.

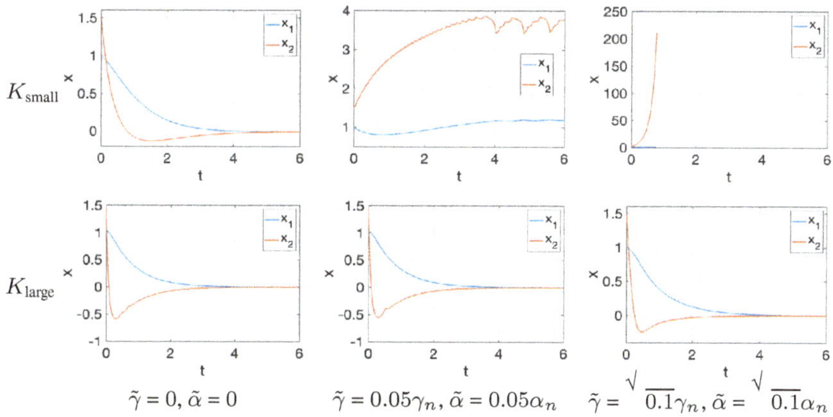

Fig. 8.3 Influence of the uncertainties $\tilde{\gamma}$ and $\tilde{\alpha}$ on the closed-loop response. Increasing values of the uncertainties are used, from left to right. The top row shows the responses with gain K_{small}. The bottom row shows the responses with gain K_{large}

Table 8.2 Influence of the ratio $\delta = \tilde{\gamma}/\gamma_n = \tilde{\alpha}/\alpha_n$ on the stability and the number of jumps ($t_{\text{final}} = 6s$)

δ	–0.1	0	0.04	0.05	0.1	$\sqrt{0.1}$	0.49	0.5
K_{small}, jumps/stab	110	149	93	Oscillating	Diverge	Diverge	Diverge	Diverge
K_{large}, jumps/stab	181	179	208	214	238	242	165	Oscillating

8.6 Conclusion

Two hybrid controller mechanisms have been developed in this chapter for fully linearizable nonlinear systems. Both of these design strategies yield globally asymptotically stable closed-loop systems with different robustness levels against model uncertainties. The plant control input is piecewise constant along the closed-loop solutions and is computed as a sample-and-hold feedback from the plant state and input and their inclusion in suitable flow and jump sets. Some numerical simulations illustrate the different robustness levels and confirm that our second main result provides a guaranteed robustness for larger uncertainties.

This work leaves some questions open. A natural extension is to overcome the need for the knowledge of the full state in the computation of the control input and in the sample or hold decisions. An observer may be useful to this end, modifying not only the control input computation but also the definition of the flow and jump sets. Another open question is the optimization of the matrix M used in the definition (8.7) of the Lyapunov function candidate. Theoretical developments are needed to complete the numerical results of Fig. 8.2 and to inspect the impact of the matrix M on the number of jumps experienced by the closed-loop solutions.

Acknowledgements This work was supported by the "Agence Nationale de la Recherche" (ANR) under Grant HANDY ANR-18-CE40-0010.

References

1. Aldhaheri, R.W., Khalil, H.K.: Effect of unmodeled actuator dynamics on output feedback stabilization of nonlinear systems. Automatica **32**(9), 1323–1327 (1996)
2. Arcak, M., Seron, M., Braslavsky, J., Kokotovic, P.: Robustification of backstepping against input unmodeled dynamics. IEEE Trans. Autom. Control **45**(7), 1358–1363 (2000)
3. Arcak, M., Teel, A., Kokotovic, P.: Robust nonlinear control of feedforward systems with unmodeled dynamics. Automatica **37**, 265–272 (2001)
4. Biannic, J.-M., Burlion, L., Tarbouriech, S., Garcia, G.: On dynamic inversion with rate limitations. In: American Control Conference, Montreal, Canada, June 2012
5. Esfandiari, F., Khalil, H.K.: Output feedback stabilization of fully linearizable systems. Int. J. Control **56**(5), 1007–1037 (1992)
6. Franco, A.L.D., Bourlès, H., De Pieri, E.R., Guillard, H.: Robust nonlinear control associating robust feedback linearization and $\mathcal{H}_\infty$ control. IEEE Trans. Autom. Control **51**(7), 1200–1207 (2006)
7. Goebel, R., Sanfelice, R.G., Teel, A.R.: Hybrid Dynamical Systems: Modeling, Stability, and Robustness. Princeton University Press (2012)
8. Goebel, R., Sanfelice, R.G., Teel, A.R.: Hybrid dynamical systems: robust stability and control for systems that combine continuous-time and discrete-time dynamics. IEEE Control. Syst. Mag. **29**(2), 28–93 (2009)
9. Herrmann, G., Menon, P., Turner, M., Bates, D., Postlethwaite, I.: Anti-windup synthesis for nonlinear dynamic inversion control schemes. Int. J. Robust Nonlinear Control **20**(13), 1465–1482 (2010)
10. Khalil, H.K.: Nonlinear Systems, 3rd edn. Prentice-Hall (2002)
11. Lavergne, F., Villaume, F., Garcia, G., Tarbouriech, S., Jeanneau, M.: Nonlinear and robust flight control laws design for the longitudinal axis of an aircraft. In: 11th Australian International Aerospace Congress (AIAC'11), Melbourne (Australia), March 2005
12. Leite, V.J.S., Tarbouriech, S., Garcia, G.: Energy-peak evaluation of nonlinear control systems under neglected dynamics. In: 9th IFAC Symposium on Nonlinear Control Systems (NOLCOS), Toulouse, France, September 2013
13. Löfberg, J.: YALMIP: a toolbox for modeling and optimization in MATLAB. In: In Proceedings of the CACSD Conference, Taipei, Taiwan (2004)
14. Menon, P.P., Herrmann, G., Turner, M.C., Lowenberg, M., Bates, D., Postlethwaite, I.: Nonlinear dynamic inversion based anti-windup—an aerospace application. In: World IFAC Congress, Seoul, Korea (2008)
15. Papageorgiou, C., Glover, K.: Robustness analysis of nonlinear flight controllers. AIAA J. Guid. Control. Dyn. **28**(4), 639–648 (2005)
16. Prieur, C., Tarbouriech, S., Zaccarian, L.: Lyapunov-based hybrid loops for stability and performance of continuous-time control systems. Automatica **49**(2), 577–584 (2013)
17. Reiner, J., Balas, G.J., Garrard, W.L.: Flight control design using robust dynamic inversion and time-scale separation. Automatica **32**(11), 1491–1625 (1996)
18. Sanfelice, R.G., Copp, D., Nanez, P.: A toolbox for simulation of hybrid systems in Matlab/Simulink: hybrid equations (HyEQ) toolbox. In: Hybrid Systems: Computation and Control Conference (2013)
19. Topcu, U., Packard, A.: Local robust performance analysis for nonlinear dynamical systems. In: American Control Conference, St. Louis, USA, June 2009, pp. 784–789

20. Wang, Q., Stengel, R.F.: Robust nonlinear flight control of a high-performance aircraft. IEEE Trans. Control. Syst. Technol. **13**(1), 15–26 (2005)
21. Wang, X., Van Kampen, E.J., Chu, Q., Lu, P.: Stability analysis for incremental nonlinear dynamic inversion control. J. Guid. Control. Dyn. **42**(5), 1116–1129 (2019)
22. Yang, J., Chen, W.-H., Li, S.: Non-linear disturbance observer-based robust control for systems with mismatched disturbances/uncertainties. IET Control. Theory & Appl. **5**(18), 2053–2062 (2011)

Chapter 9
Improved Synthesis of Saturating Sampled-Data Control Laws for Linear Plants

Arthur Scolari Fagundes, João Manoel Gomes da Silva Jr., and Marc Jungers

Abstract The focus of this chapter is the stabilization of linear systems under saturating aperiodic sampled-data control. By employing a hybrid system representation, we establish conditions for the local and global stability of the origin of the closed-loop system using a specific class of quadratic timer (clock) dependent Lyapunov functions. These conditions are formulated as Sum-of-Squares constraints within optimization problems, enabling the design of stabilizing control laws that aim to maximize an estimate of the Region of Attraction to the Origin (RAO) or to maximize the admissible interval between two sampling instants in order to ensure that a certain given set of initial conditions is included in the RAO.

9.1 Introduction

In recent decades, there has been significant research focused on sampled-data control systems [15], particularly due to their relevance in representing networked and embedded systems [1, 25]. These systems involve the connection of a digital controller to a continuous-time plant. One important aspect of such systems is the presence of aperiodic sampling, which can occur in networks as a result of jitters or packet losses. To capture this characteristic, sampled-data control systems with varying sampling intervals have been studied extensively, disregarding the assumption of a constant sampling period considered in earlier studies [2]. Another important motivation for considering the actual continuous- and discrete-time behaviors of sampled-data systems regards parametric uncertainty issues. For instance, if a linear continuous-

A. S. Fagundes · J. M. Gomes da Silva Jr. (✉)
Universidade Federal do Rio Grande do Sul (UFRGS), Porto Alegre, RS, Brazil
e-mail: jmgomes@ufrgs.br

A. S. Fagundes
e-mail: arthur.fagundes@ufrgs.br

M. Jungers
Université de Lorraine, CNRS, CRAN, 54000 Nancy, France
e-mail: marc.jungers@univ-lorraine.fr

R. Postoyan et al. (eds.), *Hybrid and Networked Dynamical Systems*,
Lecture Notes in Control and Information Sciences 493,
https://doi.org/10.1007/978-3-031-49555-7_9

time plant presents some uncertain parameters, the mapping of the uncertainties in a discrete-time system obtained from an exact discretization approach (considering periodic sampling) is not straightforward.

Several approaches to tackle sampled-data systems have been proposed in the literature. For instance, [11] considers a continuous-time system with time-varying delay on the inputs; [3] incorporates the lifting technique applied to the input signal within a continuous-time system; [13] uses a hybrid system formulation. From these models, stability analysis conditions based on Lyapunov methods have been derived using different mathematical tools. The Lyapunov–Krasovskĭi functionals approach has been employed in works such as [10, 11]. The looped-functionals approach has been utilized in publications like [21, 22]. Furthermore, clock (or timer) dependent Lyapunov functions have been employed in [4, 12]. For a comprehensive overview of the literature on sampled-data systems, readers can refer to the survey provided in [15], which offers a broader perspective on the topic.

On the other hand, actuator saturation is a prevalent characteristic observed in control systems, and similar to aperiodic sampling, it can significantly degrade the performance of closed-loop systems and even lead to instability [16, 24]. It is important to note that, in the presence of saturating inputs, ensuring global stability of the origin may not be possible, even for linear plants [18, 23]. As a result, the literature has been focused on addressing the problems of characterizing estimates of the Region of Attraction to the Origin (RAO) and designing control laws that aim to achieve larger estimates of the RAO, which are interpreted as safe operating zones for the closed-loop system.

Various approaches have been proposed to tackle the problems arising from both sampled-data control and actuator saturation. Combining these two features, the design of stabilizing control laws by using a looped-functional approach is presented in [19, 22], where a generalized sector condition is employed to handle saturation. Considering an impulsive system representation and a convenient partition of the sampling interval inspired by [5], stability and stabilization conditions based on difference inclusions defined by two set-valued maps are proposed in [17]. In [6], it is shown that this design problem can be tackled with the hybrid systems approach, leading to even larger estimates of RAO than in [22] from a problem with a lower number of decision variables than in [17].

This chapter proposes an improvement in the design of stabilizing aperiodic sampled-data control laws conceived in [6], following the hybrid system approach [13]. A more general class of Lyapunov functions, with polynomial dependence on a timer variable, is used to derive local as well as global stability conditions for linear plants under control saturation. Two Sum-of-Squares (SOS) optimization problems are formulated to compute the parameters of the control law aiming at increasing an estimate of the RAO or an estimate of the upper bound of the sampling intervals while ensuring that a given set of admissible initial conditions is included in the RAO.

The chapter is organized as follows. The characterization of the closed-loop system and the stabilization problem are given in Sect. 9.2. Preliminaries for the stabilization conditions, including a hybrid system representation of the closed-loop system, are presented in Sect. 9.3. Based on a timer-dependent Lyapunov function candidate, infinite-dimensional matrix inequalities are proposed as stabilization conditions in Sect. 9.4. Considering the candidate with a polynomial timer dependence, in Sects. 9.5 and 9.6, a Sum-of-Squares relaxation is applied to the infinite-dimensional matrix inequalities, resulting in optimization problems to maximize an estimate of the RAO or an estimate of the maximum allowable intersampling time for a set of admissible initial conditions. Numerical examples and some concluding remarks are presented in Sects. 9.7 and 9.8, respectively.

Notation. $\mathbb{N}$ is the set of natural numbers, $\mathbb{R}$ is the set of real numbers, and $\mathbb{R}_{\geq 0}$ is the set of nonnegative real numbers. The ith element of a vector v is denoted by $v_{(i)}$, while $M_{(i)}$ denotes the ith row of M. The vector v has Euclidean norm given by $|v| = \sqrt{v_{(1)}^2 + \cdots + v_{(n)}^2}$. The distance of a vector v to a closed set $\mathcal{A}$ is denoted $|v|_{\mathcal{A}}$ and is defined by $|v|_{\mathcal{A}} = \inf_{y \in \mathcal{A}} |v - y|$. $\mathbb{S}^n$ is the set of symmetric matrices of order n, and for a symmetric matrix $S \in \mathbb{S}^n$, $S > 0$ means that S is positive definite. $\mathbb{D}^n$ is the set of diagonal matrices of order n. M' denotes the transpose of M, and $\text{He}\{M\} = M' + M$. I denotes an identity matrix of appropriate dimensions. $\star$ denotes a symmetric block in matrices.

9.2 Problem Formulation

Consider the continuous-time plant described by the following model:

$$\dot{x}_p(t) = Ax_p(t) + Bu(t), \tag{9.1}$$

where $x_p \in \mathbb{R}^{n_p}$ is the state of the plant and $u \in \mathbb{R}^m$ is its input. Matrices $A \in \mathbb{R}^{n_p \times n_p}$ and $B \in \mathbb{R}^{n_p \times m}$ are constant.

The control signal $u(t)$ is given by a saturating sampled-data feedback control law, which is generically described as follows [17]:

$$u(t) = \text{sat}(K_x x_p(t_k) + K_u u(t_{k-1})), \qquad t \in [t_k \; t_{k+1}), \tag{9.2}$$

where $\text{sat}(\cdot) : \mathbb{R}^m \to \mathbb{R}^m$ is a vector valued saturation function, defined as

$$\text{sat}(v)_{(i)} \triangleq \text{sign}(v_{(i)})\min(|v_{(i)}|, 1) \;\; \forall i = 1, \ldots, m,$$

and $K_x \in \mathbb{R}^{m \times n_p}$, $K_u \in \mathbb{R}^{m \times m}$ are the controller gains. The instants $t = t_k$, with $k \in \mathbb{N}$, denote the sequence of sampling instants. We assume that $t_0 = 0$ and $u(t_{k-1}) = 0$ for $k = 0$. The term $K_u u(t_{k-1})$, which depends on the past value of the control input, adds a degree of freedom to the problem and also simplifies the stabilization conditions, as it will be clear later. Note that a pure state feedback corresponds to $K_u = 0$.

It is assumed that the sampling intervals are possibly aperiodic, ranging from a lower value $\underline{\tau}$ to an upper value $\overline{\tau}$, that is

$$0 < \underline{\tau} \leq t_{k+1} - t_k \leq \overline{\tau}. \tag{9.3}$$

Supposing that the control law $u(t)$ given in (9.2) ensures the asymptotic stability of the origin of the nonlinear closed-loop system formed by (9.1) and (9.2), the region of attraction of the origin is defined as

$$R_A = \{\xi \in \mathbb{R}^n \; : \; \lim_{t\to\infty} x_p(t) = 0 \text{ for } x_p(0) = \xi\}.$$

From the above setup, the problems we are interested in solving can be stated as follows:

P1. Given system (9.1) and the sampling interval limits $\underline{\tau}$ and $\overline{\tau}$, determine the feedback gains K_x and K_u such that the origin of the closed-loop system (9.1)–(9.2) is asymptotically stable and an estimate of R_A is maximized.

P2. Given system (9.1), the lower sampling interval limit $\underline{\tau}$ and a set of admissible initial conditions $\mathcal{E}_Q^{x_p}$, determine the feedback gains K_x and K_u such that the origin of the closed-loop system (9.1)–(9.2) is asymptotically stable, $\mathcal{E}_Q^{x_p} \subset R_A$ and an upper limit on the sampling interval $\overline{\tau}$ is maximized.

In some cases, namely when the matrix A is Hurwitz, these problems of local stability can be solved in the global sense. The global stabilization problem, i.e., the control law synthesis achieving $R_A = \mathbb{R}^{n_p}$, will be treated as a particular case of the results to solve **P1** and **P2**.

In the following section, we use a hybrid system framework to derive stability conditions in the form of infinite-dimensional matrix inequalities and thereafter finite dimensional ones with Sum-of-Squares (SOS) relaxations.

9.3 Preliminaries

9.3.1 Stability of Hybrid Systems

The closed-loop system described by (9.1) and (9.2) can be represented in the hybrid system framework proposed by [13] as a system $\mathcal{H}$, with state given by $\eta = [x' \; \tau']' \in \mathbb{R}^{n+1}$, where $x = \left[x_p' \; u'\right]' \in \mathbb{R}^n$, with $n = n_p + m$, and $\tau \in \mathbb{R}$ is a timer:

$$\mathcal{H} \begin{cases} \dot{\eta} = \begin{bmatrix} \dot{x} \\ \dot{\tau} \end{bmatrix} = f(\eta), & \forall \eta \in C \\ \eta^+ = \begin{bmatrix} x^+ \\ \tau^+ \end{bmatrix} = g(\eta), & \forall \eta \in D \end{cases} \tag{9.4}$$

The sets $C = \mathbb{R}^n \times [0, \overline{\tau}]$ and $D = \mathbb{R}^n \times [\underline{\tau}, \overline{\tau}]$ are the *flow* and the *jump sets*, respectively, while the function $f : \mathbb{R}^{n+1} \to \mathbb{R}^{n+1}$ is the *flow map* and the function $g : \mathbb{R}^{n+1} \to \mathbb{R}^{n+1}$ is the *jump map*. f and g are defined as follows:

$$f(\eta) = \begin{bmatrix} A_f x \\ 1 \end{bmatrix}, \quad g(\eta) = \begin{bmatrix} A_j x + B_j \text{sat}(K_j x) \\ 0 \end{bmatrix}$$

where A_f, $A_j \in \mathbb{R}^{n\times n}$, $B_j \in \mathbb{R}^{n\times m}$ and $K_j \in \mathbb{R}^{m\times n}$ are given by:

$$A_f = \begin{bmatrix} A & B \\ 0 & 0 \end{bmatrix}, \quad A_j = \begin{bmatrix} I & 0 \\ 0 & 0 \end{bmatrix}, \quad B_j = \begin{bmatrix} 0 \\ I \end{bmatrix}, \quad K_j' = \begin{bmatrix} K_x' \\ K_u' \end{bmatrix}.$$

The system $\mathcal{H}$ has solutions given by an hybrid arc η defined in the domain dom $\eta = \cup_{k=0}^{\infty}([t_k, t_{k+1}], k)$. In particular, dom $\eta(t, k)$ is complete and, since $\underline{\tau} > 0$ from (9.3), without Zeno behavior.

The stability of system $\mathcal{H}$ is associated to a closed set $\mathcal{A}$ containing the origin, and the domain of the timer. The set is defined as follows:

$$\mathcal{A} = \{0\} \times [0, \overline{\tau}]. \tag{9.5}$$

Note that τ is an auxiliary variable that counts the time elapsed since the last sampling instant, being reset to zero at each sampling instant. In other words, during the flow, $\tau(t, k) = t - t_k$, and, at jumps, $\tau(t_k, k) = 0$. Since we assume that $t_0 = 0$, it follows that $\tau(0, 0) = 0$. Thus, we consider that the initial conditions of system (9.4) are expressed as $\eta(0, 0) = \begin{bmatrix} x(0, 0) \\ 0 \end{bmatrix}$.

The next theorem follows from the results in [12] and [13] and provides sufficient conditions to ensure that $\mathcal{A}$ is locally asymptotically stable for system $\mathcal{H}$.

Theorem 9.3.1 *Consider the sets* $\mathcal{X}_C = \mathcal{X} \times [0, \overline{\tau}] \subset C$ *and* $\mathcal{X}_D = \mathcal{X} \times [\underline{\tau}, \overline{\tau}] \subset D$ *with* $\mathcal{X}$ *containing the origin in its interior. If there exists a function* $V : \mathbb{R}^{n+1} \to \mathbb{R}_{\geq 0}$ *and class* $\mathcal{K}_\infty$ *functions* α_1 *and* α_2 *such that*

$$\alpha_1(|\eta|_{\mathcal{A}}) \leq V(\eta) \leq \alpha_2(|\eta|_{\mathcal{A}}) \qquad \forall \eta \in (\mathcal{X}_C \cup \mathcal{X}_D) \tag{9.6}$$

$$\langle \nabla V(\eta), f(\eta) \rangle < 0 \qquad \forall \eta \in \mathcal{X}_C \backslash \mathcal{A} \tag{9.7}$$

$$V(g(\eta)) - V(\eta) < 0 \qquad \forall \eta \in \mathcal{X}_D \backslash \mathcal{A} \tag{9.8}$$

then

1. *the set* $\mathcal{A}$ *is asymptotically stable with respect to system* (9.4),
2. *for any initial condition* $\eta \in \mathcal{L}_V(\mu) = \{\eta \in \mathbb{R}^{n+1} \ : \ V(\eta) \leq \mu\} \subset \mathcal{X}_C$, *it follows that* $\eta(t, k) \to \mathcal{A}$ *as* $t + k \to \infty$.

Recalling that $\eta = [x' \ \tau']'$ and considering that $V(\eta) \triangleq V(x, \tau)$, define the set

$$\mathcal{L}_V^x(\mu) = \{x \in \mathbb{R}^n \ ; \ V(x, 0) \leq \mu\}.$$

As $\tau(0,0)=0$, the verification of Theorem 9.3.1 implies that if $x(0,0)\in\mathcal{L}_V^x$, it follows that $x(t,k)\to 0$ as $t+k\to+\infty$. Assuming that the piecewise constant function $u(t)$ is equal to 0 for $t<0$, one has that $u(0)=u(t_0)=K_x x_p(0)$ for $t=t_0=0$. Then, it follows that

$$x(0,0)=\begin{bmatrix} x_p(0) \\ \text{sat}(K_x x_p(0)) \end{bmatrix}.$$

The following corollary relates the estimates of the region of attraction of the actual sampled-data closed-loop system (9.1)–(9.2), and the ones of the hybrid system $\mathcal{H}$ in (9.4) obtained as level sets of the function V:

Corollary 9.3.2 *If the conditions of Theorem 9.3.1 hold, the set*

$$\mathcal{L}_V^{x_p}(\mu)=\{x_p\in\mathbb{R}^{n_p}\ :\ V(\begin{bmatrix} x_p \\ \text{sat}(K_x x_p) \end{bmatrix},0)\le\mu\} \tag{9.9}$$

is included in the region of attraction of the origin of the closed-loop system (9.1)–(9.2), *i.e.*, $\mathcal{L}_V^{x_p}(\mu)\subset R_A$.

Remark 9.3.3 A global stability result can be straightforwardly formulated by considering $\mathcal{X}=\mathbb{R}^n$ in Theorem 9.3.1. In this case, for any initial condition $\eta\in\mathbb{R}^{n+1}$, it follows that $\eta(t,k)\to\mathcal{A}$ as $t+k\to\infty$.

9.3.2 Stability Analysis with Saturation

Consider a deadzone function $\text{dz}(v)$ defined by:

$$\text{dz}(v)=\text{sat}(v)-v \tag{9.10}$$

that is, $\text{dz}(v)_{(i)}=\text{sign}(v_{(i)})\big(1-\max(|v_{(i)}|,1)\big)$. To handle the saturation nonlinearity in a local context and allow a relaxation of the stability conditions, we consider a generalized sector condition as stated in the following lemma:

Lemma 9.3.4 ([14]) *Consider a matrix $G_j\in\mathbb{R}^{m\times n}$ and define the set*

$$\mathcal{S}=\{x\in\mathbb{R}^n;\ |(K_{j(i)}-G_{j(i)})x|\le 1,\ i=1,\dots,m\} \tag{9.11}$$

If $x\in\mathcal{S}$, then the deadzone nonlinearity $dz(K_j x)$ defined in (9.10) *satisfies the following inequality:*

$$\text{dz}(K_j x)'T(\text{dz}(K_j x)+G_j x)\le 0 \tag{9.12}$$

for any diagonal positive definite matrix $T\in\mathbb{D}^m$.

In order to apply Lemma 9.3.4 in the stability analysis, the saturation function in $\mathcal{H}$ must be replaced by the deadzone function. In this case, $g(\eta)$ can be rewritten as follows:

$$g(\eta) = \begin{bmatrix} \mathbb{A}_j x + B_j \mathrm{dz}(K_j x) \\ 0 \end{bmatrix} \tag{9.13}$$

with $\mathbb{A}_j = A_j + B_j K_j$. The forthcoming results are applied to system $\mathcal{H}$ considering $g(\eta)$ as in (9.13).

9.4 Synthesis of Stabilizing Controller

From the generic result in Theorem 9.3.1, we derive now conditions in the form of matrix inequalities to compute the stabilizing control law (9.2), using a quadratic timer (clock) dependent Lyapunov function, i.e.,

$$V(\eta) = V(x, \tau) = x' P(\tau) x. \tag{9.14}$$

Theorem 9.4.1 *If there exists a matrix function $\tilde{P} : [0,\ \overline{\tau}] \to \mathbb{S}^n$, matrices $\tilde{K}_j\ \tilde{G}_j \in \mathbb{R}^{m\times n}$, $\tilde{N} \in \mathbb{R}^{n\times n}$, a diagonal matrix $\tilde{T} \in \mathbb{D}^m$ and a scalar β that satisfy the following matrix inequalities*

$$\tilde{P}(\tau) > 0 \quad \forall \tau \in [0,\ \overline{\tau}] \tag{9.15}$$

$$\begin{bmatrix} \dot{\tilde{P}}(\tau) + \beta \mathrm{He}\{A_f \tilde{N}'\} & \tilde{P}(\tau) - \beta \tilde{N}' + \tilde{N} A_f' \\ \star & -\mathrm{He}\{\tilde{N}\} \end{bmatrix} < 0 \quad \forall \tau \in [0,\ \overline{\tau}] \tag{9.16}$$

$$\begin{bmatrix} \tilde{P}(0) - \beta \mathrm{He}\{\tilde{N}\} & \beta \tilde{\mathbb{A}}_j - \tilde{N} & \beta B_j \tilde{T} \\ \star & \mathrm{He}\{\tilde{\mathbb{A}}_j\} - \tilde{P}(\tau) & B_j \tilde{T} - \tilde{G}_j' \\ \star & \star & -2\tilde{T} \end{bmatrix} < 0 \quad \forall \tau \in [\underline{\tau},\ \overline{\tau}] \tag{9.17}$$

$$\begin{bmatrix} \tilde{P}(\tau) & (\tilde{K}_{j(i)} - \tilde{G}_{j(i)})' \\ \star & 1 \end{bmatrix} \geq 0 \quad \begin{matrix} \forall \tau \in [\underline{\tau},\ \overline{\tau}] \\ i = 1, \ldots, m \end{matrix} \tag{9.18}$$

with $\tilde{\mathbb{A}}_j = A_j \tilde{N}' + B_j \tilde{K}_j$, then considering $K_j = \tilde{K}_j (\tilde{N}')^{-1}$ and $P(0) = \tilde{N}^{-1} \tilde{P}(0)$ $(\tilde{N}')^{-1}$, all initial conditions of the closed-loop system (9.1)–(9.2) such that $x_p(0) \in \mathcal{E}_P^{x_p}$, with[1]

$$\mathcal{E}_P^{x_p} = \{x_p \in \mathbb{R}^{n_p} \ :\ \begin{bmatrix} x_p \\ \mathrm{sat}(K_x x_p) \end{bmatrix}' P(0) \begin{bmatrix} x_p \\ \mathrm{sat}(K_x x_p) \end{bmatrix} \leq 1\},$$

result in solutions $x_p(t)$ that converge asymptotically to the origin as $t \to +\infty$.

---

[1] The level set $\mathcal{E}_P^{x_p}$ corresponds to (9.9) in the case where $V(\eta) = x' P(\tau) x$ and, since P can be normalized, μ can be considered as 1 without loss of generality in this case.

Proof From (9.16), note that $\tilde{N}' + \tilde{N} > 0$, and thus $\tilde{N}$ is non-singular. From (9.15), if $\tilde{P}(\tau) > 0$, then it follows that $\tilde{N}^{-1}\tilde{P}(\tau)(\tilde{N}')^{-1} = P(\tau) > 0$. Hence, considering $V(\eta) = x'P(\tau)x$, we have that $V(\eta)$ is a nonnegative function which satisfies $V(\eta) = 0$ for $\eta \in \mathcal{A}$ and $V(\eta) > 0$ for $\eta \notin \mathcal{A}$. Moreover, we can conclude that (9.6) is verified with $\alpha_1(|\eta|_{\mathcal{A}}) = \underline{\alpha}|\eta|^2_{\mathcal{A}}$ and $\alpha_2(|\eta|_{\mathcal{A}}) = \overline{\alpha}|\eta|^2_{\mathcal{A}}$, where $\underline{\alpha}$ and $\overline{\alpha}$ are respectively as the smallest and largest eigenvalues of $P(\tau)$ for $\tau \in [0, \overline{\tau}]$, respectively.

Recalling that $\dot{\tau} = 1$ for the flow dynamics, the condition (9.7) in Theorem 9.3.1 with $V(\eta) = x'P(\tau)x$ is equivalent to:

$$\begin{bmatrix} x \\ \dot{x} \end{bmatrix}' \begin{bmatrix} \dot{P}(\tau) & P(\tau) \\ \star & 0 \end{bmatrix} \begin{bmatrix} x \\ \dot{x} \end{bmatrix} < 0 \quad \forall \begin{bmatrix} x \\ \dot{x} \end{bmatrix} \text{ s.t. } \begin{bmatrix} A_f & -I \end{bmatrix} \begin{bmatrix} x \\ \dot{x} \end{bmatrix} = 0, \ \forall \tau \in [0, \ \overline{\tau}]. \tag{9.19}$$

where $\dot{P}(\tau) = \dfrac{\partial P(\tau)}{\partial \tau}\dot{\tau} = \dfrac{\partial P(\tau)}{\partial \tau}$.

Applying now the Finsler's lemma [9], (9.19) is satisfied if there exists a matrix $N \in \mathbb{R}^{2n+m \times 2n+m}$ such that

$$\begin{bmatrix} \dot{P}(\tau) & P(\tau) \\ \star & 0 \end{bmatrix} + \text{He}\{N \begin{bmatrix} A_f & -I \end{bmatrix}\} < 0 \ \ \forall \tau \in [0, \ \overline{\tau}]. \tag{9.20}$$

Considering a particular structure $N = \begin{bmatrix} \beta(\tilde{N}')^{-1} & (\tilde{N}')^{-1} \end{bmatrix}'$, (9.20) is rewritten as:

$$\begin{bmatrix} \dot{P}(\tau) + \text{He}\{\beta\tilde{N}^{-1}A_f\} & P(\tau) - \beta\tilde{N}^{-1} + A_f'(\tilde{N}')^{-1} \\ \star & -(\tilde{N}')^{-1} - \tilde{N}^{-1} \end{bmatrix} < 0 \qquad \forall \tau \in [0, \ \overline{\tau}]. \tag{9.21}$$

Now, by left and right-multiplying (9.21) by $\text{diag}(\tilde{N}, \tilde{N})$ and its transpose respectively, and considering the change of variables $\tilde{P}(\tau) = \tilde{N}P(\tau)\tilde{N}'$, the condition (9.16) is obtained. Thus, we conclude that (9.16) is a sufficient condition to verify (9.7) of Theorem 9.3.1, that is, $\langle \nabla V(\eta), f(\eta) \rangle < 0 \ \forall \tau \in [0, \ \overline{\tau}]$ with $V(\eta) = x'\tilde{N}^{-1}\tilde{P}(\tau)(\tilde{N}')^{-1}x = x'P(\tau)x$.

Adopting now the hybrid arc representation $\eta(t, k) = [x(t, k)' \quad \tau(t, k)']'$ and recalling that, by assumption, a sampling occurs at $t_0 = 0$ and $\tau(0, 0) = 0$, if $x_p(0) \in \mathcal{E}_P^{x_p}$ it follows that $x(0, 0) \in \mathcal{L}_V^x(1)$, that is $x'(t_0, 0)P(\tau(0, 0))x(t_0, 0) = x'(0, 0)P(0)x(0, 0) \leq 1$. Hence, during the flow between t_0 and t_1, i.e., $\forall t \in [t_0, t_1]$, the condition (9.16) ensures that

$$x'(t, 0)P(\tau(t, 0))x(t, 0) < x'(t_0, 0)P(\tau(0, 0))x(t_0, 0) \leq 1. \tag{9.22}$$

Applying Schur's complement to (9.18) and then left and right-multiplying it by $\tilde{N}^{-1}$ and its transpose respectively, considering the change of variables $G_j = \tilde{G}_j(\tilde{N}')^{-1}$ and $K_j = \tilde{K}_j(\tilde{N}')^{-1}$, it follows that:

$$P(\tau) - (K_{j(i)} - G_{j(i)})'(K_{j(i)} - G_{j(i)}) \geq 0, \quad \forall \tau \in [\underline{\tau}, \overline{\tau}], \; i = 1, \ldots, m \tag{9.23}$$

From (9.22) and (9.23), since $\tau(t_1, 0) \in [\underline{\tau}, \overline{\tau}]$, it follows that

$$x'(t_1, 0)(K_{j(i)} - G_{j(i)})'(K_{j(i)} - G_{j(i)})x(t_1, 0) \leq x(t_1, 0)'P(\tau(t_1, 0))x(t_1, 0) \leq 1 \tag{9.24}$$

for $i = 1, \ldots, m$ and thus, from (9.24), we conclude that

$$|(K_{j(i)} - G_{j(i)})x(t_1, 0)| \leq 1 \qquad i = 1, \ldots, m \tag{9.25}$$

Hence, from (9.25) and the definition of $\mathcal{S}$ in (9.11), one has that $x(t_1, 0) \in \mathcal{S}$ and from Lemma 9.3.4 it follows that $dz(K_j(x(t_1, 0))' \, T(\mathrm{dz}(K_j x(t_1, 0)) + G_j x(t_1, 0)) \leq 0$. Thus, supposing now that a new sampling (jump) occurs at $t = t_1$, and considering $\zeta = \left[x'(t_1, 1) \; x'(t_1, 0) \; \mathrm{dz}(K_j x(t_1, 0))\right]$, if

$$\zeta' \begin{bmatrix} P(\tau(t_1, 1)) & 0 & 0 \\ \star & -P(\tau(t_1, 0)) & -G_j' T \\ \star & \star & -2T \end{bmatrix} \zeta \leq 0, \quad \forall \zeta \text{ s.t. } \left[-I \; \mathbb{A}_j \; B_j\right] \zeta = 0 \tag{9.26}$$

then one obtains that $V(g(\eta)) - V(\eta) = V(x(t_1, 1), \tau(t_1, 1)) - V(x(t_1, 0), \tau(t_1, 0)) < 0$. Applying again the Finsler's lemma, (9.26) is satisfied if there exists a matrix $N_j \in \mathbb{R}^{2n+m \times 2n+m}$ such that

$$\begin{bmatrix} P(\tau(t_1, 1)) & 0 & 0 \\ \star & -P(\tau(t_1, 0)) & -G_j' T \\ \star & \star & -2T \end{bmatrix} + \mathrm{He}\{N_j \left[-I \; \mathbb{A}_j \; B_j\right]\} < 0. \tag{9.27}$$

Since τ is reset to zero at the jump instant, one has that $\tau(t_1, 1) = 0$ and that $\tau(t_1, 0) \in [\underline{\tau}, \overline{\tau}]$. Thus, condition (9.27) is rewritten as:

$$\begin{bmatrix} P(0) & 0 & 0 \\ \star & -P(\tau) & -G_j' T \\ \star & \star & -2T \end{bmatrix} + \mathrm{He}\{N_j \left[-I \; \mathbb{A}_j \; B_j\right]\} < 0 \qquad \forall \tau \in [\underline{\tau}, \overline{\tau}]. \tag{9.28}$$

Considering now $N_j = \left[\beta(\tilde{N}')^{-1} \; (\tilde{N}')^{-1} \; 0\right]'$ and pre and post-multiplying (9.28) by $\mathrm{diag}(\tilde{N}, \tilde{N}, T^{-1})$ and its transpose, respectively, the result is the following condition:

$$\begin{bmatrix} E_{11} & E_{12} & E_{13} \\ \star & E_{22}(\tau) & E_{23} \\ \star & \star & E_{33} \end{bmatrix} < 0 \qquad \forall \tau \in [\underline{\tau}, \overline{\tau}] \tag{9.29}$$

with

$$\begin{cases} E_{11} = \tilde{N}P(0)\tilde{N}' - \mathrm{He}\{\beta\tilde{N}\}; & E_{12} = \beta\mathbb{A}_j\tilde{N}' - \tilde{N} \\ E_{13} = \beta B_j T^{-1}; & E_{22}(\tau) = \mathrm{He}\{\mathbb{A}_j\tilde{N}'\} - \tilde{N}P(\tau)\tilde{N}' \\ E_{23} = B_j T^{-1} - \tilde{N}G_j'; & E_{33} = -2T^{-1} \end{cases}$$

Finally, considering the changes of variables: $\tilde{\mathbb{A}}_j = \mathbb{A}_j\tilde{N}'$, $\tilde{T} = T^{-1}$, $\tilde{P}(\tau) = \tilde{N}P(\tau)\tilde{N}'$, $\tilde{G}_j = G_j\tilde{N}'$ and $\tilde{K}_j = K_j\tilde{N}'$, one obtains that (9.29) is equivalent to (9.17). Thus, the satisfaction of (9.17) implies (9.8) and it is sufficient for the decrease of V at the jump instant, i.e., $V(x(t_1, 1), \tau(t_1, 1)) < V(x(t_1, 0), \tau(t_1, 0)) \leq 1$. Repeating the reasoning recursively for $t \geq t_1$, $k \geq 1$, leads to the fact that $\eta(t, k) \to \mathcal{A}$ and $x(t, k) \to 0$ when $t \to \infty$ and $k \to \infty$, which implies that $x_p(t) \to 0$ as $t \to \infty$ provided that $x_p(0) \in \mathcal{E}_P^{x_p}$. □

As a particular case of the previous theorem, the following corollary provides the conditions to global asymptotic stabilization of the origin of (9.1)–(9.2).

Corollary 9.4.2 *If there exist a matrix function $\tilde{P} : [0, \overline{\tau}] \to \mathbb{S}^n$, matrices $\tilde{K}_j \in \mathbb{R}^{m\times n}$, $\tilde{N} \in \mathbb{R}^{n\times n}$, a diagonal matrix $\tilde{T} \in \mathbb{D}^m$ and a scalar β that satisfy* (9.15), (9.16) *and* (9.17) *with $\tilde{K}_j = \tilde{G}_j$, then all initial conditions $x_p(0) \in \mathbb{R}^{n_p}$ of the closed-loop system* (9.1)–(9.2) *result in solutions $x_p(t)$ that converge asymptotically to the origin as $t \to +\infty$.*

Proof In the case that $\tilde{K}_j = \tilde{G}_j$, the relation (9.12) is satisfied for any $x \in \mathbb{R}^n$, which implies that the conditions of Theorem 9.3.1 are verified with $\mathcal{X} = \mathbb{R}^n$. □

9.5 Polynomial Timer-Dependent Lyapunov Function

The conditions of Theorem 9.4.1 depend on the definition of $P(\tau)$ and must be evaluated at each τ in the range of $[0\ \overline{\tau}]$ or $[\underline{\tau}\ \overline{\tau}]$. This would result in an infinite number of matrix inequalities to be verified. This limitation can be circumvented by choosing $P(\tau)$ to be a polynomial function, that is

$$P(\tau) \triangleq P_0 + \tau P_1 + \cdots + \tau^d P_d \tag{9.30}$$

where the degree d should be chosen taking into account the trade-off between conservatism and numerical complexity of the resulting problem. Considering the candidate (9.30), the next theorem is based on the SOS relaxation of the conditions presented in the previous theorem.

Theorem 9.5.1 *If there exist matrix polynomials $\tilde{P} : [0, \overline{\tau}] \to \mathbb{S}^n$, $Q_1 : [0, \overline{\tau}] \to \mathbb{S}^n$, $Q_2 : [0, \overline{\tau}] \to \mathbb{S}^{2n}$, $Q_3 : [0, \overline{\tau}] \to \mathbb{S}^{2n+m}$, $Q_4 : [0, \overline{\tau}] \to \mathbb{S}^{n+1}$, $\tilde{K}_j$, $\tilde{G}_j \in \mathbb{R}^{m\times n}$, and $\tilde{N} \in \mathbb{R}^{n\times n}$, a diagonal matrix $\tilde{T} \in \mathbb{D}^m$, a scalar β and a positive scalar γ that satisfy the following inequalities:*

$$Q_1(\tau),\, Q_2(\tau),\, Q_3(\tau),\, Q_4(\tau) \quad \text{are SOS} \tag{9.31}$$

$$\tilde{P}(\tau) - \gamma I - Q_1(\tau)\tau(\overline{\tau} - \tau) \text{ is SOS} \tag{9.32}$$

$$\begin{bmatrix} -\dot{\tilde{P}}(\tau) + \beta \mathrm{He}\{A_f \tilde{N}'\} & \tilde{P}(\tau) - \beta \tilde{N}' + \tilde{N} A_f' \\ \star & -\mathrm{He}\{\tilde{N}\} \end{bmatrix} - \gamma I - Q_2(\tau)\tau(\overline{\tau} - \tau) \text{ is SOS} \tag{9.33}$$

$$-\begin{bmatrix} \tilde{P}(0) - \mathrm{He}\{\tilde{N}\} & \beta \tilde{\mathbb{A}}_j - \tilde{N} & \beta B_j \tilde{T} \\ \star & \mathrm{He}\{\tilde{\mathbb{A}}_j\} - \tilde{P}(\tau) & B_j \tilde{T} - \tilde{G}_j' \\ \star & \star & -2\tilde{T} \end{bmatrix} - \gamma I - Q_3(\tau)(\tau - \underline{\tau})(\overline{\tau} - \tau) \text{ is SOS} \tag{9.34}$$

$$\begin{bmatrix} \tilde{P}(\tau) & (\tilde{K}_{j(i)} - \tilde{G}_{j(i)})' \\ \star & 1 \end{bmatrix} - \gamma I - Q_4(\tau)(\tau - \underline{\tau})(\overline{\tau} - \tau) \text{ is SOS for } i = 1, \ldots, m \tag{9.35}$$

with $\tilde{\mathbb{A}}_j = A_j \tilde{N}' + B_j \tilde{K}_j$, *then considering* $K_j = \tilde{K}_j (\tilde{N}')^{-1}$ *and* $P_0 = \tilde{N}^{-1} \tilde{P}(0)$ $(\tilde{N}')^{-1}$, *all initial conditions of the closed-loop system* (9.1)–(9.2) *such that* $x_p(0) \in \mathcal{E}_{P_0}^{x_p}$, *with*

$$\mathcal{E}_{P_0}^{x_p} = \{x_p \in \mathbb{R}^{n_p} : \begin{bmatrix} x_p \\ \mathrm{sat}(K_x x_p) \end{bmatrix}' P_0 \begin{bmatrix} x_p \\ \mathrm{sat}(K_x x_p) \end{bmatrix} \leq 1\},$$

result in solutions $x_p(t)$ *that converge asymptotically to the origin as* $t \to +\infty$.

Proof If the conditions in (9.31) are verified, then

$$Q_1(\tau),\, Q_2(\tau),\, Q_3(\tau),\, Q_4(\tau) \geq 0 \quad \forall \tau \in \mathbb{R}$$

Noting that $\tau(\overline{\tau} - \tau) \geq 0$ for $\tau \in [0,\ \overline{\tau}]$, provided that $Q_1(\tau) \geq 0\, \forall \tau \in \mathbb{R}$, it follows that (9.32) ensures that

$$\tilde{P}(\tau) > 0 \quad \forall \tau \in [0,\ \overline{\tau}]\,.$$

Similarly, as $Q_2(\tau) \geq 0\, \forall \tau \in \mathbb{R}$, the condition (9.33) implies (9.16).

On the other hand, noting that $(\tau - \underline{\tau})(\overline{\tau} - \tau) \succeq 0$ for $\tau \in \left[\underline{\tau},\ \overline{\tau}\right]$, provided that $Q_3(\tau)$ and $Q_4(\tau) \succeq 0\, \forall \tau \in \mathbb{R}$, it follows that (9.34) and (9.35) ensure respectively (9.17) and (9.18). □

Remark 9.5.2 If one wants to synthesize a controller that does not depend on the value of the control input in the last sampling instant (i.e., $K_u = 0$), the transformation $K_j = \tilde{K}_j (\tilde{N}')^{-1}$ must yield a controller of the form $K_j = [K_x\ 0]$. To this end, it suffices to impose the following structures for $\tilde{N}$ and $\tilde{K}_j$:

$$\tilde{N} = \begin{bmatrix} \tilde{N}_{11} & \tilde{N}_{12} \\ 0 & \tilde{N}_{22} \end{bmatrix}, \qquad \tilde{K}_j = \begin{bmatrix} \tilde{K}_x & 0 \end{bmatrix}. \tag{9.36}$$

Remark 9.5.3 As stated in the last paragraph of the introduction, the conditions obtained with the polynomial function candidate (9.30) can be seen as a generalization of the ones obtained with the affine function in [6], that is, of $P(\tau) = P_0 + \tau P_1$.

9.6 Optimization Problems

To address the problem **P1** the idea is to maximize the region $\mathcal{E}_{P_0}^{x_p}$, which is an estimate of the RAO, considering a shape set criterion. For this, we use the following lemma (adapted from [22]).

Lemma 9.6.1 *Let κ be a positive scalar, P_0 and $Q \in \mathbb{S}^n$ be positive definite matrices and $\tilde{N}$ be a non-singular matrix, and define $\tilde{P}_0 = \tilde{N} P_0 \tilde{N}'$. If the following inequality is verified*

$$\begin{bmatrix} \varepsilon Q & \kappa I \\ \kappa I & \kappa(\tilde{N} + \tilde{N}') - \tilde{P}_0 \end{bmatrix} \geq 0 \tag{9.37}$$

then $P_0 \leq \varepsilon Q$.

Proof If $P_0 > 0$, it follows that $(\kappa P_0^{-1} - \tilde{N}')' P_0 (\kappa P_0^{-1} - \tilde{N}') \geq 0$, which implies that $\kappa^2 P_0^{-1} \geq (\kappa(\tilde{N} + \tilde{N}') - \tilde{P}_0)$, or equivalently, $P_0 \leq \kappa^2(\kappa(\tilde{N} + \tilde{N}') - \tilde{P}_0)^{-1}$. Hence, if (9.37) is verified, it follows that $\varepsilon Q \geq \kappa^2(\kappa(\tilde{N} + \tilde{N}') - \tilde{P}_0)^{-1}$, which implies that $P_0 \leq \varepsilon Q$. □

Consider now a given matrix $Q = Q' > 0$ and a piecewise quadratic shape set described as

$$\mathcal{E}_Q^{x_p}(\varepsilon) = \{x_p \in \mathbb{R}^{n_p} : \begin{bmatrix} x_p \\ \text{sat}(K_x x_p) \end{bmatrix}' Q \begin{bmatrix} x_p \\ \text{sat}(K_x x_p) \end{bmatrix} \leq \frac{1}{\varepsilon}\}. \tag{9.38}$$

If condition (9.37) in Lemma 9.6.1 is verified, it follows that the inclusion $\mathcal{E}_Q^{x_p}(\varepsilon) \subset \mathcal{E}_{P_0}^{x_p}$ holds. Hence, the idea is to use (9.37) and minimize ε. This leads to the maximization of the region $\mathcal{E}_Q^{x_p}(\varepsilon)$, and, in virtue of the inclusion, indirectly leads to the maximization of $\mathcal{E}_{P_0}^{x_p}$. Observe also that Q can be chosen to influence the final shape of $\mathcal{E}_{P_0}^{x_p}$ and therefore of the estimate of the RAO.

Moreover, if β and κ are a priori fixed, then (9.31)–(9.37) can be cast as SOS constraints of the following optimization problem, which can be used to solve **P1**:

$$\begin{aligned} &\min_{\tilde{P},\tilde{G}_j,\tilde{K}_j,\tilde{T},\tilde{N},\varepsilon} \quad \varepsilon \\ &\text{subject to: (9.31), (9.32), (9.33), (9.34), (9.35) and (9.37)} \end{aligned} \tag{9.39}$$

Considering $\mathcal{E}_Q^{x_p}$ as in (9.38) given, the problem **P2**, in turn, is posed as follows:

$$\begin{aligned} &\max_{\tilde{P},\tilde{G}_j,\tilde{K}_j,\tilde{T},\tilde{N}} \quad \overline{\tau} \\ &\text{subject to: (9.31), (9.32), (9.33), (9.34), (9.35) and (9.37)} \end{aligned} \tag{9.40}$$

Hence, the optimal value of ε and $\overline{\tau}$ can be efficiently approximated by finding respectively the solution to (9.39) and (9.40) over a range of values for β and κ. In the case of (9.40), the objective variable $\overline{\tau}$ must be found by repeatedly solving the problem, each time increasing the value of $\overline{\tau}$, until the constraints become unfeasible. The next numerical examples illustrate the use of the optimization problems (9.39) and (9.40).

9.7 Numerical Examples

Consider system (9.1) with the following data:

$$A = \begin{bmatrix} 0 & 1 \\ 1 & 0 \end{bmatrix}; \quad B = \begin{bmatrix} 0 \\ -5 \end{bmatrix}; \quad \underline{\tau} = 0.05\,; \quad \overline{\tau} = 0.1\,; \tag{9.41}$$

Addressing **P1**, a feedback gain $K_j = [K_x \; K_u] = [2.6 \; 1.4 \; 0]$ was first synthesized in [22] considering the data in (9.41). The same controller was considered in the analysis of [8, 22]. The estimates of the RAO obtained in these references are plotted in Fig. 9.1, in green and dotted black. In the same figure, the estimates obtained in [17] for the analysis with the gain $K_x = [2.6 \; 1.4]$, and for the synthesis with the resulting gain $K_j = [1.13 \; 0.94 \; 0.008]$ are plotted as well, in red and blue. This example is also treated in [6], which presents stabilization conditions based on an affine function candidate described as

$$P(\tau) = P_0 + \tau P_1 = 10^{-4} \times \left(\begin{bmatrix} 752 & 750 & -464 \\ 750 & 753 & -461 \\ -464 & -461 & 1361 \end{bmatrix} + \tau \begin{bmatrix} -2443 & -2467 & 4868 \\ -2467 & -2486 & 4908 \\ 4868 & 4908 & -6761 \end{bmatrix} \right) \tag{9.42}$$

when adopting a shape set (9.38) with $Q = I$ and fixing $\beta = \kappa = 3.6$ (the optimal value). Here, the problem (9.39) is solved with the SOStools toolbox [20] version 3.03 considering $d = 2$ and the same value of $\beta = \kappa = 3.6$. We consider at first $Q = I$. In this case, the minimum ε obtained with (9.39) and in Ref. [6] are 0.2013 and 0.2227, respectively. Hence, with the more general Lyapunov function candidate (9.30), a lesser value was found in the minimization of ε. In this case, the obtained

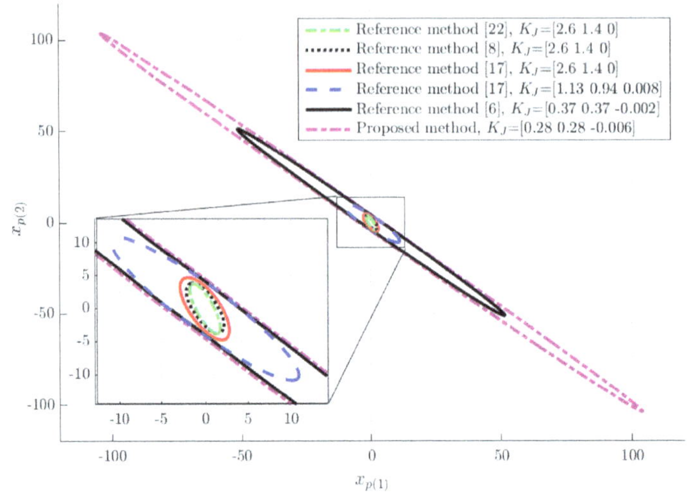

Fig. 9.1 The estimates of the RAO obtained by solving problem (9.39) compared to the ones obtained with other methods in the literature

timer-dependent Lyapunov function is defined with

$$
\begin{aligned}
P(\tau) = P_0 + \tau P_1 + \tau^2 P_2 = 10^{-4} \times & \left(\begin{bmatrix} 715 & 706 & -453 \\ 706 & 718 & -451 \\ -453 & -451 & 1033 \end{bmatrix} \right. \\
& \left. + \tau \begin{bmatrix} -887 & -907 & 3600 \\ -907 & -926 & 3655 \\ 3600 & 3655 & -5931 \end{bmatrix} + \tau^2 \begin{bmatrix} -64 & -63 & -1306 \\ -63 & -61 & -1316 \\ -1306 & -1316 & 6353 \end{bmatrix} \right)
\end{aligned}
$$

Note that when $Q = I$, the minimization of ε corresponds to the maximization of the smallest axis of the ellipsoid $\mathcal{E}_P^{x_p}$. In order to obtain an estimate of the RAO that not only surpasses on the smallest axis the one of [6], but encompasses it, the problem (9.39) is solved again, this time with $Q = P_0$ from (9.42), and the result is also shown in Fig. 9.1, in purple, together with the result from [6] in black.

For a more in-depth analysis of the results, we show now in Table 9.1 a comparison gauging the numerical complexity between the approaches as well, taking into account the number of decision variables of the proposed method and the ones in [6, 17, 22]. It should be noticed that the method in [17] requires that the interval $[\underline{\tau}, \overline{\tau}]$ is partitioned in J subintervals. In the considered example, the interval was divided in $J = 27$ partitions, which leads to 114 decision variables to be accounted for in the solution process. The method of [22], on the other hand, presents 43 decision variables, while the method proposed here presents 31 decision variables, 6 more than the 25 in [6] but leading to a greater estimate of the RAO.

Table 9.1 Comparing the numerical complexity of the methods

Number of decision variables	
Proposed	$n^2 + (n^2 + n)(d + 1)/2 + m + mn$
Reference [6]	$2n^2 + n + m + mn$
Reference [17]	$\frac{1}{2}(n^2 + n) + (mn + m)J$
Reference [22]	$\frac{1}{2}(n^2 + n) + 3n_p^2 + n_p(n_p + 1)$ $+n_p(3n_p + m) + 2n_pm + m$

Table 9.2 Results for **P2**

ε	$4\varepsilon_{\min}$	$3\varepsilon_{\min}$	$2\varepsilon_{\min}$	$\varepsilon_{\min}$
$\overline{\mathcal{T}}$ from (9.40)	0.23	0.20	0.16	0.10
$\overline{\mathcal{T}}$ from [6]	0.21	0.18	0.14	unfeas.

Considering again $Q = I$ and defining $\varepsilon_{\min} \triangleq 0.2013$, the results for problem **P2** with different values of ε are shown in Table 9.2. Note that there is a direct relationship between the value assigned to ε (which is inversely related to the "size" of the region of admissible initial conditions) and the resulting estimate of the maximum $\overline{\mathcal{T}}$ for which the stability can be guaranteed: for smaller regions of admissible initial conditions (i.e., larger values of ε), the estimate of $\overline{\mathcal{T}}$ is greater.

9.8 Concluding Remarks

In this chapter a method to design sampled-data stabilizing control laws for linear plants subject to input saturation based on an hybrid system framework has been proposed. The method consists in optimization problems obtained from a Sum-of-Squares relaxation of stability conditions based on a polynomial timer-dependent Lyapunov function candidate, whose solution provides the synthesis of a control law to maximize an estimate of the region of attraction of the origin of the sampled-data system considered. The proposed approach here showed reduced conservatism when compared to other references in the literature [6, 8, 17, 22]. Moreover, the number of decision variables is also lesser than the references (except for [6] which considers an affine time-dependence), and, in relation to [17], the problem **P2** can be solved without requiring a new partition of the range of admissible sampling intervals.

The conservatism of the estimates of the RAO provided by the method are mainly due to the particular structure of the Finsler's lemma multipliers used to obtain tractable synthesis conditions. Hence, once a stabilizing gain is computed, it is possible to obtain improved estimates by applying the analysis method proposed in [7].

The extension of the results to cope with uncertainties in the continuous-time plant is quite straightforward in the proposed framework. In this case, classical developments from robust control theory to consider polytopic or norm-bounded uncertainties can be incorporated to the presented ones in order to derive conditions to compute robust stabilizing gains and associate estimates of the RAO of the uncertain closed-loop system.

Acknowledgements This work was supported in part by the "Agence Nationale de la Recherche" (ANR, France) under Grant HANDY ANR-18-CE40-0010, by the Coordenação de Aperfeiçoamento de Pessoal de Nível Superior (CAPES, Brazil)-Finance Code 001, and by the Conselho Nacional de Desenvolvimento Cientfico e Tecnológico (CNPq, Brazil) - under Grant PQ-307449/2019-0.

References

1. Antsaklis, P., & Baillieul, J. (2007). A special issue on technology of networked control systems. *Proc. IEEE, 95*(1), 5–8.
2. Åström, K. J., & Wittenmark, B. (1984). *Computer Controlled Systems: Theory and Design*. Information and System Sciences SeriesUpper Saddle River, NJ: Prentice-Hall.
3. Bamieh, B., Pearson, J. B., Francis, B. A., & Tannenbaum, A. (1991). A lifting technique for linear periodic systems with applications to sampled-data control. *Syst. Control Lett., 17*(2), 79–88.
4. Briat, C. (2013). Convex conditions for robust stability analysis and stabilization of linear aperiodic impulsive and sampled-data systems under dwell-time constraints. *Automatica, 49*(11), 3449–3457.
5. Cloosterman, M. B. G., Hetel, L., van de Wouw, N., Heemels, W. P. M. H., Daafouz, J., & Nijmeijer, H. (2010). Controller synthesis for networked control systems. *Automatica, 46*, 1584–1594.
6. Fagundes, A.S., Gomes da Silva Jr., J.M.: Design of saturating aperiodic sampled-data control laws for linear plants: a hybrid system approach. In: 20th European Control Conference (ECC). London, UK, European Control Association (EUCA) (2022)
7. Fagundes, A.S., Gomes da Silva Jr., J.M., Jungers, M.: Stability analysis of sampled-data control systems with input saturation: a hybrid system approach. Eur. J. Control (2023)
8. Fiacchini, M., Gomes da Silva Jr., J.M.: Stability of sampled-data control systems under aperiodic sampling and input saturation. In: 57th IEEE Conference on Decision and Control, pp. 1286–1293. Miami Beach (2018)
9. Finsler, P.: Über das vorkommen definiter und semidefiniter formen in scharen quadratischer formen. Commentarii Mathematici Helvetici **9**, 188–192 (1937)
10. Fridman, E. (2010). A refined input delay approach to sampled-data control. *Automatica, 46*(2), 421–427.
11. Fridman, E., Seuret, A., & Richard, J.-P. (2004). Robust sampled-data stabilization of linear systems—an input delay approach introduction. *Automatica, 40*, 1441–1446.
12. Goebel, R., Sanfelice, R. G., & Teel, A. R. (2009). Hybrid dynamical systems. *IEEE Control Syst. Mag., 29*(2), 28–93.
13. Goebel, R., Sanfelice, R. G., & Teel, A. R. (2012). *Hybrid Dynamical Systems: Modeling, Stability and Robustness*. Princeton, NJ: Princeton University Press.
14. Gomes da Silva Jr., J.M., Tarbouriech, S.: Antiwindup design with guaranteed regions of stability: an LMI-based approach. IEEE Trans. Autom. Control **50**(1) (2005)

15. Hetel, L., Fiter, C., Omran, H., Seuret, A., Fridman, E., Richard, J.-P., & Niculescu, S. I. (2016). Recent developments on the stability of systems with aperiodic sampling: an overview. *Automatica, 76*, 309–335.
16. Hu, T., & Lin, Z. (2001). *Control Systems with Actuator Saturation: Analysis and Design*. Boston, MA: Birkhauser.
17. Huff, D.D., Fiacchini, M., Gomes da Silva Jr., J.M.: Stability and stabilization of sampled-data systems subject to control input saturation: a set invariant approach. IEEE Trans. Autom. Control **67**(3), 1423–1429 (2022)
18. Liu, W., Chitour, Y., & Sontag, E. D. (1996). On finite-gain stability of linear systems subject to input saturation. *SIAM J. Control Optim., 34*(4), 1190–1219.
19. Palmeira, A.H.K., Gomes da Silva Jr., J.M., Tarbouriech, S., Ghiggi, I.M.F.: Sampled-data control under magnitude and rate saturating actuators. Int. J. Robust Nonlinear Control **26**(15), 3232–3252 (2016)
20. Prajna, S., Papachristodoulou, A., Valmorbida, G., Anderson, J., Seiler, P., Parrilo, P.: SOSTOOLS: sum of squares optimization toolbox for MATLAB. GNU General Public License (2004)
21. Seuret, A. (2012). A novel stability analysis of linear systems under asynchrounous samplings. *Automatica, 48*(1), 177–182.
22. Seuret, A., Gomes da Silva Jr., J.M.: Taking into account period variations and actuator saturation in sampled-data systems. Syst. Control Lett. **61**, 1286–1293 (2012)
23. Sussmann, H. J., Sontag, E. D., & Yang, Y. (1994). A general result on the stabilization of linear systems using bounded controls. *IEEE Trans. Autom. Control, 39*(12), 2411–2425.
24. Tarbouriech, S., Garcia, G., Gomes da Silva Jr., J.M., Queinnec, I.: Stability and Stabilization of Linear Systems with Saturating Actuators. Springer, Berlin (2011)
25. Zhang, W., Branicky, M. S., & Phillips, S. M. (2001). Stability of networked control systems. *IEEE Control Syst. Mag., 21*, 84–99.

Part III
Emerging Trends and Approaches for Analysis and Design

Chapter 10
Trends and Questions in Open Multi-agent Systems

Renato Vizuete, Charles Monnoyer de Galland, Paolo Frasca, Elena Panteley, and Julien M. Hendrickx

Abstract This chapter presents a survey of recent trends in the analysis of *open* multi-agent systems, where the set of agents is time-varying. We first introduce the notions of arrivals, departures, and replacements of agents in the context of multi-agent systems, including several approaches for the modeling and time evolution. We then provide alternative definitions for concepts that must be adapted to conduct analyses in the open context, such as stability and convergence. We also consider some aspects of open systems that must be taken into account in the design of algorithms. Finally, some applications are presented to illustrate the current importance of open multi-agent systems as well as future perspectives in this framework.

10.1 Introduction

An increasing number of real-world problems are being understood as constantly evolving towards a networked environment where single entities interact with each other to accomplish a collective task or achieve a common objective. Systems of this type include social networks, networks of devices such as sensors or computing enti-

---

R. Vizuete (✉) · C. Monnoyer de Galland · J. M. Hendrickx
ICTEAM Institute, UCLouvain, B-1348 Louvain-la-Neuve, Belgium
e-mail: renato.vizueteharo@uclouvain.be

C. Monnoyer de Galland
e-mail: charles.monnoyer@uclouvain.be

J. M. Hendrickx
e-mail: julien.hendrickx@uclouvain.be

P. Frasca
Univ. Grenoble Alpes, CNRS, Inria, Grenoble INP, GIPSA-lab, F-38000 Grenoble, France
e-mail: paolo.frasca@gipsa-lab.fr

E. Panteley
Université Paris-Saclay, CNRS, CentraleSupélec, Laboratoire des signaux et systèmes, 91190 Gif-sur-Yvette, France
e-mail: elena.panteley@l2s.centralesupelec.fr

R. Postoyan et al. (eds.), *Hybrid and Networked Dynamical Systems*,
Lecture Notes in Control and Information Sciences 493,
https://doi.org/10.1007/978-3-031-49555-7_10

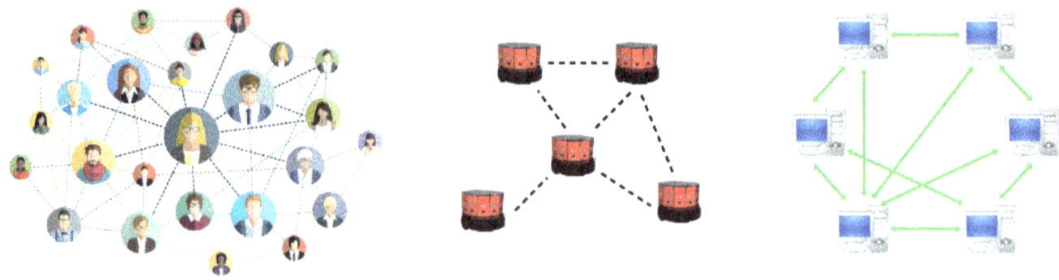

Fig. 10.1 Examples of settings which can be described as a multi-agent system (social network, group of robots, network of interconnected devices)

ties, or groups of vehicles or robots, for instance, (Fig. 10.1). They are called *multi-agent systems* (MAS) and have been extensively studied in the scientific community, considering different types of dynamics, coupling, goals, and time formulation (e.g., see the extended discussion in [1]).

Mathematically, the analysis of multi-agent systems is usually performed with a graph $\mathcal{G} = (\mathcal{V}, \mathcal{E})$ where $\mathcal{V}$ corresponds to the set of agents, and $\mathcal{E}$ represents the direct interactions between the agents (Fig. 10.2). In many scenarios, the links in $\mathcal{E}$ do not remain static and can change with time (so that $\mathcal{E}(t)$ is time-varying); many results for fixed topologies can however be extended to time-varying ones under the assumption that the network is *sufficiently connected* over time [2–4]. Yet, even if models based on time-varying graphs provide a good approach for studying the behavior of systems where communications among the agents do not remain static, they do not encompass possible changes in the *set of agents* $\mathcal{V}$ *itself.*

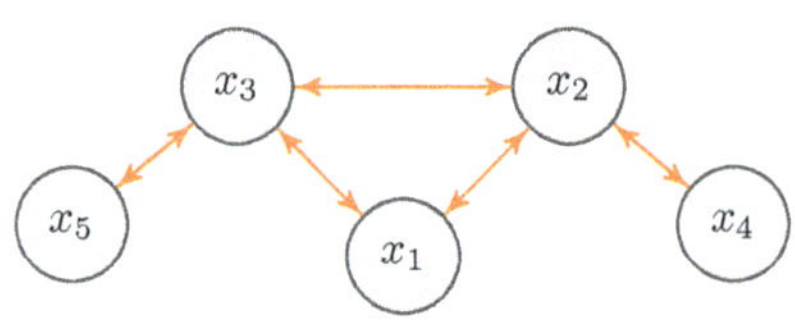

Fig. 10.2 Graph representation of a multi-agent system constituted of 5 agents, with $\mathcal{V} = \{1, \ldots, 5\}$, connected by 5 edges (in orange), where each agent $i \in \mathcal{V}$ holds a state x_i [5, Fig. 1.2]

From MAS to OMAS. While most results on multi-agent systems assume that the composition of the underlying network remains unchanged, it is indeed expected in the real world that agents join and leave the network during the process that is studied. This assumption is actually justified when arrivals and departures are rare enough as compared to the time scale of the process: the system is then expected to have the time to fully incorporate their effects before the next one occurs. Yet, it is now getting increasingly challenged with the emergence of systems reaching huge magnitudes (such as the "Internet of Things" [6] or devices in nanomedicine [7]), where the probability of agents joining and leaving grows with the system size and thus becomes significant. Similarly, systems naturally exposed to frequent arrivals

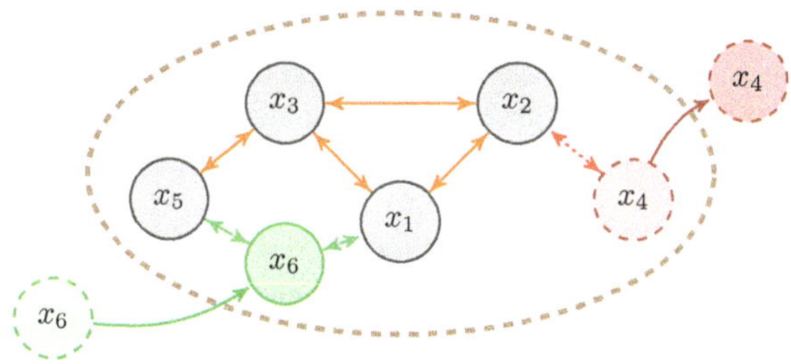

Fig. 10.3 Example of an open multi-agent system (OMAS) of initially 5 interconnected agents subject to the arrival of agent 6 (green) and the departure of agent 4 (red) [5, Fig. 1.3]

and departures such as multi-vehicles systems heading towards different directions [8] or social networks where users keep connecting and disconnecting [9] challenge it as well (the examples above notably include those of Fig. 10.1). Such systems subject to arrivals and departures of agents at a time scale similar to that of the dynamical system on the network are called an *open multi-agent systems* (OMAS). See Fig. 10.3 for an illustration.

Open Systems in Other Domains. The possibility for agents to join and leave such as in OMAS was already considered in other domains with different approaches. Typically, specific architectures such as THOMAS [10] were proposed in computer science to deploy large-scale open multi-agent systems. In the same line of work, mechanisms allowing distributed computation processes to cope with the failures of nodes or to take advantage of the arrival of new nodes also exist, such as algorithms designed to maintain network connectivity into P2P networks subject to arrivals and departures of nodes [11]. In addition, the *Plug and Play* implementation considered e.g., in [12–14] aims at designing structures robust to (un)plugging subsystems.

Openness was also empirically considered, e.g., with the model FIRE [15] in the context of trust and reputation computation to evaluate the reliability of joining agents. In the field of Game Theory, researchers have considered dynamic populations where players can exit and be replaced by new participants [16].

Finally, in the more general context of function computation, self-stabilization protocols [17, 18] consider agents undergoing temporary or permanent failures, which can be assimilated to departures. The objective of such protocols is to ensure asymptotic stability if arrivals and departures were to eventually stop at some point. There is however no guarantee about the transient performance of such protocols, for which they can even create important disruptions when arrivals and departures become frequent (as discussed e.g., in [19]).

Challenges in OMAS. Considering arrivals and departures has a significant impact on the analysis and design of algorithms around MAS, resulting in new challenges. Firstly, every arrival and departure results in an instantaneous change in both the system size and state, making the evolution of the system challenging to analyze on the one hand, and preventing classical asymptotic convergence on the other hand, so that algorithms in open systems cannot be expected to provide "exact" results. Moreover, these events amount to perturbations for the system, which can sometimes even modify the objective pursued by the agents. In particular, at each departure and depending on the nature of the problem, it may be necessary for algorithms to deal with either (i) outdated information related to agents that left the system and that is thus not relevant anymore and should be erased, or (ii) losses of information about agents that left and that should be remembered. Furthermore, preliminary studies have shown that algorithms designed for closed systems do not easily extend to open systems. In addition, (naive) correction mechanisms for handling few arrivals and departures, or guaranteeing convergence if they were to eventually stop, can be counterproductive if these events keep regularly occurring, see e.g., [19–21].

Hence, algorithms designed for OMAS must be robust to repeated arrivals and departures of agents, and able to cope with potentially variable objectives. Moreover, they generally cannot be simply extended from algorithms designed in closed systems. The conception of new algorithms tailored for open systems is therefore becoming necessary nowadays. As an example, being able to handle failures of connected basic devices such as servers for instance is more interesting economically than investing in sophisticated expensive devices that (almost) never fail, as failures are expected to ultimately happen nonetheless. Similarly, tailored representations and analysis techniques are necessary as well to model and study OMAS in general.

Outline. In this chapter, we provide an overview of some current trends and questions related to OMAS. After introducing some illustrative examples in Sect. 10.2, we present different ways of modeling OMAS both in terms of arrivals and departures and of the way OMAS evolves with time in Sect. 10.3. Section 10.4 then highlights how some classical tools for MAS are no longer appropriate for OMAS and presents existing alternatives for these. In Sect. 10.5, we focus on the definition of different objectives over OMAS, and the fundamental performance limitations that can be used for the design of algorithms. Finally, we go over some applications which are currently being studied under the scope of OMAS in Sect. 10.6, as well as possible perspectives for the future of this rather young framework in Sect. 10.7.

10.2 Illustrative Examples

We define here some standard problems in OMAS; we will refer to these throughout this chapter as illustrative examples in order to introduce and explain general concepts around OMAS.

10.2.1 Consensus

Consensus is a well-known basic problem in the context of multi-agent systems, which consists of making all the agents in the system agree on a specific state or quantity based on local interactions. More formally, in a closed system, let $\mathcal{V}$ denote the set of agents in the system, and let $x_i(t) \in E$ denote the time-varying state held

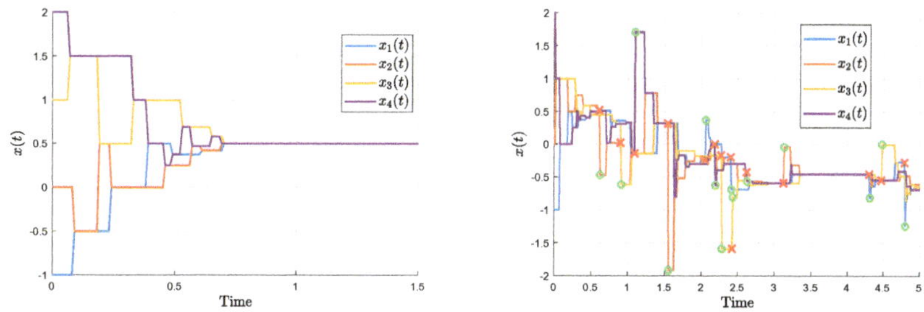

Fig. 10.4 Realization of the Gossip algorithm (see [25] for details) in a system of 4 fully connected agents, where each line is the estimate of one agent. The system is closed on the left and achieves consensus on the average of the values held by the agents. On the right, a replacement happens on average once every ten interactions (represented by the red cross and green circle, respectively pointing out the value of the replaced agent before and after being replaced), preventing convergence

by agent $i \in \mathcal{V}$ at time t for some state space E (typically $\mathbb{R}^d$ for some d). Then, *consensus* is said to be reached on some $y \in E$ if for all $i \in \mathcal{V}$, we have

$$\lim_{t \to \infty} \|x_i(t) - y\| = 0. \tag{10.1}$$

Let $z_i \in E$ denote the initial state held by $i \in \mathcal{V}$ (such that $x_i(0) = z_i$). The *agreement state* y sometimes consists of a specific function of the initial states, i.e., $y = f(\{z_i \in \mathcal{V}\})$, such as the average, maximal or median value.

Consensus problems typically appear in applications such as decentralized optimization where consensus and optimization steps typically alternate [22]; in vehicle coordination where, e.g., consensus must be reached on the distance between vehicles [23], or in sociology [24]. An example of a system reaching consensus on the average of the values initially held by the agents is illustrated in Fig. 10.4 (left).

In open systems, the size and state of the system regularly suffer variations, so that tracking the system state, and therefore, the analysis of algorithms, become challenging. Moreover, the permanent disruptions in the system prevent convergence and stability to even make sense, as each perturbation potentially drives the system state away from the agreement state, which can be impacted as well. This is illustrated in Fig. 10.4 (right), which depicts exactly the same setting as for the closed system (left), where one agent gets replaced once every ten interactions on average.

10.2.2 Opinion Dynamics

In many social interactions, the group of individuals does not remain constant since new agents may join the system and others leave. This happens especially in interactions taking place over online platforms (*Facebook*, *Twitter*, etc.), where agents can connect and disconnect in an easy manner (see Fig. 10.5). The importance of

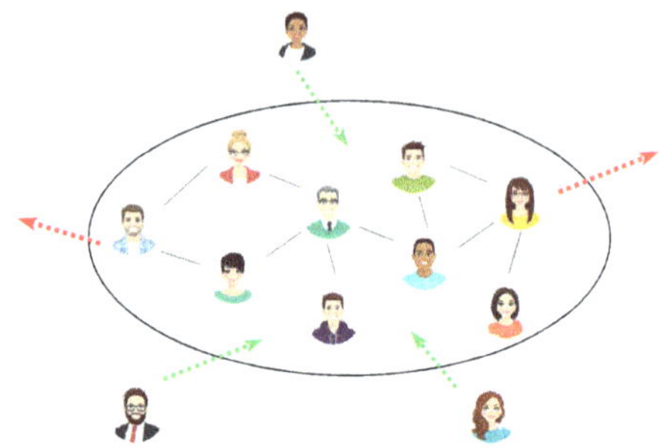

Fig. 10.5 A dynamic social network where new agents join the discussions and other agents stop interacting

these phenomena has been highlighted in [26, 27] as an important feature that must be taken into account in the analysis of dynamic social interactions.

Preliminary works have considered the analysis of an open Hegselmann–Krause (HK) model [28] where for a population of n agents indexed in a set $\mathcal{I} = \{1, \ldots, n\}$, each agent interacts through:

$$\dot{x}_i(t) = \sum_{j:|x_i(t)-x_j(t)|<1} (x_j(t) - x_i(t)), \quad \text{for all } i \in \mathcal{I}. \tag{10.2}$$

For this dynamics, the graph topology is automatically defined by the states of the agents, which overcomes the problem of the definition of connections of new agents. In [24, 29], the authors only studied the particular case of replacements where the opinion of agents change abruptly to mimic dynamic groups [24, 29]. In [30], some Lyapunov functions are proposed for the analysis of an open HK dynamics subject to decoupled arrivals and departures. Regarding stability, the authors in [31] have considered the addition of new agents only as perturbations to analyze the stability of existing clusters (connected components).

10.2.3 Epidemics

One of the most popular models in the spread of epidemics is the SIS (Susceptible-Infected-Susceptible) model, which is used for diseases where the agents, after being recovered, can become infected again. The networked version of the SIS model is given by [32]:

$$\dot{x}_i(t) = -\delta x_i(t) + \sum_{j=1}^{n} a_{ij}\beta x_j(t)\,(1 - x_i(t))\,, \tag{10.3}$$

where $x_i(t) \in [0, 1]$ is the probability of infection of an individual at time t, δ is the recovery rate, β is the infection rate, $a_{ij} = 1$ if there is an interaction between agents i and j, and $a_{ij} = 0$ otherwise. Most of the stability results derived for the SIS model

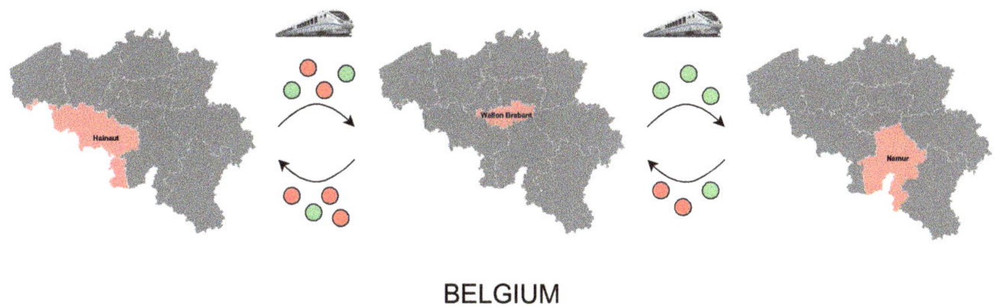

Fig. 10.6 Due to modern transportation infrastructure, susceptible individuals (green) and infected individuals (red) are constantly moving between different regions during an epidemic, modifying the evolution of the disease

have considered an ideal scenario where the composition of the system is fixed [32]. Nevertheless, in more realistic environments, individuals do not stay in a fixed place for a long time and are constantly moving between different locations. It is clear that mobility is one of the most important factors that need to be considered in the analysis of modern spreads of diseases [33], and many works begin to consider this aspect to obtain more accurate results [34, 35]. Traditionally, time-varying networks have been used to model to some degree the mobility of agents by changing the interactions among them [36, 37]. However, this type of model might fail to represent modern networks where due to the development of transportation infrastructures, agents are constantly entering and leaving towns, cities and countries, which implies not only changes of the connections a_{ij}, but also modifications of the infection probabilities x_i associated with the agents (see Fig. 10.6).

At the level of large populations of individuals (e.g., countries, cities), it is common to have similar rates of arrivals and departures such that the variations of the size of the populations are almost negligible. However, even if the total number of individuals in a specific place could remain almost invariant, it is unrealistic to assume that exactly the same agents stay in a fixed location during the evolution of the epidemic. If we consider a similar rate of departures and arrivals, which preserve the number of individuals in a specific place, epidemics under these conditions can be modeled as a dynamical system subject to replacements of individuals, where the infection probability of an agent i is modified during a replacement event [38]. Nevertheless, depending on the size of the populations and the time scale of the analysis, the introduction of decoupled arrivals and departures might be necessary in some scenarios, to model the influence of external individuals on the evolution of the disease in the population.

10.3 Modeling OMAS

This section is dedicated to the way arrivals and departures can be modelled, and how they can be incorporated to study the time evolution of OMAS.

10.3.1 Modeling Arrivals and Departures

By definition, OMAS are characterized by the possibility for agents to join and leave the system, so that their composition varies with time. In addition, we can also consider *replacements*, which consist of the simultaneous occurrence of a departure and an arrival. The way these three types of events are modelled and incorporated in the standard framework of multi-agent systems constitute the first step toward the definition of the theory around OMAS, and can already have major implications on their analysis. Different ways to model arrivals and departures in OMAS can therefore be used depending on the type of application and on the information in the system; we present three of the most standard ones below.

10.3.1.1 Time-Invariant Finite Superset

Let us consider a network of devices such as sensors. In such system, it can happen that some of the devices are unavailable, out of reach, or suffering from a failure at some time, so that only a portion of the devices is actually active in the system at that time. This means that the number of devices that can join and leave the network is finite, and that a disconnecting device is expected to eventually come back.

The first model we present is based on the idea described above, i.e., that the OMAS is composed by finitely many agents (a *finite superset*) and that, at any given time, only a subset of that superset is in the system, while the others are out of it [39, 40]. With this model, the openness is characterized by the activation and deactivation of the agents in the superset, meaning that, at each time instant, a different number of agents is effectively present in the system. It moreover means that the agents joining and leaving the system are always the same (i.e., no agents ever fully disappears nor is completely new). In that sense, the finite superset model constitutes an under-approximation of OMAS. This model is represented in Fig. 10.7.

This finite superset setting admits a convenient mathematical representation: we can associate to the state vector $x(t) \in \mathbb{R}^n$ an activation vector $a(t) \in \{0, 1\}^n$ such that $a_i(t) = 1$ indicates that agent i is active at t. One can then formulate the dynamics of the system as:

$$\dot{x}(t) = f\big(x(t), a(t)\big). \tag{10.4}$$

Since the dimension of the global system (i.e., the superset) is constant with this model, standard tools from dynamical systems can be used for the analysis. However, this approach is restricted to systems where a full knowledge of all the possible agents is available such that unexpected agents cannot join the network. For this reason, this type of systems can be considered only as *pseudo open multi-agent systems*.

10.3.1.2 Time-Varying Composition

By contrast with the previous model, in some applications, it is not possible to know in advance all the agents that can be in the system at any time. Consider, for instance, the case of opinion dynamics in social networks: Since most modern social interactions take place in online environments, not only is it hard to know *a priori* the number of agents that could take part in a discussion, but it also cannot be expected that the agents leaving and joining the network remain the same, i.e., new users might join in the process and users might leave forever. Models based on a superset are therefore not suited for such systems, and a different type of modeling is needed.

Hence, the second model we present and which we call "*Time-Varying Composition*" relies on the following assumptions:

(i) every joining agent is new, i.e., they have never been in the system before;
(ii) every departure is permanent, i.e., a leaving agent never comes back,

so that it is not limited in the number of agents that can join and leave the system, and its dimension possibly changes. We illustrate it in Fig. 10.8.

This formulation implies that the information held by leaving agents is poised to never come back and might be lost or become irrelevant forever, and that an agent joining the system supposedly has no information about it at its arrival. It also means that the way a joining agent connects to the system is determined at its arrival, and that the edges connecting a leaving agent disappear at its departure. Observe that, as opposed to the superset formulation, this setting is less accessible to classical tools, and depending on the properties of the system, one might need to introduce new tools tailored for this type of systems. Some of them are detailed below.

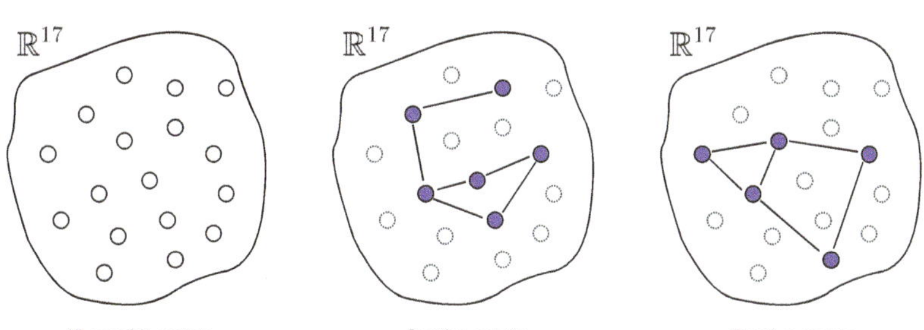

Fig. 10.7 Time-invariant finite superset of agents characterized by different active agents [38, Fig. 1.2]

Fixed State-Space. When an OMAS is subject only to replacements, the dimension of the system does not change with time. In this particular setting, an agent leaving the network is immediately replaced by a joining agent, which can be assumed to retrieve its label so that the state of the system remains in the same state space all the time, even though replacements happen.

In this specific case, trajectories are well-defined and standard concepts of dynamical systems can be applied, but jumps may appear in the trajectories due to the replacements. It proves handy for first studies or to analyze the impact of openness without adding the complexity of time-varying size. However, it is restricted to replacements only and analyses conducted with this assumption often cannot be straightforwardly adapted to the case of decoupled arrivals and departures.

This type of model has been mainly applied in the field of distributed optimization where due to the complexity of the local cost functions, the occurrence of replacements can have a serious impact on the stability and variation of the minimizer [41].

Switching Systems. In the more general case, when the OMAS is subject to decoupled arrivals and departures, switching systems between spaces of different dimensions can be a useful tool for the modeling. Preliminary works in this field can be found in [42, 43], where the switching between a fixed number of spaces of different dimensions is studied. The family of possible vector spaces to which the state belongs is defined as a discrete vector bundle and concepts of Topology are required for the analysis. For this framework, the concept of trajectories is meaningful and a global equilibrium for the system can be defined, allowing the use of some tools of hybrid systems [44]. However, the classical concept of stability could be difficult to apply, and the approach can be rigid for many applications since it is required to know all the possible vector spaces to define the discrete vector bundle.

The switching between spaces of different dimensions can notably be studied by using the well-known theory of hybrid dynamical systems. We consider a discrete vector bundle, composed by a family of vector spaces (fibers) $\mathbb{R}^{n_q}$ for $q \in Q$. The vector $x^{(q)}(t) \in \mathbb{R}^{n_q}$ is the partial state at t and $q(t) \in Q$ is the hybrid state which determines the mode [43]:

$$\mathbf{x} = (q; x^{(1)}, x^{(2)}, \ldots, x^{(N)}) \in \mathbf{X} = Q \times (\mathbb{R}^{n_1} \oplus \cdots \oplus \mathbb{R}^{n_N}).$$

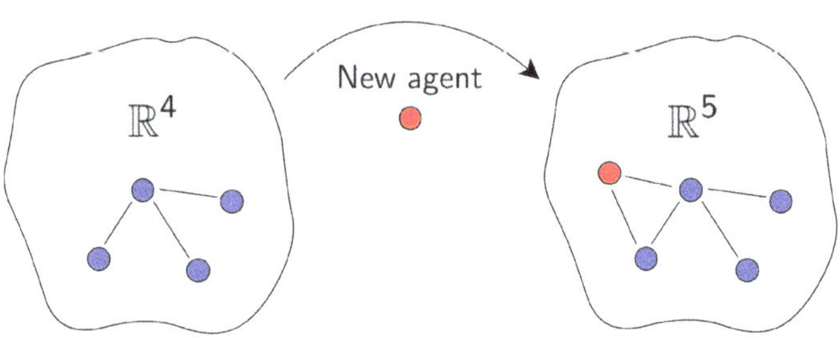

Fig. 10.8 Time-varying composition representation of an OMAS [38, Fig. 1.3]

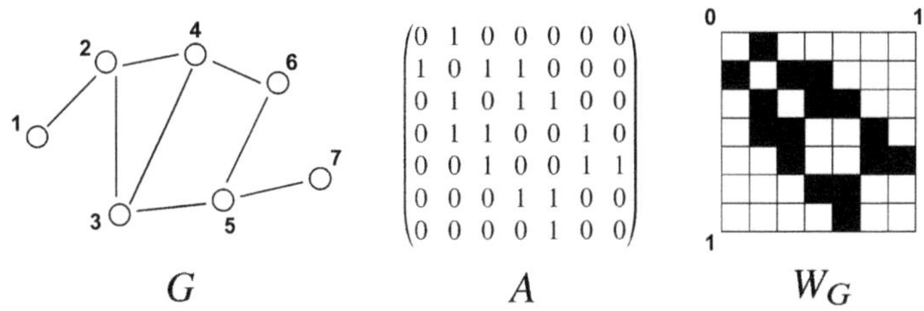

Fig. 10.9 Graph G, adjacency matrix A and step graphon W_G

When $q = i$, only the n_i dimensional state component x_i is relevant.

10.3.1.3 Infinite-Dimensional Dynamical System

Infinite-dimensional dynamical systems may be used as an over-approximation of OMAS and encompass the two previous approaches. In this case, the problem can be formulated as a dynamical system in a space of measurable functions by using a graphon [45]. Graphons can be considered as a continuous version of the adjacency matrix of a graph as we can see in Fig. 10.9.

In addition, graphons can also be used to generate sampled graphs of a desired size through sampling procedures [46]. In the case of OMAS, the connections of new agents should follow the graphon structure (see Fig. 10.10), and the dynamics can be analyzed over a graphon differential equation of the form [47]:

$$\dot{x}(t) = T_W x(t),$$

where $x(0) \in L^2[0, 1]$ and $T_W : L^2[0, 1] \to L^2[0, 1]$ is an integral operator associated with a graphon $W : [0, 1] \to [0, 1]$.

The objective would be to solve the problem in infinite dimensions and then use approximation techniques to obtain the solution to the case of OMAS with a certain margin of approximation. The main advantages of this approach is that all the possible agents can be considered in the analysis of the system and advanced tools from infinite-dimensional systems can be used. However, the definition of appropriate mappings from the finite network to the infinite system and vice versa could make it difficult to use this approach.

Due to the close relationship with the adjacency matrix, graphons have been especially used in the analysis of SIS epidemics given by (10.3). The formulation of an infinite SIS epidemic through a graphon has been formulated in [48] and a preliminary application to the particular case of replacements has been provided in [38]. Nevertheless, the extension to decoupled arrivals and departures is still under development.

Alternatively, the authors in [49] have proposed the spaces of sequences ℓ^p as a potential tool to analyze OMAS. In this case, the topology of the network is defined

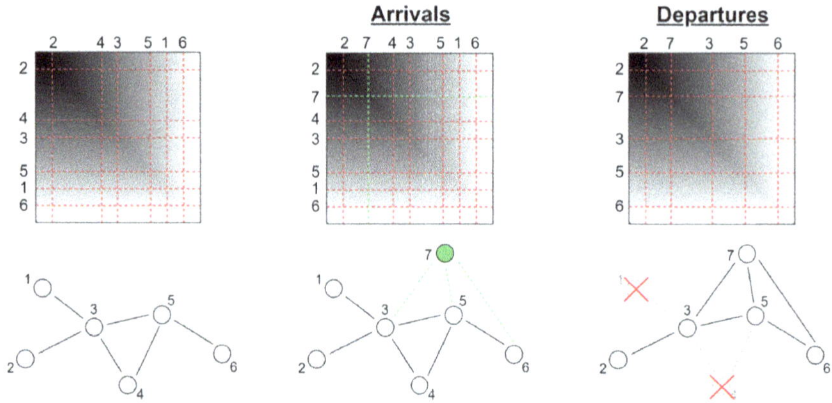

Fig. 10.10 Graphons can be used to analyze open multi-agent systems in an infinite-dimensional setting

by an infinite matrix and vectors are infinite sequences such that all possible agents in OMAS can be taking into account through an infinite vector. The dynamics can be expressed as an ODE in Banach spaces of infinite sequences of the form [49]:

$$\dot{x}(t) = f(x(t)),$$

where $x(t) \in \ell^p(\mathbb{N})$. Notice that this formulation in terms of spaces of sequences could be considered as a more complex version of the time-invariant model Sect. 10.3.1.1 when the superset is infinite.

10.3.2 Initial State of New Agents

Another important aspect in the modeling of OMAS is the determination of the initial state of new agents, which depends on the applications. It is clear that new agents cannot join the network with arbitrary states since this could lead to instabilities in the system. Therefore, in simple cases, we assume some homogeneity in the pool of potential new agents. This type of restrictions could be enough in OMAS where it is assumed that a leaving agent cannot join the network again. However, in other types of applications where several agents are constantly entering and leaving the system, several problems can appear regarding the identification of the state when a re-arrival takes place. Depending on the type of application, the state of an agent that joins the system could be the same as that it had when leaving it before (e.g., the position of a damaged robot) or different (e.g., the opinion of an individual). The latter could have a serious impact, especially in pseudo open multi-agent systems where a model based on time-invariant finite superset is used for the analysis. In this case, it would be necessary to track the behavior of active and inactive agents, which limits the advantages of the model. However, even in general open multi-agent systems, the change of the state of an agent between a departure and a re-arrival can have an impact since several control and optimization algorithms rely on past information and outdated data could be detrimental for the efficiency and even the stability.

10.3.3 Modeling Time Evolution

The same way as in usual multi-agent systems, the agents constituting an OMAS evolve according to their own dynamics, which we call the *agent dynamics*. By contrast however, OMAS are in addition characterized by a second dynamics which defines the occurrence of arrivals, departures and replacements, which we call the *network dynamics* [50, 51]. For instance, in consensus problems, the agent dynamics would correspond to the interactions between agents, whereas the network dynamics would define when an agent joins or leaves the system; an algorithm would thus exploit the former to achieve consensus while dealing with the trouble brought by the latter. Observe that no specific assumption is made on the network dynamics, so that departures could, for instance, be endogenous.

These two dynamics, which can, in some cases, depend on each other, can moreover have different time scales and natures and impact the system in different ways.

As previously established, when these two dynamics have similar time scales, they must be considered simultaneously, and properly defining how they impact the system and how they influence each other becomes crucial. In this section, we show how the evolution of OMAS with the time can be modelled with respect to these dynamics.

10.3.3.1 Event-Based Formulation for Arrivals and Departures

A perfect identification of the time instants at which arrivals, departures, and replacements take place does not seem plausible in a real application, especially in large systems. In addition, it is possible that some results that are valid for a specific series of events, could be non robust to changes. While the use of dwell times can guarantee some lower bounds on the time interval between the occurrence of two distinct events, the use of stochastic processes could model richer behaviors of OMAS that may be closer to real applications. In this case, arrivals, departures, and replacements actually amount to punctual *events* happening at random times and triggering instantaneous modifications of the system size and state; and as far as the network dynamics alone is concerned, the system remains unchanged in between. As such, one way to characterize the evolution of the system with respect to these is to consider an *event-based* formulation, which we describe below.

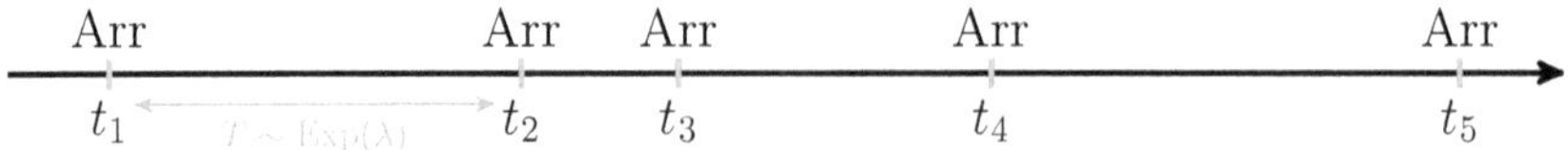

Fig. 10.11 Timeline describing the time instants at which modifications (i.e., arrivals denoted Arr) happen in the only-growing system described above. The time T separating the two consecutive arrivals (at times t_1 and t_2 in the figure) is randomly drawn from an exponential distribution with parameter λ, following the properties of the Poisson clock defining their occurrences [5, Fig. 2.10]

Example 10.3.1 (Only-Growing System) To introduce and illustrate the event-based formulation described in this section, let us first focus on the specific and simple case of only-growing systems, i.e., where only arrivals of single agents happen at random times determined with a Poisson clock of fixed rate λ. Neither departure nor any other type of interaction happens in that system. Its evolution is solely defined by the activation at specific random time instants t_k of the Poisson clock characterizing the arrivals, as illustrated in Fig. 10.11.

The evolution of the system described above can therefore be studied using a Continuous-Time Markov Chain (CTMC), whose jumps are triggered by the occurrences of an arrival in the system, which are themselves defined by a Poisson clock of parameter λ. Here, arrivals are, therefore, characterized by a counting process which triggers the evolution of the system.

General Description. In general OMAS, other processes can be involved in parallel of arrivals, such as departures and replacements. In our event-based formulation, each of these processes is given its own clock (i.e., its counting process), which determines its occurrence, and which can depend on that of the other processes. The state of the system is, therefore, a CTMC whose jumps are triggered by the activation of these different clocks.

For the simplicity of the interpretation, and with a slight abuse of language, we can refer to the activation of a given clock ϵ as an "*event of* ϵ". This terminology can actually be assimilated to the notion of "probabilistic event" from probability theory, as ϵ defines the type of modification experienced by the system when a clock ticks, which can be seen as the outcome of an experiment.

The Poisson process is certainly the most standard counting process that is used to characterize the occurrences of the events. With this process, arrivals, departures, and replacements are memoryless and independent of each other, and have numerous useful properties for analyses. This is, however, not the only possibility, and more elaborate processes could be considered.

History of an OMAS. With the *event-based* formulation described above, one can define the *event set* of the corresponding OMAS as the set containing all the existing clocks which trigger modifications of the system state. The time evolution of an OMAS is then characterized by the different events, i.e., by the activation at random times of the different clocks in the event set of the OMAS, as illustrated in Fig. 10.12.

Fig. 10.12 Timeline describing the time instants at which a modification of an OMAS happen. Each time t_i corresponds to the time instant at which an event of a clock ξ_i in the event set of the system happens, triggering its modification [5, Fig. 2.11]

Based on this, we can now build the *history* of the process up to some time t as the sequence of all the events that happened in the system and the time instant at which they happened until that time:

$$\mathcal{H}(t) = \{(t_1, \xi_1), \ldots, (t_k, \xi_k)\}, \tag{10.5}$$

where $t_i \leq t$ and $\xi_i \in \Xi$ for all $i = 1, \ldots, k$.

Moreover, this formulation typically allows discretizing the time at the specific time instants where events take place (up to the assumption that two distinct events never happen simultaneously). In that case, one can then alternatively define the history of the process in discrete time up to the k-th time step as

$$\mathcal{H}^k := \{(1, \xi_1), \ldots, (k, \xi_k)\}, \tag{10.6}$$

where $\xi_i \in \Xi$ for all $i = 1, \ldots, k$. Under the additional assumption that the different processes (and hence the clocks) are independent of each other and of the system states (past or present), one can also define the probabilities for each clock to tick at a given time step, so that the effects of the different clocks can be decoupled.

The concepts and definitions introduced above can be useful for conducting analyses in settings where the effect of each type of event can be characterized and studied separately, such as in [50] in the context of consensus or in [41, 52] for decentralized optimization.

10.3.3.2 Hybrid Formulation for Arrivals and Departures

In some cases, the agent dynamics is discrete, such as in the context of, e.g., some discrete consensus algorithms which rely on instantaneous interactions happening at random times in the network, such as Gossip interactions [25]. In that case, the agent dynamics can directly be incorporated in the event-based formulation discussed above by considering the interactions in the network as an additional event which can be added to the event set, and studied alongside with arrivals, departures, and replacements. This was done, for instance, in [41], and allows among other to discretize the time at the occurrences of either arrivals, departures, and replacements, or of instances of the dynamics of the agents. This is, however, not so simple when one considers continuous-time agent dynamics like in the case of oscillators [53].

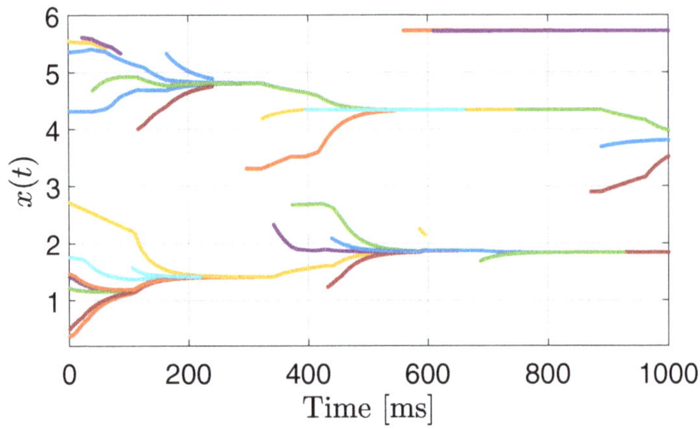

Fig. 10.13 Evolution of the opinions for an open HK dynamics introduced in Sect. 10.2.2

In this case, the state of the agents may change between the occurrence of events linked to the openness of the system, but the occurrence of one of these events does not necessarily affect the state of all the agents in the system. Nevertheless, at a global level, the evolution of the system in continuous time and the changes due to arrivals, departures and replacements become a stochastic hybrid system [54]. This behavior is more easily visualized when the OMAS is analyzed with the use of descriptors, which are introduced in Sect. 10.4.3.

10.4 Analysis of OMAS

In this section, we illustrate several phenomena that make the analysis of OMAS more challenging, and we introduce alternative mathematical tools that can be used for the study of this type of systems.

10.4.1 Definition of Trajectories

The concepts of state spaces and trajectories are not convenient in the analysis of OMAS due to the changes of dimension and complexity that require the use of non-standard mathematical spaces. For instance, during the removal of an agent, its associated trajectory is completely removed while when an agent joins the system, a new trajectory is created. Even if only replacements occur in the system, trajectories experiment jumps, which can drastically change the behavior of the current state.

For instance, let us consider the case of the HK dynamics introduced in Sect. 10.2.2, where the arrivals and departures are determined by homogeneous Poisson processes.

Figure 10.13 presents a realization of this dynamics where 19 new agents join the system and 20 departures take place (see [30] for details). We can observe the time instants of arrivals where new trajectories appear and the time instants of departures corresponding to the disapeareance of current trajectories. In addition, notice how arrivals and departures affect the creation and merging of clusters[1] where initially there was only two clusters and at the end and additional cluster was generated.

10.4.2 Stability and Convergence

Since an OMAS is constantly being perturbed by the addition and removal of agents, several classical definitions of stability and convergence might not be well suited for this type of systems. Alternative concepts like practical stability have been used in some works in OMAS, where certain level of error is tolerated in the case of consensus [55]. However, the time-varying number of agents could lead to new definitions of distance, stability and convergence, that exhibit interesting characteristics and are more appropriate for OMAS. One of the first proposal is provided in [51] where a stability radius for open system is introduced, which depends on the open distance $d_O : \mathbb{R}^{\mathcal{V}_1} \times \mathbb{R}^{\mathcal{V}_2} \to \mathbb{R}_{\geq 0}$:

$$d_O(x, y) = \sqrt{\sum_{v \in \mathcal{V}_1 \cap \mathcal{V}_2} (x_v - y_v)^2 + \sum_{v \in \mathcal{V}_1 \setminus \mathcal{V}_2} (x_v)^2 + \sum_{v \in \mathcal{V}_2 \setminus \mathcal{V}_1} (y_v)^2},$$

where $\mathcal{V}_1$ and $\mathcal{V}_2$ are two finite set of agents, $x \in \mathbb{R}^{\mathcal{V}_1}$ and $y \in \mathbb{R}^{\mathcal{V}_2}$. However, this open distance is only one possibility and other formulations could lead to different results. For instance, an alternative distance could involve only the states of agents in $\mathcal{V}_1 \cap \mathcal{V}_2$ or give more priority to these agents.

In addition to previous works, in this chapter we would like to introduce a normalized norm inspired by the concept of consensus that could be useful for OMAS. For each ℓ_p-norm we consider the following normalization:

$$|||x(t)|||_p := \frac{\|x(t)\|_p}{\|\mathbb{1}_{n(t)}\|_p} = \frac{\|x(t)\|_p}{\sqrt[p]{n(t)}},$$

where $\mathbb{1}_{n(t)} \in \mathbb{R}^{n(t)}$ is a vector of ones.

The choice of this normalization factor[2] is justified by the fact that if we analyze the value of the normalized norm for a vector x of dimension n with the same component x_i (i.e., consensus), it would be more appropriate to have the same value for the normalized norm independent of the choice of the norm. Notice that standard

---

[1] Connected components of the graph associated with the HK dynamics.

[2] This normalized norm is equivalent to the *generalized mean* defined for sets of numbers.

stability definitions can be formulated in terms of this normalized norm, which could lead to alternative approaches for the analysis of OMAS.

10.4.3 Scale-Independent Quantities

As discussed above, the varying size and abrupt changes in the state of the system make OMAS difficult to analyze, in particular because directly tracking the system state becomes infeasible. As an alternative, the concept of *descriptors* was introduced in [21, 56] and extended in [50] as an alternative to handle these challenges. These descriptors are essentially scale-independent quantities, i.e., which do not scale with the system size, and which characterize its state or some properties which one aims to study. The goal is then to analyze the evolution of these quantities instead of that of the system directly. Of course, the choice of the descriptors is central and must be appropriate regarding the nature of the analysis that is conducted. Furthermore, since in stochastic settings, the system could be continuously perturbed by arrivals, departures and replacements, an analysis of the statistical properties of the descriptors like the expectation, could be more appropriate for OMAS in certain applications.

Discrete Time. The case of agents evolving in discrete time has been considered e.g., in [50] where the disagreement between the agents subject to pairwise Gossip interactions is studied. In order to perform the analysis, the *squared mean* $\bar{x}^2(t)$ and the *mean of squares* $\overline{x^2}(t)$ are chosen as descriptors. Their characterization then allows that of the Variance $\text{Var}(x(t)) := \overline{x^2}(t) - \bar{x}^2(t)$, which characterizes the level of disagreement in the system.

In their setting, the authors of [50] show that the evolution of the descriptors can be characterized with an affine dynamical system by using standard tools from Markov Theory. Eventually, they characterize exactly the evolution of the expected disagreement in systems of fixed size (see Fig. 10.14), and obtain upper bounds on the asymptotic expected disagreement for systems of varying size.

Continuous Time. When the dynamics of the agents evolve in continuous time and the system is analyzed with the use of a descriptor $V(x(t))$ (e.g., Lyapunov functions), it is common that the evolution of $V(x(t))$ is determined by a stochastic differential equation (SDE) of the form:

$$dV(x(t)) = f(V(x(t)))dt + g_a(V(x(t)))dN_a + g_d(V(x(t)))dN_d + g_r(V(x(t)))dN_r, \tag{10.7}$$

where x is the state vector, f determines the evolution of $V(x(t))$ in continuous time, g_a, g_d and g_r determine the changes of $V(x(t))$ during arrivals, departures and replacements respectively, and N_a, N_d and N_r are Poisson processes. The SDE (10.7) is also known as a *Poisson Counter driven Stochastic Differential Equation* [57–59] and the solution is a cadlag function associated with a Piecewise-deterministic Markov Process (PDMP) and can also be studied using the theory of PDMP [60].

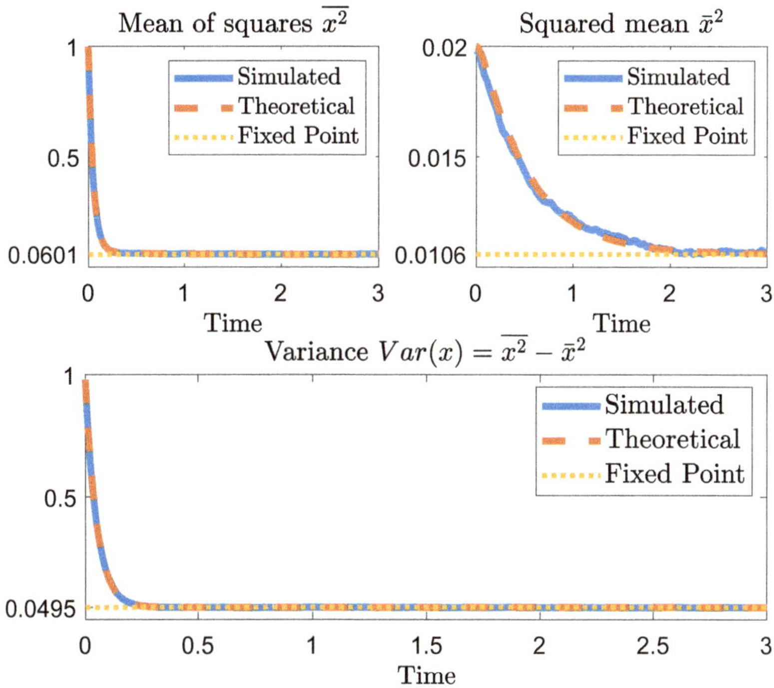

Fig. 10.14 Comparison of the simulated (plain blue) and theoretical (dashed red) expected evolution of the mean of squares (top left), squared mean (top right) and variance (bottom) of a system of 50 agents subject to pairwise Gossip interactions and of replacements on average once every 20 interactions. See [50] for details

Even if the solution of (10.7) is unique, every realization of the stochastic process can be completely different and an analysis of the asymptotic behavior is not appropriate because the limit $\lim_{t\to\infty} V(x(t))$ could not exist due to the ongoing occurrence of arrivals, departures and replacements. For this reason, an analysis of the expectation of the descriptor $\mathbb{E}[V(x(t))]$, seems more suitable in continuous time. The advantage of this type of formulation is that further moments (e.g., variance) of the descriptor $V(x(t))$ can be studied by the application of the Itô lemma in (10.7), which allows us to obtain further information about the OMAS.

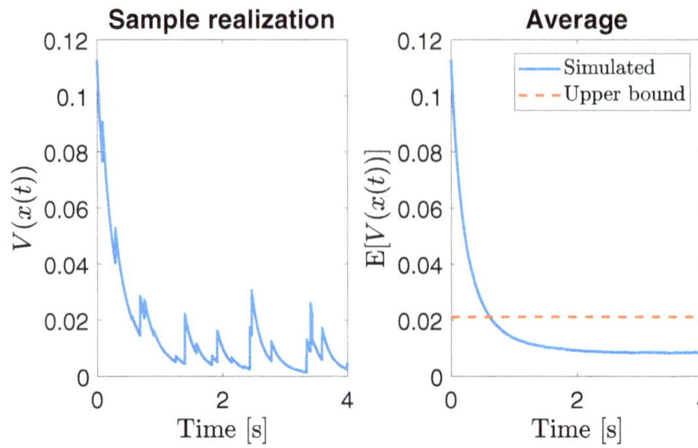

Fig. 10.15 Evolution of the descriptor $V(x(t))$ defined in (10.8) for a SIS epidemic subject to replacements. See [38] for details

This approach has been used in [38] for the analysis of the SIS epidemic model presented in Sect. 10.2.3 under the particular case of replacements. In this case, the descriptor used for the analysis is the Lyapunov function $V : [0, 1]^n \to \mathbb{R}_{\geq 0}$:

$$V(x(t)) = \frac{1}{n} \|x(t)\|^2, \tag{10.8}$$

where $\|\cdot\|$ is the Euclidean norm. The replacement instants are determined by a homogeneous Poisson process and the new infection value of the replaced agent is taken from a continuous distribution. Figure 10.15 presents the evolution of $V(x(t))$ for the SIS epidemic subject to replacement where in the left plot, we can observe the evolution in continuous time of $V(x(t))$ and the jumps during the replacements. The upper bound in the right plot has been obtained by using a SDE of the form (10.7) and allow us to estimate how far we are from the stability point (i.e., no infection) due to the replacements.

10.5 Towards Algorithm Design in OMAS

We now turn to several aspects which must be taken into account in the design of algorithms in OMAS, and which are direct consequences of how openness directly impacts the definition and resolution of problems in that context.

10.5.1 Different Types of Objectives in OMAS

When designing algorithms in OMAS, it is important to take into account the *nature* of the objective of the problem it is trying to solve, as it can have a significant impact on the behavior of algorithms. We distinguish two main types of objectives in OMAS: *intrinsic* and *external*, which we describe below.

To illustrate this distinction, we consider as an example the *max-consensus problem*, which is a specific instance of the consensus problem defined in Sect. 10.2.1 where agents aim at agreeing on the maximal value in the system, i.e., (with the notation of Sect. 10.2.1):

$$y = \max \{z_i \mid i \in \mathcal{V}\}. \tag{10.9}$$

In closed system this problem is therefore clearly defined. In OMAS however, as we describe below, it is necessary to properly characterize how arrivals and departures impact this objective, yielding different versions of the max-consensus problem.

10.5.1.1 Intrinsic and External Objectives

A first way to define objectives in OMAS is to consider they directly rely on the agents *presently in the system at that time*. Such objectives, which we call *intrinsic*, thus vary as agents join and leave, and therefore require algorithms to take it into account to respectively incorporate and eliminate the information related to agents joining and leaving the system.

An *intrinsic* version of the max-consensus in OMAS would read the following, where $\mathcal{V}(t)$ is the set of agents presently in the system at time t:

$$y(t) := \max \{z_i \mid i \in \mathcal{V}(t)\}. \tag{10.10}$$

This definition implies that the objective $y(t)$ changes as agents join and leave the system, so that agents trying to solve this problem must adapt consequently. Namely:

- if the agent holding the largest value leaves, then the other agents must potentially eliminate (or "forget") its value, as the current maximum has changed;
- if a joining agent holds the new largest value, then the other agents must learn it.

This is illustrated in Fig. 10.16.

The second type of objectives, which contrast with the previous one, is called *external*. This type of objective is defined *outside* the system, independently of its composition. The state held by an agent joining the system typically depends on the current value of the objective; it can thus be exploited to solve the problem, and remembering the state af agents that left the system can become crucial.

In the context of the max-consensus problem in OMAS, let us assume that at all time t each agent $i \in \mathcal{V}(t)$ holds a nonnegative value $z_i \in \mathbb{R}_+$ satisfying $z_i \leq M$ for some $M \in \mathbb{R}_+$. Then an *external* version of the max-consensus objective in OMAS could read

$$y(t) := M. \tag{10.11}$$

With this definition, the objective is not impacted by arrivals and departures of agents. Instead, it contributes to defining the value they hold when they join the system. This is illustrated in Fig. 10.17.

In order to estimate M, one strategy for the agents is to collect as many information about all the agents that have been in the system since the beginning of the process, so that they can compute the following approximation of the actual objective:

$$\tilde{y}(t) = \max \{z_i \mid i \in \mathcal{V}(s),\ \forall s \leq t\}. \tag{10.12}$$

10.5.1.2 Impact on Algorithms

The two types of objectives described above can have a significant impact on the efficiency of an algorithm. To illustrate that, let us consider the well-known *Flooding*

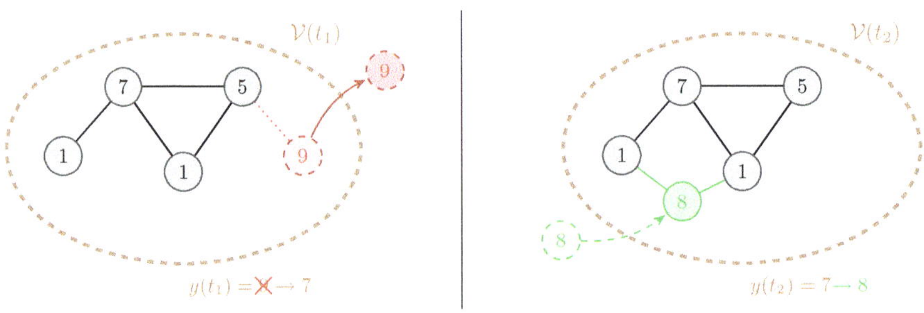

Fig. 10.16 Intrinsic max-consensus : the objective changes whenever the agent holding the largest value leaves the system (left) and whenever a new agent holding the new largest value joins (right) [5, Fig. 2.13]

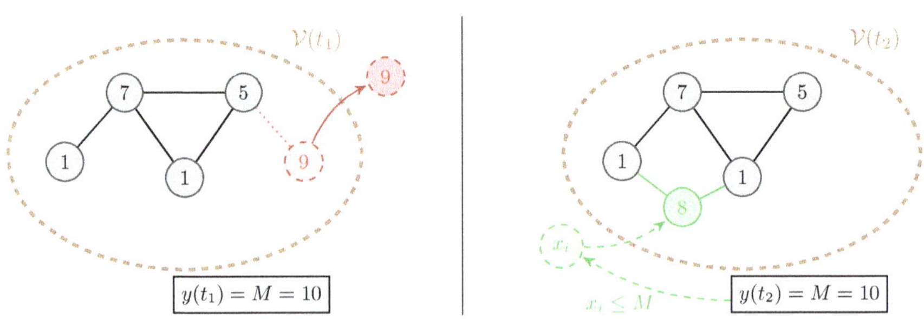

Fig. 10.17 External max-consensus: the objective is not impacted by departures (left) and arrivals (right) of agents, even if they hold the largest value in the system, but contributes to define that value when they join (right) [5, Fig. 2.14]

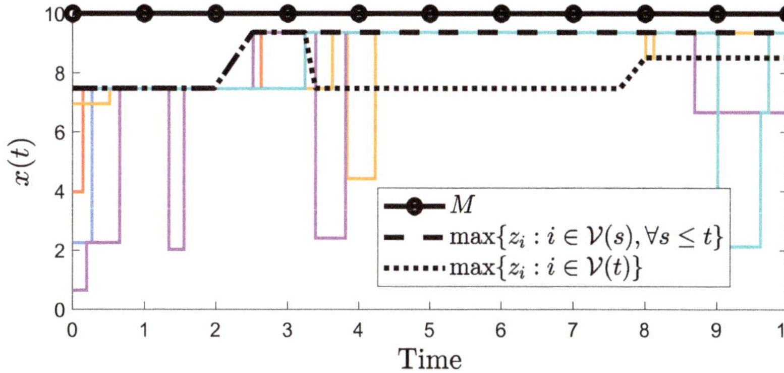

Fig. 10.18 Behavior of the Flooding algorithm described above with respect to the different variations of the max-consensus in OMAS discussed in this section, in a system of 5 agents subject to either pairwise interactions or replacements a single agents, the latter happening once every 5 occurrences of the former on average. Each plain thin line is the estimate held by an agent, and the black thick lines correspond to the different objectives defined above, i.e., the dotted one to (10.10), the circled one to (10.11) and the dashed one to (10.12) [5, Fig. 2.15]

algorithm, which is a simple strategy building on pairwise interactions, and which is known to solve the max-consensus problem in closed systems as long as the system is connected. In open system, it can be defined as follows:

- Initialization (at $t = 0$ or when an agent i joins the system): $x_i = z_i$;
- Upon interaction between agents i and j: $x_i^+ = x_j^+ = \max(x_i, x_j)$.

The behavior of this algorithm in our setting with respect to the different versions of the max-consensus objective discussed above is illustrated in Fig. 10.18.

It appears that, for the considered setting, the Flooding algorithm is not efficient to handle the intrinsic objective (10.10) as it is not designed to eliminate the outdated information corresponding to agents that already left the system. In opposition, and for the exact same reason, it behaves rather well with respect to the external objective (10.11), and in particular to compute the approximated external objective $\tilde{y}(t)$ from (10.12).

The distinction between *intrinsic* and *external* objectives discussed in this section already exists in the context of closed systems, yet its impact is rather limited. By contrast, as the example above clearly illustrates, it can significantly influence the efficiency of an algorithm in OMAS, and should definitely be taken into account during their design. In particular:

- Since an intrinsic objective depends on the composition of the system, the information related to agents having left the system is most of the time irrelevant and must consequently be eliminated. This can be a serious challenge for algorithm design, as it somehow requires keeping track of the origin and validity of the information held by the agents. This is rather clear from the example above, and strategies for

this problem can be found e.g., in [20] where information is given a life span and must be eliminated once it is exceeded in the context of max-consensus in OMAS.
- Since the states of the agents typically depend on an external objective, which is independent of the system composition, the information corresponding to any agent that has been or will be in the system is potentially relevant, and should be remembered. The challenge is then for algorithms to gather all the information without losing any, and for analyses to somehow quantify the information having been in the system at any time. The design of algorithms for external consensus problems of this type in OMAS was recently explored in a master thesis [61].

10.5.2 Fundamental Performance Limitations

As already discussed in Sect. 10.4, because of the ever-changing nature of OMAS, algorithms are poised to never achieve convergence. This is even more clear for intrinsic objectives (see Sect. 10.5.1), which rely on and therefore vary with the composition of the system. This observation has two main implications: (i) it means that there tends to always remain some residual error that algorithms cannot get rid of when working with OMAS, and (ii) it implies that convergence is not a suited criterion to evaluate the performance of algorithms in OMAS.

An alternative to evaluate the performance of algorithms in such settings is the derivation of *fundamental performance limitations*. For a given problem and setting in OMAS, and for some arbitrarily chosen performance metric, those take the form of lower bounds on the performance that can be achieved by any algorithm for that problem in that setting. These bounds then constitute a relevant quality criterion to measure the performance of algorithms for that problem in that setting.

The Intrinsic Average Consensus Case. Fundamental performance limitations are typically obtained by analyzing a conceptual algorithm which achieves optimal performance in a setting that is more favorable than that which is actually considered for the problem. As an illustration, let us consider the intrinsic average consensus problem, which we remind is an instance of the consensus problem defined in Sect. 10.2.1 where the agreement state is the average of the values held by the agents in the system at that time.

In [62], performance limitations were obtained for that problem in OMAS of fixed size. Since their derivation requires the characterization and analysis of an optimal algorithm, an improved setting was considered by assuming agents have access to more information they actually have, such as the system size, the dynamics of the system and the distribution of the values they hold. Then, the concept of *knowledge set* was introduced to model *all the information* that an agent might have possibly gathered at some time given a sequence of replacements and interactions. These knowledge sets therefore model the maximum amount of information an agent might have access to at some time about the system at that time. Hence, by properly characterizing how interactions and replacements happen in the system and how

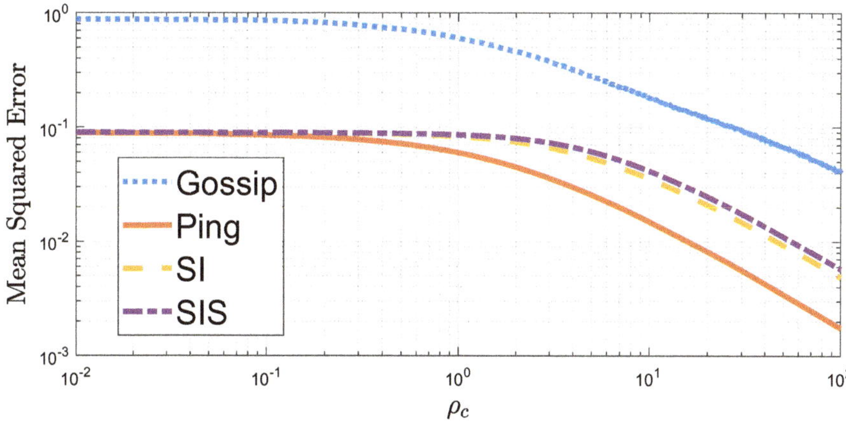

Fig. 10.19 Comparison between the expected Mean Squared Error achieved by the Gossip algorithm in a system of 10 agents (dotted blue) and three different valid lower bounds (plain red, dashed yellow and dash-dotted purple) with respect to the amount of interactions happening on average between two replacements

information propagates in the system, these assumptions and definitions allowed fully characterizing the performance of an optimal algorithm, yielding the desired fundamental performance limitations for that problem.

Typically, the analysis conducted in [62] yielded several lower bounds on the Mean Squared Error achieved by algorithms for that problem, depending on the definition of the interactions in the system. These bounds are shown and compared to the performance of the Gossip algorithm (see [25] for details) in Fig. 10.19. We refer to [62] for additional details on their derivation and use.

10.6 Further Applications

In addition to the examples previously mentioned, in this section, we present additional applications of OMAS in different fields.

10.6.1 Consensus

Since consensus is one of the most important problems associated with multi-agent systems, the preliminary works in OMAS were focused on the extension of classical results for open systems. Throughout this chapter, we presented several works focused on the consensus problem with pairwise interactions in a complete graph.

However, other works involving the consensus problem have also been developed considering different scenarios.

Regarding deterministic formulations of OMAS, one of the preliminary works is [51], where authors study the dynamic consensus in a contractive OMAS under the assumption that the arrival process is bounded. Other works related to the analysis of the dynamic consensus in OMAS are [63, 64] where the median value and the max value are the object of the study respectively and the assumption of a dwell time is used for the derivation of the results. In [55], the authors analyze the stability of multi-dimensional switched systems, where restrictions are imposed on the switching signal to guarantee practical consensus.

Other works include the analysis of OMAS under stochastic interactions in a finite superset of agents where the activation of agents are determined by Bernoulli random variables [39]. In the same framework, the deviation from consensus in randomized interactions subject to noise is analyzed in [65].

10.6.2 Distributed Optimization and Learning

The problem of distributed optimization in OMAS is mainly associated with the introduction of new cost functions carried by joining agents that can have an impact on the stability and location of the minimizer of the system. Mathematically, this optimization problem is generally formulated as:

$$\min_{x(t)\in X(t)} f^t(x(t)) = \sum_{i\in\mathcal{V}(t)} f_i^t(x_i(t)), \tag{10.13}$$

where $x(t)$ is a vector containing the states of all the agents in $\mathcal{V}(t)$, f_i^t is the cost function at time t associated with the agent i in $\mathcal{V}(t)$, and $X(t)$ is the feasible set at time t. Problems of this type can be observed for instance at the level of servers, where every unit can easily connect to a network to share loads and perform tasks more efficiently, and disconnect when its objectives are achieved. Two frameworks are currently being used for the analysis, depending on the objectives of the optimization algorithms.

The first approach is based on *time-varying optimization* [66], where the objective is to analyze the stability and convergence of the minimizer through the metric:

$$\left\| x(t) - x^*(t) \right\|,$$

where $x^*(t)$ is the minimizer of (10.13) at time t. This approach is better suited to optimization problems where the goal is to be as close as possible to the instantaneous minimizer at all time instants. In this type of applications, to keep uniformity of the potential cost functions, a common assumption is to consider that all the possible local minimizers are located inside a ball. This approach has been used in the analysis of the stability of the gradient descent algorithm in [67] and the analysis of convergence of the random coordinate descent algorithm in [41]. The influence of packet losses in open scenarios has been investigated in [68] where it has been shown that some performance metrics might diverge in OMAS due to changes of the cost functions. In the case of federated learning, the framework of time-varying optimization has been used in [69] for the analysis of stability of the minimizer.

The second approach is based on the well-known *online optimization framework* [70, 71] where the goal is to obtain a sublinear *regret*:

$$\mathrm{Reg}_T := \sum_{t=1}^{T} \left(f^t(x(t)) - f^t(x^*(t)) \right).$$

For instance, the resource allocation problem is open scenarios is investigated in [52] where several performance indices inspired from online optimization are introduced. In [40], an algorithm based on dual averaging was proposed to minimize a global cost function that depends on a time-varying set of active agents in a fixed size network. The problem of cooperative learning for adversarial bandit in open networks have been addressed in [72] where an upper bound for a pseudo regret is derived.

10.6.3 Vehicles and Robots

Transportation systems are inherently subject to frequent arrivals and departures due to the different destinations and starting points. For instance, cars keep joining and leaving a stretch of road [73]. In this context, frameworks similar to OMAS have been considered in [74] where Virtual Traffic Lights (VTL) are proposed for autonomous cars to deal with crossroads. In the case of transportation in highways, the size of the platoon change along the route depending on the particular destinations of the vehicles. In this case, since the communications of a vehicle are usually established only with the direct neighbors, properties of path graphs of different sizes play an important role in the analysis of open platoons. It is clear that for more complex communications patterns, other graph topologies must be studied, but they will be usually sparse due to the distances between vehicles.

In the case of robots, the problem of large groups of robots with different sizes have been extensively studied in the field of swarm robotics where the objective is the emergence of a desired collective behavior [75, 76]. However, from a dynamical point of view where each robot has a dynamics, there are only few works where an analysis of the stability and behavior of groups of robots with different sizes due to arrivals or departures is performed. For example, in [77] the consensus of a network of robots with potential arrivals is analyzed using a framework of switching systems.

10.7 Future Work

The framework of OMAS is extremely rich and dense. Even though it started receiving increasing attention recently, it remains largely unexplored, and many questions and problems are still open. We provide below several perspectives.

Models. Most of the existing analyses on OMAS rely on the use of either time-invariant finite supersets (e.g., [39, 65]) or fixed state-spaces subject to only replacements (e.g., [41, 50]). Dynamical compositions dealing with varying-size systems also start receiving attention with e.g., [50, 51], but these are rather recent and further work can still be envisioned to extend such modeling to more complex problems. Finally, recent works involving infinite-dimensional systems through the use of graphons can be found e.g., in [78] for the study of epidemics and in [79] for the analysis of the spectrum of the Laplacian matrix corresponding to sampled graphs.

More generally, since most of the dynamics in multi-agent systems are commonly encapsulated in the Laplacian matrix, extending the theory of graphons for this matrix is a promising line of work regarding infinite-dimensional models, which would allow us to use graphons for dynamics based on the Laplacian matrix. In this line of research, the authors in [80] studied the case of consensus in first-order graphon models, while in [81], the authors analyzed the spectrum of the Laplacian matrix for particular types of graphons. However, the relation between the spectrum of the Laplacian operator associated with an arbitrary graphon and the eigenvalues of the Laplacian matrix of a sampled graph is still under analysis.

Topology. The question of the topology of the underlying network of an OMAS is rather critical in some applications and can be the source of many issues in terms of connectivity. In particular, in OMAS it is crucial to properly define how arrivals and departures of agents impact the topology of the network, as these could break its connectivity.

So far, works in OMAS considering specific network topologies rely on the assumption that it remains fixed (with replacements only, such as in [41]). There are few specific cases where the network topologies follow a specific pattern that can be completely extended for any number of agents such as line, cycle, grids, complete or bipartite graphs [82]. However, it is unlikely that real networks will follow such restrictive fixed or pattern-based topologies. The possibility to extend analyses to (time-varying) graph topologies subject to arrivals and departures thus remains an open question, and the use of random graphs [83], graphons [45] and algorithms inspired from plug-and-play applications [84] could allow us to model more realistic interactions.

Moreover, most works on OMAS are centered around the analysis of undirected graphs. In [55], the authors investigate the case of switching between multi-agent systems of different dimensions where agents interact through digraphs. Their approach however disregards the impact of arrivals, departures and replacements, and the possibility to consider digraphs in OMAS remains open. In fact, since most analyses exploit the symmetry of the Laplacian matrix, it is likely that the extension to digraphs might require introducing new techniques.

Decoupled Arrivals and Departures. In a similar line, many works on OMAS focus on the specific case where the system size is fixed, as replacements already represent a significant challenge with many interesting properties. A natural yet complex extension for these work is therefore to consider decoupled arrivals and departures, and hence varying-size systems. As discussed previously, this extension requires properly defining how the connectivity of the underlying network is impacted, but also how to represent and analyze the system itself, as its dimension keeps changing. Tools regarding these questions were introduced in [50], but their application to other analyses might require adapting several aspects in a non-trivial way.

Also, the case of arrivals and departures linked to the current state of the agents or the system is an interesting field of research that need to be explored. Consider, for example, the case of robotic systems where the number of required units depend on the state of the mission, or the case of platoons where a vehicle leaves the system when it arrives to the desired destination.

Heterogeneity. An important and largely unexplored aspect of OMAS is *heterogeneity*. In simple settings, it is often assumed that the agents constituting a system all have the same dynamics. This is however not always the case regarding e.g., the case of robotics where different types of robots need to interact to achieve a desired goal [85]. Several approaches could be considered for such analysis, including for instance the study of emergent dynamics [86, 87], which has provided promising results in the case of switching systems with heterogeneous agents [88].

Agents Dynamics. Most studies in OMAS focus on rather simple agent dynamics in order to reduce the mathematical complexity brought by the introduction of arrivals and departures in the analysis. Therefore, a future line of work for OMAS is to consider agents with more general interaction schemes and dynamics, such as double integrators or nonlinear dynamics (coupled oscillators, Lotka–Volterra, etc.). Approaches based on the use of descriptors might be promising in this line of research, where Lyapunov functions could be used to study the dynamics in terms of a SDE. Alternatively, the theory of stochastic hybrid systems appears as an important framework that can be used for the analysis of OMAS when the objective is the determination of conditions to guarantee the stability of OMAS [54, 89].

Advanced Problems and Algorithm Design. Finally, a natural next step for analyses around OMAS is to extend them to more complex and advanced tasks, and to exploit them in order to design models and algorithms tailored for open systems. So far, many analyses focus on simple problems in order to emphasize the characteristics and properties of OMAS, and exploiting these to tackle complex real-life problems remains a promising line of work.

Typically, as it was remarked in [26, 27], OMAS look as a promising tool to handle opinion dynamics where agents are constantly entering and leaving the system as in the case of social networks, forums, discussions, etc. Models based on opinion-dependent connectivity like the HK model and k-nearest neighbors [90] seem to be more suitable for OMAS where the interactions are determined only by the values of the opinions. This type of interconnections simplifies the definition of the network topology in OMAS in the case of arrivals and departures, since it is not necessary to define the connections of new agents or reconfigure the connections of the network when agents leave the system.

10.8 Conclusions

Open multi-agent systems (OMAS) allow to model more realistic dynamic networks where agents can enter, leave or being replaced in the system. Unlike single changes in the network topology, these variations in the set of agents make the analysis more complex since the dimension of the system may change and traditional concepts associated to dynamical systems might not be applied in many cases. In this chapter, we introduced several notions and concepts associated with OMAS that make the analysis tractable. In addition, we presented current applications for OMAS and future perspectives to highlight the importance of these systems in the actual world where the networks become more dynamic and the common concept of *closed system* starts to become a strong assumption in most of the analyses.

Acknowledgements This work was supported by F.R.S.-FNRS via the *KORNET* project and via the Incentive Grant for Scientific Research (MIS) *Learning from Pairwise Comparisons*, and by the *RevealFlight* Concerted Research Action (ARC) of the Fédération Wallonie-Bruxelles and in part by the "Agence Nationale de la Recherche" (ANR) under Grant HANDY ANR-18-CE40-0010. R. Vizuete is a FNRS Postdoctoral Researcher – CR.

References

1. Bullo, F.: Lectures on Network Systems, 1st ed. Kindle Direct Publishing (2022)
2. Moreau, L.: Stability of continuous-time distributed consensus algorithms. In: IEEE Conference on Decision and Control, vol. 4, pp. 3998–4003 (2004)
3. Moreau, L.: Stability of multiagent systems with time-dependent communication links. IEEE Trans. Autom. Control **50**(2), 169–182 (2005)
4. Lin, Z., Francis, B., Maggiore, M.: State agreement for continuous-time coupled nonlinear systems. SIAM J. Control Optim. **46**(1), 288–307 (2007)
5. Monnoyer de Galland, C.: Open multi-agent systems: representation, limitations and decentralized optimization. Ph.D. dissertation, UCLouvain (2022)
6. Staedter, T.: 100,000 IoT sensors monitor a 1,400-km canal in China. IEEE Spectr. TechTalk Blog (2018)
7. Hauert, S., Bhatia, S.N.: Mechanisms of cooperation in cancer nanomedicine: towards systems nanotechnology. Trends Biotechnol. **32**(9), 448–455 (2014)
8. Claes, R., Holvoet, T., Weyns, D.: A decentralized approach for anticipatory vehicle routing using delegate multiagent systems. IEEE Trans. Intell. Transp. Syst. **12**(2), 364–373 (2011)
9. Proskurnikov, A., Tempo, R.: A tutorial on modeling and analysis of dynamic social networks. Part I. Ann. Rev. in Control **43**, 65–79 (2017). (05)

10. Argente, E., Botti, V., Carrascosa, C., Giret, A., Julián, V., Rebollo, M.: An abstract architecture for virtual organizations: the Thomas approach. Knowl. Inf. Syst. **29**, 379–403 (2011). (11)
11. Giret, A., Julián, V., Rebollo, M., Argente, E., Carrascosa, C., Botti, V.: An open architecture for service-oriented virtual organizations. Lect. Notes Comput. Sci. **5919**(09), 118–132 (2010)
12. Riverso, S., Farina, M., Ferrari-Trecate, G.: Plug-and-play decentralized model predictive control. In: IEEE Conference on Decision and Control, pp. 4193–4198 (2012)
13. Riverso, S., Farina, M., Ferrari-Trecate, G.: Plug-and-play state estimation and application to distributed output-feedback model predictive control. Eur. J. Control **25**, 04 (2015)
14. Farina, M., Carli, R.: Partition-based distributed Kalman filter with plug and play features. IEEE Trans. Control Netw. Syst. **5**(1), 560–570 (2018)
15. Huynh, T.D., Jennings, N.R., Shadbolt, N.R.: An integrated trust and reputation model for open multi-agent systems. Auton. Agents Multi-Agent Syst. **13**(2), 119–154 (2006)
16. Lykouris, T., Syrgkanis, V., Tardos, É.: Learning and efficiency in games with dynamic population. In: Proceedings of the Twenty-seventh Annual ACM-SIAM Symposium on Discrete Algorithms. SIAM, pp. 120–129 (2016)
17. Delporte-Gallet, C., Fauconnier, H., Guerraoui, R., Ruppert, E.: When birds die: making population protocols fault-tolerant. In: International Conference on Distributed Computing in Sensor Systems. Springer pp. 51–66 (2006)
18. Angluin, D., Aspnes, J., Fischer, M.J., Jiang, H.: Self-stabilizing population protocols. In: Principles of Distributed Systems, pp. 103–117. Springer, Berlin Heidelberg (2006)
19. Brasseur, B.: Algorithms in open multi-agent systems: gossiping with random replacements. In: Hendrickx, J.M. (ed.) Ecole polytechnique de Louvain, Université catholique de Louvain, 2018, prom (2018). http://hdl.handle.net/2078.1/thesis:16470
20. Abdelrahim, M., Hendrickx, J.M., Heemels, W.: Max-consensus in open multi-agent systems with gossip interactions. In: IEEE Conference on Decision and Control, pp. 4753–4758 (2017)
21. Hendrickx, J.M., Martin, S.: Open multi-agent systems: gossiping with deterministic arrivals and departures. In: 54th Annual Allerton Conference on Communication, Control, and Computing (Allerton), pp. 1094–1101 (2016)
22. Nedić, A., Olshevsky, A., Rabbat, M.G.: Network topology and communication-computation tradeoffs in decentralized optimization. Proc. IEEE **106**, 953–976 (2018)
23. Molinari, F., Raisch, J.: Efficient consensus-based formation control with discrete-time broadcast updates. In: IEEE Conference on Decision and Control, pp. 4172–4177 (Dec 2019)
24. Török, J., Iñiguez, G., Yasseri, T., San Miguel, M., Kaski, K., Kertész, J.: Opinions, conflicts, and consensus: modeling social dynamics in a collaborative environment. Phys. Rev. Lett. **110**(8), 088701 (2013)
25. Boyd, S., Ghosh, A., Prabhakar, B., Shah, D.: Randomized gossip algorithms. IEEE Trans. Inf. Theory **52**(6), 2508–2530 (2006)
26. Ravazzi, C., Dabbene, F., Lagoa, C., Proskurnikov, A.V.: Learning hidden influences in large-scale dynamical social networks: a data-driven sparsity-based approach, in memory of Roberto tempo. IEEE Control Syst. Mag. **41**(5), 61–103 (2021)
27. Proskurnikov, A.V., Tempo, R.: A tutorial on modeling and analysis of dynamic social networks. Part II. Ann. Rev. Control **45**, 166–190 (2018)
28. Hegselmann, R., Krause, U.: Opinion dynamics and bounded confidence models, analysis, and simulation. J. Artif. Soc. Soc. Simul. **5**(3) (2002)
29. Grauwin, S., Jensen, P.: Opinion group formation and dynamics: structures that last from nonlasting entities. Phys. Rev. E **85**, 066113 (Jun 2012)
30. Vizuete, R., Frasca, P., Panteley, E.: On Lyapunov functions for open Hegselmann–Krause dynamics (2023). arXiv:2303.07074
31. Blondel, V.D., Hendrickx, J.M., Tsitsiklis, J.N.: Continuous-time average-preserving opinion dynamics with opinion-dependent communications. SIAM J. Control Opti. **48**(8), 5214–5240 (2010)
32. Nowzari, C., Preciado, V.M., Pappas, G.J.: Analysis and control of epidemics: a survey of spreading processes on complex networks. IEEE Control Syst. Mag. **36**(1), 26–46 (2016)

33. Alamo, T., Millán, P., Reina, D.G., Preciado, V.M., Giordano, G.: Challenges and future directions in pandemic control. IEEE Control Syst. Lett. **6**, 722–727 (2022)
34. Tizzoni, M., Bajardi, P., Decuyper, A., Kon Kam King, G., Schneider, C.M., Blondel, V., Smoreda, Z., González, M.C., Colizza, V.: On the use of human mobility proxies for modeling epidemics. PLoS Comput. Biol. **10**(7), e1003716 (2014)
35. Hazarie, S., Soriano-Paños, D., Arenas, A., Gómez-Gardeñes, J., Ghoshal, G.: Interplay between population density and mobility in determining the spread of epidemics in cities. Commun. Phys. **4**(1), 1–10 (2021)
36. Ogura, M., Preciado, V.M.: Stability of spreading processes over time-varying large-scale networks. IEEE Trans. Netw. Sci. Eng. **3**(1), 44–57 (2016)
37. Paré, P.E., Beck, C.L., Nedić, A.: Epidemic processes over time-varying networks. IEEE Trans. Control Netw. Syst. **5**(3), 1322–1334 (2018)
38. Vizuete, R.: Contributions to open multi-agent systems: consensus, optimization and epidemics. Ph.D. dissertation, Université Paris-Saclay (2022)
39. Varma, V.S., Morărescu, I.-C., Nešić, D.: Open multi-agent systems with discrete states and stochastic interactions. IEEE Control Syst. Lett. **2**(3), 375–380 (2018)
40. Hsieh, Y.-G., Iutzeler, F., Malick, J., Mertikopoulos, P.: Optimization in open networks via dual averaging. In: IEEE Conference on Decision and Control, pp. 514–520 (2021)
41. Monnoyer de Galland, C., Vizuete, R., Hendrickx, J.M., Panteley, E., Frasca, P.: Random coordinate descent for resource allocation in open multi-agent systems (2022). arXiv:2205.10259
42. Verriest, E.I.: Multi-mode multi-dimensional systems. In: Proceedings of the 17th International Symposium on Mathematical Theory of Networks and Systems, pp. 1268–1274 (2006)
43. Verriest, E.I.: Pseudo-continuous multi-dimensional multi-mode systems. Discrete Event Dyn. Syst. **22**(1), 27–59 (2012)
44. Goebel, R., Sanfelice, R.G., Teel, A.R.: Hybrid Dynamical Systems. Princeton University Press (2012)
45. Lovász, L.: Large Networks and Graph Limits. American Mathematical Society, vol. 60 (2012)
46. Avella-Medina, M., Parise, F., Schaub, M.T., Segarra, S.: Centrality measures for graphons: accounting for uncertainty in networks. IEEE Trans. Netw. Sci. Eng. **7**(1), 520–537 (2020)
47. Gao, S., Caines, P.E.: Graphon control of large-scale networks of linear systems. IEEE Trans. Autom. Control **65**(10), 4090–4105 (2020)
48. Delmas, J.-F., Dronnier, D., Zitt, P.-A.: An infinite-dimensional metapopulation sis model. J. Differ. Equ. **313**, 1–53 (2022)
49. Noroozi, N., Mironchenko, A., Kawan, C., Zamani, M.: Small-gain theorem for stability, cooperative control and distributed observation of infinite networks (2020). arXiv:2002.07085
50. Monnoyer de Galland, C., Martin, S., Hendrickx, J.M.: Modelling gossip interactions in open multi-agent systems (2020). arXiv:2009.02970
51. Franceschelli, M., Frasca, P.: Stability of open multi-agent systems and applications to dynamic consensus. IEEE Trans. Autom. Control **66**(5), 2326–2331 (2021)
52. Vizuete, R., Monnoyer de Galland, C., Hendrickx, J.M., Frasca, P., Panteley, E.: Resource allocation in open multi-agent systems: an online optimization analysis. In: IEEE Conference on Decision and Control, pp. 5185–5191 (2022)
53. Dörfler, F., Bullo, F.: Synchronization in complex networks of phase oscillators: a survey. Automatica **50**(6), 1539–1564 (2014)
54. Teel, A.R., Hespanha, J.P.: Stochastic hybrid systems: a modeling and stability theory tutorial. In: IEEE Conference on Decision and Control, pp. 3116–3136 (2015)
55. Xue, M., Tang, Y., Ren, W., Qian, F.: Stability of multi-dimensional switched systems with an application to open multi-agent systems. Automatica **146**, 110644 (2022)
56. Hendrickx, J.M., Martin, S.: Open multi-agent systems: gossiping with random arrivals and departures. In: IEEE Conference on Decision and Control, pp. 763–768 (2017)
57. Brockett, R.: Stochastic Control. Harvard University, Lecture Notes (2009)
58. Brockett, R., Gong, W., Guo, Y.: Stochastic analysis for fluid queueing systems. In: IEEE Conference on Decision and Control, vol. 3, pp. 3077–3082 (1999)
59. Privault, N.: Stochastic Finance: An Introduction with Market Examples. CRC Press (2013)

60. Jacobsen, M.: Point Process Theory and Applications: Marked Point and Piecewise Deterministic Processes. Birkhäuser Boston (2006)
61. Colla, S.: Data fusion in open multi-agent systems for decentralized estimation. Hendrickx, J.M. (ed.) Ecole polytechnique de Louvain, Université catholique de Louvain, 2020, prom (2020). https://dial.uclouvain.be/memoire/ucl/object/thesis:25154
62. Monnoyer de Galland, C., Hendrickx, J.M.: Fundamental performance limitations for average consensus in open multi-agent systems. IEEE Trans. Autom. Control (2022)
63. Sanai Dashti, Z.A.Z., Seatzu, C., Franceschelli, M.: Dynamic consensus on the median value in open multi-agent systems. In: IEEE Conference on Decision and Control, pp. 3691–3697 (2019)
64. Deplano, D., Franceschelli, M., Giua, A.: Dynamic min and max consensus and size estimation of anonymous multiagent networks. IEEE Trans. Autom. Control **68**(1), 202–213 (2023)
65. Vizuete, R., Frasca, P., Panteley, E.: On the influence of noise in randomized consensus algorithms. IEEE Control Syst. Lett. **5**(3), 1025–1030 (2021)
66. Simonetto, A., Dall'Anese, E., Paternain, S., Leus, G., Giannakis, G.B.: Time-varying convex optimization: time-structured algorithms and applications. Proc. IEEE **108**(11), 2032–2048 (2020)
67. Hendrickx, J.M., Rabbat, M.G.: Stability of decentralized gradient descent in open multi-agent systems. In: IEEE Conference on Decision and Control, pp. 4885–4890 (2020)
68. Vizuete, R., Frasca, P., Panteley, E.: Gradient descent for resource allocation with packet loss. IFAC-PapersOnLine **55**(13), 109–114 (2022)
69. Sun, Y., Fernando, H., Chen, T., Shahrampour, S.: On the stability analysis of open federated learning systems. In: American Control Conference, pp. 867–872. IEEE (2023)
70. Hazan, E.: Introduction to online convex optimization. Found. Trends Opti. **2**, 157–325 (2016)
71. Bubeck, S.: Introduction to online optimization (Dec 2011). https://www.microsoft.com/en-us/research/publication/introduction-online-optimization/
72. Nakamura, T., Hayashi, N., Inuiguchi, M.: Cooperative learning for adversarial multi-armed bandit on open multi-agent systems. IEEE Control Syst. Lett. **7**, 1712–1717 (2023)
73. Claes, R., Holvoet, T., Weyns, D.: A decentralized approach for anticipatory vehicle routing using delegate multiagent systems. IEEE Trans. Intell. Transp. Syst. **12**(2), 364–373 (2011)
74. Tonguz, O.K.: Red light, green light-no light: tomorrow's communicative cars could take turns at intersections. IEEE Spectr. **55**(10), 24–29 (2018)
75. Hamann, H.: Swarm Robotics: A Formal Approach, vol. 221. Springer (2018)
76. Valentini, G.: Achieving Consensus in Robot Swarms, vol. 706. Springer (2017)
77. Restrepo, E., Loría, A., Sarras, I., Marzat, J.: Consensus of open multi-agent systems over dynamic undirected graphs with preserved connectivity and collision avoidance. In: IEEE Conference on Decision and Control, pp. 4609–4614 (2022)
78. Vizuete, R., Frasca, P., Garin, F.: Graphon-based sensitivity analysis of SIS epidemics. IEEE Control Syst. Lett. **4**(3), 542–547 (2020)
79. Vizuete, R., Garin, F., Frasca, P.: The Laplacian spectrum of large graphs sampled from graphons. IEEE Trans. Netw. Sci. Eng. **8**(2), 1711–1721 (2021)
80. Bonnet, B., Duteil, N.P., Sigalotti, M.: Consensus formation in first-order graphon models with time-varying topologies. In: Mathematical Models and Methods in Applied Sciences (2022)
81. Bramburger, J., Holzer, M.: Pattern formation in random networks using graphons. SIAM J. Math. Anal. **55**(3), 2150–2185 (2023)
82. Fagnani, F., Frasca, P.: Introduction to Averaging Dynamics over Networks. Springer, Cham, Switzerland (2018)
83. Bollobás, B.: Random Graphs, 2nd ed. Cambridge University Press (2001)
84. Stüdli, S., Yan, Y., Seron, M.M., Middleton, R.H.: Plug-and-play network reconfiguration algorithms to maintain regularity and low network reconfiguration needs. IEEE Control Syst. Lett. **6**, 3451–3456 (2022)
85. Abad Torres, J., Cruz, P.J., Vizuete, R., Fierro, R.: On resilience and heterogeneity in robotic networks. In: Cooperative Localization and Navigation, pp. 141–158. CRC Press (2019)

86. Panteley, E., Loría, A.: Synchronization and dynamic consensus of heterogeneous networked systems. IEEE Trans. Autom. Control **62**(8), 3758–3773 (2017)
87. Maghenem, M., Panteley, E., Loría, A.: Singular-perturbations-based analysis of dynamic consensus in directed networks of heterogeneous nonlinear systems (2022). arXiv:2205.15646
88. Adhikari, B., Morărescu, I.-C., Panteley, E.: An emerging dynamics approach for synchronization of linear heterogeneous agents interconnected over switching topologies. IEEE Control Syst. Lett. **5**(1), 43–48 (2021)
89. Teel, A.R., Subbaraman, A., Sferlazza, A.: Stability analysis for stochastic hybrid systems: a survey. Automatica **50**(10), 2435–2456 (2014)
90. Rossi, W.S., Frasca, P.: Asynchronous opinion dynamics on the k-nearest-neighbors graph. In: IEEE Conference on Decision and Control, pp. 3648–3653 (2018)

Chapter 11
Layers Update of Neural Network Control via Event-Triggering Mechanism

Sophie Tarbouriech, Carla De Souza, and Antoine Girard

Abstract The chapter deals with the design of event-triggering mechanisms (ETM) for discrete-time linear systems stabilized by neural network controllers. The proposed event-triggering mechanism is based on the use of local sector conditions related to the activation functions, to reduce the computational cost associated with the neural network evaluation. Such a mechanism avoids redundant computations by updating only a portion of the layers instead of evaluating periodically the complete neural network. Sufficient matrix inequality conditions are provided to design the parameters of the event-triggering mechanism and compute an inner-approximation of the region of attraction for the feedback system. The theoretical conditions are obtained by using a quadratic Lyapunov function and an adequate abstraction of the activation functions via generalised sector condition to decide whether the outputs of the layers should be transmitted through the network or not. Convex optimisation procedures can be associated to the theoretical conditions in order to maximise the approximation of the region of attraction or to minimise the number of updates. The advantages and the drawbacks of our approach are illustrated in an example borrowed from the literature, namely the nonlinear inverted pendulum system stabilized by a trained neural network.

This work has been supported by ANR under the project HANDY number 18-CE40-0010.

S. Tarbouriech (✉)
LAAS-CNRS, Université de Toulouse CNRS, Toulouse, France
e-mail: tarbour@laas.fr

C. De Souza
Leyfa Measurement une filiale SNCF, Aucamville, France
e-mail: carla.souza93@hotmail.com

A. Girard
Université Paris-Saclay, CNRS, CentraleSupélec Laboratoire des Signaux et Systèmes, 91190 Gif-sur-Yvette, France
e-mail: antoine.girard@l2s.centralesupelec.fr

R. Postoyan et al. (eds.), *Hybrid and Networked Dynamical Systems*,
Lecture Notes in Control and Information Sciences 493,
https://doi.org/10.1007/978-3-031-49555-7_11

11.1 Introduction

The use of neural networks (NNs) for control purpose is attracting a lot of works, in particular, due to recent advancements in deep learning. Indeed, NNs may replace some existing computationally expensive controllers typically as model predictive controllers (MPCs), requiring online solutions to an optimal control problem. Thus, NNs allow for an efficient and inexpensive embedded implementation [9, 13, 28]. However, NN controllers lack guarantees, which typically restrict their use in safety-critical applications such as for example in surgical procedures, and medical support systems [24]. Therefore, the ability to develop tools that can provide useful certificates of stability, safety, and robustness for NN controlled systems appears as a crucial direction of research. One can cite several works studying different aspect as verifying such systems. References [5, 14, 15, 17, 21].

From a control point of view, the structures of NNs are interesting due to various types of nonlinear activation functions, potentially numerous layers, and a large number of hidden neurons. Furthermore, the adaptation of Lyapunov theory and the development of diverse ways to embed the nonlinear activation functions is a very exciting prospective. Hence, the use of adequate quadratic constraints (QCs) allowing to handle the nonlinear activation functions has been considered in different contexts: [4] checked the safety of NNs by using QCs to outer-bound their outputs given a set of inputs. Reference [25] combined Lyapunov theory with local sector QCs to analyze the stability of feedback systems with neural network controllers. Reference [16] analyzed stability in offset-free setpoint tracking with a piecewise constant reference. Reference [12] proposed QCs based on partial gradients to certify the input-output stability of reinforcement-learning (RL) controlled systems.

Moreover, in the context of network control systems it is important not only to satisfy some control requirements but also to alleviate the communication network consumption. A pertinent alternative control paradigm to reduce the computational overhead and communication traffic, thus saving the limited network resources consists in event-triggered control (ETC) strategy. Such a strategy allows to update the control only when necessary (corresponding to a predefined mechanism), which is different from executing control tasks at periodic intervals. Then, the ETC approach describes a mechanism to determine the sampling instants without compromising the desired system performance (see, for example, [3, 7, 8, 20]). In the particular context of neural networks, the ETC technique has been mainly used to transmit states during the learning process, as, for example, studied in [6, 18, 23, 29].

Taking inspiration from our previous work [1], in this chapter, we consider that the neural network has already been trained and we design an event-triggering mechanism (ETM) to transmit the output of the layers, to reduce the computational cost associated with the control law evaluation. The ETM we investigate in this chapter uses the knowledge on the nonlinear activation functions, which are supposed to be saturation functions. We then take advantage of less conservative conditions to deal with saturation functions [22] and we propose an ETM based on a local sector conditions satisfied by the activation functions to decide whether the outputs of the layers should be transmitted through the network or not. We use quadratic constraints (QCs) not only to abstract the nonlinear activation functions but also to model the event-triggering rule. Hence, considering discrete-time linear time-invariant (LTI) plants stabilized by known neural network controllers, combining Lyapunov theory and generalized sector condition, we develop sufficient matrix inequalities conditions to compute the triggering parameters and characterize an estimate of the domain of attraction for the feedback system. An optimization scheme incorporating these conditions is also established to effectively reduce the update of the neural network layers. An example borrowed from the literature is used to illustrate the effectiveness of the proposal.

Notation. $\mathbb{N}$, $\mathbb{R}^n$, $\mathbb{R}^{n\times m}$ denote respectively the sets of nonnegative integers, n-dimensional vectors and $n \times m$ matrices. $\mathbb{D}^n_{>0}$ (respectively $\mathbb{D}^n_{\geq 0}$) is the set of diagonal positive definite (respectively, positive semi-definite) matrices of dimension n. For any matrix A, $A^\top$ denotes its transpose. For any square matrix A, $\mathtt{trace}(A)$ denotes its trace and $He\{A\} = A + A^\top$. $\mathtt{diag}(A_1, A_2)$ is a block-diagonal matrix with block diagonal matrices A_1 and A_2. For two symmetric matrices of same dimensions, A and B, $A > B$ means that $A - B$ is symmetric positive definite. $\mathbf{I}$ and $\mathbf{0}$ stand respectively for the identity and the null matrix of appropriate dimensions. For a partitioned matrix, the symbol $\star$ stands for symmetric blocks. For any vector $x \in \mathbb{R}^n$ and any symmetric positive definite (or semi-positive definite) matrix, $\|x\|^2_Q$ denotes the quadratic form $x^\top Qx$.

11.2 Modeling and Problem Statement

In this chapter, we consider a nonlinear control system described by the connection of a linear plant and an event-triggered neural network controller π_{ETM}. The objective is then to design an event-triggering mechanism to reduce the computational cost associated with the neural network evaluation. The following section presents the complete system under consideration.

11.2.1 Model Description

The plant under consideration is a discrete-time linear time-invariant (LTI) system defined by:

$$x(k+1) = A_p x(k) + B_p u(k), \tag{11.1}$$

where $x(k) \in \mathbb{R}^{n_p}$ is the state vector and $u(k) \in \mathbb{R}^{n_u}$ is the control input. Matrices A_p and B_p of appropriate dimensions are supposed to be constant and known. π_{ETM} is an ℓ-layer, event-triggered feedforward neural network (NN) described by:

$$
\begin{aligned}
\hat{\omega}^0(k) &= x(k), \\
\nu^i(k) &= W^i \hat{\omega}^{i-1}(k) + b^i, \ i \in \{1, \ldots, \ell\}, \\
\omega^i(k) &= \mathtt{sat}(\nu^i(k)), \\
u(k) &= W^{\ell+1} \hat{\omega}^\ell(k) + b^{\ell+1} + Kx(k),
\end{aligned}
\tag{11.2}
$$

where $\nu^i \in \mathbb{R}^{n_i}$ is the input to the ith activation function, $\omega^i \in \mathbb{R}^{n_i}$ and $\hat{\omega}^i \in \mathbb{R}^{n_i}$ are the current output and the last forwarded output from the ith layer, respectively. The activation function is a saturation map $\mathtt{sat}(\cdot)$, which is applied element-wise and corresponds to the classical decentralized and symmetric saturation function:

$$
\mathtt{sat}(\nu_j^i(k)) = \mathtt{sign}(\nu_j^i(k)) \min(|\nu_j^i(k)|, \bar{\nu}_j^i), \tag{11.3}
$$

where ν_j^i is the jth component of ν^i, and $\bar{\nu}_j^i > 0$, $i \in [1, \ell]$, $j \in [1, n_i]$, is the level of the saturation ($\underline{\nu}_j^i = -\bar{\nu}_j^i$). As in [2], the operations for each layer are defined by a weight matrix $W^i \in \mathbb{R}^{n_i \times n_{i-1}}$, a bias vector $b^i \in \mathbb{R}^{n_i}$, and the activation function $\mathtt{sat}(.)$. Differently from [2], a linear term $Kx(k)$ is added to the output of the NN and we will study its potential later in the chapter.

We can rewrite the closed-loop system (11.1)–(11.2) in a compact form with isolated nonlinear functions to separate the linear operations from the nonlinear ones in the neural network [4, 25]. Then, we consider the augmented vectors

$$
\begin{aligned}
\nu_\phi &= \begin{bmatrix} \nu^{1\top} & \ldots & \nu^{\ell\top} \end{bmatrix}^\top, \\
\omega_\phi &= \begin{bmatrix} \omega^{1\top} & \ldots & \omega^{\ell\top} \end{bmatrix}^\top, \\
\hat{\omega}_\phi &= \begin{bmatrix} \hat{\omega}^{1\top} & \ldots & \hat{\omega}^{\ell\top} \end{bmatrix}^\top \in \mathbb{R}^{n_\phi},
\end{aligned}
$$

with $n_\phi = \sum_{i=1}^{\ell} n_i$. The ith element of ν_ϕ, for instance, is denoted by $\nu_{\phi,i}$. A combined nonlinearity $\mathtt{sat}(\cdot) : \mathbb{R}^{n_\phi} \to \mathbb{R}^{n_\phi}$ can also be obtained by stacking all saturation functions, i.e.,

$$
\omega_\phi(k) = \mathtt{sat}(\nu_\phi) = \begin{bmatrix} \mathtt{sat}(\nu^1)^\top & \ldots & \mathtt{sat}(\nu^\ell)^\top \end{bmatrix}^\top \in \mathbb{R}^{n_\phi}, \tag{11.4}
$$

The neural network control policy can therefore be described as follows:

$$
\begin{bmatrix} u(k) \\ \nu_\phi(k) \end{bmatrix} = N \begin{bmatrix} x(k) \\ \hat{\omega}_\phi(k) \\ 1 \end{bmatrix}, \tag{11.5}
$$

with

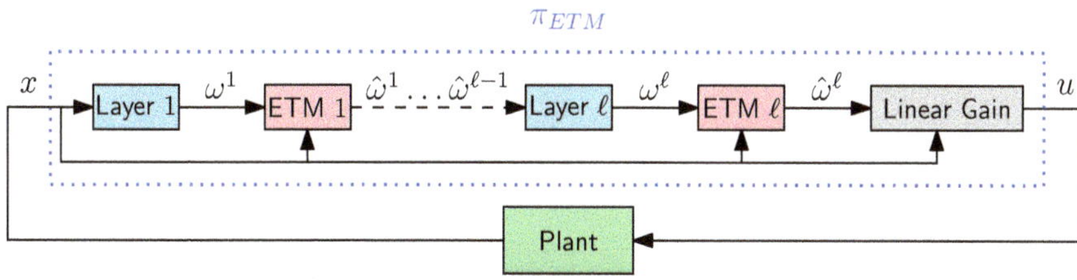

Fig. 11.1 Control system with event-triggered network controller π_{ETM}

$$N = \left[\begin{array}{c|cccc|c} K & \mathbf{0} & \dots & \mathbf{0} & W^{\ell+1} & b^{\ell+1} \\ \hline W^1 & \mathbf{0} & \dots & \mathbf{0} & \mathbf{0} & b^1 \\ \mathbf{0} & W^2 & \dots & \mathbf{0} & \mathbf{0} & b^2 \\ \vdots & \vdots & \ddots & \vdots & \vdots & \vdots \\ \mathbf{0} & \mathbf{0} & \dots & W^\ell & \mathbf{0} & b^\ell \end{array}\right] = \begin{bmatrix} N_{ux} & N_{u\omega} & N_{ub} \\ N_{\nu x} & N_{\nu\omega} & N_{\nu b} \end{bmatrix}. \tag{11.6}$$

Remark 11.2.1 Differently from [2], it should be noted that $N_{ux} = K \neq \mathbf{0}$ due to the presence of the control gain $K \in \mathbb{R}^{n_u \times n_p}$.

The relation between the current and transmitted outputs $\omega_\phi(k)$ and $\hat{\omega}_\phi(k)$ is described by event-triggering mechanisms (ETMs) located at each layer of the NN: if an event is triggered at instant k at layer i then $\hat{\omega}^i(k) = \omega^i(k)$, otherwise $\hat{\omega}^i(k) = \hat{\omega}^i(k-1)$. The conditions under which events are triggered will be formally introduced in Sect. 11.2.3. The overall architecture of our control system is shown in Fig. 11.1.

11.2.2 Preliminary Model Analysis

It is clear from (11.4)–(11.5), that, whenever an event is triggered at layer i, the definition of ν_ϕ presents a nonlinear algebraic loop due to the presence of the saturation ($\hat{\omega}^i = \mathtt{sat}(\nu^i)$) in the right-hand term. With $N_{\nu\omega} = \mathbf{0}$ one retrieves the classical linear state-feedback case without algebraic loop (see, for example, [22]). Then, let us discuss the aspects related to the nonlinear algebraic loop when $N_{\nu\omega} \neq \mathbf{0}$. Necessary and sufficient well-posedness conditions can be, for instance, derived from [11, Claim 2]. If the algebraic loop is well-posed then one can guarantee the existence of a (piecewise affine) solution to ν_ϕ in (11.4)–(11.5). In the particular case considered here, the lower triangular structure of $N_{\nu\omega}$ ensures that the algebraic loop, and thus the control system, are well posed. Then, we can state the following result regarding the uniform boundedness of the trajectories of the control system:

Lemma 11.2.2 *Assume that the matrix $A_p + B_p K$ is Schur–Cohn. Then, there exist $\beta, \gamma_x, \gamma_\omega > 0$ such that for all trajectories of the control system, it holds*

$$\|x(k)\| \leq \beta + \gamma_x \|x(0)\| + \gamma_\omega \|\hat{\omega}^\ell(0)\|, \ \forall k \in \mathbb{N}. \tag{11.7}$$

Proof From the use of bounded activation functions given by saturation maps and of ETMs, we have that for all $k \in \mathbb{N}$, $|\hat{\omega}^\ell(k)| \leq \max(|\hat{\omega}^\ell(0)|, \bar{\nu}^\ell)$, which implies from (11.2) that

$$\|u(k) - Kx(k)\| \leq \left\| \, |W^{\ell+1}| \max(|\hat{\omega}^\ell(0)|, \bar{\nu}^\ell) + |b^\ell| \, \right\|, \ \forall k \in \mathbb{N}.$$

Then, (11.7) follows from the previous inequality and the fact that the matrix $A_p + B_p K$ is Schur–Cohn. □

Let us now discuss briefly the existence of equilibrium points of (11.1) and (11.5)–(11.6), that we denote $(x_*, u_*, \nu_*, \omega_*, \hat{\omega}_*)$ with $\omega_* = \hat{\omega}_*$. Then, the following conditions must be satisfied

$$\begin{aligned} x_* &= A_p x_* + B_p u_*, \ \omega_*^0 = x_*, \\ \begin{bmatrix} u_* \\ \nu_* \end{bmatrix} &= N \begin{bmatrix} x_* \\ \omega_* \\ 1 \end{bmatrix}, \\ \omega_* &= \mathtt{sat}(\nu_*). \end{aligned} \tag{11.8}$$

Then, consider the matrices:

$$\begin{aligned} R \ &= (\mathbf{I} - N_{\nu\omega})^{-1} \\ R_\omega &= N_{ux} + N_{u\omega} R N_{\nu x} = K + N_{u\omega} R N_{\nu x} \\ R_b &= N_{u\omega} R N_{\nu b} + N_{ub}. \end{aligned} \tag{11.9}$$

Note that the lower triangular structure of $N_{\nu\omega}$ ensures that $\mathbf{I} - N_{\nu\omega}$ is always invertible. Then, let us make the following assumptions:

Assumption 11.2.3 The matrix $\mathbf{I} - A_p - B_p R_\omega$ is invertible.

Assumption 11.2.4 $-\bar{\nu} \leq R N_{\nu x} [\mathbf{I} - A_p - B_p R_\omega]^{-1} B_p R_b + R N_{\nu b} \leq \bar{\nu}$.

Then, we can state the following result.

Lemma 11.2.5 *Under Assumptions 11.2.3 and 11.2.4, there exists a unique equilibrium* $(x_*, u_*, \nu_*, \omega_*, \hat{\omega}_*)$ *such that* $-\bar{\nu} \leq \nu_* \leq \bar{\nu}$. *This equilibrium point is given by*

$$\begin{aligned} x_* &= [\mathbf{I} - A_p - B_p R_\omega]^{-1} B_p R_b, \\ u_* &= R_\omega x_* + R_b, \\ \nu_* &= R N_{\nu x} x_* + R N_{\nu b}, \\ \omega_* &= \hat{\omega}_* = \nu_*, \end{aligned} \tag{11.10}$$

with the matrices R, R_ω *and* R_b *defined in (11.9).*

Proof Let us look for an equilibrium point such that $-\bar{\nu} \leq \nu_* \leq \bar{\nu}$, i.e., such that $\mathtt{sat}(\nu_*) = \nu_*$. Then, it follows from (11.6)–(11.8) and Assumption 11.2.3 that such an equilibrium point is unique and given by (11.10). Then, Assumption 11.2.4 ensures that $-\bar{\nu} \leq \nu_* \leq \bar{\nu}$ indeed holds. □

In the rest of the chapter, we will assume that Assumptions 11.2.3 and 11.2.4 hold.

11.2.3 Event-Triggering Mechanism

To reduce the computational cost incurred by the evaluation of the neural network at each time step, the NN architecture is equipped with ETMs located at the outputs of each layer, as shown in Fig. 11.1. The ETMs decide whether or not to update the current outputs through the neural network.

In this work, we propose to use ETMs given by rules of the following form for $i \in \{1, \ldots, \ell\}$:

$$\hat{\omega}^i(k) = \begin{cases} \omega^i(k), & \text{if } f^i(\hat{\omega}^i(k-1), \nu^i(k), x(k)) > 0, \\ \hat{\omega}^i(k-1), & \text{otherwise,} \end{cases} \tag{11.11}$$

with $f^i(\hat{\omega}^i(k-1), \nu^i(k), x(k))$ defined by

$$\begin{aligned} f^i(\hat{\omega}^i(k-1),&\nu^i(k), x(k)) = \\ &\left[\nu^i(k) - \hat{\omega}^i(k-1)\right]^\top T^i \left[G^i(x(k) - x_*) - (\hat{\omega}^i(k-1) - \omega^i_*)\right] \end{aligned} \tag{11.12}$$

where $T^i \in \mathbb{R}^{n_i \times n_i}$ is a diagonal positive definite matrix and $G^i \in \mathbb{R}^{n_i \times n_p}$.

We can now provide a statement of the problem under consideration in this paper:

Problem 11.2.6 Consider the NN controller π_{ETM} (11.4)–(11.5) that stabilizes the plant (11.1). Design ETMs, according to (11.11)–(11.12), i.e., design the matrices T^i, G^i, to reduce the computational cost associated with the evaluation of the neural network while preserving the stability of the control system.

The presence of the saturation functions implies that the control system is nonlinear and asymptotic stability of the equilibrium point x_* can be ensured globally (that is for any initial condition $x(0) \in \mathbb{R}^n$) or only locally (that is, only for initial conditions in a neighborhood of x_*). Regional (local) asymptotic stability holds if and only if $A_p + B_p R_\omega$ is Schur, but in that case, the characterization of a large inner approximation of the basin of attraction requires nontrivial derivations (see, e.g., [10, 22, 27]). That constitutes an implicit problem in Problem 11.2.6. Hence, a classical trade-off can be considered between the size of the estimate of the region of attraction and the update saving.

Fundamental limitations of bounded stabilization of linear systems [19] imply that when $A_p + B_P K$ has unstable eigenvalues (i.e., outside the unit circle), the maximal set of initial conditions providing solutions that converge to x_* (corresponding to the basin of attraction of x_*) is bounded. The exact characterization of the basin of attraction remains an open problem in general. Hence, a challenging problem consists in computing accurate approximations. A guarantee on the size of the basin of attraction is typically done through an inner approximation, which corresponds to study the regional asymptotic stability.

Let us briefly discuss another important point related to an effective implementation of ETMs (11.11)–(11.12). In comparison to [2] or [1] where the ETM at layer i uses only local information (i.e., values of $\hat{\omega}^{i-1}$, $\hat{\omega}^i$, ν_i or ω^i), ETM (11.11)–(11.12) uses non-local information since the knowledge of the value of $x(k)$ is required. For the ETM at layer i, we define variables $q_i(k) \in \mathbb{R}^{n_p}$, $r_i(k) \in \mathbb{R}$ given by

$$q_i(k) = G^{i\top} T^i \left[\nu^i(k) - \hat{\omega}^i(k)\right],$$

and

$$r_i(k) = \left[\nu^i(k) - \hat{\omega}^i(k)\right]^\top T^i \left[G^i x_* + (\hat{\omega}^i(k) - \omega_*^i)\right].$$

Let us remark, that if at time k, an event is not triggered at layer $i - 1$, then

$$f^i(\hat{\omega}^i(k-1), \nu^i(k), x(k)) = q_i(k-1)^\top x(k) - r_i(k-1).$$

The triggering condition (11.11) can then be checked much more efficiently than using (11.12) particularly when the dimension of the state n_p is much smaller than the number of neurons in the layer n_i. Moreover, it is to be noted that if no event is triggered at layer $i - 1$ nor at layer i then we have $q_i(k) = q_i(k-1)$ and $r_i(k) = r_i(k-1)$. Thus, these variables only need to be updated when events are triggered at layer $i - 1$ or at layer i.

11.3 LMI-Based Design of ETM

In this section, we propose an approach to solve Problem 11.2.6 based on solving an optimization problem given by linear matrix inequalities (LMIs). We first show how the nonlinear activation functions can be abstracted by quadratic constraints. Then, we establish sufficient conditions for stability under the form of LMIs. Then, we discuss the optimization procedure required to compute a solution to these LMIs.

11.3.1 Activation Functions as Quadratic Constraints

We consider activation functions given by saturation maps. Thus, we have the following property adapted from [22, Lemma 1.6, p. 43], see also [2].

Lemma 11.3.1 *For $i \in \{1, \ldots, \ell\}$, consider a matrix $G^i \in \mathbb{R}^{n_i \times n_p}$. If $x(k)$ belongs to the set $\mathcal{S}^i$ defined by*

$$\mathcal{S}^i = \left\{x(k) \in \mathbb{R}^{n_p} : -\bar{\nu}^i - \nu_*^i \leq G^i(x(k) - x_*) \leq \bar{\nu}^i - \nu_*^i\right\}$$

then, the following quadratic constraint holds:

$$\left[\nu^i(k) - \omega^i(k)\right]^\top T^i \left[G^i(x(k) - x_*) - (\omega^i(k) - \omega_*^i)\right] \leq 0 \tag{11.13}$$

for any diagonal positive definite matrix $T^i \in \mathbb{R}^{n_i \times n_i}$.

Proof If $x(k)$ is an element of $\mathcal{S}^i$, it follows that for all $j \in \{1, \ldots, n_i\}$, $G_j^i(x(k) - x_*) - (-\bar{\nu}_j^i - \nu_{*,j}^i) \geq 0$ and $G_j^i(x(k) - x_*) - (\bar{\nu}_j^i - \nu_{*,j}^i) \leq 0$. Consider now the three cases below.

- Case 1: $\nu_j^i(k) \geq \bar{\nu}_j^i$. It follows that $\nu_j^i(k) - \omega_j^i(k) = \nu_j^i(k) - \bar{\nu}_j^i \geq 0$ and that $G_j^i(x(k) - x_*) - (\omega_j^i(k) - \omega_{*,j}^i) = G_j^i(x(k) - x_*) - (\bar{\nu}_j^i - \nu_{*,j}^i) \leq 0$. Then, one gets

$$\left[\nu_j^i(k) - \omega_j^i(k)\right] \times \left[G_j^i(x(k) - x_*) - (\omega_j^i(k) - \omega_{*,j}^i)\right] \leq 0. \tag{11.14}$$

- Case 2: $\nu_j^i(k) \leq -\bar{\nu}_j^i$. It follows that $\nu_j^i(k) - \omega_j^i(k) = \nu_j^i(k) + \bar{\nu}_j^i \leq 0$ and that $G_j^i(x(k) - x_*) - (\omega_j^i(k) - \omega_{*,j}^i) = G_j^i(x(k) - x_*) - (-\bar{\nu}_j^i - \nu_{*,j}^i) \geq 0$. Then, one gets (11.14).
- Case 3: $-\bar{\nu}_i < \nu_{\phi,i} < \bar{\nu}_i$. It follows that $\nu_j^i(k) - \omega_j^i(k) = 0$ and one gets (11.14).

Since (11.14) holds for all $j \in \{1, \ldots, n_i\}$, we get that (11.13) holds for all diagonal positive definite matrix T^i. □

Remark 11.3.2 Note that relation (11.13) can be rewritten equivalently as:

$$\left[(\nu^i(k) - \nu_*^i) - (\omega^i(k) - \omega_*^i)\right]^\top T^i \left[G^i(x(k) - x_*) - (\omega^i(k) - \omega_*^i)\right] \leq 0 \tag{11.15}$$

since by Lemma 11.2.5 one gets $\nu_* = \omega_*$.

From the previous lemma, we can deduce that a quadratic constraint is enforced by the ETM (11.11)–(11.12).

Proposition 11.3.3 *Consider matrices $G = \left[G^{1\top} \ldots G^{\ell\top}\right]^\top$ with $G^i \in \mathbb{R}^{n_i \times n_p}$, $i \in \{1, \ldots, \ell\}$ and $T = \mathtt{diag}(T^1 \ldots, T^\ell)$ with $T^i \in \mathbb{R}^{n_i \times n_i}$, $i \in \{1, \ldots, \ell\}$ diagonal positive definite matrices. Consider NN controller π_{ETM} (11.4)–(11.5) with ETM (11.11)–(11.12). Under Assumption 11.2.3, if $x(k)$ belongs to the set $\mathcal{S}$ defined by*

$$\mathcal{S} = \left\{x(k) \in \mathbb{R}^{n_p} : -\bar{\nu} - \nu_* \leq G(x(k) - x_*) \leq \bar{\nu} - \nu_*\right\}$$

then, the following quadratic constraint holds:

$$\left[\nu_\phi(k) - \hat{\omega}_\phi(k)\right]^\top T\left[G(x(k) - x_*) - (\hat{\omega}_\phi(k) - \omega_*)\right] \leq 0. \tag{11.16}$$

Proof Since T is a diagonal matrix, it follows that (11.16) holds if and only if the following holds for all $i \in \{1, \ldots, \ell\}$

$$\left[\nu^i(k) - \hat{\omega}^i(k)\right]^\top T^i\left[G^i(x(k) - x_*) - (\hat{\omega}^i(k) - \omega^i_*)\right] \leq 0. \tag{11.17}$$

Then, let us consider two possible cases according to (11.11). If $\hat{\omega}^i(k) = \omega^i(k)$, then since $x(k) \in \mathcal{S} \subseteq \mathcal{S}^i$, we obtain (11.17) from (11.13) in Lemma 11.3.1. If $\hat{\omega}^i(k) \neq \omega^i(k)$ then we get from (11.11)–(11.12) that $\hat{\omega}^i(k) = \hat{\omega}^i(k-1)$ and that

$$\left[\nu^i(k) - \hat{\omega}^i(k-1)\right]^\top T^i\left[G^i(x(k) - x_*) - (\hat{\omega}^i(k-1) - \omega^i_*)\right] \leq 0,$$

which gives us (11.17). □

Here, it should be highlighted that the particular form of the ETM (11.11)–(11.12) has been chosen such that the quadratic constraint used to abstract the nonlinear activation functions are satisfied at all time. This particular construction of ETMs has already been considered in the case of different activation functions in [1].

11.3.2 Sufficient Conditions for Stability

In this section, we establish sufficient conditions under the form of LMIs that guarantee the regional (local) asymptotic stability of the control system and provide an estimate of the basin of attraction of the equilibrium point.

For further discussions, we introduce the auxiliary variable $\psi_\phi(k) = \nu_\phi(k) - \hat{\omega}_\phi(k)$. Let us remark that at the equilibrium $\psi_* = \nu_* - \hat{\omega}_* = 0$. Moreover, from Proposition 11.3.3, we get that if $x(k)$ belongs to the set $\mathcal{S}$, then

$$\psi_\phi(k)^\top T\left[G(x(k) - x_*) + (\psi_\phi(k) - \nu_\phi(k) - \psi_* + \nu_*)\right] \leq 0. \tag{11.18}$$

Moreover, we can write the control system in the following compact form:

$$\begin{array}{ll} x(k+1) = (A_p + B_p R_\omega)x(k) - B_p N_{u\omega} R\psi_\phi(k) + B_p R_b \\ \nu_\phi(k) \quad = RN_{\nu x}x(k) + (\mathbf{I} - R)\psi_\phi(k) + RN_{\nu b} \end{array} \tag{11.19}$$

with matrices R, R_ω and R_b defined in (11.9). Let us define the matrices

$$R_\phi = \begin{bmatrix} \mathbf{I} & \mathbf{0} \\ RN_{\nu x} & \mathbf{I} - R \\ \mathbf{0} & \mathbf{I} \end{bmatrix}, \bar{A}_p = A_p + B_p R_\omega = A_p + B_p K + B_p N_{u\omega} R N_{\nu x}. \tag{11.20}$$

Theorem 11.3.4 *Consider the control system given by the plant* (11.1)*, the NN controller* π_{ETM} (11.4)–(11.5)*. Assume that there exist a symmetric positive definite matrix* $P \in \mathbb{R}^{n_p \times n_p}$*, a diagonal positive definite matrix* $T = \mathtt{diag}(T^1 \ldots, T^\ell)$ *with* $T^i \in \mathbb{R}^{n_i \times n_i}$*,* $i \in \{1, \ldots, \ell\}$*, a matrix* $Z = \begin{bmatrix} Z^{1\top} & \ldots & Z^{\ell\top} \end{bmatrix}^\top$*,* $Z^i \in \mathbb{R}^{n_i \times n_p}$*,* $i \in \{1, \ldots, \ell\}$*, and a scalar* $0 < \alpha \in \mathbb{R}$*, such that the following matrix inequalities hold*

$$\begin{bmatrix} \bar{A}_p^\top \\ (-B_p N_{u\omega} R)^\top \end{bmatrix} P \begin{bmatrix} \bar{A}_p & -B_p N_{u\omega} R \end{bmatrix} - \begin{bmatrix} P & \mathbf{0} \\ \mathbf{0} & \mathbf{0} \end{bmatrix} - R_\phi^\top \begin{bmatrix} \mathbf{0} & Z^\top \\ \mathbf{0} & -T \\ \mathbf{0} & T \end{bmatrix} - \begin{bmatrix} \mathbf{0} & \mathbf{0} & \mathbf{0} \\ Z & -T & T \end{bmatrix} R_\phi < \mathbf{0}, \tag{11.21}$$

$$\begin{bmatrix} P & Z_j^{i\top} \\ \star & 2\alpha T_{j,j}^i - \alpha^2 (\hat{\nu}_j^i)^{-2} \end{bmatrix} \geq \mathbf{0}, \ \forall i \in \{1, \ldots, \ell\}, \ j \in \{1, \ldots, n_i\} \tag{11.22}$$

where $\hat{\nu}_j^i = \min\left(|-\bar{\nu}_j^i - \nu_{*,j}^i|, |\bar{\nu}_j^i - \nu_{*,j}^i|\right)$*. Let the parameters of the ETMs* (11.11)–(11.12) *be given by* T *and* $G_j = (T_{j,j})^{-1} Z_j$*. Then:*

1. *the control system is locally stable around* x_**,*
2. *the set* $\mathcal{E}(P, x_*) = \{x \in \mathbb{R}^{n_F} : (x - x_*)^\top P (x - x_*) \leq 1\}$ *is an estimate of the domain of attraction for the control system.*

Proof The proof mimics that one in [2]. Consider the candidate Lyapunov function as $V(x(k)) = (x(k) - x_*)^\top P (x(k) - x_*)$ with $\mathbf{0} < P = P^\top \in \mathbb{R}^{n_p \times n_p}$. First, note that from the definition of R_ϕ in (11.20), it follows that

$$\begin{bmatrix} x(k) - x_* \\ \nu_\phi(k) - \nu_* \\ \psi_\phi(k) - \psi_* \end{bmatrix} = R_\phi \begin{bmatrix} x(k) - x_* \\ \psi_\phi(k) - \psi_* \end{bmatrix}$$

which allows us to re-write condition (11.18) as

$$\begin{bmatrix} x(k) - x_* \\ \psi_\phi(k) - \psi_* \end{bmatrix}^\top \begin{bmatrix} \mathbf{0} \\ \mathbf{I} \end{bmatrix} T \begin{bmatrix} G & -\mathbf{I} & \mathbf{I} \end{bmatrix} R_\phi \begin{bmatrix} x(k) - x_* \\ \psi_\phi(k) - \psi_* \end{bmatrix} \leq 0 \tag{11.23}$$

Assume the feasibility of relation (11.22), use the inequality $[\alpha \hat{\nu}_i^{-2} - T_{i,i}] \hat{\nu}_i^2 [\alpha \hat{\nu}_i^{-2} - T_{i,i}] \geq 0$ or equivalently $T_{i,i}^2 \hat{\nu}_i^2 \geq 2\alpha T_{i,i} - \alpha^2 \hat{\nu}_i^{-2}$ and consider the change of variables $G_j = (T_{j,j})^{-1} Z_j$. Then relation (11.22) reads

$$\begin{bmatrix} P & G_j^\top T_{j,j} \\ \star & T_{j,j}^2 \hat{\nu}_j^2 \end{bmatrix} \geq \mathbf{0}, \ \forall j \in \{1, \ldots, n_i\}.$$

That corresponds to ensure

$$(x - x_*)^\top G_i^\top \hat{\nu}_j^{-2} G_j (x - x_*) \leq (x - x_*)^\top P (x - x_*)$$

which ensures that $\mathcal{E}(P, x_*) \subseteq \mathcal{S}$, consequently, Lemma 11.3.1 applies. As the bounds of the set $\mathcal{S}$ are asymmetric, one considers the minimal bound in absolute value: $\hat{\nu}_i = \min(|-\bar{\nu}_i - \nu_{*,i}|, |\bar{\nu}_i - \nu_{*,i}|)$.

Consider that condition (11.21) holds, then there exists a scalar $\epsilon > 0$ such that pre and post multiplying (11.21) by $\left[(x(k)-x_*)^\top \; (\psi_\phi(k)-\psi_*)^\top\right]$ and its transpose, respectively, it holds

$$\begin{bmatrix} x(k)-x_* \\ \psi_\phi(k)-\psi_* \end{bmatrix}^\top \left(\begin{bmatrix} \bar{A}_p^\top \\ (-B_p N_{u\omega} R)^\top \end{bmatrix}^\top P \begin{bmatrix} \bar{A}_p & -B_p N_{u\omega} R \end{bmatrix} - \begin{bmatrix} P & \mathbf{0} \\ \mathbf{0} & \mathbf{0} \end{bmatrix} \right.$$
$$\left. - R_\phi^\top \begin{bmatrix} \mathbf{0} & Z^\top \\ \mathbf{0} & -T \\ \mathbf{0} & T \end{bmatrix} - \begin{bmatrix} \mathbf{0} & \mathbf{0} & \mathbf{0} \\ Z & -T & T \end{bmatrix} R_\phi \right) \begin{bmatrix} x(k)-x_* \\ \psi_\phi(k)-\psi_* \end{bmatrix} < -\epsilon \|x(k)-x_*\|^2. \tag{11.24}$$

Then, replacing Z by TG, and rearranging terms by taking into account the definition of the closed-loop in (11.19), yields:

$$\Delta V(x(k)) - 2 \begin{bmatrix} x(k)-x_* \\ \psi_\phi(k)-\psi_* \end{bmatrix}^\top \begin{bmatrix} \mathbf{0} \\ \mathbf{I} \end{bmatrix} T \begin{bmatrix} G & -\mathbf{I} & \mathbf{I} \end{bmatrix} R_\phi \begin{bmatrix} x(k)-x_* \\ \psi_\phi(k)-\psi_* \end{bmatrix} < -\epsilon \|x(k)-x_*\|^2 \tag{11.25}$$

with $\Delta V(x(k)) = V(x(k+1)) - V(x(k)) = (x(k+1)-x_*)^\top P (x(k+1)-x_*) - (x(k)-x_*)^\top P (x(k)-x_*)$.

By invoking (11.23), we can conclude that if $x(k)$ belongs to the set $\mathcal{S}$, then

$$\Delta V(x(k)) \leq \Delta V(x(k)) - 2 \begin{bmatrix} x(k)-x_* \\ \psi_\phi(k)-\psi_* \end{bmatrix}^\top \begin{bmatrix} \mathbf{0} \\ \mathbf{I} \end{bmatrix} T \begin{bmatrix} G & -\mathbf{I} & \mathbf{I} \end{bmatrix} R_\phi \begin{bmatrix} x(k)-x_* \\ \psi_\phi(k)-\psi_* \end{bmatrix} < -\epsilon \|x(k)-x_*\|^2. \tag{11.26}$$

Inequality (11.26) implies that if $x(k) \in \mathcal{E}(P, x_*)$ then $x(k+1) \in \mathcal{E}(P, x_*)$ (i.e., $\mathcal{E}(P, x_*)$ is an invariant set) and $x(k)$ converges to the equilibrium point x_*. Therefore, $\mathcal{E}(P, x_*)$ is an inner-approximation of the domain of attraction of x_* for the closed-loop system. The proof is complete. □

If matrix $A_p + B_p K$ is Schur–Cohn, Theorem 11.3.4 may also be considered in a global context.

Theorem 11.3.5 *Consider the control system given by the plant* (11.1), *the NN controller* π_{ETM} (11.4)–(11.5). *Assume that there exist a symmetric positive definite matrix* $P \in \mathbb{R}^{n_p \times n_p}$, *a diagonal positive definite matrix* $T = \mathtt{diag}(T^1 \ldots, T^\ell)$ *with* $T^i \in \mathbb{R}^{n_i \times n_i}$, $i \in \{1, \ldots, \ell\}$ *such that the following matrix inequality holds*

$$\begin{bmatrix} \bar{A}_p^\top \\ (-B_p N_{u\omega} R)^\top \end{bmatrix} P \begin{bmatrix} \bar{A}_p & -B_p N_{u\omega} R \end{bmatrix} - \begin{bmatrix} P & \mathbf{0} \\ \mathbf{0} & \mathbf{0} \end{bmatrix} - R_\phi^\top \begin{bmatrix} \mathbf{0} & \mathbf{0} \\ \mathbf{0} & -T \\ \mathbf{0} & T \end{bmatrix} - \begin{bmatrix} \mathbf{0} & \mathbf{0} & \mathbf{0} \\ \mathbf{0} & -T & T \end{bmatrix} R_\phi < \mathbf{0}. \tag{11.27}$$

Then the equilibrium point x_* is globally asymptotically stable for the closed-loop.

Proof The proof follows the same reasoning as that one of Theorem 11.3.4. In the global case Lemma 11.3.1 is applied by choosing $G = \mathbf{0}$ [22]. Hence, relation (11.22) disappears and relation (11.21) is rewritten as in (11.27). The satisfaction of this condition means that one gets

$$\begin{aligned} &\Delta V(x(k)) \leq \\ &\Delta V(x(k)) - 2 \begin{bmatrix} x(k) - x_* \\ \psi_\phi(k) - \psi_* \end{bmatrix}^\top \begin{bmatrix} \mathbf{0} \\ \mathbf{I} \end{bmatrix} T \begin{bmatrix} \mathbf{0} & -\mathbf{I} & \mathbf{I} \end{bmatrix} R_\phi \begin{bmatrix} x(k) - x_* \\ \psi_\phi(k) - \psi_* \end{bmatrix} \\ &\qquad < -\epsilon \|x(k) - x_*\|^2. \end{aligned} \tag{11.28}$$

In other words, $x(k)$ globally converges to the equilibrium point x_* and the proof is complete. □

In order to better understand the necessary assumption on the stability of matrix $A = A_p + B_p K$, one can study the feasibility of condition (11.27). To do this one can rewrite condition (11.27) as follows:

$$\begin{bmatrix} \bar{A}_p^\top \\ (-B_p N_{u\omega} R)^\top \end{bmatrix} P \begin{bmatrix} \bar{A}_p & -B_p N_{u\omega} R \end{bmatrix} - \begin{bmatrix} P & \mathbf{0} \\ \mathbf{0} & \mathbf{0} \end{bmatrix} + He\{\begin{bmatrix} \mathbf{0} \\ \mathbf{I} \end{bmatrix} T \begin{bmatrix} R N_{\nu x} & -R \end{bmatrix}\} < \mathbf{0}. \tag{11.29}$$

By pre- and post-multiplying (11.29) by $\begin{bmatrix} \mathbf{I} & N_{\nu x}^\top \\ \mathbf{0} & \mathbf{I} \end{bmatrix}$ and $\begin{bmatrix} \mathbf{I} & \mathbf{0} \\ N_{\nu x} & \mathbf{I} \end{bmatrix}$, one gets

$$\begin{bmatrix} (A_p + B_p K)^\top \\ (-B_p N_{u\omega} R)^\top \end{bmatrix} P \begin{bmatrix} (A_p + B_p K) & -B_p N_{u\omega} R \end{bmatrix} - \begin{bmatrix} P & \mathbf{0} \\ \mathbf{0} & \mathbf{0} \end{bmatrix} + He\{\begin{bmatrix} N_{\nu x}^\top \\ \mathbf{I} \end{bmatrix} T \begin{bmatrix} \mathbf{0} & -R \end{bmatrix}\} < \mathbf{0}.$$

Note that the block (1, 1) of the expression above is $(A_p + B_p K)^\top P (A_p + B_p K) - P$. Hence, it is clear that a necessary condition for the feasibility of (11.29), or equivalently for (11.27), is to have $(A_p + B_p K)^\top P (A_p + B_p K) - P < \mathbf{0}$. That corresponds to have the assumption that the matrix $A_p + B_p K$ is Schur–Cohn.

11.3.3 Optimization Procedure

Let us first provide some comments regarding conditions (11.21) and (11.22). Condition (11.21) is linear in the decision variables P, T and Z, whereas condition (11.22) is only linear in P but not in T and α. However, condition (11.22) will become linear if we fix α.

In this section, an optimization scheme is proposed to reduce the evaluation activity in the neural network, but taking into account the trade-off between the update saving and the size of the inner approximation of the region of attraction of x_*. Then, consider $\mathcal{X}_0 = \mathcal{E}(P_0, x_*) = \{x \in \mathbb{R}^{n_p} : (x - x_*)^\top P_0(x - x_*) \leq 1\}$, with $\mathbf{0} < P_0^\top = P_0 \in \mathbb{R}^{n_p \times n_p}$, a set to be contained in the region of attraction of the closed-loop system, i.e., $\mathcal{X}_0 \subseteq \mathcal{E}(P, x_*)$. Then the inclusion is obtained by imposing

$$\begin{bmatrix} P_0 & P \\ P & P \end{bmatrix} \geq \mathbf{0}. \tag{11.30}$$

Then, by taking inspiration of [2], the following optimization scheme that can help to enlarge such an estimate, while reducing the amount of computation:

$$\begin{array}{l} \min \rho \\ \text{subject to } (21), (22), (30),\ M_1 \geq -\rho \mathbf{I} \end{array} \tag{11.31}$$

with $\rho > 0$ and M_1 is the matrix defined in the left-hand term of (11.21).

By considering the optimization problem (11.31) we are trying to indirectly (1) maximizing the inner approximation of the region of attraction, and (2) approaching the infeasibility of the stability condition.

11.4 Simulations

Consider the inverted pendulum system with mass $m = 0.15$ kg, length $l = 0.5$ m, and friction coefficient $\mu = 0.5$ Nms/rad. The following discrete-time model describes its dynamics:

$$\begin{bmatrix} x_1(k+1) \\ x_2(k+1) \end{bmatrix} = \begin{bmatrix} 1 & \delta \\ \dfrac{g\delta}{l} & 1 - \dfrac{\delta\mu}{ml^2} \end{bmatrix} \begin{bmatrix} x_1(k) \\ x_2(k) \end{bmatrix} + \begin{bmatrix} 0 \\ \dfrac{\delta}{ml^2} \end{bmatrix} u(k), \tag{11.32}$$

where the states $x_1(k)$ and $x_2(k)$ represent respectively the angular position (rad) and velocity (rad/s), $u(k)$ is the control input (Nm) and $\delta = 0.02$ is the sampling time [26].

To stabilize (11.32), we have designed a controller π under the form of a 2-layer, feedforward neural network, with $n_1 = n_2 = 32$, the saturation as the activation function for both layers, and a linear term $Kx(k)$ in the output of the neural network.

Table 11.1 Comparison of update rates (columns 2 and 3) and $\mathcal{E}(P, x_*)$ areas (column 4) achieved with optimization procedure (11.31) for different gains K

Gains	1st layer (%)	2nd layer (%)	Area
$K = \begin{bmatrix}-0.1 & 0\end{bmatrix}$	23.05	22.61	111.22
$K = \begin{bmatrix}-0.3 & 0\end{bmatrix}$	21.20	20.67	126.17
$K = \begin{bmatrix}-0.5 & 0\end{bmatrix}$	18.61	17.97	135.45

For training purpose only, we replaced the saturation by its smooth approximation provided by tanh, making it possible to rely on MATLAB®'s Reinforcement Learning toolbox. During the training, the agent's decision is characterized by a Gaussian distribution probability with mean $\pi(x(k))$ and standard deviation σ. In addition, to illustrate the applicability of our condition to equilibrium points different from 0, we have not set the bias in neural network to zero during training. After training, the policy mean π is used as the deterministic controller $u(k) = \pi(x(k))$ with the saturation as the activation function.

First, we have designed the ETMs (11.11)–(11.12) for 3 different linear gains K to show the influence of these gains on the transmission saving and the size of the estimate of the region of attraction. In this case, by setting $\bar{\nu} = -\underline{\nu} = 1 \times 1_{64\times 1}$, we use the optimization procedure (11.31) with

$$P_0 = \begin{bmatrix} 0.2916 & 0.0054 \\ \star & 0.0090 \end{bmatrix}, \qquad (11.33)$$

chosen based on the maximum estimate of the region of attraction obtained without trigger and $\alpha = 9 \times 10^{-4}$. Then, we simulate the feedback system for 100 initial conditions belonging to the estimates of the region of attraction. The results are shown in Table 11.1.

To illustrate, considering the first gain, $K = \begin{bmatrix}-0.1 & 0\end{bmatrix}$, we have plotted in Fig. 11.2 the state trajectories and the control signals of the feedback system for the 100 initial condition belonging to the boundary of the estimate of region of attraction. Also, by considering the initial condition $x_0 = \begin{bmatrix}-1.8341 & -6.9559\end{bmatrix}^\top$, we have plotted in the top of Fig. 11.3, the inter-event time of both layers ω^1 and ω^2 and in the middle and bottom, the outputs of the first neuron of each layer, respectively. We can see that all curves converge to their respective equilibrium points with an update rate of 23.71% and 22.29% for ω^1 and ω^2, respectively, thus reducing the computational cost associated with the control law evaluation. In addition, it is possible to see that ω_1^2 saturates in the first instants of simulation.

Moreover, Figs. 11.4, 11.5 and 11.6, depict respectively for each gain in Table 11.1, the estimate of the region of attraction $\mathcal{E}(P, x_*)$ (blue solid line) and the set of

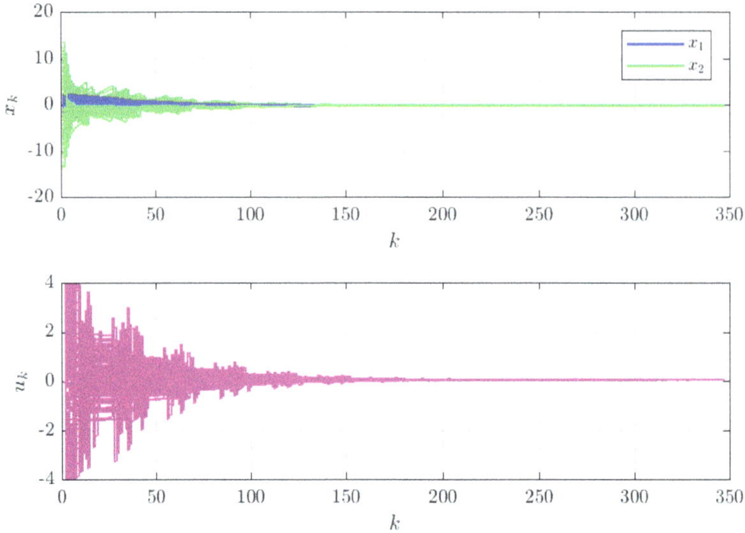

Fig. 11.2 The feedback response of the system using $K = [-0.1 \quad 0]$ for different initial conditions in the border of the $\mathcal{E}(P, x_*)$. Top: system' states; bottom: control signal computed (blue-line)

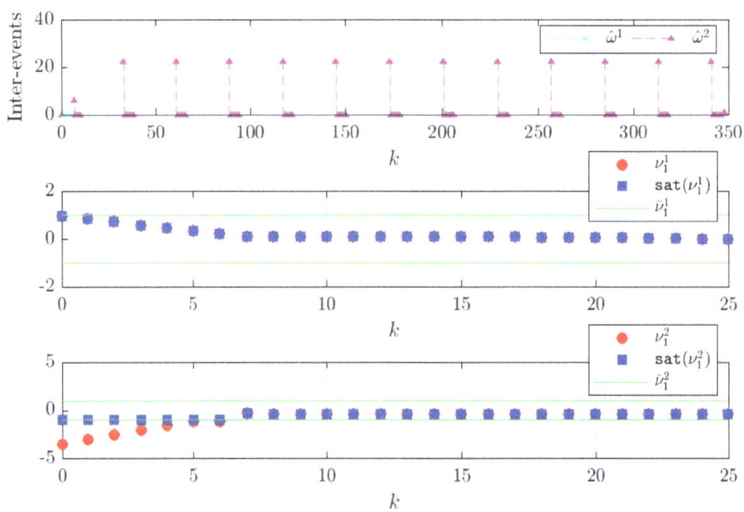

Fig. 11.3 Top: Inter-event time of both layers ω^1 and ω^2; Middle and bottom: Outputs of the first neuron of each layer; for $K = [-0.1 \quad 0]$ and $x_0 = [-1.8341 \quad -6.9559]^\top$

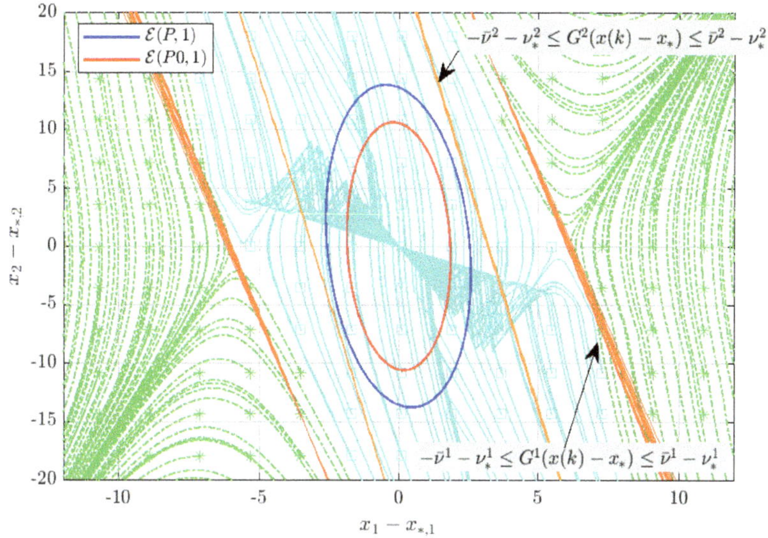

Fig. 11.4 Estimate of the region of attraction for the feedback system in the plan $(x_1 - x_{*,1} \times x_2 - x_{*,2})$ for $K = [-0.1\ \ 0]$

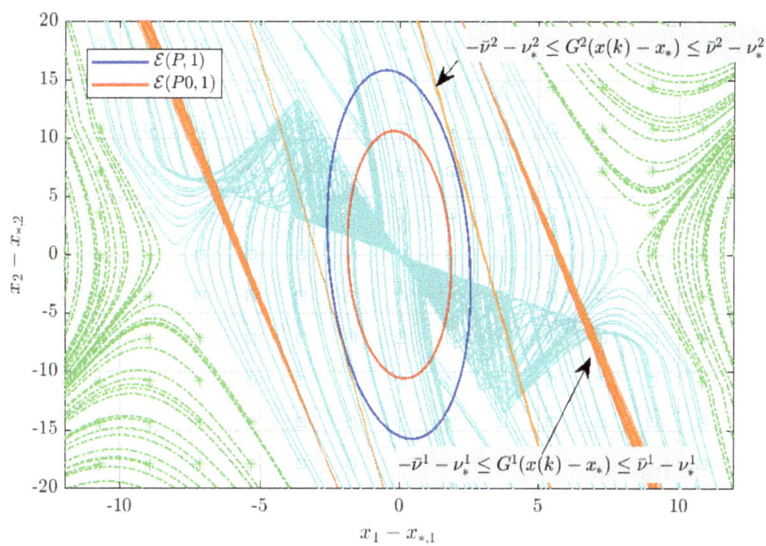

Fig. 11.5 Estimate of the region of attraction for the feedback system in the plan $(x_1 - x_{*,1} \times x_2 - x_{*,2})$ for $K = [-0.3\ \ 0]$

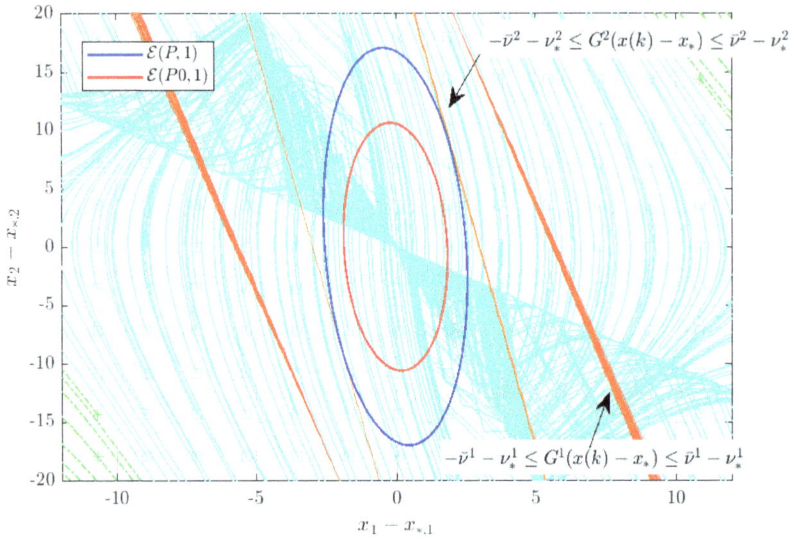

Fig. 11.6 Estimate of the region of attraction for the feedback system in the plan $(x_1 - x_{*,1} \times x_2 - x_{*,2})$ for $K = [-0.5 \;\; 0]$

admissible initial states $\mathcal{E}(P_0, x_*)$ (red solid line). Some convergent and divergent trajectories are also shown in cyan dashed lines and green dash-dotted lines, starting from the points marked with cyan circles and green asterisks, respectively. Note that in all cases the estimate of the region of attraction does not overstep the bounds $\{-\bar{\nu} - \nu_* \leq G(x - x_*) \leq \bar{\nu} - \nu_*\}$ (orange lines) for both layers, as expected. Also, we can verify that as the absolute value of the gain increases, the size of the estimate of the region of attraction increases, which can be verified in the last column of Table 11.1, which shows the area of the ellipses $\mathcal{E}(P, x_*)$. Note that the estimate of the region of attraction increased 13.44% and 21.78%, respectively, in relation to the first case.

11.5 Conclusion

This chapter proposed an event-triggering strategy, based on (local) sectors conditions related to the activation functions in order to decide whether the outputs of the layers should be transmitted through the network or not. This chapter can be viewed as an extension of our previous works [1, 2]. Theoretical conditions allowed to design the event-triggering mechanism and to estimate the region of attraction, while preserving the stability of the closed-loop system. Simulations have illustrated the effectiveness of the event-triggering scheme, showing a significant reduction of the transmission activity in the neural network even if the number of layer increase.

The results open the doors for future works as studying other event-triggering structures and different abstractions in order to reduce the conservatism of the conditions. Another interesting direction could be to study co-design problem, that is, design in the same time the neural network and the event-triggering mechanism.

Acknowledgements This work was supported by the "Agence Nationale de la Recherche" (ANR) under Grant HANDY ANR-18-CE40-0010.

References

1. De Souza, C., Girard, A., Tarbouriech, S.: Event-triggered neural network control using quadratic constraints for perturbed systems. Automatica (2023). To appear
2. de Souza, C., Tarbouriech, S., Girard, A.: Event-triggered neural network control for lti systems. IEEE Control Syst. Lett. **7**, 1381–1386 (2023)
3. de Souza, C., Tarbouriech, S., Leite, V.J.S., Castelan, E.B.: Co-design of an event-triggered dynamic output feedback controller for discrete-time LPV systems with constraints. J. Franklin Inst. (2020). Submitted
4. Fazlyab, M., Morari, M., Pappas, G.J.: Safety verification and robustness analysis of neural networks via quadratic constraints and semidefinite programming. IEEE Trans. Autom, Control (2020)
5. Fazlyab, M., Robey, A., Hassani, H., Morari, M., Pappas, G.: Efficient and accurate estimation of lipschitz constants for deep neural networks. Adv. Neural Inf. Process. Syst. 32 (2019)
6. Gao, Y., Guo, X., Yao, R., Zhou, W., Cattani, C.: Stability analysis of neural network controller based on event triggering. J. Franklin Inst. **357**(14), 9960–9975 (2020)
7. Girard, A.: Dynamic triggering mechanisms for event-triggered control. IEEE Trans. Autom. Control **60**(7), 1992–1997 (2014)
8. Heemels, W.P.M.H., Donkers, M.C.F., Teel, A.R.: Periodic event-triggered control for linear systems. IEEE Trans. Autom. Control **58**(4), 847–861 (2012)
9. Hertneck, M., Köhler, J., Trimpe, S., Allgöwer, F.: Learning an approximate model predictive controller with guarantees. IEEE Control Syst. Lett. **2**(3), 543–548 (2018)
10. Hu, T., Lin, Z.: Control Systems with Actuator Saturation: Analysis and Design. Birkhäuser, Boston (2001)
11. Hu, T., Teel, A.R., Zaccarian, L.: Stability and performance for saturated systems via quadratic and nonquadratic Lyapunov functions. IEEE Trans. Autom. Control **51**(11), 1770–1786 (2006)
12. Jin, M., Lavaei, J.: Stability-certified reinforcement learning: a control-theoretic perspective. IEEE Access **8**, 229086–229100 (2020)
13. Karg, B., Lucia, S.: Efficient representation and approximation of model predictive control laws via deep learning. IEEE Trans. Cybern. **50**(9), 3866–3878 (2020)
14. Kim, K.-K.K., Patrón, E.R., Braatz, R.D.: Standard representation and unified stability analysis for dynamic artificial neural network models. Neural Netw. **98**, 251–262 (2018)
15. Pauli, P., Koch, A., Berberich, J., Kohler, P., Allgöwer, F.: Training robust neural networks using lipschitz bounds. IEEE Control Syst. Lett. **6**, 121–126 (2021)
16. Pauli, P., Köhler, J., Berberich, J., Koch, A., Allgöwer, F.: Offset-free setpoint tracking using neural network controllers. In: Learning for Dynamics and Control, pp. 992–003. PMLR (2021)
17. Revay, M., Wang, R., Manchester, I.R.: Lipschitz bounded equilibrium networks. arXiv preprint arXiv:2010.01732 (2020)
18. Sahoo, A., Xu, H., Jagannathan, S.: Neural network-based event-triggered state feedback control of nonlinear continuous-time systems. IEEE Trans. Neural Netw. Learn. Syst. **27**(3), 497–509 (2015)

19. Schmitendorf, W.E., Barmish, B.R.: Null controllability of linear systems with constrained controls. SIAM J. Contr. Opt. **18**(4), 327–345 (1980)
20. Tabuada, P.: Event-triggered real-time scheduling of stabilizing control tasks. IEEE Trans. Autom. Control **52**(9), 1680–1685 (2007)
21. Tanaka, K.: An approach to stability criteria of neural-network control systems. IEEE Trans. Neural Netw. **7**(3), 629–642 (1996)
22. Tarbouriech, S., Garcia, G., Gomes da Silva Jr. J.M., Queinnec, I.: Stability and Stabilization of Linear Systems with Saturating Actuators. Springer (2011)
23. Vamvoudakis, K.G.: Event-triggered optimal adaptive control algorithm for continuous-time nonlinear systems. IEEE/CAA J. Autom. Sin. **1**(3), 282–293 (2014)
24. Xiang, W., Musau, P., Wild, A.A., Lopez, D.M., Hamilton, N., Yang, X., Rosenfeld, J., Johnson, T.T.: Verification for machine learning, autonomy, and neural networks survey. arXiv preprint arXiv:1810.01989 (2018)
25. Yin, H., Seiler, P., Arcak, M.: Stability analysis using quadratic constraints for systems with neural network controllers. IEEE Trans. Autom, Control (2021)
26. Yin, H., Seiler, P., Jin, M., Arcak, M.: Imitation learning with stability and safety guarantees. IEEE Control Syst. Lett. **6**, 409–414 (2021)
27. Zaccarian, L., Teel, A.R.: Modern Anti-windup Synthesis: Control Augmentation for Actuator Saturation, vol. 36. Princeton University Press (2011)
28. Zhang, X., Bujarbaruah, M., Borrelli, F.: Safe and near-optimal policy learning for model predictive control using primal-dual neural networks. In: American Control Conference, pp. 354–359. IEEE (2019)
29. Zhong, X., Ni, Z., He, H., Xu, X., Zhao, D.: Event-triggered reinforcement learning approach for unknown nonlinear continuous-time system. In: International Joint Conference on Neural Networks (IJCNN), pp. 3677–3684. IEEE (2014)

Chapter 12
Data-Driven Stabilization of Nonlinear Systems via Taylor's Expansion

Meichen Guo, Claudio De Persis, and Pietro Tesi

Abstract Lyapunov's indirect method is one of the oldest and most popular approaches to model-based controller design for nonlinear systems. When the explicit model of the nonlinear system is unavailable for designing such a linear controller, finite-length off-line data is used to obtain a data-based representation of the closed-loop system, and a data-driven linear control law is designed to render the considered equilibrium locally asymptotically stable. This work presents a systematic approach for data-driven linear stabilizer design for continuous-time and discrete-time general nonlinear systems. Moreover, under mild conditions on the nonlinear dynamics, we show that the region of attraction of the resulting locally asymptotically stable closed-loop system can be estimated using data.

12.1 Introduction

Most control approaches of nonlinear systems are based on well-established models of the system constructed by prior knowledge or system identification. When the models are not explicitly constructed, nonlinear systems can be directly controlled using input–output data. The problem of controlling a system via input–output data without explicitly identifying the model has been gaining more and more attentions for both linear and nonlinear systems. An early survey of data-driven control

M. Guo (✉)
Delft Center for Systems and Control, Delft University of Technology, 2628 CD Delft, The Netherlands
e-mail: m.guo@tudelft.nl

C. De Persis
Engineering and Technology Institute Groningen, University of Groningen, Nijenborgh 4, 9747 Groningen, The Netherlands
e-mail: c.de.persis@rug.nl

P. Tesi
Department of Information Engineering, University of Florence, 50139 Florence, Italy
e-mail: pietro.tesi@unifi.it

R. Postoyan et al. (eds.), *Hybrid and Networked Dynamical Systems*,
Lecture Notes in Control and Information Sciences 493,
https://doi.org/10.1007/978-3-031-49555-7_12

methods can be found in [17]. More recently, the authors of [32] developed an online control approach, the work [18] utilized the dynamic linearization data models for discrete-time non-affine nonlinear systems, the authors of [13, 30] considered feedback linearizable systems, and the works [2, 20] designed data-driven model predictive controllers. Inspired by Willems et al.'s fundamental lemma, [11] proposed data-driven control approaches for linear and nonlinear discrete-time systems. Using a matrix Finsler's lemma, [34] applied data-driven control to Lur'e systems. The authors of [10] used state-dependent representation and proposed an online optimization method for data-driven stabilization of nonlinear dynamics. For polynomial systems, [14] designed global stabilizers using noisy data, and [23] synthesized data-driven safety controllers. The recent work [22] investigated dissipativity of nonlinear systems based on polynomial approximation.

Related works. Some recent works related to nonlinear data-driven control and the region of attraction (RoA) estimation are discussed in what follows.

Deriving a data-based representation of the dynamics is one of the important steps in data-driven control of unknown nonlinear systems. If the controlled systems are of certain classes, such as polynomial systems having a known degree, the monomials of the state can be chosen as basis functions to design data-driven controllers such as presented in [14, 15]. By integrating noisy data and side information, [1] showed that unknown polynomial dynamics can be learned via semi-definite programming. When the nonlinearities satisfy quadratic constraints, data-driven stabilizer was developed in [21]. With certain knowledge and assumptions on the nonlinear basis functions, systems containing more general types of nonlinearities have also been studied in recent works. For instance, under suitable conditions, some nonlinear systems can be lifted into polynomial systems in an extended state for control, such as the results shown in [29, Sect. IV] and [19, Sect. 3.2]. Using knowledge of the basis functions, [24] designed data-driven controllers by (approximate) cancellation of the nonlinearity. When the system nonlinearities cannot be expressed as a combination of known functions, [24] presented data-driven local stabilization results by choosing basis functions carefully such that the neglected nonlinearities are small in a known set of the state. On the other hand, if the knowledge on the basis functions is not available, approximations of the nonlinear systems are often involved. The previous work [11] tackled the nonlinear data-driven control problem by linearizing the dynamics around the known equilibrium and obtaining a local stability result. According to these existing results, it is clear that the efficiency and the performance of data-driven controllers can be improved via prior knowledge such as specific classes of the systems or the nonlinear basis functions. Nonetheless, there is still a lack of comprehensive investigation of the more general case where the nonlinear basis functions cannot be easily and explicitly obtained.

The RoA estimation is another relevant topic in nonlinear control. For general nonlinear systems, it is common that the designed controllers only guarantee local stability. Hence, it is of importance to obtain the RoA of the closed-loop systems for the purpose of theoretical analysis as well as engineering applications. Unfortunately, for general nonlinear systems, it is extremely difficult to derive the exact RoA even when the model is explicitly known. A common solution is to estimate the RoA based on Lyapunov functions. Using Taylor's expansion and considering the worst-case remainders, [7] estimated the RoA of uncertain non-polynomial systems via linear matrix inequality (LMI) optimizations. RoA analysis for polynomial systems was presented in [31] using polynomial Lyapunov functions and sum of squares (SOS) optimizations. The authors of [33] studied uncertain nonlinear systems subject to perturbations in certain forms and used the SOS technique to compute invariant subsets of the RoA. It is noted that, in these works, the RoA estimation winds up in solving bilinear optimization problems, and techniques such as bisection or special bilinear inequality solver are required to find the solutions. For nonlinear systems without explicit models, there are also efforts devoted to learning the RoA by various approaches. The authors of [6] developed a sampling-based approach for a class of piecewise continuous nonlinear systems that verifies stability and estimates the RoA using Lyapunov functions. Based on the converse Lyapunov theorem, [9] processed system trajectories to lift a Lyapunov function whose level sets lead to an estimation of the RoA. Using the properties of recurrent sets, [27] proposed an approach that learns an inner approximation of the RoA via finite-length trajectories. It should be pointed out that, all the aforementioned works focus on stability analysis of autonomous nonlinear systems, i.e., the control design is not considered.

Contributions. For general nonlinear systems, this chapter presents a data-driven approach to simultaneously obtaining a Lyapunov function and designing a state feedback stabilizer that renders the known equilibrium locally asymptotically stable. Specifically, the unknown dynamics are approximated by linear dynamics with an approximation error. Then, linear stabilizers are designed for the approximated models using finite-length input-state data collected in an off-line experiment. To handle the approximation error we conduct the experiment close to the known equilibrium such that the approximation error is small with a known bound. An over-approximation of all the feasible dynamics is then found using the collected data, and Petersen's lemma [25] is used for the controller design. The data-driven stabilizer design can be seen as a generalization of the nonlinear control result in the previous work [11, Sect. V.B]. Specifically, this work considers the linear approximations of both continuous-time and discrete-time systems.

For estimating the RoA with the designed data-driven controller, we first derive an estimation of the approximation error by assuming a known bound on the derivatives of the unknown functions. With the help of the Positivstellensatz [28] and the SOS relaxations, we derive data-driven conditions that verify whether a given sublevel set of the obtained Lyapunov function is an invariant subset of the RoA. This is achieved by finding a sufficient condition for the negativity of the derivative of the Lyapunov function based on the estimation of the approximation error. The conditions are derived via data and some prior knowledge on the dynamics, and can be easily solved by software such as MATLAB®. The estimated RoA gives insights to the closed-loop system under the designed data-driven controller and is relevant for both theoretical and practical purposes. Note that alternatively, the RoA can be estimated based on other data-driven methods, such as the ones developed in [6, 9, 27]. Simulations results on the inverted pendulum show the applicability of the design and estimation approach.

The rest of the chapter is arranged as follows. Section 12.2 formulates the nonlinear data-driven stabilization problem. Data-based descriptions of feasible dynamics consistent with the collected data and the over-approximation of the feasible set is presented in Sect. 12.3. Data-driven control designs for both continuous-time and discrete-time systems are presented in Sect. 12.4. The data-driven characterization of the RoA is derived in Sect. 12.5. Numerical results and analysis on an example are illustrated in Sect. 12.6. Finally, Sect. 12.7 summarizes this chapter.

Notation. Throughout the chapter, $A \succ (\succeq) 0$ denotes that matrix A is positive (semi-) definite, and $A \prec (\preceq) 0$ denotes that matrix A is negative (semi-) definite. For vectors $a, b \in \mathbb{R}^n$, $a \preceq b$ means that $a_i \leq b_i$ for all $i = 1, \ldots, n$. $\|\cdot\|$ denotes the Euclidean norm.

12.2 Data-Driven Nonlinear Stabilization Problem

Consider a general nonlinear continuous-time system

$$\dot{x} = f(x, u) \tag{12.1}$$

or a nonlinear discrete-time system

$$x^+ = f(x, u), \tag{12.2}$$

where the state $x \in \mathbb{R}^n$ and the input $u \in \mathbb{R}^m$. Assume that (x_e, u_e) is a known equilibrium of the system to be stabilized. For simplicity and without loss of generality, in this chapter we let $(x_e, u_e) = (0, 0)$, as any known equilibrium can be converted to the origin by a change of coordinates.

To gather information regarding the nonlinear dynamics, E experiments are performed on the system, where E is an integer satisfying $1 \leq E \leq T$ and T is the total number of collected samples. On one extreme, one could perform 1 single experiment during which a total of T samples are collected. On the other extreme, one could perform T independent experiments, during each one of which a single sample is collected. The advantage of short multiple experiments is that they allow information collection about the system at different points in the state space without incurring problems due to the free evolution of the system.

Either way, a data set $\mathcal{DS}_c := \{\dot{x}(t_k); x(t_k); u(t_k)\}_{k=0}^{T-1}$ for the continuous-time system, or $\mathcal{DS}_d := \{x(t_k); u(t_k)\}_{k=0}^{T}$ for the discrete-time system can be obtained. Organize the data collected in the experiment(s) as

$$
\begin{aligned}
X_0 &= \left[x(t_0) \ldots x(t_{T-1})\right], \\
U_0 &= \left[u(t_0) \ldots u(t_{T-1})\right], \\
X_1 &= \left[\dot{x}(t_0) \ldots \dot{x}(t_{T-1})\right] \text{ for continuous-time systems,} \\
X_1 &= \left[x(t_1) \ldots x(t_T)\right] \text{ for discrete-time systems.}
\end{aligned}
$$

Remark 12.2.1 (*On the experimental data*) To avoid further complicating the problem and focus on the unknown general nonlinear dynamics, we do not consider external disturbances in the dynamics or measurement noise in the data. In the case of disturbed dynamics or noisy data, if some bounds on the disturbance/noise data are known then we can follow the same approach used in this work to design a data-driven controller, *cf.* [24].

Data-driven Stabilization Problem

Use the data set $\mathcal{DS}_c$ ($\mathcal{DS}_d$) to design a feedback controller $u = Kx$ for the system (12.1) ((12.2)) such that the origin is locally asymptotically stable for the closed-loop system, and an estimation of the RoA with respect to the origin is derived.

12.3 Data-Based Feasible Sets of Dynamics

As limited information is available for the nonlinear dynamics, a natural way to deal with the unknown model is to approximate it as a linear model in a neighborhood of the considered equilibrium. Due to the approximation error, true dynamics of the linearized models cannot be uniquely determined by the collected data. Instead of explicitly identifying the linearized model from the data set $\mathcal{DS}_c$ or $\mathcal{DS}_d$, in this section, we find feasible sets that contain all linear dynamics that are consistent with the collected data. To achieve this, we will use the approach proposed in [3], where an over-approximation of the feasible set is found by solving an optimization problem depending on the data set and the bound of the approximation error during the experiment.

In what follows, we will address the linear approximations of the continuous-time system (12.1) and the discrete-time system (12.2), and the over-approximation of the feasible sets of dynamics.

For the system (12.1), denote each element of f as f_i, and let $f_i \in C^1(\mathbb{R}^n \times \mathbb{R}^m)$, $i = 1, \ldots, n$. The linear approximation of (12.1) at the origin is

$$
\dot{x} = Ax + Bu + R(x, u), \tag{12.3}
$$

where $R(x, u)$ denotes the approximation error and

$$A = \left.\frac{\partial f(x,u)}{\partial x}\right|_{(x,u)=(0,0)}, \quad B = \left.\frac{\partial f(x,u)}{\partial u}\right|_{(x,u)=(0,0)}.$$

Using the same definitions, the linear approximation of the discrete-time nonlinear systems are given as

$$x^+ = Ax + Bu + R(x,u). \tag{12.4}$$

For both types of systems, one can treat the approximation error $R(x,u)$ as a disturbance that affects the data-driven characterization of the unknown dynamics, and focus on controlling the approximated linear dynamics to obtain a locally stabilizing controller. Then, by tackling the impact of $R(x,u)$, the RoA of the closed-loop system can also be characterized.

Assumption 12.3.1 (*Bound on the approximation error*) For $k = 0, \ldots, T$ and a known γ,

$$R(x(t_k), u(t_k))^\top R(x(t_k), u(t_k)) \le \gamma^2. \tag{12.5}$$

Remark 12.3.2 (*Bound on the approximation error*) Assumption 12.3.1 gives an *instantaneous* bound on the maximum amplitude of the approximation error during the experiment. The bound can be obtained by prior knowledge of the model, such as the physics of the system. If such knowledge is unavailable, one may resort to an over-estimation of γ.

Next, we determine the feasible set of dynamics that are consistent with the data sets.

Denote $S = [B \;\; A]$. Based on the dynamics (12.3) and (12.4), at each time t_k, $k = 0, \ldots, T-1$, the collected data satisfies

$$\dot{x}(t_k) = S \begin{bmatrix} u(t_k) \\ x(t_k) \end{bmatrix} + R\big(x(t_k), u(t_k)\big)$$

for continuous-time systems and

$$x(t_{k+1}) = S \begin{bmatrix} u(t_k) \\ x(t_k) \end{bmatrix} + R\big(x(t_k), u(t_k)\big)$$

for discrete-time systems. Under Assumption 12.3.1, at each time $t_k, k = 0, \ldots, T-1$, the matrices $\widehat{S} = [\widehat{B} \;\; \widehat{A}]$ consistent with the data belongs to the set

$$\mathcal{C}_k = \left\{\widehat{S} : \mathbf{C}_k + \widehat{S}\mathbf{B}_k + \mathbf{B}_k^\top \widehat{S}^\top + \widehat{S}\mathbf{A}_k\widehat{S}^\top \preceq 0\right\}, \tag{12.6}$$

where

$$\mathbf{A}_k = l(t_k)l(t_k)^\top, \quad \mathbf{B}_k = -l(t_k)\dot{x}(t_k)^\top,$$
$$\mathbf{C}_k = \dot{x}(t_k)\dot{x}(t_k)^\top - \gamma^2 I, \quad l(t_k) = \begin{bmatrix} u(t_k) \\ x(t_k) \end{bmatrix}$$

for continuous-time systems, and the same definitions except for

$$\mathbf{B}_k = -l(t_k)x(t_{k+1})^\top, \quad \mathbf{C}_k = x(t_{k+1})x(t_{k+1})^\top - \gamma^2 I$$

hold for discrete-time systems. Then, the feasible set of matrices $\widehat{S}$ that is consistent with all data collected in the experiment(s) is the intersection of all the sets $\mathcal{C}_k$, i.e., $\mathcal{I} = \bigcap_{k=0}^{T-1} \mathcal{C}_k$.

Though the exact set $\mathcal{I}$ is difficult to obtain, an over-approximation of $\mathcal{I}$ in the form of a matrix ellipsoid and of minimum size can be computed. Denote the over-approximation set as

$$\overline{\mathcal{I}} := \left\{ \widehat{S} : \overline{\mathbf{C}} + \widehat{S}\overline{\mathbf{B}} + \overline{\mathbf{B}}^\top \widehat{S}^\top + \widehat{S}\overline{\mathbf{A}}\widehat{S}^\top \preceq 0 \right\}, \tag{12.7}$$

where $\overline{\mathbf{A}} = \overline{\mathbf{A}}^\top \succ 0$, $\overline{\mathbf{C}}$ is set as $\overline{\mathbf{C}} = \overline{\mathbf{B}}^\top \overline{\mathbf{A}}^{-1} \overline{\mathbf{B}}^\top - \delta I$ and $\delta > 0$ is a constant fixed arbitrarily. Following [3, Sect. 5.1], the set $\overline{\mathcal{I}}$ can be found by solving the optimization problem

$$\begin{aligned}
&\underset{\overline{\mathbf{A}},\overline{\mathbf{B}},\overline{\mathbf{C}}}{\text{minimize}} && -\log\det(\overline{\mathbf{A}}) \\
&\text{subject to} && \overline{\mathbf{A}} = \overline{\mathbf{A}}^\top \succ 0 \\
& && \tau_k \geq 0,\; k = 0, \ldots, T-1 \\
& && \begin{bmatrix} -\delta I - \sum_{k=0}^{T-1} \tau_k \mathbf{C}_k & \overline{\mathbf{B}}^\top - \sum_{k=0}^{T-1} \tau_k \mathbf{B}_k^\top & \overline{\mathbf{B}}^\top \\ \overline{\mathbf{B}} - \sum_{k=0}^{T-1} \tau_k \mathbf{B}_k & \overline{\mathbf{A}} - \sum_{k=0}^{T-1} \tau_k \mathbf{A}_k & 0 \\ \overline{\mathbf{B}} & 0 & -\overline{\mathbf{A}} \end{bmatrix} \preceq 0.
\end{aligned} \tag{12.8}$$

Proposition 12.3.3 (Over-approximated feasible set) *Consider the data set $\mathcal{DS}_c$ ($\mathcal{DS}_d$) collected from the dynamics* (12.3) *((*12.4*)), which satisfies Assumption 12.3.1. If the optimization problem* (12.8) *is feasible for $\mathbf{A}_k$, $\mathbf{B}_k$, and $\mathbf{C}_k$ defined in* (12.6)*, then the set $\overline{\mathcal{I}}$ defined in* (12.7) *is an over-approximation set of all dynamics that is consistent with the data set $\mathcal{DS}_c$ ($\mathcal{DS}_d$).*

Remark 12.3.4 (*Persistency of excitation*) As pointed out in [3, Sect. 3.1], if the collected data is rich enough, i.e., $\begin{bmatrix} U_0 \\ X_0 \end{bmatrix}$ has full row rank, then the intersection set $\mathcal{I}$ is bounded, which allows the optimization problem (12.8) to have a solution. Hence, $\begin{bmatrix} U_0 \\ X_0 \end{bmatrix}$ having full row rank implies the feasibility of (12.8).

12.4 Data-Driven Controller Design

Stabilizing the linear approximation of the unknown system renders the origin locally asymptotically stable as the approximation error $R(x, u)$ contains higher-order terms and converges to the origin faster than the linear part in a neighborhood of the origin. Hence, the objective of the controller design is to stabilize the origin for all dynamics belonging to the over-approximation set $\overline{\mathcal{I}}$. This can be achieved in the same manner as done in [4] via Petersen's lemma; see Appendix 12.8.1 for background material. For the completeness of this work, we include the following results on designing data-driven local stabilizers using Petersen's lemma.

Theorem 12.4.1 (Data-driven controller design for continuous-time systems) *Under Assumption 12.3.1, given a constant $w > 0$, if there exist matrices Y and $P = P^\top$ such that*

$$\begin{bmatrix} wP - \overline{\mathbf{C}} & \overline{\mathbf{B}}^\top - \begin{bmatrix} Y \\ P \end{bmatrix}^\top \\ \overline{\mathbf{B}} - \begin{bmatrix} Y \\ P \end{bmatrix} & -\overline{\mathbf{A}} \end{bmatrix} \preceq 0 \qquad P \succ 0, \tag{12.9}$$

then the origin is a locally asymptotically stable equilibrium for the closed-loop system composed of (12.1) *and the control law $u = YP^{-1}x$.*

Proof To make the origin a locally asymptotically stable, we look for a control gain K such that the linear part of the closed-loop dynamics $(\widehat{A} + \widehat{B}K)$ is Hurwitz for all $\begin{bmatrix}\widehat{A} & \widehat{B}\end{bmatrix} \in \overline{\mathcal{I}}$. For this purpose, it suffices to find a $P = P^\top \succ 0$ such that

$$P^{-1}(\widehat{A} + \widehat{B}K) + (\widehat{A} + \widehat{B}K)^\top P^{-1} + wP^{-1} \preceq 0 \quad \forall \begin{bmatrix}\widehat{B} & \widehat{A}\end{bmatrix} \in \overline{\mathcal{I}} \tag{12.10}$$

for any fixed $w > 0$. Recalling that $\widehat{S} = \begin{bmatrix}\widehat{B} & \widehat{A}\end{bmatrix}$, and left- and right- multiplying P to both sides of the inequality gives

$$\widehat{S}\begin{bmatrix} K \\ I \end{bmatrix} P + P \begin{bmatrix} K \\ I \end{bmatrix}^\top \widehat{S}^\top + wP \preceq 0 \quad \forall \widehat{S} \in \overline{\mathcal{I}}. \tag{12.11}$$

Following the description (12.7) of set $\overline{\mathcal{I}}$ and the definition of $\overline{\mathbf{C}}$, it holds that

$$\begin{aligned} &\overline{\mathbf{C}} + \widehat{S}\overline{\mathbf{B}} + \overline{\mathbf{B}}^\top \widehat{S}^\top + \widehat{S}\overline{\mathbf{A}}\widehat{S}^\top \\ &= \left(\widehat{S}^\top + \overline{\mathbf{A}}^{-1}\overline{\mathbf{B}}\right)^\top \overline{\mathbf{A}} \left(\widehat{S}^\top + \overline{\mathbf{A}}^{-1}\overline{\mathbf{B}}\right) - \overline{\mathbf{B}}^\top \overline{\mathbf{A}}^{-1} \overline{\mathbf{B}} + \overline{\mathbf{C}} \\ &= \left(\widehat{S}^\top + \overline{\mathbf{A}}^{-1}\overline{\mathbf{B}}\right)^\top \overline{\mathbf{A}} \left(\widehat{S}^\top + \overline{\mathbf{A}}^{-1}\overline{\mathbf{B}}\right) - \delta I \preceq 0. \end{aligned}$$

Define $\Delta = \overline{\mathbf{A}}^{1/2}\left(\widehat{S}^\top + \overline{\mathbf{A}}^{-1}\overline{\mathbf{B}}\right)$, and it follows that $\Delta^\top \Delta \preceq \delta I$ and

$$\widehat{S}^\top = -\overline{\mathbf{A}}^{-1}\overline{\mathbf{B}} + \overline{\mathbf{A}}^{-1/2}\Delta.$$

Defining $Y = KP$ we can rewrite (12.11) as

$$\begin{aligned}
&\widehat{S}\begin{bmatrix} K \\ I \end{bmatrix} P + P \begin{bmatrix} K \\ I \end{bmatrix}^\top \widehat{S}^\top + wP \\
&= \left(-\overline{\mathbf{A}}^{-1}\overline{\mathbf{B}}\right)^\top \begin{bmatrix} Y \\ P \end{bmatrix} + \left(\overline{\mathbf{A}}^{-1/2}\Delta\right)^\top \begin{bmatrix} Y \\ P \end{bmatrix} + (\star)^\top + wP \preceq 0 \quad \forall\, \Delta^\top\Delta \preceq \delta I.
\end{aligned}$$

By Petersen's lemma, the inequality above holds if and only if there exists $\epsilon > 0$ such that

$$\left(-\overline{\mathbf{A}}^{-1}\overline{\mathbf{B}}\right)^\top \begin{bmatrix} Y \\ P \end{bmatrix} + \begin{bmatrix} Y \\ P \end{bmatrix}^\top \left(-\overline{\mathbf{A}}^{-1}\overline{\mathbf{B}}\right) + wP + \epsilon \begin{bmatrix} Y \\ P \end{bmatrix}^\top \overline{\mathbf{A}}^{-1} \begin{bmatrix} Y \\ P \end{bmatrix} + \epsilon^{-1}\delta I \preceq 0, \tag{12.12}$$

which is equivalent to

$$\left(-\overline{\mathbf{A}}^{-1}\overline{\mathbf{B}}\right)^\top \begin{bmatrix} \epsilon Y \\ \epsilon P \end{bmatrix} + \begin{bmatrix} \epsilon Y \\ \epsilon P \end{bmatrix}^\top \left(-\overline{\mathbf{A}}^{-1}\overline{\mathbf{B}}\right) + \epsilon w P + \begin{bmatrix} \epsilon Y \\ \epsilon P \end{bmatrix}^\top \overline{\mathbf{A}}^{-1} \begin{bmatrix} \epsilon Y \\ \epsilon P \end{bmatrix} + \delta I \preceq 0.$$

As Y and P are unknown variables, we can neglect ϵ and obtain the inequality

$$\begin{aligned}
&\left(-\overline{\mathbf{A}}^{-1}\overline{\mathbf{B}}\right)^\top \begin{bmatrix} Y \\ P \end{bmatrix} + \begin{bmatrix} Y \\ P \end{bmatrix}^\top \left(-\overline{\mathbf{A}}^{-1}\overline{\mathbf{B}}\right) + wP + \begin{bmatrix} Y \\ P \end{bmatrix}^\top \overline{\mathbf{A}}^{-1} \begin{bmatrix} Y \\ P \end{bmatrix} + \delta I \\
&= \left(\overline{\mathbf{B}} - \begin{bmatrix} Y \\ P \end{bmatrix}\right)^\top \overline{\mathbf{A}}^{-1} \left(\overline{\mathbf{B}} - \begin{bmatrix} Y \\ P \end{bmatrix}\right) + wP - \overline{\mathbf{C}} \preceq 0.
\end{aligned}$$

By the Schur complement, the inequality above is equivalent to (12.9), and the matrices Y and P satisfies (12.9) renders $(\widehat{A} + \widehat{B}YP^{-1})$ Hurwitz.

Recall that the closed-loop dynamics is $\dot{x} = (\widehat{A} + \widehat{B}YP^{-1})x + R(x, u)$, where the approximation error $R(x, u)$ contains high-order terms and converges to the origin faster than the linear part for all x in a neighborhood of the origin. Therefore, the origin is a locally asymptotically stable equilibrium for the closed-loop system under the designed linear controller. □

A similar result is obtained for discrete-time systems as shown in the following theorem. The proof follows that of Theorem 12.4.1 and thus is omitted.

Theorem 12.4.2 (Data-driven controller design for discrete-time systems) *Under Assumption 12.3.1, given a* $w \in (0, 1)$, *if there exist matrices* Y *and* $P = P^\top$ *such that*

$$\begin{bmatrix} -(1-w)P & 0 & \begin{bmatrix} Y \\ P \end{bmatrix}^\top \\ 0 & -P-\overline{\boldsymbol{C}} & -\overline{\boldsymbol{B}}^\top \\ \begin{bmatrix} Y \\ P \end{bmatrix} & -\overline{\boldsymbol{B}} & -\overline{\boldsymbol{A}} \end{bmatrix} \preceq 0 \tag{12.13}$$

$$P \succ 0, \tag{12.14}$$

then the origin is a locally asymptotically stable equilibrium for the closed-loop system composed of (12.2) *and the control law* $u = YP^{-1}x$.

Remark 12.4.3 (*Enforcing a decay rate via* w) Consider the Lyapunov function $V(x) = x^\top P^{-1}x$. The controller designed by Theorem 12.4.1 guarantees that the derivative of $V(x)$ along the trajectory of the closed-loop system $\dot{x} = \widehat{A}x + \widehat{B}u$ with the designed control law $u = YP^{-1}x$ satisfies that

$$\dot{V}(x) \leq -wV(x)$$

for any given $w > 0$ and any $\begin{bmatrix} \widehat{B} & \widehat{A} \end{bmatrix} \in \overline{\mathcal{I}}$. Hence, by choosing the value of w, a certain decay rate of the closed-loop solution is enforced. Similarly, for the discrete-time system, Theorem 12.4.2 leads to a closed-loop system $x^+ = \widehat{A}x + \widehat{B}u$ with the designed control law $u = YP^{-1}x$ such that

$$V(x^+) - V(x) \leq -wV(x)$$

for any given $w \in (0, 1)$ and $\begin{bmatrix} \widehat{B} & \widehat{A} \end{bmatrix} \in \overline{\mathcal{I}}$.

Remark 12.4.4 (*Data-driven control design via high-order approximation*) Besides approximating the nonlinear dynamics (12.1) and (12.2) as linear systems with approximated errors, one can also perform high-order approximation using Taylor's expansion. In particular, consider the continuous-time input-affine system

$$\dot{x} = f(x) + g(x)u, \tag{12.15}$$

where $x \in \mathbb{R}^n$, $u \in \mathbb{R}^m$, and $f(0) = 0$. Under certain continuity assumptions on the functions in f and g, one can write the nonlinear dynamics into a polynomial system having a linear-like form, i.e.,

$$\dot{x} = AZ(x) + BW(x)u + R(x, u), \tag{12.16}$$

where $Z(x)$ and $W(x)$ are a vector and a matrix containing monomials in x, respectively, and $R(x, u)$ is the approximation error. If $R(x, u)$ satisfies Assumption 12.3.1, then an over-approximated set of the feasible dynamics can also be found in a similar manner as shown in Sect. 12.3. Based on the over-approximation of the feasible set and using Petersen's lemma, a *nonlinear* data-driven control law can be designed by solving SOS conditions. This approach is also applicable to discrete-time nonlinear dynamics. Detailed results on nonlinear data-driven control design via high-order approximation for continuous-time and discrete-time systems can be found in our work [16].

12.5 RoA Estimation

In the previous section, we have shown that data-driven stabilizers can be designed for unknown nonlinear systems using linear approximations. The resulting controllers make the origin locally asymptotically stable. Besides this property, it is of paramount importance to estimate the RoA of the closed-loop system. The definition of the RoA is given as follows.

Definition 12.5.1 (*Region of attraction*) For the system $\dot{x} = f(x)$ or $x^+ = f(x)$, if for every initial condition $x(t_0) \in \mathcal{R}$, it holds that $\lim_{t\to\infty} x(t) = 0$, then $\mathcal{R}$ is a region of attraction of the system with respect to the origin. If there exists a $\mathcal{C}^1$ function $V : \mathbb{R}^n \to \mathbb{R}$ and a positive constant c such that

$$\Omega_c := \{x \in \mathbb{R}^n : V(x) \leq c\}$$

is bounded and

$$V(0) = 0, \quad V(x) > 0 \ \forall x \in \mathbb{R}^n$$

$\{x \in \mathbb{R}^n : V(x) \leq c, x \neq 0\} \subseteq \{x \in \mathbb{R}^n : \dot{V}(x) < 0\}$ for the continuous system, or

$$\{x \in \mathbb{R}^n : V(x) \leq c, x \neq 0\} \subseteq \{x \in \mathbb{R}^n : V(x^+) - V(x) < 0\}$$

for the discrete-time system, then Ω_c is an invariant subset, or called an estimation, of the RoA.

In this section, for the designed data-driven controllers in Sect. 12.4, we derive data-driven conditions to determine whether a given sublevel set of the Lyapunov function is an invariant subset of the RoA. The derived conditions are data-driven because they are obtained using the over-approximated set $\overline{\mathcal{I}}$. We note that once the controller is computed, it is possible to use any other data-driven method to estimate the RoA, see for example [6, 9, 27].

By the controller design method in Sect. 12.4, the Lyapunov function $V(x)$, and thus the set Ω_c with a given c, are available for analysis. To characterize the set $\{x \in \mathbb{R}^n : \dot{V}(x) < 0\}$, we need a bound on the approximation error for all x in a neighborhood of the origin. This is achievable by posing the following assumption on the partial derivative of each f_i.

Assumption 12.5.2 For all $z \in \mathbb{D} \subseteq \mathbb{R}^{n+m}$, where $\mathbb{D}$ is a star-convex neighborhood of the origin, f_i is continuously differentiable and

$$\left|\frac{\partial f_i}{\partial z_j}(z) - \frac{\partial f_i}{\partial z_j}(0)\right| \leq L_i \|z\| \quad \forall j = 1, \ldots, m+n \tag{12.17}$$

for $i = 1, \ldots, n$ with known $L_i > 0$.

Under Assumption 12.5.2, a bound on the approximation error can be obtained by the following lemma.

Lemma 12.5.3 *Under Assumption 12.5.2, the approximation errors $R(x, u)$ in* (12.3) *and* (12.4) *satisfy*

$$|R_i(x, u)| \leq \frac{\sqrt{n+m} L_i}{2} \|(x, u)\|^2 \quad \forall (x, u) \in \mathbb{D}, \tag{12.18}$$

where $R_i(x, u)$, $i = 1, \ldots, n$, is such that $R(x, u) = \begin{bmatrix} R_1(x, u) & \cdots & R_n(x, u) \end{bmatrix}^\top$.

The proof of Lemma 12.5.3 can be found in Appendix 12.8.2.

Remark 12.5.4 (*Existence of L_i*) Assumption 12.5.2 is the weakest condition needed for deriving a bound on the approximation error using Lemma 12.5.3. A stronger condition, such as the Lipschitz continuity of $\frac{\partial f_i}{\partial z_j}$, guarantees the existence of L_i. It is also noted that L_i can be estimated using a data-based bisection procedure as shown in [22, Sect. III.C].

Remark 12.5.5 (*On Assumptions* 12.3.1 *and* 12.5.2) Under Assumption 12.5.2, using Lemma 12.5.3, the bound γ^2 in Assumption 12.3.1 can be derived for the experimental data. During the experiment(s), suppose that the smallest ball containing $(x(t), u(t))$ has radius R_e, i.e., $\|(x(t_k), u(t_k))\| \leq R_e$ for all $k = 0, \ldots, T$. Then, for $k = 0, \ldots, T$,

$$\begin{aligned}
& R(x(t_k), u(t_k))^\top R(x(t_k), u(t_k)) \\
&= \sum_{i=1}^{n} R_i(x(t_k), u(t_k))^2 \\
&\leq \sum_{i=1}^{n} \frac{(m+n)L_i^2}{4} \|x(t_k), u(t_k)\|^4 \\
&\leq \sum_{i=1}^{n} \frac{(m+n)L_i^2}{4} R_e^4.
\end{aligned}$$

Hence, if some prior knowledge on the dynamics is known such that Assumption 12.5.2 holds, γ^2 can be chosen as $\sum_{i=1}^{n} \frac{(m+n)L_i^2}{4} R_e^4$ to satisfy Assumption 12.3.1.

12.5.1 Continuous-Time Systems

We now characterize the time derivative of $V(x)$ along the trajectory of the continuous-time closed-loop system with the designed controller, and analyze the RoA of the closed-loop system.

Lemma 12.5.6 *Consider system* (12.1) *and the linear controller* $u = YP^{-1}x$, *where* Y *and* P *are designed to satisfy* (12.9) *with any given constant* $w > 0$. *Under Assumption 12.5.2, the derivative of the Lyapunov function* $V(x) = x^\top P^{-1}x$ *along the trajectory of the closed-loop system with the controller* $u = YP^{-1}x$ *satisfies, for all* $x \in \mathbb{D}$,

$$\dot{V}(x) \leq -wx^\top P^{-1}x + 2\kappa(x)\rho(x), \tag{12.19}$$

where

$$\kappa(x) := \left[x^\top Q_1 \|(x, Kx)\|^2 \cdots x^\top Q_n \|(x, Kx)\|^2\right], \tag{12.20}$$

Q_i *is the* i*th column of* P^{-1}, *and the vector* $\rho(x)$ *is contained in the polytope*

$$\mathcal{H} := \{\varrho : -\bar{h} \preceq \varrho \preceq \bar{h}\} \tag{12.21}$$

with

$$\bar{h} = \left[\bar{h}_1 \cdots \bar{h}_n\right]^\top = \left[\tfrac{\sqrt{m+n}L_1}{2} \cdots \tfrac{\sqrt{m+n}L_n}{2}\right]^\top.$$

The proof of Lemma 12.5.6 can be found in Appendix 12.8.3.

Denote the number of distinct vertices of $\mathcal{H}$ as ν and each vertex of $\mathcal{H}$ as h_k, $k = 1, \ldots, \nu$. Using SOS techniques (see Appendix 12.8.4 for background material) and the Positivstellensatz (Appendix 12.8.5), we have the following result.

Proposition 12.5.7 *Suppose that the controller* $u = Kx$ *renders the origin a locally asymptotically stable equilibrium for* (12.1) *with the Lyapunov function* $V(x) = x^\top P^{-1}x$. *Under Assumption 12.5.2, given a* $c > 0$ *such that* $\Omega_c = \{x \in \mathbb{R}^n : V(x) \leq c\} \subseteq \mathbb{D}$, *if there exist SOS polynomials* s_{1k}, s_{2k} *in* x, $k = 1, \ldots, \nu$ *such that*

$$-\left[s_{1k}(c - V(x)) + s_{2k}\left(-wx^\top P^{-1}x + 2\kappa(x)h_k\right) + x^\top x\right] \tag{12.22}$$

is SOS, where $\kappa(x)$ *is defined as in* (12.20) *and* h_k *are the distinct vertices of the polytope* $\mathcal{H}$ *defined in* (12.21), *then* Ω_c *is an invariant subset of the RoA of the system* $\dot{x} = f(x, Kx)$ *relative to the equilibrium* $x = 0$.

Proof According to [5, p. 87], the polytope $\mathcal{H}$ can be expressed as

$$\mathcal{H} = \left\{\varrho = \sum_{k=1}^{\nu} \lambda_k(x)h_k,\ \sum_{k=1}^{\nu} \lambda_k(x) = 1,\ \lambda_k(x) \geq 0\right\}$$

for any fixed $x \in \Omega_c$. Then, the derivative of the Lyapunov function satisfies that

$$
\begin{aligned}
\dot{V}(x) &\le -wx^\top P^{-1}x + 2\kappa(x)\rho(x) \\
&= -wx^\top P^{-1}x + 2\kappa(x)\sum_{k=1}^{\nu}\lambda_k(x)h_k \\
&= \sum_{k=1}^{\nu}\lambda_k(x)(-wx^\top P^{-1}x) + \sum_{k=1}^{\nu}\lambda_k(x)2\kappa(x)h_k \\
&= \sum_{k=1}^{\nu}\lambda_k(x)\left(-wx^\top P^{-1}x + 2\kappa(x)h_k\right).
\end{aligned}
$$

As $\lambda_k(x) \ge 0$ and $\sum_{k=1}^{\nu}\lambda_k(x) = 1$, if

$$
-wx^\top P^{-1}x + 2\kappa(x)h_k < 0
$$

holds for all $k = 1, \ldots, \nu$, then $\dot{V}(x) < 0$.

By Lemma 12.8.4 in Appendix 12.8.5, for each $k = 1, \ldots, \nu$, if there exist SOS polynomials s_{1k}, s_{2k} such that

$$
-\left[s_{1k}(c - V(x)) + s_{2k}\left(-wx^\top P^{-1}x + 2\kappa(x)h_k\right) + x^\top x\right]
$$

is SOS, then the set inclusion condition

$$
\{x \in \mathbb{R}^n : V(x) \le c,\ x \ne 0\} \subseteq \{x \in \mathbb{R}^n : -wx^\top P^{-1}x + 2\kappa(x)h_k < 0\}
$$

holds. This leads to the set inclusion condition

$$
\{x \in \mathbb{R}^n : V(x) \le c,\ x \ne 0\} \subseteq \{x \in \mathbb{R}^n : \dot{V}(x) < 0\},
$$

and hence Ω_c is an inner estimate of the RoA. □

12.5.2 Discrete-Time Systems

Similarly, for discrete-time systems satisfying Assumption 12.5.2, we can describe the difference of the Lyapunov functions of the closed-loop system with the designed data-driven controller.

Lemma 12.5.8 *Consider system* (12.2) *and the linear controller* $u = YP^{-1}x$, *where* Y *and* P *are designed to satisfy* (12.13) *and* (12.14) *with any given* $w \in (0, 1)$. *Under Assumption 12.5.2, the difference of the Lyapunov function* $V(x) = x^\top P^{-1}x$ *along the trajectory of the closed-loop system with the controller* $u = YP^{-1}x$ *satisfies that, for all* $x \in \mathbb{D}$ *and any* $\varepsilon > 0$,

$$
V(x^+) - V(x) \le -x^\top\left(wP^{-1} - \varepsilon^{-1}r_1^2 I\right)x + \left(\varepsilon + \|P^{-1}\|\right)\widehat{\kappa}(x)\widehat{\rho}(x), \tag{12.23}
$$

where

$$r_1 := \left\| -P^{-1}\overline{\boldsymbol{B}}^\top \overline{\boldsymbol{A}}^{-1} \begin{bmatrix} K \\ I \end{bmatrix} \right\| + \sqrt{\delta} \left\| P^{-1} \right\| \left\| \overline{\boldsymbol{A}}^{-1/2} \begin{bmatrix} K \\ I \end{bmatrix} \right\|, \tag{12.24}$$

$$\widehat{\kappa}(x) := \left[x^\top Q_1 \|(x, Kx)\|^2 \cdots x^\top Q_n \|(x, Kx)\|^2 \right], \tag{12.25}$$

Q_i is the ith column of P^{-1}, and the vector $\widehat{\rho}(x)$ is contained in the polytope

$$\widehat{\mathcal{H}} := \{\varrho : 0 \preceq \varrho \preceq \widehat{h}\} \tag{12.26}$$

with

$$\widehat{h} = \begin{bmatrix} \frac{(m+n)L_1^2}{4} & \cdots & \frac{(m+n)L_n^2}{4} \end{bmatrix}^\top .$$

The proof of Lemma 12.5.8 can be found in Appendix 12.8.6.

Let $\widehat{\nu}$ denote the number of distinct vertices of $\widehat{\mathcal{H}}$, and let $\widehat{h}_k, k = 1, \ldots, \widehat{\nu}$, denote a vertex of $\widehat{\mathcal{H}}$. It can be proved that if the set inclusion condition

$$\begin{aligned} &\{x \in \mathbb{R}^n : V(x) \leq c,\ x \neq 0\} \\ &\subseteq \left\{x \in \mathbb{R}^n : -x^\top \left(wP^{-1} - \varepsilon^{-1} r_1^2 I\right) x + \left(\varepsilon + \|P^{-1}\|\right) \widehat{\kappa}(x)\widehat{h}_k < 0\right\} \end{aligned}$$

holds for $k = 1, \ldots, \widehat{\nu}$, then the origin is a locally asymptotically stable equilibrium of the closed-loop system and Ω_c is an invariant subset of the RoA. Using Lemma 12.8.4, we derive a result similar to Proposition 12.5.7 for discrete-time systems.

Proposition 12.5.9 *Suppose that the $u = Kx$ renders the origin a locally asymptotically stable equilibrium for* (12.2) *with the Lyapunov function $V(x) = x^\top P^{-1} x$. Under Assumption 12.5.2, given a $c > 0$ such that $\Omega_c \subseteq \mathbb{D}$, if there exist SOS polynomials s_{1k}, s_{2k} in x, $k = 1, \ldots, \nu$ such that*

$$-\left\{ s_{1k}(c - V(x)) + s_{2k} \left[-x^\top \left(wP^{-1} - \varepsilon^{-1} r_1^2 I \right) x + \left(\varepsilon + \|P^{-1}\| \right) \widehat{\kappa}(x)\widehat{h}_k \right] + x^\top x \right\}$$

is SOS, where $\widehat{\kappa}(x)$ is defined as in (12.25) *and $\widehat{h}_k$ are the distinct vertices of polytope $\widehat{\mathcal{H}}$ defined in* (12.26)*, then Ω_c is an invariant subset of the RoA of the system $x^+ = f(x, Kx)$ relative to the equilibrium $x = 0$.*

Remark 12.5.10 (*Alternative methods for RoA estimation*) In [24], the RoA is estimated by a numerical method, i.e., a sufficient condition of $V(x^+) - V(x) < 0$ is found using data, and the grids in a compact region are tested to see whether the sufficient condition is satisfied so that the sublevel sets of $V(x)$ can be found as the RoA estimation. In this work, we derive SOS conditions for the RoA estimation that is an alternative to the mesh method used in [24].

Remark 12.5.11 (*Data-driven RoA estimation via high-order approximation*) When the data-driven controller is designed via high-order approximation, the RoA of the resulting closed-loop system can also be estimated using data. In the case of high-order approximation, an assumption on the high-order derivative similar to Assumption 12.5.2 is needed to characterize the time derivative or difference of the Lyapunov function via data. For details on the RoA estimation via high-order approximation, the reader is referred to our work [16].

12.6 Example

In this section, we illustrate the proposed nonlinear data-driven stabilizing design by presenting simulation results on a system with $x = \begin{bmatrix} x_1 & x_2 \end{bmatrix} \in \mathbb{R}^2$ and $u \in \mathbb{R}$. The true dynamics of the system is the inverted pendulum given in Appendix 12.8.7, on which an open-loop experiment is conducted to obtain the data matrices.

12.6.1 Continuous-Time Systems

To collect the data, an experiment is conducted with $x(0) = \begin{bmatrix} 0.01 & -0.01 \end{bmatrix}^\top$ and $u = 0.1\sin(t)$ during the time interval [0, 0.5]. The data is sampled with fixed sampling period $T_s = 0.05$s. We collect the data and arrange them into a data set with length $T = 10$.

We assume that the bound on $R(x, u)$ is over-approximated by 100%; in other words, the bound γ is twice the largest instantaneous norm of $R(x, u)$ during the experiment. Then, for the experimental data, Assumption 12.3.1 holds with

$$\gamma = 3.3352 \cdot 10^{-6}.$$

Setting $\delta = 0.01$, we first solve the optimization problem (12.8) to find $\overline{\mathcal{I}}$, and then apply Theorem 12.4.1 with $w = 1$. The solution found by CVX is

$$\begin{aligned} P &= 10^3 \cdot \begin{bmatrix} 1.0152 & -1.3289 \\ -1.3289 & 1.7727 \end{bmatrix}, \\ u &= -12.0432x_1 - 8.887x_2. \end{aligned} \tag{12.27}$$

To estimate the RoA for the closed-loop system with the designed data-driven controller, we need to first find L_1 and L_2 satisfying Assumption 12.5.2. From physical considerations we can argue that the x_1-subsystem is linear, which leads to $L_1 = 0$. As for L_2, by over-estimating the true bound we let $L_2 = 1.4697$.

By applying Proposition 12.5.7, the largest c found for the controller in (12.27) is $c^* = 7.58 \cdot 10^{-4}$.

Figure 12.1 illustrates the estimated RoA using different approaches. In particular, the light gray area is obtained by checking every point in an meshed area, the medium

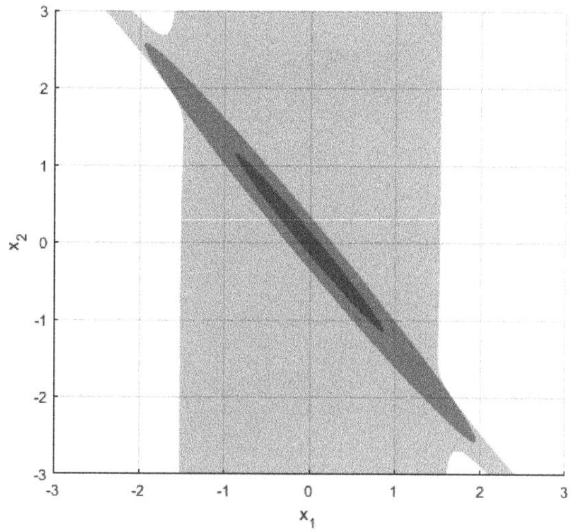

Fig. 12.1 Estimations of the RoA using the Lyapunov function by linear approximation. The light gray area is the estimated RoA by checking the sign of the derivative of the Lyapunov function for the grids in a compact region with explicit known dynamics; the medium gray area is the largest sublevel set of the Lyapunov function contained in the light gray estimated RoA; the dark gray area is the estimated RoA obtained by Proposition 12.5.7

gray area is the largest sublevel set of the obtained Lyapunov function contained in the light gray RoA, and the dark gray area is the estimated RoA found using Proposition 12.5.7.

12.6.2 Discrete-Time Systems

For the discrete-time system, we consider the Euler discretization of the true dynamics used for the continuous-time case with the sampling time $T_s = 0.1$. The true dynamics used for data generation can be found in Appendix 12.8.7. The initial condition and the input are randomly chosen in the interval $[-1, 1]$. In particular, the initial condition and input used in the experiment is $x(0) = \begin{bmatrix}-0.196 & 0.0395\end{bmatrix}$ and

$$U_0 = [0.0724 \quad -0.0960 \quad 0.0720 \quad 0.0118 \quad -0.0194 \quad 0.0517 \quad 0.0434 \quad 0.0975 \quad -0.0444 \quad -0.0992].$$

Same as the continuous-time case, we collect and arrange the data into a data set with length $T = 10$. To obtain a reasonable estimation of the bound on $R(x, u)$ in Assumption 12.3.1, we compare the data generated by the true dynamics and the data generated by the linear approximation and over-estimate the difference by 100%. Then, Assumption 12.3.1 holds for this example with

$$\gamma^2 = 3.3646 \cdot 10^{-8}.$$

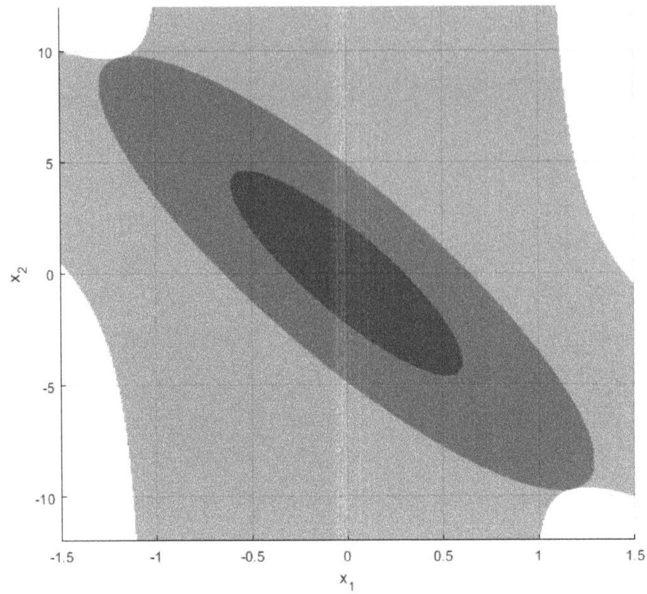

Fig. 12.2 Estimations of the RoA using the Lyapunov function by linear approximation. The light gray area is the estimated RoA by checking the sign of the derivative of the Lyapunov function for the grids in a compact region with explicit known dynamics; the medium gray area is the largest sublevel set of the Lyapunov function contained in the light gray estimated RoA; the dark gray area is the estimated RoA obtained by Proposition 12.5.9

The over-estimated bound is used to obtain the over-approximation of the feasible set by Proposition 12.3.3. Then, applying Theorem 12.4.2 with $\delta = 10^{-2}$ gives the solution

$$P = \begin{bmatrix} 5.9860 & -45.3511 \\ -45.3511 & 448.1767 \end{bmatrix},$$
$$u = -9.2787x_1 - 1.8095x_2.$$

To estimate the RoA of the closed-loop system under the designed controllers, we choose L_1 and L_2 satisfying Assumption 12.5.2 as $L_1 = 0$ and $L_2 = 2.1213$. These parameters are chosen by similar arguments made for the continuous-time case. Applying Proposition 12.5.9, the largest c found for the controller is $c^* = 2.3 \cdot 10^{-3}$.

To evaluate the RoA estimation, we plot in Fig. 12.2 the RoA estimation using Proposition 12.5.9 (darkest area), the estimated RoA by checking point-by-point of a mesh of initial conditions using explicit dynamics (light gray area), and the largest sublevel set of the Lyapunov function contained in the estimated RoA (medium dark area).

12.7 Summary

Without any model information on the unknown general nonlinear dynamics, this work proposes data-driven stabilizer designs and RoA analysis by approximating the unknown functions using linear approximation. Using finite-length input-state data, linear stabilizers are designed for both continuous-time and discrete-time systems that make the known equilibrium locally asymptotically stable. Then, by estimating a bound on the approximation error, data-driven conditions are given to find an invariant subset of the RoA. Simulation results on the inverted pendulum illustrate the designed data-driven controllers and the RoA estimations. Topics such as enlarging the RoA estimation and case studies on more complicated nonlinear benchmarks are all interesting directions to be considered in future works.

12.8 Appendix

12.8.1 Petersen's Lemma

In the section for data-driven controller design, Petersen's lemma is essential for deriving the sufficient condition characterizing the controller. Due to the space limit, the proof of the lemma is omitted and one may refer to works such as [4, 25, 26] for more details.

Lemma 12.8.1 (Petersen' s lemma [25]) *Consider matrices* $\mathcal{G} = \mathcal{G}^\top \in \mathbb{R}^{n\times n}$, $\mathcal{M} \in \mathbb{R}^{n\times m}$, $\mathcal{M} \neq 0$, $\mathcal{N} \in \mathbb{R}^{p\times n}$, $\mathcal{N} \neq 0$, *and a set* F *defined as*

$$F = \{\mathcal{F} \in \mathbb{R}^{m\times p} : \mathcal{F}^\top \mathcal{F} \preceq \overline{\mathcal{F}}\},$$

where $\overline{\mathcal{F}} = \overline{\mathcal{F}}^\top \succeq 0$. *Then, for all* $\mathcal{F} \in F$,

$$\mathcal{G} + \mathcal{M}\mathcal{F}\mathcal{N} + \mathcal{N}^\top \mathcal{F}^\top \mathcal{M}^\top \preceq 0$$

if and only if there exists $\mu > 0$ *such that*

$$\mathcal{G} + \mu\mathcal{M}\mathcal{M}^\top + \mu^{-1}\mathcal{N}^\top \overline{\mathcal{F}}\mathcal{N} \preceq 0.$$

12.8.2 Proof of Lemma 12.5.3

Let $z = \begin{bmatrix} x^\top & u^\top \end{bmatrix}^\top$. Each element $f_i(z)$, $i = 1, \ldots, n$ in $f(z)$ can bc written as

$$f_i(z) = \sum_{j=1}^{n+m} \frac{\partial f_i}{\partial z_j}(0)z_j + R_i(z).$$

On the other hand, as shown in [12], the function $f_i(z)$ can be expressed as

$$f_i(z) = \sum_{j=1}^{n+m} z_j \int_0^1 \frac{\partial f_i}{\partial z_j}(tz)dt.$$

As a consequence, one can write $R_i(z)$ as

$$\begin{aligned} R_i(z) &= \sum_{j=1}^{n+m} z_j \int_0^1 \frac{\partial f_i}{\partial z_j}(tz)dt - \sum_{j=1}^{n+m} \frac{\partial f_i}{\partial z_j}(0)z_j \\ &= \sum_{j=1}^{n+m} z_j \int_0^1 \left(\frac{\partial f_i}{\partial z_j}(tz) - \frac{\partial f_i}{\partial z_j}(0) \right) dt. \end{aligned}$$

Under Assumption 12.5.2, one has

$$\left| \frac{\partial f_i}{\partial z_j}(tz) - \frac{\partial f_i}{\partial z_j}(0) \right| \leq L_i \|tz\|, \quad t \in (0, 1).$$

Then, it holds that

$$\begin{aligned} |R_i(z)| &\leq \sum_{j=1}^{n+m} |z_j| \int_0^1 L_i \|z\| \cdot |t| dt \\ &= L_i \|z\| \int_0^1 |t| dt \cdot \sum_{j=1}^{n+m} |z_j|. \end{aligned}$$

By the fact that $\int_0^1 |t|dt = \frac{1}{2}$ and $|z_1 + \cdots z_{n+m}| \leq \sqrt{n+m}\|z\|$, it holds that

$$|R_i(z)| \leq \frac{\sqrt{n+m}L_i}{2} \|z\|^2.$$

The proof is complete.

12.8.3 *Proof of Lemma 12.5.6*

For the closed-loop system with the controller $u = Kx$ designed via Theorem 12.4.1, the derivative of the Lyapunov function $V(x) = x^\top P^{-1} x$ satisfies

$$\begin{aligned}\dot{V}(x) &= x^\top P^{-1}(A+BK)x + x^\top (A+BK)^\top P^{-1}x + 2x^\top P^{-1}R(x, Kx) \\ &\le -wx^\top P^{-1}x + 2x^\top P^{-1}R(x, Kx).\end{aligned}$$

Under Assumption 12.5.2, for all $x \in \mathbb{D}$ and $i = 1, \ldots, n$, the bounds on the approximation error can be found as

$$|R_i(x, Kx)| \le \frac{\sqrt{m+n}L_i}{2}\|(x, Kx)\|^2.$$

Hence, for all $x \in \mathbb{D}$, there exists a continuous $\rho_i(x)$ for each $i = 1, \ldots, n$ such that for each $x \in \mathbb{D}$

$$\begin{aligned}R_i(x, Kx) &= \rho_i(x)\|(x, Kx)\|^2, \\ \rho_i(x) &\in \left[-\frac{\sqrt{m+n}L_i}{2}, \frac{\sqrt{m+n}L_i}{2}\right].\end{aligned}$$

Define $\rho(x) = [\rho_1(x) \ \ldots \ \rho_n(x)]^\top$. By the definition of polytope [5, Definition 3.21], the vector $\rho(x)$ belongs to the polytope

$$\mathcal{H} = \{\varrho : -\bar{h} \preceq \varrho \preceq \bar{h}\},$$

where

$$\bar{h} = [\bar{h}_1 \ \cdots \ \bar{h}_n]^\top = \left[\tfrac{\sqrt{m+n}L_1}{2} \ \ldots \ \tfrac{\sqrt{m+n}L_n}{2}\right]^\top.$$

Denote Q_i as the ith column of P^{-1}. It holds that

$$\begin{aligned}&2x^\top P^{-1}R(x, Kx) \\ &= 2\left[x^\top Q_1 \cdots x^\top Q_n\right]\begin{bmatrix}\rho_1(x)\|(x, Kx)\|^2 \\ \vdots \\ \rho_n(x)\|(x, Kx)\|^2\end{bmatrix} \\ &= 2\sum_{i=1}^n x^\top Q_i\rho_i(x)\|(x, Kx)\|^2 \\ &= 2\sum_{i=1}^n x^\top Q_i\|(x, Kx)\|^2 \cdot \rho_i(x) \\ &= 2\left[x^\top Q_1\|(x, Kx)\|^2 \cdots x^\top Q_n\|(x, Kx)\|^2\right]\rho(x).\end{aligned}$$

Denote

$$\kappa(x) = \left[x^\top Q_1\|(x, Kx)\|^2 \cdots x^\top Q_n\|(x, Kx)\|^2\right].$$

Then, the derivative of the Lyapunov function satisfies for all $x \in \mathbb{D}$

$$\dot{V}(x) \leq -wx^\top P^{-1}x + 2\kappa(x)\rho(x),$$

where $\rho(x) \in \mathcal{H}$.

12.8.4 Sum of Squares Relaxation

As solving positive conditions of multivariable polynomials is in general NP-hard, the SOS relaxations are often used to obtain sufficient conditions that are tractable. The SOS polynomial matrices are defined as follows.

Definition 12.8.2 (*SOS polynomial matrix* [8]) $M : \mathbb{R}^n \to \mathbb{R}^{\sigma\times\sigma}$ is an SOS polynomial matrix if there exist $M_1, \dots, M_k : \mathbb{R}^n \to \mathbb{R}^{\sigma\times\sigma}$ such that

$$M(x) = \sum_{i=1}^{k} M_i(x)^\top M_i(x) \;\; \forall x \in \mathbb{R}^n. \tag{12.28}$$

Note that when $\sigma = 1$, $M(x)$ becomes a scalar SOS polynomial.

It is straightforward to see that if a matrix $M(x)$ is an SOS polynomial matrix, then it is positive semi-definite, i.e., $M(x) \succeq 0 \, \forall x \in \mathbb{R}^n$. Relaxing the positive polynomial conditions into SOS polynomial conditions makes the conditions tractable and easily solvable by common software.

12.8.5 Positivstellensatz

In the RoA analysis, we need to characterize polynomials that are positive on a semialgebraic set, and the Positivstellensatz plays an important role in this characterization.

Let $p_1, \dots, p_k$ be polynomials. The multiplicative monoid, denoted by $\mathcal{S}_M(p_1, \dots, p_k)$, is the set generated by taking finite products of the polynomials $p_1, \dots, p_k$. The cone $\mathcal{S}_C(p_1, \dots, p_k)$ generated by the polynomials is defined as

$$\mathcal{S}_C(p_1, \dots, p_k)$$
$$= \{s_0 + \sum_{i=1}^{j} s_i q_i : s_0, \dots, s_j \text{ are SOS polynomials}, q_1, \dots, q_j \in \mathcal{S}_M(p_1, \dots, p_k)\}.$$

The ideal $\mathcal{S}_I(p_1, \dots, p_k)$ generated by the polynomials is defined as

$$\mathcal{S}_I(p_1, \dots, p_k) = \left\{ \sum_{i=1}^{k} r_i p_i : r_1, \dots, r_k \text{ are polynomials} \right\}.$$

Stengle's Positivstellensatz [28] is presented as follows in [8].

Theorem 12.8.3 (Positivstellensatz) *Let $f_1, \dots, f_k$, $g_1, \dots, g_l$, and $h_1, \dots, h_m$ be polynomials. Define the set*

$$\begin{aligned} \mathcal{X} = \{x \in \mathbb{R}^n : \; & f_1(x) \geq 0, \dots, f_k(x) \geq 0, \\ & g_1(x) = 0, \dots, g_l(x) = 0, \\ & \textit{and } h_1(x) \neq 0, \dots, h_m(x) \neq 0\}. \end{aligned}$$

Then, $\mathcal{X} = \emptyset$ if and only if

$$\exists f \in \mathcal{S}_C(f_1, \dots, f_k), \;\; g \in \mathcal{S}_I(g_1, \dots, g_l), \;\; h \in \mathcal{S}_M(h_1, \dots, h_m)$$

such that

$$f(x) + g(x) + h(x)^2 = 0.$$

For the subsequent RoA analysis, we will use the following result derived from the Positivstellensatz.

Lemma 12.8.4 *Let φ_1 and φ_2 be polynomials in x. If there exist SOS polynomials s_1 and s_2 in x such that*

$$-\left(s_1\varphi_1(x) + s_2\varphi_2(x) + x^\top x\right) \textit{ is SOS } \forall x \in \mathbb{R}^n \tag{12.29}$$

then the set inclusion condition

$$\{x \in \mathbb{R}^n : \varphi_1(x) \geq 0, x \neq 0\} \subseteq \{x \in \mathbb{R}^n : \varphi_2(x) < 0\} \tag{12.30}$$

holds.

Proof The set inclusion condition (12.30) can be equivalently written as

$$\{x \in \mathbb{R}^n : \varphi_1(x) \geq 0, \varphi_2(x) \geq 0, x \neq 0\} = \emptyset.$$

By Theorem 12.8.3, we know that this is true if and only if there exist $\varphi(x) \in \mathcal{S}_C(\varphi_1, \varphi_2)$ and $\zeta(x) \in \mathcal{S}_M(x)$, such that

$$\varphi(x) + \zeta(x)^2 = 0. \tag{12.31}$$

Let

$$\varphi = s_0 + s_1\varphi_1 + s_2\varphi_2,$$

where s_j, $j = 0, 1, 2$ are SOS polynomials. By the definition of the cone $\mathcal{S}_C$, one has that $\varphi \in \mathcal{S}_C(\varphi_1, \varphi_2)$. Choosing $\zeta(x)^2 = x^\top x$, we write the condition (12.31) as

$$s_0 + s_1\varphi_1 + s_2\varphi_2 + x^\top x = 0. \tag{12.32}$$

As $s_0 = -(s_1\varphi_1 + s_2\varphi_2 + x^\top x)$ from (12.32), if there exist SOS polynomials s_1 and s_2 such that the SOS condition (12.29) holds, then there exist SOS polynomials s_j, $j = 0, 1, 2$ such that (12.32) is true, and hence the set inclusion condition (12.30) holds. □

12.8.6 Proof of Lemma 12.5.8

For the closed-loop system with the controller $u = Kx$ designed via Theorem 12.4.2, the difference between the Lyapunov functions $V(x^+) = (x^+)^\top P^{-1}x^+$ and $V(x) = x^\top P^{-1}x$ is

$$\begin{aligned}
& V(x^+) - V(x) \\
&= [(A + BK)x + R(x, Kx)]^\top P^{-1}[(A + BK)x + R(x, Kx)] - x^\top P^{-1}x \\
&= x^\top \left[(A + BK)^\top P^{-1}(A + BK) - P^{-1}\right]x + 2R(x, Kx)^\top P^{-1}(A + BK)x \\
&\quad + R(x, Kx)^\top P^{-1}R(x, Kx).
\end{aligned}$$

Observe that

$$A + BK = \left(-\overline{\mathbf{A}}^{-1}\overline{\mathbf{B}} + \overline{\mathbf{A}}^{-1/2}\Delta\right)^\top \begin{bmatrix} K \\ I \end{bmatrix}$$

with $\Delta\Delta^\top \preceq \delta I$. Then, it holds that

$$\left\|P^{-1}(A + BK)\right\| \le \left\|-P^{-1}\overline{\mathbf{B}}^\top\overline{\mathbf{A}}^{-1}\begin{bmatrix} K \\ I \end{bmatrix}\right\| + \sqrt{\delta}\left\|P^{-1}\right\|\left\|\overline{\mathbf{A}}^{-1/2}\begin{bmatrix} K \\ I \end{bmatrix}\right\| = r_1.$$

For any $\varepsilon > 0$, it holds that

$$\begin{aligned}
& 2R(x, Kx)^\top P^{-1}(A + BK)x \\
&\le \varepsilon R(x, Kx)^\top R(x, Kx) + \varepsilon^{-1}\left\|P^{-1}(A + BK)\right\|^2 x^\top x \\
&\le \varepsilon\|R(x, Kx)\|^2 + \varepsilon^{-1}r_1^2 x^\top x.
\end{aligned}$$

Recall that, by Theorem 12.4.2, $(A + BK)^\top P^{-1}(A + BK) - P^{-1} \preceq -wP^{-1}$. Hence, one has that

$$V(x^+) - V(x) = -x^\top\left(wP^{-1} - \varepsilon^{-1}r_1^2 I\right)x + \left(\varepsilon + \|P^{-1}\|\right)\|R(x, Kx)\|^2.$$

Under Assumption 12.5.2,

$$\|R(x, Kx)\|^2 = \sum_{i=1}^{n} R_i(x, Kx)^2 \leq \sum_{i=1}^{n} \frac{(m+n)L_i^2}{4} \|(x, Kx)\|^4.$$

If we write $\|R(x, Kx)\|^2$ as

$$\|R(x, Kx)\|^2 = \begin{bmatrix} \|(x, Kx)\|^4 & \cdots & \|(x, Kx)\|^4 \end{bmatrix} \begin{bmatrix} \widehat{\rho}_1(x) \\ \vdots \\ \widehat{\rho}_n(x) \end{bmatrix},$$

then the scalars $\widehat{\rho}_i(x)$ are such that $\widehat{\rho}_i(x) \in \left[0, \frac{(m+n)L_i^2}{4}\right]$, $i = 1, \ldots, n$ for all $x \in \mathbb{D}$. Defining

$$\widehat{\kappa}(x) = \begin{bmatrix} \|(x, Kx)\|^4 & \cdots & \|(x, Kx)\|^4 \end{bmatrix} \quad \text{and} \quad \widehat{\rho}(x) = \begin{bmatrix} \widehat{\rho}_1(x) & \cdots & \widehat{\rho}_n(x) \end{bmatrix}^\top$$

gives $\|R(x, Kx)\|^2 = \widehat{\kappa}(x)\widehat{\rho}(x)$, and for any $x \in \mathbb{D}$ the vector $\widehat{\rho}(x)$ is contained in the polytope $\widehat{\mathcal{H}}$ defined in (12.26).

12.8.7 Dynamics Used for Data Generation in the Example

The dynamics used for data generation in Sect. 12.6 is the inverted pendulum written as

$$\begin{aligned} \dot{x}_1 &= x_2, \\ \dot{x}_2 &= \frac{mgl}{J}\sin(x_1) - \frac{r}{J}x_2 + \frac{l}{J}\cos(x_1)u, \end{aligned} \tag{12.33}$$

where $m = 0.1$, $g = 9.8$, $r = l = J = 1$. For the discrete-time case, consider the Euler discretization of the inverted pendulum, i.e.,

$$\begin{aligned} x_1^+ &= x_1 + T_s x_2, \\ x_2^+ &= \frac{T_s g}{l}\sin(x_1) + \left(1 - \frac{T_s r}{ml^2}\right)x_2 + \frac{T_s}{ml^2}\cos(x_1)u, \end{aligned}$$

where $m = 0.1$, $g = 9.8$, $T_s = 0.1$, $l = 1$, and $r = 0.01$.

References

1. Ahmadi, A.A., El Khadir, B.: Learning dynamical systems with side information. In: Conference on Learning for Dynamics and Control, vol. 120, pp. 718–727 (2020)
2. Berberich, J., Köhler, J., Müller, M.A., Allgöwer, F.: Linear tracking MPC for nonlinear systems—part II: the data-driven case. IEEE Trans. Autom. Control **67**(9), 4406–4421 (2022)
3. Bisoffi, A., De Persis, C., Tesi, P.: Trade-offs in learning controllers from noisy data. Syst. Control Lett. **154**, 104985(1)–104985(8) (2021)
4. Bisoffi, A., De Persis, C., Tesi, P.: Data-driven control via Petersen's lemma. Automatica **145**, 110537(1)–110537(14) (2022)
5. Blanchini, F., Miani, S.: Set-Theoretic Methods in Control. In: Systems & Control: Foundations & Applications. Birkhäuser Boston (2008)
6. Bobiti, R., Lazar, M.: Automated sampling-based stability verification and DOA estimation for nonlinear systems. IEEE Trans. Autom. Control **63**(11), 3659–3674 (2018)
7. Chesi, G.: Estimating the domain of attraction for non-polynomial systems via LMI optimizations. Automatica **45**(6), 1536–1541 (2009)
8. Chesi, G.: LMI techniques for optimization over polynomials in control: a survey. IEEE Trans. Autom. Control **55**(11), 2500–2510 (2010)
9. Colbert, B.K., Peet, M.M.: Using trajectory measurements to estimate the region of attraction of nonlinear systems. In: IEEE Conference on Decision and Control, pp. 2341–2347. Miami, FL, USA (2018)
10. Dai, T., Sznaier, M.: Nonlinear data-driven control via state-dependent representations. In: IEEE Conference on Decision and Control, pp. 5765–5770. Austin, TX, USA, 13–15 Dec 2021
11. De Persis, C., Tesi, P.: Formulas for data-driven control: stabilization, optimality, and robustness. IEEE Trans. Autom. Control **65**(3), 909–924 (2020)
12. Folland, G.B.: Higher-order derivatives and Taylor's formula in several variables
13. Fraile, L., Marchi, M., Tabuada, P.: Data-driven stabilization of SISO feedback linearizable systems (2021). arXiv:2003.14240
14. Guo, M., De Persis, C., Tesi, P.: Data-driven stabilization of nonlinear polynomial systems with noisy data. IEEE Trans. Autom. Control **67**(8), 4210–4217 (2022)
15. Guo, M., De Persis, C., Tesi, P.: Learning control for polynomial systems using sum of squares relaxations. In: IEEE Conference on Decision and Control, pp. 2436–2441. Jeju Island, Republic of Korea (2020)
16. Guo, M., De Persis, C., Tesi, P.: Data-based stabilizer design and closed-loop analysis of unknown general nonlinear systems (2022). arXiv:2209.01071
17. Hou, Z., Wang, Z.: From model-based control to data-driven control: survey, classification and perspective. Inf. Sci. **235**, 3–35 (2013)
18. Hou, Z., Xiong, S.: On model-free adaptive control and its stability analysis. IEEE Trans. Autom. Control **64**(11), 4555–4569 (2019)
19. Kaiser, E., Kutz, J.N., Brunton, S.L.: Data-driven discovery of Koopman eigenfunctions for control. Mach. Learn.: Sci. Technol. **2**(3), 035023(1)–035023(31) (2021)
20. Liu, W., Sun, J., Wang, G., Bullo, F., Chen, J.: Data-driven self-triggered control via trajectory prediction. IEEE Trans. Autom. Control **68**(11), 6951–6958 (2023)
21. Luppi, A., De Persis, C., Tesi. P.: On data-driven stabilization of systems with nonlinearities satisfying quadratic constraints. Syst. Control Lett. **163**, 105206(1)–105206(11) (2022)
22. Martin, T., Allgöwer, F.: Data-driven system analysis of nonlinear systems using polynomial approximation (2022). arXiv:2108.11298
23. Nejati, A., Zhong, B., Caccamo, M., Zamani, M.: Data-driven controller synthesis of unknown nonlinear polynomial systems via control barrier certificates. In: Proceedings of The 4th Annual Learning for Dynamics and Control Conference, vol. 168, pp. 763–776 (2022)
24. De Persis, C., Rotulo, M., Tesi, P.: Learning controllers from data via approximate nonlinearity cancellation. IEEE Trans. Autom. Control **68**(10), 6082–6097 (2023)

25. Petersen, I.R.: A stabilization algorithm for a class of uncertain linear systems. Syst. Control Lett. **8**(4), 351–357 (1987)
26. Shcherbakov, P.S., Topunov, M.V.: Extensions of Petersen's lemma on matrix uncertainty. IFAC Proc. Vol. **41**(2), 11385–11390 (2008)
27. Shen, Y., Bichuch, M., Mallada, E.: Model-free learning of regions of attraction via recurrent sets. In: IEEE Conference on Decision and Control, pp. 4714–4719. Cancún, Mexico (2022)
28. Stengle, G.: A nullstellensatz and a positivstellensatz in semialgebraic geometry. Math. Ann. **207**, 87–97 (1974)
29. Strasser, R., Berberich, J., Allgöwer, F.: Data-driven control of nonlinear systems: beyond polynomial dynamics. In: IEEE Conference on Decision and Control, pp. 4344–4351. Austin, TX, USA (2021)
30. Tabuada, P., Loong Ma, W., Grizzle, J., Ames, A.D.: Data-driven control for feedback linearizable single-input systems. In: Proceedings of the 56th IEEE Conference on Decision and Control, pp. 6265–6270. Melbourne, VIC, Australia (2017)
31. Tan, W., Packard, A.: Stability region analysis using polynomial and composite polynomial Lyapunov functions and sum-of-squares programming. IEEE Trans. Autom. Control **53**(2), 565–571 (2008)
32. Tanaskovic, M., Fagiano, L., Novara, C., Morari, M.: Data-driven control of nonlinear systems: an on-line direct approach. Automatica **75**, 1–10 (2017)
33. Topcu, U., Packard, A.: Local stability analysis for uncertain nonlinear systems. IEEE Trans. Autom. Control **54**(5), 1042–1047 (2009)
34. van Waarde, H.J., Camlibel, M.K.: A matrix Finsler's lemma with applications to data-driven control. In: Proceedings of the 60th IEEE Conference on Decision and Control, pp. 5777–5782. Austin, TX, USA (2021)

Chapter 13
Harmonic Modeling and Control

Flora Vernerey, Pierre Riedinger, and Jamal Daafouz

Abstract Harmonic modeling involves transforming a periodic system into an equivalent time-invariant model of infinite dimension. The states of this model, also referred to as phasors, are represented by coefficients obtained through a sliding Fourier decomposition process. This chapter aims to present a unified and coherent mathematical framework for harmonic modeling and control. By adopting this framework, the analysis and design processes become significantly simplified, as all the methods established for time-invariant systems can be directly applied. Within this framework, we explore the application of these methods to tackle the task of designing control laws based on harmonic pole placement for Linear Time-Periodic (LTP) systems. Additionally, we delve into the computational aspects associated with these control designs.

13.1 Introduction

Harmonic modeling and control play a crucial role in a wide range of application domains, including energy management and embedded systems, among others [1, 2, 5–8, 11, 12, 16, 17, 19, 21, 24]. Recently, a groundbreaking development in this field has emerged with the introduction of a unified and comprehensive mathematical framework for harmonic modeling and control [4]. This framework establishes an equivalent time-invariant model of infinite dimension, where the coefficients, referred to as phasors, are derived through a sliding Fourier decomposition. Importantly, a rigorous equivalence is established between this harmonic model and the conventional periodic system model. By adopting this proposed framework, the analysis

F. Vernerey (✉) · P. Riedinger · J. Daafouz
Université de Lorraine, CNRS, CRAN, Nancy, France
e-mail: flora.vernerey@univ-lorraine.fr

P. Riedinger
e-mail: pierre.riedinger@univ-lorraine.fr

J. Daafouz
e-mail: jamal.daafouz@univ-lorraine.fr

R. Postoyan et al. (eds.), *Hybrid and Networked Dynamical Systems*,
Lecture Notes in Control and Information Sciences 493,
https://doi.org/10.1007/978-3-031-49555-7_13

and design processes are significantly simplified as well-established methods for time-invariant systems can be directly applied. However, a key challenge arises due to the infinite-dimensional nature of the harmonic time-invariant model obtained. The associated infinite-dimensional harmonic state matrix comprises a combination of a (block) Toeplitz matrix and a diagonal matrix. Consequently, stability analysis or control design within this context requires solving algebraic Lyapunov and Riccati equations of infinite dimension. This introduces computational complexities that must be addressed in order to effectively utilize the harmonic modeling and control framework.

Despite the existence of extensive literature on infinite-dimensional Toeplitz matrices and their solutions to quadratic matrix equations, these findings cannot be directly applied to the field of harmonic control design. Additionally, approaches that rely on Floquet factorization [9, 12, 17, 19, 23] pose challenges when it comes to extending them for control design purposes. The reason is that Floquet factorization, while providing an existence result, lacks constructiveness [8, 9]. However, a recent study addresses these limitations by leveraging the spectral properties of the harmonic state operator [14]. In this work, efficient algorithms are proposed for solving infinite-dimensional harmonic algebraic Lyapunov and Riccati equations. These algorithms allow for practical implementations and overcome the computational complexities associated with harmonic control design. In this context, we provide an overview of the harmonic control methodology and elucidate the process of designing state feedback harmonic control with pole placement for continuous-time periodic systems in both their general and harmonic formulations. By incorporating the advancements highlighted in [14], the presented methodology enables the practical realization of harmonic control strategies, bridging the gap between theory and application in this field.

Notation. The transpose of a matrix A is denoted A' and A^* denotes its complex conjugate transpose $A^* = \bar{A}'$. The n-dimensional identity matrix is denoted Id_n. The infinite identity matrix is denoted $\mathcal{I}$. $A \otimes B$ is the Kronecker product of two matrices A and B. L^p (resp. ℓ^p) denotes the Lebesgues spaces of $p-$integrable functions (resp. $p-$summable sequences) for $1 \leq p \leq \infty$. L^p_{loc} is the set of locally $p-$integrable functions, i.e., on any compact set. The notation $f(t) = g(t)$ $a.e.$ means almost everywhere in t or for almost every t. We denote by $col(X)$ the vectorization of a matrix X, formed by stacking the columns of X into a single column vector. Finally, $< \cdot, \cdot >$ refers to the scalar product in ℓ^2.

13.2 Preliminaries

13.2.1 Toeplitz Block Matrices

For harmonic modeling purposes, let us introduce the following definition:

Definition 13.2.1 The Toeplitz transformation of a T-periodic function $a \in L^2([0\ T], \mathbb{R})$, denoted $\mathcal{T}(a)$, defines a constant Toeplitz and infinite-dimensional matrix as follows:

$$\mathcal{T}(a) = \begin{bmatrix} \ddots & & \vdots & & \iddots \\ & a_0 & a_{-1} & a_{-2} & \\ \cdots & a_1 & a_0 & a_{-1} & \cdots \\ & a_2 & a_1 & a_0 & \\ \iddots & & \vdots & & \ddots \end{bmatrix},$$

where $(a_k)_{k\in\mathbb{Z}}$ is the Fourier coefficient sequence of a.

From the subsequence $a^+ = (a_k)_{k>0}$ and $a^- = (a_k)_{k<0}$ of $(a_k)_{k\in\mathbb{Z}}$, we also define the semi-infinite Hankel matrices:

$$\mathcal{H}(a^+) = (a_{i+j-1})_{i,j>0}, \quad \mathcal{H}(a^-) = (a_{-i-j+1})_{i,j>0}.$$

We denote by $\mathcal{H}_{(p,q)}(a^+)$(resp. $\mathcal{H}_{(p,q)}(a^-)$) for any $p, q > 0$, the $(2p+1) \times (2q+1)$ Hankel matrix obtained by selecting the first $(2p+1)$ rows and $(2q+1)$ columns of $\mathcal{H}(a^+)$(resp. $\mathcal{H}(a^-)$).

The Toeplitz block transformation of a T-periodic $n \times n$ matrix function $A = (a_{ij})_{i,j=1,\ldots,n} \in L^2([0\ \ T], \mathbb{R}^{n\times n})$, denoted $\mathcal{A} = \mathcal{T}(A)$, defines a constant $n \times n$ Toeplitz Block (TB) and infinite-dimensional matrix:

$$\mathcal{A} = \begin{pmatrix} \mathcal{A}_{11} & \cdots & \mathcal{A}_{1n} \\ \vdots & \ddots & \vdots \\ \mathcal{A}_{n1} & \cdots & \mathcal{A}_{nn} \end{pmatrix}, \tag{13.1}$$

where $\mathcal{A}_{ij} = \mathcal{T}(a_{ij})$, $i, j = 1, \ldots, n$.

Similarly, the $n \times n$ Hankel block matrices $\mathcal{H}(A^+)$, $\mathcal{H}(A^-)$ are also defined, respectively, by $\mathcal{H}(A^+)_{ij} = \mathcal{H}(a^+_{ij})$ and $\mathcal{H}(A^-)_{ij} = \mathcal{H}(a^-_{ij})$ for $i, j = 1, \ldots, n$. Their principal submatrices $\mathcal{H}(A^+)_{(p,q)}$, $\mathcal{H}(A^-)_{(p,q)}$ for $p > 0$, $q > 0$ are obtained by considering the principal submatrices of the entries $\mathcal{H}(a^+_{ij})_{(p,q)}$ and $\mathcal{H}(a^-_{ij})_{(p,q)}$ for $i, j = 1, \ldots, n$.

In the sequel, to distinguish a time-varying matrix function A, its sliding Fourier decomposition $\mathcal{F}(A)$ and its Toeplitz transformation $\mathcal{T}(A)$, we use the notation $A(t)$, $\mathbf{A} := \mathcal{F}(A)$ and $\mathcal{A} := \mathcal{T}(A)$.

Definition 13.2.2 The left $m-$truncation (resp. right $m-$truncation) of a $n \times n$ block Toeplitz matrix $\mathcal{A}$ with blocks of infinite size is given by

$$\mathcal{A}_{m^+} := \begin{pmatrix} \mathcal{A}_{11_{m^+}} & \mathcal{A}_{12_{m^+}} & \cdots & \mathcal{A}_{1n_{m^+}} \\ \mathcal{A}_{21_{m^+}} & \mathcal{A}_{22_{m^+}} & & \vdots \\ \vdots & & \ddots & \vdots \\ \mathcal{A}_{n1_{m^+}} & \cdots & \mathcal{A}_{n(n-1)_{m^+}} & \mathcal{A}_{nn_{m^+}} \end{pmatrix}$$

(resp. $\mathcal{A}_{m^-}$), where $\mathcal{A}_{ij_{m^+}}$, $i, j := 1, \ldots, n$ are obtained by suppressing in the infinite matrices $\mathcal{A}_{ij}$ all the columns and lines having an index strictly smaller than $-m$ (respectively, strictly greater than m). Finally, the $m-$truncation is obtained by applying successively a left and a right $m-$truncations and is denoted by $\mathcal{A}_m$ (or equivalently $\mathcal{T}_m(A)$.

Finally we introduce symbol definition useful to compute product of two TB matrices.

Definition 13.2.3 The symbol matrix $A(z)$ associated with a $n \times n$ block Toeplitz matrix is given by

$$A(z) := \begin{pmatrix} a_{11}(z) & \cdots & a_{1n}(z) \\ \vdots & \ddots & \vdots \\ a_{n1}(z) & \cdots & a_{nn}(z) \end{pmatrix}, \tag{13.2}$$

where $a_{ij}(z)$ is the symbol associated to a_{ij} and is defined by $a_{ij}(z) = \sum_{k \in \mathbb{Z}} a_{ij,k} z^k$, for $i, j = 1, \ldots, n$.

Recall that the product of two TB matrices is a TB matrix only in infinite dimension. In finite dimension, we have the following result:

Theorem 13.2.4 *Let $\mathcal{A}$ and $\mathcal{B}$ be two infinite-dimensional TB matrices with, respectively, associated symbol $A(z)$ and $B(z)$. Then, the symbol associated to the product $\mathcal{C} := \mathcal{A}\mathcal{B}$ is $C(z) := A(z)B(z)$ and*

$$\begin{aligned} \mathcal{A}_m \mathcal{B}_m &= \mathcal{C}_m - \mathcal{H}_{(m,\eta)}(A^+)\mathcal{H}_{(\eta,m)}(B^-) \\ &\quad - \mathcal{J}_{n,m}\mathcal{H}_{(m,\eta)}(A^-)\mathcal{H}_{(\eta,m)}(B^+)\mathcal{J}_{n,m}, \end{aligned} \tag{13.3}$$

where J_m is the $(2m+1) \times (2m+1)$ flip matrix having 1 on the anti-diagonal and zeros elsewhere, $\mathcal{J}_{n,m} := Id_n \otimes J_m$ and η is such that $2\eta \geq \min d^o(A(z), B(z))$.

An illustration of the above theorem is given in Fig. 13.1 for $n := 1$ with $a(z)$ and $b(z)$ Laurent polynomials of degree much less than m so that $\mathcal{T}_m(a)$ and $\mathcal{T}_m(b)$ are banded. If $a(z) := \sum_{i=-k}^{k} a_i z^i$ and $b(z) := \sum_{i=-k}^{k} b_i z^i$ with k much smaller than m, then the matrices $E^+ := \mathcal{H}_m(a^+)\mathcal{H}_m(b^-)$ and $E^- := J_m\mathcal{H}_m(a^-)\mathcal{H}_m(b^+)J_m$ have disjoint supports located in the upper leftmost corner and in the lower rightmost corner, respectively. As a consequence, $\mathcal{T}_m(a)\mathcal{T}_m(b)$ can be represented as the sum of the Toeplitz matrix associated with $c(z)$ and two correcting terms E^+ and E^-.

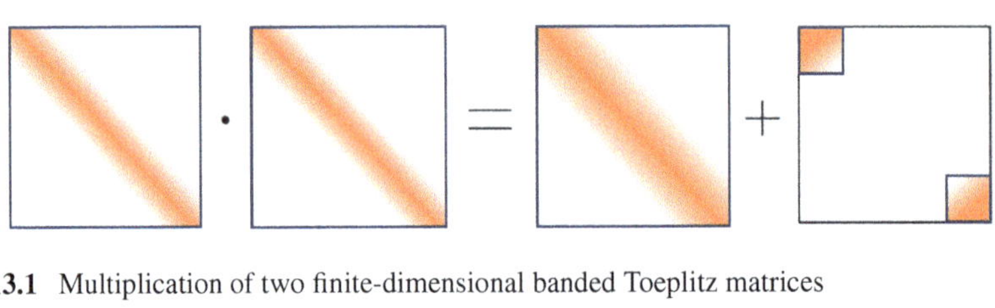

Fig. 13.1 Multiplication of two finite-dimensional banded Toeplitz matrices

13.2.2 Sliding Fourier Decomposition

We recall the definition of the sliding Fourier decomposition over a window of length T and the so-called "Coincidence Condition" introduced in [4].

Definition 13.2.5 The sliding Fourier decomposition over a window of length T of a complex valued function of time $x \in L^2_{loc}(\mathbb{R}, \mathbb{C})$ is defined by the time-varying infinite sequence $X = \mathcal{F}(x) \in C^a(\mathbb{R}, \ell^2(\mathbb{C}))$ (see [4]) whose components satisfy

$$X_k(t) = \frac{1}{T}\int_{t-T}^{t} x(\tau)e^{-\mathrm{j}\omega k\tau}d\tau$$

for $k \in \mathbb{Z}$, with $\omega = \frac{2\pi}{T}$.

If $x = (x_1, \ldots, x_n) \in L^2_{loc}(\mathbb{R}, \mathbb{C}^n)$ is a complex valued vector function, then

$$X = \mathcal{F}(x) = (\mathcal{F}(x_1), \ldots, \mathcal{F}(x_n)).$$

The vector $X_k = (X_{1,k}, \ldots, X_{n,k})$ with

$$X_{i,k}(t) = \frac{1}{T}\int_{t-T}^{t} x_i(\tau)e^{-\mathrm{j}\omega k\tau}d\tau$$

is called the $k-$th phasor of X.

Definition 13.2.6 We say that X belongs to H if X is an absolutely continuous function (i.e., $X \in C^a(\mathbb{R}, \ell^2(\mathbb{C}^n))$ and fulfills for any k the following condition:

$$\dot{X}_k(t) = \dot{X}_0(t)e^{-j\omega kt} \; a.e. \tag{13.4}$$

Similarly to the Riesz–Fischer theorem which establishes a one-to-one correspondence between the spaces L^2 and ℓ^2, the following "Coincidence Condition" establishes a one-to-one correspondence between the spaces L^2_{loc} and H.

Theorem 13.2.7 (Coincidence Condition [4]) *For a given $X \in L^\infty_{loc}(\mathbb{R}, \ell^2(\mathbb{C}^n))$, there exists a representative $x \in L^2_{loc}(\mathbb{R}, \mathbb{C}^n)$ of X, i.e., $X = \mathcal{F}(x)$, if and only if X belongs to H.*

13.3 Harmonic Modeling

Under the "Coincidence Condition" of Theorem 13.2.7, it is established in [4] that any periodic system having solutions in Carathéodory sense can be transformed by a sliding Fourier decomposition into a time-invariant system. For instance, consider T-periodic functions $A(\cdot)$ and $B(\cdot)$, respectively, of class $L^2([0\ T], \mathbb{C}^{n\times n})$ and $L^\infty([0\ T], \mathbb{C}^{n\times m})$ and let the linear time periodic system:

$$\dot{x}(t) = A(t)x(t) + B(t)u(t) \quad x(0) := x_0. \tag{13.5}$$

If, x is a solution (in Carathéodory sense) associated to the control $u \in L^2_{loc}(\mathbb{R}, \mathbb{C}^m)$ of the linear time periodic system (13.5) then, $X := \mathcal{F}(x)$ is a solution of the linear time-invariant system:

$$\dot{X}(t) = (\mathcal{A} - \mathcal{N})X(t) + \mathcal{B}U(t), \quad X(0) := \mathcal{F}(x)(0), \tag{13.6}$$

where $\mathcal{A} := \mathcal{T}(A)$, $\mathcal{B} := \mathcal{T}(B)$ and

$$\mathcal{N} := Id_n \otimes diag(j\omega k,\ k \in \mathbb{Z}). \tag{13.7}$$

Reciprocally, if $X \in H$ is a solution of (13.6) with $U \in H$, then its representative x (i.e., $X = \mathcal{F}(x)$) is a solution of (13.5). Moreover, for any $k \in \mathbb{Z}$, the phasors $X_k \in C^1(\mathbb{R}, \mathbb{C}^n)$ and $\dot{X}_k \in C^0(\mathbb{R}, \ell^\infty(\mathbb{C}^n))$. As the solution x is unique for the initial condition x_0, X is also unique for the initial condition $X(0) := \mathcal{F}(x)(0)$. In addition, it is proved in [4] that one can reconstruct time trajectories from harmonic ones, that is

$$x(t) = \mathcal{F}^{-1}(X)(t) := \sum_{k=-\infty}^{+\infty} X_k(t)e^{j\omega kt} + \frac{T}{2}\dot{X}_0(t). \tag{13.8}$$

In the same way, a strict equivalence between a periodic differential Lyapunov equation and its associated harmonic algebraic Lyapunov equation is also proved [4]. Namely, let $Q \in L^\infty([0\ T])$ be a T-periodic symmetric and positive definite matrix function. P is the unique T-periodic symmetric positive definite solution of the periodic differential Lyapunov equation:

$$\dot{P}(t) + A'(t)P(t) + P(t)A(t) + Q(t) = 0,$$

if and only if $\mathcal{P} := \mathcal{T}(P)$ is the unique Hermitian and positive definite solution of the harmonic algebraic Lyapunov equation:

$$\mathcal{P}(\mathcal{A} - \mathcal{N}) + (\mathcal{A} - \mathcal{N})^*\mathcal{P} + \mathcal{Q} = 0, \tag{13.9}$$

where $\mathcal{Q} := \mathcal{T}(Q)$ is Hermitian positive definite and $\mathcal{A} := \mathcal{T}(A)$. Moreover, $\mathcal{P}$ is a bounded operator on ℓ^2 and P is an absolutely continuous function.

Instead of dealing with a periodic differential Lyapunov equation, solving an algebraic Lyapunov equation proves to be advantageous for analysis and control design. However, it poses a challenge due to the infinite dimension nature of Eq. (13.9), specifically concerning the non-Toeplitz and non-compact nature of the diagonal matrix $\mathcal{N}$ defined by (13.7). In [14], it is demonstrated that the solution to the infinite-dimensional harmonic Lyapunov equation (13.9) can be approximated by solving a finite-dimensional problem with a controllable error margin. This practical result circumvents the need for a Floquet factorization computation and significantly reduces the computational burden. Furthermore, these findings extend to harmonic Riccati equations encountered in periodic optimal control. The basis of these results lies in the spectral characterization of the harmonic state operator $(\mathcal{A}-\mathcal{N})$, which plays a vital role in developing the algorithms presented in [14]. Firstly, it is proven that the spectrum of $(\mathcal{A}-\mathcal{N})$ is both unbounded and discrete. In fact, it is given by the unbounded and discrete set:

$$\sigma(\mathcal{A}-\mathcal{N}) := \{\lambda_p + \mathrm{j}\omega k : k \in \mathbb{Z},\ p := 1, \ldots, n\},$$

where λ_p, $p := 1, \ldots, n$ are not necessarily distinct eigenvalues. Secondly, analyzing the impact of an $m-$truncation on the spectrum of $(\mathcal{A}_m - \mathcal{N}_m)$ allows to derive efficient algorithms that can be used for harmonic control design. Therefore, one has to analyze the alterations in the spectrum of $(\mathcal{A}_m - \mathcal{N}_m)$ compared to that of $(\mathcal{A}-\mathcal{N})$. Henceforth, it is assumed that the T-periodic matrix function $A(t)$ belongs to $L^\infty(\mathbb{R}, \mathbb{R}^{n\times n})$, or equivalently, $\mathcal{A}$ is a bounded operator on ℓ^2. This assumption facilitates the development of algorithms that provide guarantees with arbitrarily small error tolerances when an m-truncation is applied. For the sake of simplicity, we present the results for the case where the operator $(\mathcal{A}-\mathcal{N})$ is non-defective. However, it is important to note that these results hold true in a general setting.

Theorem 13.3.1 *Assume that $A(t) \in L^\infty([0\ T])$ and $(\mathcal{A}-\mathcal{N})$ is non-defective. Denote by $\sigma := \{\lambda_p + j\omega k : k \in \mathbb{Z},\ p := 1, \ldots, n\}$ the spectrum of $(\mathcal{A}-\mathcal{N})$. Let $(\mathcal{A}_{m^+} - \mathcal{N}_{m^+})$ be a left $m-$truncation of $(\mathcal{A}-\mathcal{N})$ according to Definition 13.2.2 and assume that it is non-defective, with an eigenvalues set denoted by Λ_{m^+}.*

1. *For $\epsilon > 0$, there exists an index j_0 such that for any eigenvalue $\lambda \in \Lambda_1^+(m) := \{\lambda_p + j\omega k : k \in \mathbb{Z}, k \le m+1-j_0,\ p := 1, \ldots, n\} \subset \sigma$:*

$$\|(\mathcal{A}_{m^+} - \mathcal{N}_{m^+} - \lambda\mathcal{I}_{m^+})V|_{m^+}\|_{\ell^2} < \epsilon, \tag{13.10}$$

   where $V|_{m^+}$ is the left $m-$truncation of the eigenvector associated with λ.
2. *For $\epsilon > 0$, an approximation of the set Λ_{m^+} is provided by*

$$\Lambda_{m^+} \approx \Lambda_1^+(m) \cup \Lambda_2^+(m),$$

   where $\Lambda_1^+(m)$ is defined by (13.10) and where $\Lambda_2^+(m)$ is a finite subset of Λ_{m^+}. Moreover, any eigenvalue λ_{m+1} which belongs to the set $\Lambda_2^+(m+1)$ is obtained by the relation: $\lambda_{m+1} = \lambda_m + j\omega$, where λ_m belongs to $\Lambda_2^+(m)$.

Theorem 13.3.2 *Assume that the matrix $A(t) \in L^\infty([0\ T])$ and that $(\mathcal{A}-\mathcal{N})$ is non-defective with $\sigma := \{\lambda_p + j\omega k : k \in \mathbb{Z},\ p := 1, \ldots, n\}$ its spectrum. Assume that the $m-$truncation $(\mathcal{A}_m - \mathcal{N}_m)$ is non-defective with its eigenvalues set denoted by Λ_m.*

For $\epsilon > 0$, there exists an m_0 such that for $m \ge m_0$:

1. *there exists an index j_0 such that for any eigenvalue $\lambda_1 \in \Lambda_1(m)$ defined by the subset $\{\lambda_p + j\omega k : |k| \leq j_0, p := 1, \ldots, n\}$ of σ, the following relation is satisfied:*

$$\|(\mathcal{A}_m - \mathcal{N}_m - \lambda_1 \mathcal{I}_m) V_1|_m\|_{\ell^2} < \epsilon, \tag{13.11}$$

   where $V_1|_m$ is the $m-$truncation of the eigenvector associated with λ_1.
2. *for any eigenvalue $\lambda_2 \in \Lambda_2(m^+)$ or in $\Lambda_2(m^-) := \bar{\Lambda}_2(m^+)$ with $\Lambda_2(m^+)$ defined in Theorem 13.3.1, the following relation is satisfied:*

$$\|(\mathcal{A}_m - \mathcal{N}_m - \lambda_2 \mathcal{I}_m) V_2|_m\|_{\ell^2} < \epsilon,$$

   where $V_2|_m$ is the $m-$truncation of the eigenvector associated with λ_2.

Then, the set Λ_m can be approximated by the union of the sets $\Lambda_1(m)$, $\Lambda_2^-(m)$ and $\Lambda_2^+(m)$ that is

$$\Lambda_m \approx \Lambda_1(m) \cup \Lambda_2^-(m) \cup \Lambda_2^+(m).$$

Corollary 13.3.3 *Assume that the matrix $A(t) \in L^\infty([0\ T])$. If $(\mathcal{A} - \mathcal{N})$ is invertible, there exists an $m_0 > 0$ such that for any $m \geq m_0$, the matrix $(\mathcal{A}_m - \mathcal{N}_m)$ is invertible. Moreover, $\|(\mathcal{A}_m - \mathcal{N}_m)^{-1}\|_{\ell^2}$ is uniformly bounded, i.e., $\sup_{m \geq m_0} \|(\mathcal{A}_m - \mathcal{N}_m)^{-1}\|_{\ell^2} < +\infty$.*

To highlight the fact that even if $(\mathcal{A} - \mathcal{N})$ is Hurwitz, attempting to solve a Lyapunov equation using a truncated version $(\mathcal{A}_m - \mathcal{N}_m)$ would never yield a positive definite solution, regardless of the value of m, let us consider a 2×2 block Toeplitz matrix:

$$\mathcal{A} := \begin{pmatrix} \mathcal{A}_{11} & \mathcal{A}_{12} \\ \mathcal{A}_{21} & \mathcal{A}_{22} \end{pmatrix},$$

where the Toeplitz matrices $\mathcal{A}_{ij}$ are characterized by

$$\begin{aligned}
a_{11} &:= (0.5, 0.6 - \mathrm{j}, \underline{-1}, 0.6 + \mathrm{j}, 0.5) \\
a_{12} &:= (1.3 - 0.4\mathrm{j}, -2.2 + 0.5\mathrm{j}, \underline{-0.4}, -2.2 - 0.5\mathrm{j}, 1.3 + 0.4\mathrm{j}) \\
a_{21} &:= (-0.3 - 0.6\mathrm{j}, 0.4 + 0.7\mathrm{j}, \underline{-0.1}, 0.4 - 0.7\mathrm{j}, -0.3 + 0.6\mathrm{j}) \\
a_{22} &:= (-1.3 - 1.8\mathrm{j}, 1.4 - 1.6\mathrm{j}, \underline{-2}, 1.4 + 1.6\mathrm{j}, -1.3 + 1.8\mathrm{j})
\end{aligned}$$

with the underlined terms corresponding to the index 0. The eigenvalues of $(\mathcal{A}_m - \mathcal{N}_m)$ are depicted in Fig. 13.2 for $m := 20$ (blue circles) and $m := 40$ (red stars). We clearly observe the sets $\Lambda_1(m)$, $\Lambda_2^-(m)$ and $\Lambda_2^+(m)$. Notice that $\Lambda_1(m)$ is defined by two eigenvalues (as expected) satisfying $R(\lambda_1) \approx -0.3$ and $R(\lambda_2) \approx -2.7$ as shown by the alignment of the eigenvalues along these vertical axes. Thus, $(\mathcal{A} - \mathcal{N})$ is Hurwitz while $(\mathcal{A}_m - \mathcal{N}_m)$ is never Hurwitz for all m since $\Lambda_2^-(m)$ and $\Lambda_2^+(m)$ have eigenvalues with positive real parts. As mentioned, we see that the set $\Lambda_2^+(40)$ is obtained from the set $\Lambda_2^+(20)$ by a translation of $\mathrm{j}\omega 20$ $(\omega := 1)$.

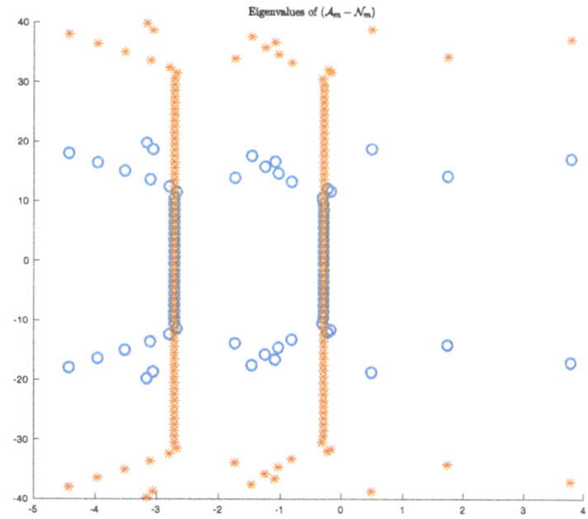

Fig. 13.2 Eigenvalues of $(\mathcal{A}_m - \mathcal{N}_m)$ for $m := 20$ (blue circles) or 40 (red stars)

The following proposition introduces the symbol Lyapunov equation.

Proposition 13.3.4 *Assume that $A(t) \in L^\infty([0\ T], \mathbb{R}^{n\times n})$. The symbol harmonic Lyapunov equation is given by*

$$A(z)'P(z) + P(z)A(z) + (\mathbb{1}_{n,n} \otimes N(z)) \cdot P(z) + Q(z) = 0, \tag{13.12}$$

where the symbol matrices $A(z)$, $Q(z)$, $P(z)$ are given by (13.2) *and where $N(z) := \sum_{k=-\infty}^{+\infty} j\omega k z^k$.*

Upon examining the previous symbolic Lyapunov equation, it becomes evident that factorizing $P(z)$ to obtain a solution is not feasible. The subsequent theorem demonstrates that if we attempt to solve a truncated version of (13.9), the resulting solution does not possess the Toeplitz property. This property is crucial in the infinite-dimensional case, thereby leading to an important practical consequence: the absence of the temporal counterpart $P(t)$, and its inability to be reconstructed using (13.8). The following theorem reveals that the solution P_m derived from solving the truncated harmonic Lyapunov equation deviates from the solution of the infinite-dimensional harmonic Lyapunov equation by a corrective term ΔP_m.

Theorem 13.3.5 *Consider finite dimension Toeplitz matrices $\mathcal{A}_m := \mathcal{T}_m(A)$ and $\mathcal{Q}_m := \mathcal{T}_m(Q)$. The solution P_m of the Lyapunov equation*

$$(\mathcal{A}_m - \mathcal{N}_m)^* P_m + P_m(\mathcal{A}_m - \mathcal{N}_m) + \mathcal{Q}_m = 0 \tag{13.13}$$

is given by $P_m := \mathcal{P}_m + \Delta P_m$, where $\mathcal{P}_m := \mathcal{T}_m(P)$ with $P(z)$ solution of (13.12) *and ΔP_m satisfies*

$$(\mathcal{A}_m - \mathcal{N}_m)^* \Delta P_m + \Delta P_m(\mathcal{A}_m - \mathcal{N}_m) = E^+ + E^-$$

with

$$
\begin{aligned}
E^{+} &:= \mathcal{H}_{(m,\eta)}(A^{+})\mathcal{H}_{(\eta,m)}(P^{-}) + \mathcal{H}_{(m,\eta)}(P^{+})\mathcal{H}_{(\eta,m)}(A^{-}) \\
E^{-} &:= \mathcal{J}_{n,m}(\mathcal{H}_{(m,\eta)}(A^{-})\mathcal{H}_{m}(P^{+}) \\
&\qquad\qquad + \mathcal{H}_{(m,\eta)}(P^{-})\mathcal{H}_{(\eta,m)}(A^{+}))\mathcal{J}_{n,m}
\end{aligned}
$$

and $2\eta \geq \min d^{o}(A(z), P(z))$.

In practical terms, extracting the Toeplitz component $\mathcal{P}_m$ from P_m is not straightforward, as the symbol $P(z)$ is implicitly defined by (13.12). Furthermore, it can be proven that this linear problem is rank deficient and possesses an infinite number of solutions. Therefore, the objective is to establish that $\mathcal{P}$ can be determined with an arbitrarily small error. This necessity to determine $\mathcal{P}$ instead of P_m with high accuracy is crucial for demonstrating the stability of $(\mathcal{A} - \mathcal{N})$. This is primarily because, even if $(\mathcal{A} - \mathcal{N})$ is Hurwitz, the matrix $(\mathcal{A}_m - \mathcal{N}_m)$ would never exhibit Hurwitz properties for any value of m.

Theorem 13.3.6 *Assume that $(\mathcal{A} - \mathcal{N})$ is invertible. The phasor $\mathbf{P} := \mathcal{F}(P)$ associated with the solution $\mathcal{P} := \mathcal{T}(P)$ of the infinite-dimensional harmonic Lyapunov equation (13.9) is given by*

$$\mathrm{col}(\mathbf{P}) := -(Id_n \otimes (\mathcal{A} - \mathcal{N})^{*} + Id_n \circ \mathcal{A}^{*})^{-1}\mathrm{col}(\mathbf{Q}), \tag{13.14}$$

where

$$Id_n \circ \mathcal{A} := \begin{pmatrix} Id_n \otimes \mathcal{A}_{11} & Id_n \otimes \mathcal{A}_{12} & \cdots & Id_n \otimes \mathcal{A}_{1n} \\ Id_n \otimes \mathcal{A}_{21} & Id_n \otimes \mathcal{A}_{22} & & \vdots \\ \vdots & & \ddots & \vdots \\ Id_n \otimes \mathcal{A}_{n1} & \cdots & \cdots & Id_n \otimes \mathcal{A}_{nn} \end{pmatrix} \tag{13.15}$$

with $\mathcal{N}$ given by (13.7) *and where the matrix $\mathbf{Q} := \mathcal{F}(Q)$.*

To state the next result, define for any given m the $m-$truncated solution as

$$\mathrm{col}(\tilde{\mathbf{P}}_m) := -(Id_n \otimes (\mathcal{A}_m - \mathcal{N}_m)^{*} + Id_n \circ \mathcal{A}_m^{*}))^{-1}\mathrm{col}(\mathbf{Q}|_m) \tag{13.16}$$

with $\mathcal{A}_m := \mathcal{T}_m(A), \mathcal{N}_m := Id_n \otimes \mathrm{diag}(\mathrm{j}\omega k,\ |k| \leq m)$ and where $Id_n \circ \mathcal{A}$ is defined by (13.15). The components of the $m-$truncated matrix $\mathbf{Q}|_m$ are given by

$$(\mathbf{Q}|_m)_{ij} := \mathcal{F}|_m(q_{ij}),\ i, j := 1, \ldots, n$$

with $\mathcal{F}|_m(q_{ij})$ the $m-$truncation of $\mathcal{F}(q_{ij})$ obtained by suppressing all phasors of order $|k| > m$.

Theorem 13.3.7 *Assume that $A(t) \in L^{\infty}([0\ T], \mathbb{R}^{n\times n})$ and $(\mathcal{A} - \mathcal{N})$ is invertible. For any given $\epsilon > 0$, there exists m_0 such that for any $m \geq m_0$:*

$$\|\mathrm{col}(\mathbf{P} - \tilde{\mathbf{P}}_m)\|_{\ell^2} < \epsilon,$$

where **P**, *given by* (13.14), *is the solution of the infinite-dimensional problem. Moreover,*

$$\|\mathcal{P} - \tilde{\mathcal{P}}_m\|_{\ell^2} < \epsilon$$

with $\mathcal{P} := \mathcal{T}(P)$ *and* $\tilde{\mathcal{P}}_m := \mathcal{T}(\tilde{P}_m)$.

By employing the symbolic equation to derive an approximate solution, a substantial reduction in computational burden can be achieved. This is due to the fact that the linear problem defined by (13.16) has a dimension of $n(2m+1)$, whereas the one defined by (13.13) has a dimension of $n^2(2m+1)^2$. Consequently, utilizing the symbolic equation allows for a significant decrease in computational complexity. The following corollary holds significant practical implications for accurately determining a solution to the infinite harmonic Lyapunov equation from (13.16). Specifically, by setting a desired tolerance level $\epsilon > 0$, one can simply increase the value of m in (13.16) until the condition (13.17) is met. This provides a practical guideline for achieving an accurate solution to the infinite harmonic Lyapunov equation within the desired error margin.

Corollary 13.3.8 *For a given* $\epsilon > 0$, *there exists* m_0 *such that for any* $m \geq m_0$, *the symbol* $\tilde{P}_m(z)$ *associated with* $\tilde{\mathbf{P}}_m$ *satisfies*

$$\|A(z)'\tilde{P}_m(z) + \tilde{P}_m(z)A(z) + (\mathbb{1}_n \otimes N(z)) \cdot \tilde{P}_m(z) + Q(z)\|_{\ell^2} < \epsilon. \tag{13.17}$$

13.3.1 Harmonic Control and Pole Placement

In this section, we show how one can design an harmonic state feedback control law of the form $U = -\mathcal{K}X$ that assigns the poles of the closed-loop harmonic model to some desired locations and provide the corresponding representative in the time domain $u(t) = -K(t)x(t)$. To this end, we need the following notion of controllability.

Definition 13.3.9 (see [13, 15]) System (13.6) is said to be exactly controllable at time t, if for all $X_0, X_1 \in \ell^2$, there exists a control $U(t) \in L^2([0,t], \ell^2)$ such that the corresponding solution satisfies $X(0) = X_0$ and $X(t) = X_1$.

In this setting, exact controllability can be characterized as follows.

Proposition 13.3.10 *System* (13.6) *is exactly controllable at time t if and only if*

$$\exists \delta > 0 : \forall x \in \ell^2, \int_0^t \|\mathcal{B}^* e^{(\mathcal{A}-\mathcal{N})^*\tau} x\|^2 d\tau \geq \delta \|x\|^2.$$

Proof See [13]. □

The next theorem provides a way to design a state feedback harmonic control law that assigns the poles of the closed-loop system to some specified location.

Theorem 13.3.11 *Assume that the pair* $(\mathcal{A}-\mathcal{N}, \mathcal{B})$ *is exactly controllable. Let* $G(t)$ *be a* T*-periodic function in* L^∞ *and define* $\mathcal{G} := \mathcal{T}(G)$. *Consider a* $n-$*dimensional Jordan normal form* Λ *such that the spectrum of* $\mathcal{A}-\mathcal{N}$ *and* $\Lambda \otimes \mathcal{I} - \mathcal{N}$ *have no common value, i.e.,* $\sigma(\mathcal{A}-\mathcal{N}) \cap \sigma(\Lambda \otimes \mathcal{I} - \mathcal{N}) = \emptyset$.

Then, the bounded operator (on ℓ^2*)* $\mathcal{P}$ *is the unique solution of the harmonic Sylvester equation:*

$$(\mathcal{A}-\mathcal{N})\mathcal{P} - \mathcal{P}(\Lambda \otimes \mathcal{I} - \mathcal{N}) = \mathcal{B}\mathcal{G} \tag{13.18}$$

if and only if $P := \mathcal{T}^{-1}(\mathcal{P})$ *is the* T*-periodic solution (in Carathéodory sense) of the differential Sylvester equation*

$$\dot{P}(t) = A(t)P(t) - P(t)\Lambda - B(t)G(t). \tag{13.19}$$

If $\mathcal{P}$ *is invertible, then* $\mathcal{P}^{-1}$ *is also a bounded operator on* ℓ^2 *and* P^{-1} *is a* T*-periodic solution in Carathéodory sense of the differential Sylvester equation:*

$$\dot{P}^{-1}(t) = -P^{-1}(t)(A(t) - B(t)K(t)) + \Lambda P^{-1}(t), \tag{13.20}$$

where $K(t) := G(t)P^{-1}(t) \in L^\infty$. *Moreover, the harmonic state feedback* $U := -\mathcal{K}X$ *with* $\mathcal{K} := \mathcal{G}\mathcal{P}^{-1}$, *is a bounded operator on* ℓ^2 *and assigns the poles of the closed-loop harmonic system, that is*

$$\mathcal{P}^{-1}(\mathcal{A}-\mathcal{N}-\mathcal{B}\mathcal{K})\mathcal{P} = (\Lambda \otimes \mathcal{I} - \mathcal{N}). \tag{13.21}$$

In addition, taking $z(t) := P^{-1}(t)x(t)$ *transforms the closed-loop LTP system* (13.5) *with* $u(t) := -K(t)x(t)$ *and* $K(t) := G(t)P^{-1}(t) \in L^\infty$ *into the LTI system*

$$\dot{z} = \Lambda z \ a.e. \tag{13.22}$$

Proof If $\mathcal{P}$ solves (13.18) then $\mathcal{P}$ belongs trivially to H, and $P := \mathcal{T}^{-1}(\mathcal{P})$ satisfies the differential Sylvester equation (13.19) and reciprocally. This equivalence is obtained using similar steps to the proof of Theorem 5 in [4]. As the product BG belongs to L^∞, we also prove that $\mathcal{P}$ is a bounded operator on ℓ^2 (see the proof of Theorem 5 in [4]).

If $\mathcal{P}$ is invertible, then so is P. Thus, P and $P^{-1} := \mathcal{T}^{-1}(\mathcal{P}^{-1})$ are both in L^∞ (see Theorem 13.6.2). As $G \in L^\infty$, the product GP^{-1} belongs to L^∞ and $\mathcal{K} := \mathcal{G}\mathcal{P}^{-1} = \mathcal{T}(G)\mathcal{T}(P^{-1}) = \mathcal{T}(GP^{-1})$ is a (block) Toeplitz and bounded operator on ℓ^2.

From (13.18), we have

$$\mathcal{P}^{-1}(\mathcal{A} - \mathcal{B}\mathcal{K} - \mathcal{N}) - (\Lambda \otimes \mathcal{I} - \mathcal{N})\mathcal{P}^{-1} = 0,$$

which implies (using similar steps as the proof of Theorem 5 in [4]) that $P^{-1} := \mathcal{T}^{-1}(\mathcal{P}^{-1})$ satisfies

$$\dot{P}^{-1}(t) = -P^{-1}(t)(A(t) - B(t)K(t)) + \Lambda P^{-1}(t) \; a.e.$$

with $K(t) := G(t)P^{-1}(t) \in L^{\infty}$.

Furthermore, the control $U := -\mathcal{K}X$ allows to assign the poles of the closed-loop harmonic system to the set $\sigma := \{\lambda + j\omega k : k \in \mathbb{Z}, \lambda \in diag(\Lambda)\}$ according to the following relation: $\mathcal{P}^{-1}(\mathcal{A} - \mathcal{N} - \mathcal{BK})\mathcal{P} = (\Lambda \otimes \mathcal{I} - \mathcal{N})$.

Finally, taking $z(t) := P^{-1}(t)x(t)$, we have

$$\begin{aligned} \dot{z} &= \dot{P}^{-1}(t)x + P^{-1}(t)\dot{x} \; a.e. \\ &= \Lambda z(t) \; a.e. \end{aligned}$$

□

The significance of Theorem 13.3.11 becomes clearer with further elaboration on the selection of pole locations associated with Λ. This limited choice within an infinite space directly stems from the spectral properties of the harmonic state operator $\mathcal{A} - \mathcal{N}$. Recall that the spectrum of this operator constitutes a discrete and unbounded set, consisting solely of eigenvalues. Consequently, there exist only n locations to be determined for pole placement. The aforementioned theorem demonstrates that this selection determines the dynamics of $z := P(t)^{-1}x(t)$, emphasizing the importance of these choices in shaping the system's behavior. To illustrate the impact of pole placement in Linear Time-Periodic (LTP) systems, let's consider a one-dimensional system. After applying pole placement techniques, the Eq. (13.22) can be expressed as $\dot{z} = -\alpha z$, and $\dot{x} = (a(t) - b(t)k(t))x(t)$, where $x(t) := P(t)z(t)$. Since $P(t)$ is periodic, we can express, for $k = 1, 2, \ldots$:

$$x(t_0 + kT) = P(t_0)e^{-\alpha Tk}z(t_0),$$

where $t_0 \in [0, T]$. It can be observed that the sequence $x(t_0 + kT)$, for any $t_0 \in [0, T]$ and $k = 1, 2, \ldots$, exhibits a decay rate $\gamma := e^{-\alpha T}$. In contrast to the classical finite-dimensional pole placement problem, where ensuring the invertibility of the Sylvester equation solution requires an observability property [18, 20], such a condition is not sufficient in the case of infinite dimensions. To illustrate this, we provide a counterexample toward the end of this chapter, highlighting the limitations of relying solely on observability as a guarantee for the solution's invertibility. In practical terms, to verify the invertibility of $\mathcal{P}$, we refer to Theorem 13.6.4. This involves checking the existence of a positive constant $c > 0$ such that

$$|\mathcal{P}x|\ell^2 \geq c|x|\ell^2, \; x \in \ell^2. \tag{13.23}$$

Alternatively, due to the continuity of $P(t)$, we can verify

$$|\det P(t)| \neq 0 \text{ for any } t \in [0\; T]. \tag{13.24}$$

Moreover, we propose a sufficient condition that ensures the invertibility of $\mathcal{P}$ for a specific choice of pole locations. For this purpose, we consider $(V(t), J)$ as the Floquet factorization [8, 22] of $A(t)$, such that $V^{-1}(t)A(t)V(t) = J$. Here, J represents the Jordan normal form, and $V(t)$ is a T-periodic, invertible, and absolutely continuous matrix function. It is important to note that such a Floquet factorization always exists and can be explicitly computed using the closed-form formula provided by Theorem 6 in [14].

Theorem 13.3.12 *Consider a Floquet factorization given by $(V(t), J)$ and let $\mathcal{V} := \mathcal{T}(V)$. The solution $\mathcal{P}$ of the Sylvester equation* (13.18) *is invertible if $\mathcal{G}$ is set to $\mathcal{G} := \mathcal{B}^*\mathcal{V}^{*-1}$ in* (13.18) *and the pole locations correspond to $\Lambda := -J^* - \alpha Id_n$ for any real number α.*

Proof To ease the proof, we assume that J has a spectrum located in the open right half plane and we assume that Λ has a spectrum located in the open left half plane. Using the Floquet factorization, and taking $w(t) := V^{-1}x(t)$, system (13.6) can be rewritten as

$$\dot{W}(t) = (J \otimes \mathcal{I} - \mathcal{N})W(t) + \mathcal{V}^{-1}\mathcal{B}U(t), \tag{13.25}$$

where $W := \mathcal{F}(w)$. Thus, the goal now is to solve the Sylvester equation

$$(J \otimes \mathcal{I} - \mathcal{N})\mathcal{R} - \mathcal{R}(\Lambda \otimes \mathcal{I} - \mathcal{N}) = \mathcal{V}^{-1}\mathcal{B}\mathcal{G}. \tag{13.26}$$

Following the assumption concerning the spectra of $(J \otimes \mathcal{I} - \mathcal{N})$ and $(\Lambda \otimes \mathcal{I} - \mathcal{N})$, the solution of (13.26) is provided by (see [3] for more detail):

$$\mathcal{R} := \int_0^{+\infty} e^{-(J\otimes\mathcal{I}-\mathcal{N})t}\mathcal{V}^{-1}\mathcal{B}\mathcal{G}e^{(\Lambda\otimes\mathcal{I}-\mathcal{N})t}dt.$$

For any $w \in \ell^2$ and taking $\mathcal{G} := \mathcal{B}^*\mathcal{V}^{*-1}$, we have

$$< w, \mathcal{R}w > = \int_0^{+\infty} w^* e^{-(J\otimes\mathcal{I}-\mathcal{N})t}\mathcal{G}^*\mathcal{G}e^{(\Lambda\otimes\mathcal{I}-\mathcal{N})t}wdt.$$

As $\mathcal{N}^* = -\mathcal{N}$, we get

$$e^{(\Lambda\otimes\mathcal{I}-\mathcal{N})t} = e^{-(J\otimes\mathcal{I}-\mathcal{N})^*t}e^{((J^*+\Lambda)\otimes\mathcal{I})t},$$

and it follows that

$$< w, \mathcal{R}w >= \int_0^{+\infty} w^* e^{-(J\otimes\mathcal{I}-\mathcal{N})t}\mathcal{G}^*\mathcal{G}e^{-(J\otimes\mathcal{I}-\mathcal{N})^*t} e^{((J^*+\Lambda)\otimes\mathcal{I})t}wdt.$$

Taking $\Lambda := -J^* - \alpha Id_n$ with $\alpha \in \mathbb{R}$ such that the spectrum $\sigma(\Lambda)$ is located in the open left half plane, we get

$$< w, \mathcal{R}w >= \int_0^{+\infty} w^* e^{-(J\otimes\mathcal{I}-\mathcal{N})t}\mathcal{G}^*\mathcal{G}e^{-(J\otimes\mathcal{I}-\mathcal{N})^*t} w\, e^{-\alpha t}dt.$$

As the pair $(\mathcal{A}-\mathcal{N}, \mathcal{B})$ is assumed to be exactly controllable, the pair $((J\otimes\mathcal{I}-\mathcal{N}), \mathcal{G}^*)$ is also exactly controllable. It follows that for any $t > 0$ and any $w \neq 0$

$$\int_0^{t} w^* e^{-(J\otimes\mathcal{I}-\mathcal{N})t}\mathcal{G}^*\mathcal{G}e^{-(J\otimes\mathcal{I}-\mathcal{N})^*t} w dt > 0.$$

Consequently, $< w, \mathcal{R}w >$ is strictly positive with $w \neq 0$ and hence $\mathcal{R}$ is positive definite and invertible.

To conclude that $\mathcal{P}$, solution of (13.18), is invertible, it is sufficient to see that $\mathcal{P}$ satisfies

$$\begin{aligned}
\mathcal{P} &:= \int_0^{+\infty} e^{-(A-\mathcal{N})t}\mathcal{B}\mathcal{G}e^{(\Lambda\otimes\mathcal{I}-\mathcal{N})t}dt \\
&:= \int_0^{+\infty} \mathcal{V}e^{-(J\otimes\mathcal{I}-\mathcal{N})t}\mathcal{V}^{-1}\mathcal{B}\mathcal{G}e^{(\Lambda\otimes\mathcal{I}-\mathcal{N})t}dt \\
&:= \mathcal{V}\mathcal{R}.
\end{aligned}$$

Then, $\mathcal{P}$ is invertible as $\mathcal{V}$ and $\mathcal{R}$ are invertible. □

13.3.2 *Solving Harmonic Sylvester Equation*

The harmonic Sylvester equation (13.18) can be effectively solved with the same approach as employed previously for solving harmonic Lyapunov equations, ensuring an arbitrarily small error in the solution. Let us define the product, denoted by $\circ$, of a $n \times n$ block Toeplitz matrix $\mathcal{A}$ with a matrix B as follows:

$$B \circ \mathcal{A} := \begin{pmatrix} B\otimes\mathcal{A}_{11} & B\otimes\mathcal{A}_{12} & \cdots & B\otimes\mathcal{A}_{1n} \\ B\otimes\mathcal{A}_{21} & B\otimes\mathcal{A}_{22} & & \vdots \\ \vdots & & \ddots & \vdots \\ B\otimes\mathcal{A}_{n1} & \cdots & \cdots & B\otimes\mathcal{A}_{nn} \end{pmatrix}. \tag{13.27}$$

Theorem 13.3.13 *The phasor* $\mathbf{P} := \mathcal{F}(P)$ *of the solution* $\mathcal{P} := \mathcal{T}(P)$ *of the infinite-dimensional Sylvester equation (13.18) is given by*

$$col(\mathbf{P}) := (Id_n \otimes (\mathcal{A}-\mathcal{N}) - Id_n \circ (\Lambda\otimes\mathcal{I})^*)^{-1} col(\mathbf{Q}) \tag{13.28}$$

with $\mathcal{N}$ given by (13.7) *and where* $\mathbf{Q} := \mathcal{F}(BG)$.

Proof The proof is similar to the proof of Theorem 10 in [14]. □

In practice, due to the infinite dimension nature of our problem, a truncation is necessary. Define for any given m:

1. the $m-$truncation of the $n \times n$ block Toeplitz matrix $\mathcal{A}$ obtained by applying a $m-$truncation of all of its blocks $\mathcal{T}(a_{ij})_m, i, j := 1, \ldots, n$, where $\mathcal{T}(a_{ij})_m$ is the $(2m+1) \times (2m+1)$ leading principal submatrices of $\mathcal{T}(a_{ij})$.
2. the $m-$truncation $\mathbf{Q}|_m$ of $\mathbf{Q} := \mathcal{F}(Q)$ is obtained by suppressing all phasors (components) of order $|k| > m$.

We define also the $m-$truncated solution as

$$col(\tilde{\mathbf{P}}_m) := (Id_n \otimes (\mathcal{A}_m - \mathcal{N}_m) - Id_n \circ (\Lambda \otimes \mathcal{I}\lceil_m)^*)^{-1} col(\mathbf{Q}|_m), \qquad (13.29)$$

where $\mathcal{N}_m := Id_n \otimes diag(j\omega k,\ |k| \leq m)$. The next theorem shows that an approximated solution to the harmonic Sylvester equation can always be determined up to an arbitrarily small error.

Theorem 13.3.14 *Under assumption of Theorem 13.3.11 and assuming that* $A(t) \in L^\infty([0\ T])$, *for any given* $\epsilon > 0$, *there exists* m_0 *such that for any* $m \geq m_0$*:*

$$\|col(\mathbf{P} - \tilde{\mathbf{P}}_m)\|_{\ell^2} < \epsilon,$$

where $\mathbf{P}$ *is the solution of the infinite-dimensional problem* (13.14). *Moreover,*

$$\|\mathcal{P} - \tilde{\mathcal{P}}_m\|_{\ell^2} < \epsilon$$

with $\mathcal{P} := \mathcal{T}(P)$ *and* $\tilde{\mathcal{P}}_m := \mathcal{T}(\tilde{P}_m)$.

13.4 Illustrative Example

Consider a LTP system defined by

$$\dot{x} = \begin{pmatrix} a_{11}(t) & a_{12}(t) \\ a_{21}(t) & a_{22}(t) \end{pmatrix} x + \begin{pmatrix} b_{11}(t) \\ 0 \end{pmatrix} u, \qquad (13.30)$$

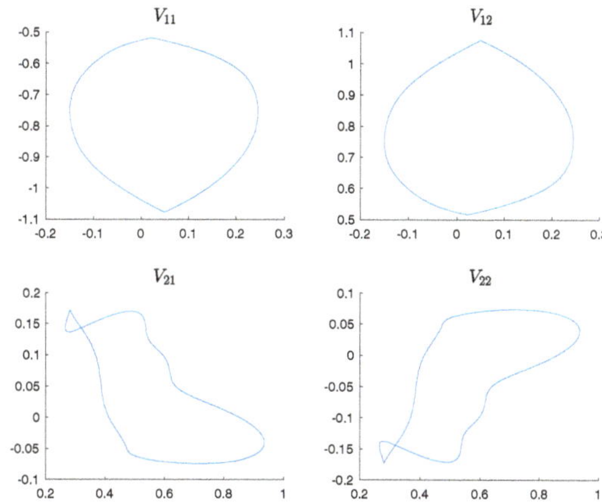

Fig. 13.3 Complex components $V_{ij}(t)$, $i, j = 1, 2$ of $V(t)$ for $t = [0\ T]$

where

$$a_{11}(t) := 1 + \frac{4}{\pi}\sum_{k=0}^{\infty}\frac{1}{2k+1}\sin(\omega(2k+1)t),$$

$$a_{12}(t) := 2 + \frac{16}{\pi^2}\sum_{k=0}^{\infty}\frac{1}{(2k+1)^2}\cos(\omega(2k+1)t),$$

$$a_{21}(t) := -1 + \frac{2}{\pi}\sum_{k=1}^{\infty}\frac{(-1)^k}{k}\sin(\omega kt + \frac{\pi}{4}),$$

$$a_{22}(t) := 1 - 2\sin(2\pi t) - 2\sin(6\pi t) + 2\cos(6\pi t) + 2\cos(10\pi t),$$

$$b_{11}(t) := 1 + 2\cos(2\omega t) + 4\sin(3\omega t) \text{ with } \omega := 2\pi.$$

Observe that a_{11}, a_{12} and a_{21} are, respectively, square, triangular, and sawtooth signals and include an offset part. The associated Toeplitz matrix has an infinite number of phasors and is not banded.

We first compute the Floquet factorization $(V(t), J)$ (see [14]). We find for J a diagonal matrix defined by two complex conjugate eigenvalues $\{1 \pm j1.64\}$ and the components of the T-periodic, absolutely continuous and complex valued matrix function $V(t)$ is shown on Fig. 13.3. Thus, system (13.30) can be rewritten as

$$\dot{z} = Jz + V(t)^{-1}B(t)u$$

and one can see that this LTP system is unstable. Moreover, the spectrum of $\mathcal{A} - \mathcal{N}$ is given by $\sigma := \{\lambda_p + j\omega k : k \in \mathbb{Z}, p := 1, 2\}$ with $\lambda_{1,2} := \{1 \pm j1.64\}$ (see [14]).

Using Theorem 13.3.12, we set $G(t) := B(t)^* V(t)^{*-1}$ and fix the poles of the closed loop by choosing $\Lambda := -J^* - Id$ ($\alpha = 1$). We solve the Harmonic Sylvester equation by computing $\tilde{\mathbf{P}}_m$ (see (13.16)) for $m := 4, 6, 8, 10$. Figure 13.4 shows the components of complex valued T-periodic and absolutely continuous matrix function $\tilde{P}_m := \mathcal{F}^{-1}(\tilde{\mathbf{P}}_m)$ while Fig. 13.5 shows the phasor modulus of each entries of the

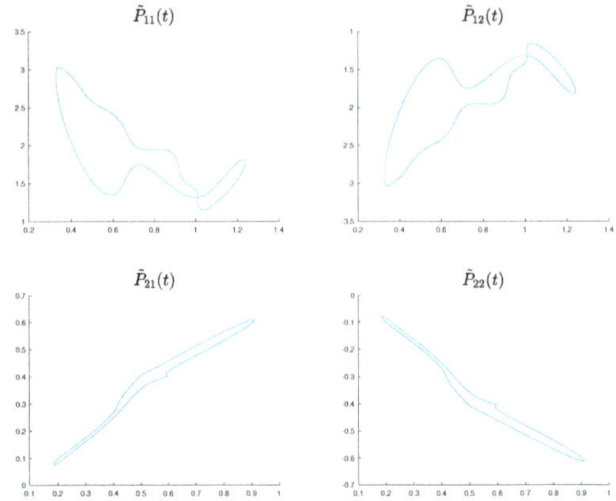

Fig. 13.4 Complex components $\tilde{P}_{ij}(t)$, $i, j = 1, 2$ of $\tilde{P}_m(t)$ for $t = [0\ T]$ and $m = 10$

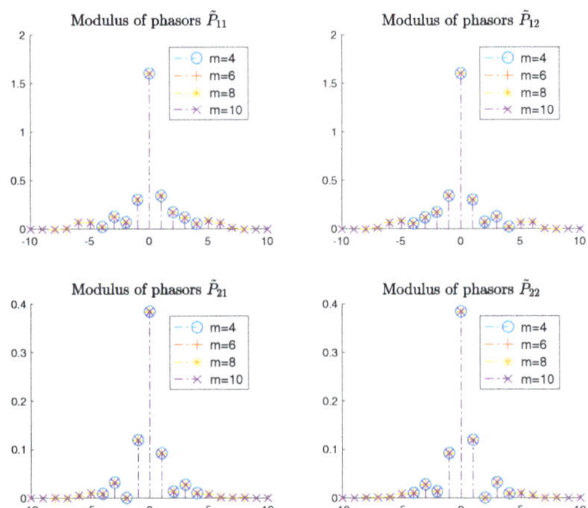

Fig. 13.5 $|\tilde{\mathbf{P}}_m|$ for $m := 4, 6, 8, 10$

matrix $\tilde{\mathbf{P}}_m$ with respect to m. We see that significant values are obtained for $m \leq 8$ (Fig. 13.5).

We deduce the T-periodic and real gain matrix $K(t) := G(t)\tilde{P}_m^{-1}(t)$ for $m := 4, 6, 8, 10$ as shown on Fig. 13.6 for $m = 10$.

Now, it is straightforward to show that the control

$$u(t) := -K(t)(x(t) - x_{ref}(t)) + u_{ref}(t)$$

stabilizes globally and asymptotically the unstable LTP system (13.30) on any T-periodic trajectory $x_{ref}(t) := \mathcal{F}^{-1}(X_{ref})$ and $u_{ref}(t) := \mathcal{F}^{-1}(U_{ref})$, where the pair (X_{ref}, U_{ref}) satisfies the harmonic equilibrium equation

$$0 = (\mathcal{A} - \mathcal{N})X_{ref} + \mathcal{B}U_{ref}. \tag{13.31}$$

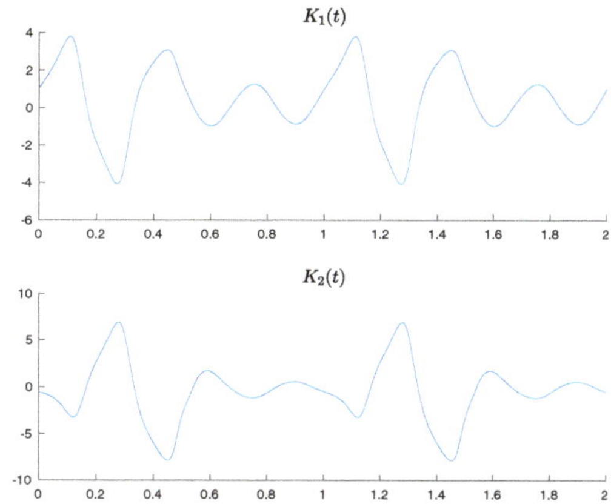

Fig. 13.6 $K(t) = [K_1(t)\ K_2(t)] = G(t)\tilde{P}_m^{-1}(t)$ for $m = 10$, $t \in [0\ 2T]$

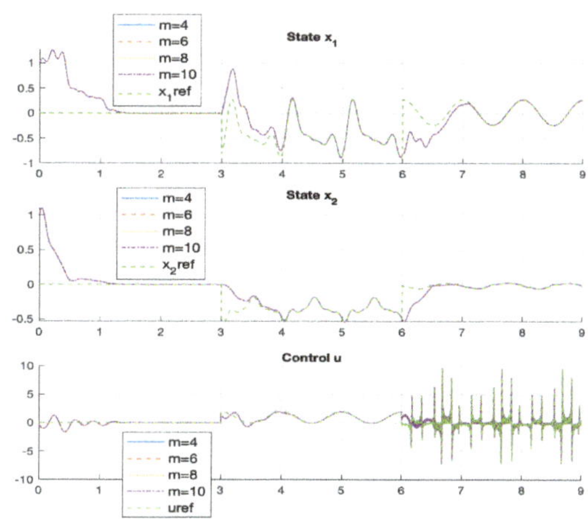

Fig. 13.7 Closed-loop response with $u(t) := -K(t)(x(t) - x_{ref}(t)) + u_{ref}(t)$ and for $K(t) = \sum_{k=-m}^{m} K_k e^{j\omega kt}$ and $m := 4, 6, 8, 10$ and $\Lambda := -J^* - Id$ and $J := diag(1 \pm j1.64)$

To illustrate this, we plot on Fig. 13.7 the closed-loop response for three T-periodic reference trajectories (x_{ref}, u_{ref}). We start by $u_{ref}(t) := 0$, for $t < 3$, then $u_{ref} := 1 + \cos(2\pi t)$ for $3 \leq t < 6$ and for $t \geq 6$ we consider a desired steady state X_d given by $\mathcal{F}^{-1}(X_d)(t) := (\frac{1}{4}\cos(2\pi t), 0)$ and look for the nearest harmonic equilibrium, solution of the minimization problem $\min_{U_{ref}} \|X_d - X_{ref}\|^2$ subject to (13.31). Clearly it can be observed on Fig. 13.7 that the provided state feedback allows to track any T-periodic trajectory corresponding to any equilibrium of (13.31) even if a relative small number m of involved harmonics are considered.

Now, it can be noticed that Theorem 13.3.12 provides only a sufficient condition for the invertibility of $\mathcal{P}$. In fact, this invertibility can be obtained for many other pole placements. For example, if we set $\Lambda := diag(-10, -12)$ instead of $\Lambda := -J^* - \alpha Id$, from $\tilde{P}_m(t) := \mathcal{F}^{-1}(\tilde{\mathbf{P}}_m) = \sum_{k=-m}^{m} \tilde{\mathbf{P}}_k e^{j\omega kt}$ which is a T-periodic and continuous function, it is easy to check that $|\det \tilde{P}_m(t)| > 0$ for any $t \in [0\ 1]$ $(T = 1)$ and thus $\tilde{P}_m(t)$ is invertible. The corresponding results depicted in Fig. 13.8 illustrate the improvement of the transient.

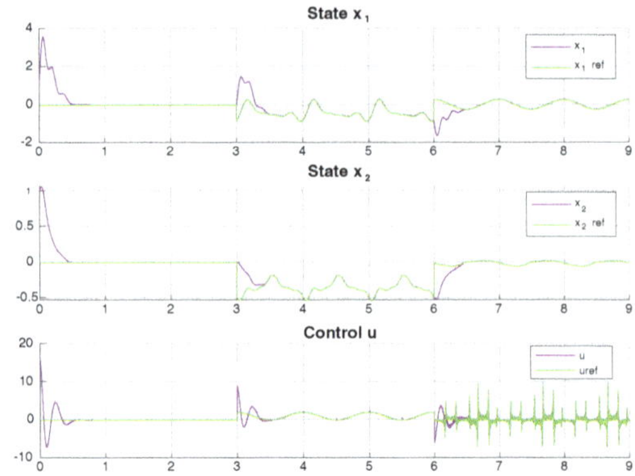

Fig. 13.8 Closed-loop response with $u(t) := -K(t)(x(t) - x_{ref}(t)) + u_{ref}(t)$ for $K(t) = \sum_{k=-m}^{m} K_k e^{j\omega kt}$, $m := 10$ and $\Lambda := diag(-10, -12)$

13.5 Conclusion

This chapter offers an overview of the harmonic control methodology, focusing on the design of highly effective harmonic control laws specifically tailored for Linear Time-Periodic (LTP) systems. A key aspect addressed in this methodology is the incorporation of performance considerations by leveraging the pole placement property. By carefully selecting the pole locations, the resulting closed-loop system, provided the solution to the harmonic Sylvester equation is invertible, exhibits the desired pole placement characteristic. From a practical perspective, the challenge lies in solving the infinite-dimensional Sylvester equation. However, through the proposed approach, a successful resolution of this equation is achieved, yielding a solution with an arbitrarily small error. This significant accomplishment enables the precise determination of the harmonic control parameters, leading to the desired performance features in the closed-loop system.

Acknowledgements This work was supported by the "Agence Nationale de la Recherche" (ANR) under Grant HANDY ANR-18-CE40-0010.

13.6 Appendix

We provide here some results concerning operator norms used in this chapter. The missing proofs can be found in [10] (Part V, pp. 562–574).

The norm of an operator M from ℓ^p to ℓ^q is given by

$$\|M\|_{\ell^p,\ell^q} := \sup_{\|X\|_{\ell^p}=1} \|MX\|_{\ell^q}.$$

This operator norm is sub-multiplicative, i.e., if $M : \ell^p \to \ell^q$ and $N : \ell^q \to \ell^r$ then $\|NM\|_{\ell^p,\ell^r} \leq \|M\|_{\ell^p,\ell^q}\|N\|_{\ell^q,\ell^r}$. If $p = q$, we use the notation: $\|M\|_{\ell^p} := \|M\|_{\ell^p,\ell^p}$.

Definition 13.6.1 Consider a vector $x(t) \in L^2([0\ T], \mathbb{C}^n)$ and define $X := \mathcal{F}(x)$ with its symbol $X(z)$. The ℓ^2-norm of $X(z)$ is given by

$$\|X(z)\|_{\ell^2} := \|X\|_{\ell^2},$$

where $\|X\|_{\ell^2} := \left(\sum_{k\in\mathbb{Z}} |X_k|^2\right)^{\frac{1}{2}}$.

Theorem 13.6.2 *Let $A \in L^2([0\ T], \mathbb{C}^{n\times m})$. Then, $\mathcal{A} := \mathcal{T}(A)$ is a bounded operator on ℓ^2 if and only if $A \in L^\infty([0\ T], \mathbb{C}^{n\times m})$. Moreover, we have*

1. *the operator norm induced by the ℓ^2-norm satisfies*

$$\|A(z)\|_{\ell^2} = \|\mathcal{A}\|_{\ell^2} = \|A\|_{L^\infty}$$

2. *the operator norm of the semi-infinite Toeplitz matrix satisfies: $\|\mathcal{T}_s(A)\|_{\ell^2} = \|\mathcal{A}\|_{\ell^2}$*
3. *the operator norm of the Hankel operators $\mathcal{H}(A^+)$, $\mathcal{H}(A^-)$ satisfies $\|\mathcal{H}(A^-)\|_{\ell^2} \leq \|A\|_{L^\infty}$ and $\|\mathcal{H}(A^+)\|_{\ell^2} \leq \|A\|_{L^\infty}$*
4. *the operator norm related to the left and right $m-$truncations satisfies: $\|\mathcal{A}_{m^+}\|_{\ell^2} = \|\mathcal{A}_{m^-}\|_{\ell^2} = \|\mathcal{A}\|_{\ell^2} = \|A\|_{L^\infty}$.*

Proposition 13.6.3 *Let $P(\cdot)$ be a matrix function in $L^\infty([0\ T], \mathbb{C}^{n\times n})$. Define $\mathbf{P} := \mathcal{F}(P)$ and $\mathcal{P} := \mathcal{T}(P)$. If $\|\text{col}(\mathbf{P})\|_{\ell^2} \leq \epsilon$ then $\|\mathcal{P}\|_{\ell^2} \leq \epsilon$.*

Proof Using Riesz–Fischer theorem, we have

$$\|\text{col}(\mathbf{P})\|_{\ell^2} = \|\text{col}(P)\|_{L^2} = (\sum_{i,j=1}^{n} \|P_{ij}\|_{L^2}^2)^{1/2} = \|P\|_F,$$

where $\|P(t)\|_F$ stands for the Frobenius norm. As $P \in L^\infty([0\ T], \mathbb{C}^{n\times n})$, Hölder's inequality implies $Px \in L^2([0\ T], \mathbb{C}^n)$ for any $x \in L^2([0\ T], \mathbb{C}^n)$. Thus, the result follows from the following relations between operator norms:

$$\begin{aligned}
\|\mathcal{P}\|_{\ell^2} &= \sup_{\|X\|_{\ell^2}=1} (< \mathcal{P}X, \mathcal{P}X >_{\ell^2})^{1/2} \\
&= \sup_{\|x\|_{L^2}=1} (< Px, Px) >_{L^2})^{1/2} \\
&\leq (\text{trace}(P^* P))^{1/2} = \|P\|_F,
\end{aligned}$$

where $< \cdot, \cdot >$ stands for the scalar product. □

Theorem 13.6.4 *Let* $A(t) \in L^{\infty}([0\ T], \mathbb{C}^{n\times n})$. *$\mathcal{A}$ is invertible if and only if there exists* $\gamma > 0$ *such that the set* $\{t : |\det(A(t))| < \gamma\}$ *has measure zero. The inverse* $\mathcal{A}^{-1}$ *is determined by* $\mathcal{T}(A^{-1})$. *In addition,* $\mathcal{A}$ *is invertible if and only if* $\mathcal{A}$ *is a Fredholm operator [10], or equivalently in this setting if and only if there exists* $c > 0$ *such that*

$$\|\mathcal{A}x\|_{\ell^2} > c\|x\|_{\ell^2},\ \textit{for any } x \in \ell^2.$$

References

1. Almer, S., Jonsson, U.: Dynamic phasor analysis of periodic systems. IEEE Trans. Autom. Control **54**(8), 2007–2012 (2009)
2. Almér, S., Mariéthoz, S., Morari, M.: Dynamic phasor model predictive control of switched mode power converters. IEEE Trans. Control Syst. Technol. **23**(1), 349–356 (2014)
3. Bhatia, R., Rosenthal, P.: How and why to solve the operator equation ax- xb= y. Bull. Lond. Math. Soc. **29**(1), 1–21 (1997)
4. Blin, N., Riedinger, P., Daafouz, J., Grimaud, L., Feyel, P.: Necessary and sufficient conditions for harmonic control in continuous time. IEEE Trans. Autom. Control **67**(8), 4013–4028 (2021)
5. Bolzern, P., Colaneri, P.: The periodic Lyapunov equation. SIAM J. Matrix Anal. Appl. **9**(4), 499–512 (1988)
6. Chavez, J.J., Ramirez, A.: Dynamic harmonic domain modeling of transients in three-phase transmission lines. IEEE Trans. Power Deliv. **23**(4), 2294–2301 (2008)
7. Demiray, T.: Simulation of power system dynamics using dynamic phasor models. Ph.D. thesis, ETH Zurich (2008)
8. Farkas, M.: Periodic Motions, vol. 104. Springer Science & Business Media (2013)
9. Floquet, G.: Sur les équations linéaires a coefficients périodiques. Ann. Sci. Ecole Norm. Sup. **12**, 47–88 (1883)
10. Gohberg, I., Goldberg, S., Kaashoek, M.A.: Classes of Linear Operators, vol. 63. Birkhäuser (2013)
11. Mattavelli, P., Verghese, G.C., Stankovic, A.M.: Phasor dynamics of thyristor-controlled series capacitor systems. IEEE Trans. Power Syst. **12**(3), 1259–1267 (1997)
12. Montagnier, P., Spiteri, R.J., Angeles, J.: The control of linear time-periodic systems using Floquet-Lyapunov theory. Int. J. Control **77**(5), 472–490 (2004)
13. Rabah, R., Sklyar, G., Barkhayev, P.: Exact null controllability, complete stabilizability and continuous final observability of neutral type systems. Int. J. Appl. Math. Comput. Sci. **27**(3), 489–499 (2017)
14. Riedinger, P., Daafouz, J.: Solving infinite-dimensional harmonic Lyapunov and Riccati equations. IEEE Trans. Autom. Control (2022)
15. Rodman, L.: On exact controllability of operators. Rocky Mt. J. Math. **20**(2), 549–560 (1990)
16. Sandcrs, S.R., Noworolski, J.M., Liu, X.Z., Verghese, G.C.: Generalized averaging method for power conversion circuits. IEEE Trans. Power Electron. **6**(2), 251–259 (1991)
17. Sinha, S.C., Pandiyan, R., Bibb, J.S.: Liapunov-Floquet transformation: computation and applications to periodic systems. J. Vib. Acoust. 209–219 (1996)
18. Varga, A.: Robust pole assignment via Sylvester equation based state feedback parametrization. In: IEEE International Symposium on Computer-Aided Control System Design, pp. 13–18. IEEE (2000)
19. Wereley, N.M.: Analysis and control of linear periodically time varying systems. Ph.D. thesis, Massachusetts Institute of Technology (1990)
20. Zbigniew, E.: Infinite-dimensional Sylvester equations: basic theory and application to observer design. Int. J. Appl. Math. Comput. Sci. **22**(78), 245–257 (2012)

21. Zhou, B., Duan, G.-R.: Periodic Lyapunov equation based approaches to the stabilization of continuous-time periodic linear systems. IEEE Trans. Autom. Control **57**(8), 2139–2146 (2011)
22. Zhou, J.: Harmonic Lyapunov equations in continuous-time periodic systems: solutions and properties. IET Control Theory Appl. **1**(4), 946–954 (2007)
23. Zhou, J., Hagiwara, T., Araki, M.: Spectral characteristics and eigenvalues computation of the harmonic state operators in continuous-time periodic systems. Syst. Control Lett. **53**(2), 141–155 (2004)
24. Zhou, Jun: Derivation and solution of harmonic Riccati equations via contraction mapping theorem. Trans. Soc. Instrum. Control Eng. **44**(2), 156–163 (2008)

Index

R. Postoyan et al. (eds.), *Hybrid and Networked Dynamical Systems*, Lecture Notes in Control and Information Sciences 493, https://doi.org/10.1007/978-3-031-49555-7

L

M

N

O

P

Q